Computational Fluid and Solid Mechanics

Series Editor

K.J. Bathe
Massachusetts Institute of Technology, Cambridge, MA, USA

For other titles published in this series, go to
http://www.springer.com/series/4449

Miguel Luiz Bucalem · Klaus-Jürgen Bathe

The Mechanics of Solids and Structures – Hierarchical Modeling and the Finite Element Solution

Springer

Prof. Miguel Luiz Bucalem
Escola Politécnica da Universidade
de São Paulo
São Paulo, Brazil
mlbucale@usp.br

Prof. Klaus-Jürgen Bathe
Massachusetts Institute of Technology
Cambridge, MA
USA
kjb@mit.edu

ISSN 1860-482X
e-ISSN 1860-4838
ISBN 978-3-540-26331-9
e-ISBN 978-3-540-26400-2
DOI 10.1007/978-3-540-26400-2
Springer Heidelberg Dordrecht London New York

Cover design: deblik, Berlin

Printed on acid-free paper

Springer is part of Springer Science+Business Media (www.springer.com)

Preface

Our main objective in this book is to provide a rational, structured and modern framework for the modeling and analysis of engineering structures.

Although engineering structures have been modeled and analyzed for centuries, the mathematical models that could actually be solved were relatively simple. This situation has dramatically changed during the last decades. Today, with powerful computers and reliable finite element procedures widely available, very complex models of solids and structures can be solved, and consequently the range and complexity of analyses has drastically increased.

In any finite element analysis, the first step for an analyst is to choose an *appropriate mathematical model*, and the second step is to solve that model using *finite element procedures*. In almost all analyses, the first step is most important and also most difficult. In order to choose an appropriate model, the analyst must be familiar with the basic mathematical models that are available, and in particular know the hierarchy of such models. Only if the analyst is deeply familiar with the various mathematical models available, their hierarchy, and reliable finite element procedures, can the analyst choose the most effective model, perform an efficient analysis, and properly interpret the analysis results.

Many books on finite element methods have been published; however, our aim in this book is broader. We aim to present in one treatise – both – the basic mathematical models of solid and structural mechanics and modern reliable finite element procedures for the solution of these models. The book can be used for teaching, in a modern way, structural and solid mechanics, finite element methods, and for self-study – from elementary to quite advanced material.

We draw in this treatise heavily from the material published in the books Theory of Elasticity, by S. P. Timoshenko and J. N. Goodier (1970), and Theory of Plates and Shells, by S. P. Timoshenko and S. Woinowsky-Krieger (1959), regarding the mechanics of solids and structures, and from the material in the book Finite Element Procedures, by K. J. Bathe (1996), regarding finite element formulations and solution techniques. In essence, we try to synthesize the presentation of the models and methods of classical mechanics with the procedures of finite element analysis, add new insights, and show

how to solve the classical general models of elasticity in a modern – effective and reliable – manner.

Towards that aim, we first develop from elementary concepts the basic mathematical models of solids and structures, then we present modern finite element methods for solution, and finally we give examples of applications using the finite element program ADINA. Emphasis is given to the hierarchical nature of the mathematical models, from simple to complex, and on choosing the simplest reliable and effective model for analysis, see Chapter 1. The process of hierarchical modeling and finite element solution, with the benefits reached, is finally illustrated in the examples of Chapter 7. These broadly indicate how we recommend modern analysis to be conducted.

To perform an effective analysis is an art. In many cases, the analyst tries to look into the future by asking how the design of a structure will perform; in other cases, the analyst tries to understand a phenomenon of nature to be able to predict when and how this phenomenon will occur, how it can be affected, and possibly be remedied. All these analyses are based on present knowledge and the analyst endeavors to predict the future by means of computational modeling and simulations. This clearly cannot be an easy task.

Since any analysis depends on the knowledge of the analyst, performing in the *art of analysis* is most stimulating and requires constant learning. We hope that this book will be valuable in this learning process.

While we focus on the linear elastic static analysis of solids and structures, with only a short introduction to nonlinear analysis in Chapter 8, the general concepts that we present regarding modeling and analysis are equally applicable in much more complex cases, including fluids and multi-physics phenomena. Hence, this book not only gives valuable information regarding the linear analysis of solids, but also demonstrates universal concepts of analysis that are generally applicable.

The writing of this book required a very large effort, and we would like to thank all of those who have supported us in this endeavor.

Miguel Luiz Bucalem is very thankful to the Department of Structural and Geotechnical Engineering of the University of São Paulo. He was very fortunate to first learn structural engineering from great engineers and professors, specifically Decio Leal de Zagottis and Henrique Lindenberg Neto. Decio Leal de Zagottis was his first mentor whose brilliant teaching and encouragement attracted him to the field. Since he joined the Faculty of the Department in 1985, he has had an excellent environment for his teaching and research and for writing this book. The interactions with his colleagues Carlos Eduardo Nigro Mazzilli, João Cyro André and Sergio Cifu in teaching subjects related to this book were very rich and helped him shape his thoughts about some of the contents in this book. He is also very grateful to have been the doctoral student of Klaus-Jürgen Bathe at M.I.T. Finally, he would like to thank his students who have been a source of inspiration for his work and who have helped with this book, Daniel Carvalho Lepkison de Oliveira, Christian Hideki Furukawa, Ivan José de Godoy Mazella, Estela Mari Ricetti Bueno, Carlos Alberto Medeiros, Alex Neves Júnior, Krishna Martins Leão, and Lucas Anastasi Fiorani.

Klaus-Jürgen Bathe is very thankful to the Department of Mechanical Engineering, M.I.T. for having provided to him an excellent environment for his teaching, research and scholarly writing. He has been fortunate to work with many brilliant students and colleagues, including Miguel Luiz Bucalem, who have much contributed to his research and teaching. The idea to write this book was born when he was asked to teach the undergraduate course in solid mechanics and he interacted with Miguel. It became apparent that new, exciting and modern ways to teach mechanics and analysis are needed, and we believe that this book offers one such option. K. J. Bathe is also thankful to his students Daniel Payen and Seounghyun Ham for their help in the proof-reading of the manuscript, and would like to acknowledge that for his teaching and research, and work on this book, his involvement in ADINA R & D, Inc. has been very valuable.

Finally, we both would like to express our unbounded thanks to our wives Tamara and Zorka, and our children, for their encouragement and unconditional support over many years in all our work endeavors, including the writing of this book.

M. L. Bucalem and K. J. Bathe

Contents

1. Mathematical models and the finite element solution. Hierarchical modeling

The objective of this chapter is to introduce a modern approach for the modeling of structures (and in fact any physical system). The hierarchical modeling process is the central concept. We discuss the need for such a process, detail it in a demonstrative problem and examine the important benefits associated with its usage in engineering. Mathematical modeling and finite element methods are introduced as natural ingredients of the hierarchical modeling procedure.

While the discussion in this chapter and in those to follow focuses on the analysis of solids and structures, we also point out that the basic ideas of the hierarchical modeling approach are directly applicable to the analysis of fluid flows, to the solution of multi-physics and multi-scale problems, and in fact to the analysis of any problem encountered in engineering and in the sciences.

1.1 Introductory remarks

Historically, the design of engineering structures has progressed through a number of stages. Engineering structures have been built much before any rational method of analysis was available. Then a structural design was based on trial and error and as experience was gathered for a particular type of structure, practical guidelines were developed which helped new designs.

The development of mathematical methods and experimental approaches have drastically altered and improved structural design methodologies. Nowadays, the design process is aided by several factors: a vast experience with previous designs of similar structures, modern methods of analysis and, of course, still experiments.

It is easy to recognize that structural design is a vast field, and involves many architectural, engineering and societal issues that we encounter in the design of bridges, buildings, airplanes, motor cars, ships, bio-medical devices, any household appliances, power plants, jet engines, and so forth. However, in order to focus the discussion on the issues that we would like to address in this chapter, we assume that a design for a structure has been proposed. Of course, many issues should have been addressed before a structural design is reached. The major considerations are linked to structural safety, operational requirements, construction/manufacturing feasibility and cost. During this design process, the most important challenge that the structural engineer faces is to give reliable answers to the following questions:

- Is the structure safe to be used?
- Will the structure perform in operation as required?

Of course, additional questions arise depending on the type of structure considered. Nevertheless, the questions above highlight the main concern of a structural engineer.

If we consider the broad spectrum of engineering structures, to build prototypes and small-scale models for experiments can be very inefficient. Of course experimental structural analysis is a very important field and physical experiments in the laboratory or on prototype structures constitute usually an important step to assure that a design is safe and functional. However, to reach a satisfactory design it is, in many industries, common practice today, to perform finite element analyses of proposed designs – prior to actually building the structure or a physical (laboratory) model thereof.

1.2 Mathematical models

This book is concerned with methods that permit to predict the behavior of a structure without requiring that the structure has already been built. Even if the structure already exists, such methods do not demand physical experimentation with the structure. It is important to appreciate that to have a method with this characteristic constitutes a major step since it implicitly assumes that we are able to construct an *abstract representation* of a structure which can capture its structural behavior. In other words, for a given structure we need to represent its geometry, materials, the loads acting on it and the governing relations linking these quantities, all of which allows us to simulate its structural behavior. In general, we are interested in predicting the deformations and internal forces of the structure subjected to various loadings and whether it would "break" or not.

The abstract representation of the structure which allows us to make predictions of its behavior is called a *mathematical model* of the structure at hand. We summarize the process described above in Figure 1.1. Note that we intentionally refer to a physical problem in this figure to place mathematical modeling into a wide perspective. Our physical problem will be a structural problem and its modeling constitutes the main objective of this book. However, the discussion that we will undertake is also valid in the broad context of predicting the behavior of other, very general, physical systems, for example involving also fluids and the interactions of structures and fluids.

To further consider this point, the structures that we primarily think of analyzing are of course the traditional structures like bridges, motor cars,

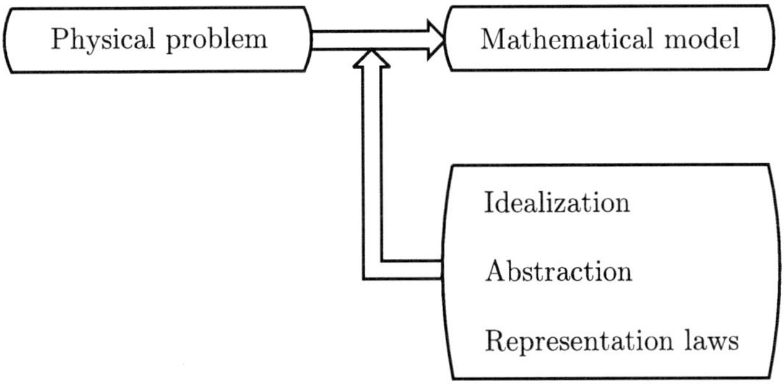

Fig. 1.1. Concepts involved in the construction of a mathematical model

airplanes, that we see around us in everyday life. We are aiming to predict the stiffness and strength characteristics of these structures. However, our discussion is also directly applicable to the analysis of structures of much smaller and much larger scales, see e.g. Bathe, 2005. For example, on the much smaller scale, we may design a new composite material based on micromechanical considerations, or analyze, on the nano-scale, cells moving in the human body. On the much larger scale, geological problems like the eroding of beaches and the construction of tunnels through mountains may be analyzed, or in weather forecasting, the development and movement of storms and their effect on structures may be simulated. It is apparent that all these "structures" can be exposed to important multi-physics events, that may involve fluid flows, and thermal, chemical and electro-magnetic effects. In such cases, the full coupling between very different media, that need to be modeled at different scales, might need to be considered.

While we do not discuss in this book the analysis of fluid flows, multi-physics and multi-scale problems, it is important to note, though, that the basic concepts of hierarchical modeling – as we expose them in detail in this book – are directly applicable to the analysis of such problems as well. Hence the basic modern approach of analysis presented here is very general and indeed can be followed in any field of analysis.

In order to further develop our discussion of modeling, we now will focus on an example analysis that we use to exemplify the basic concepts of hierarchical modeling.

1.2.1 A demonstrative problem – a carabiner

A photograph of the chosen "structure" is shown in Figure 1.2. It is called a carabiner and it is used in rock climbing. A schematic representation of its usage is shown in Figure 1.3a. From a structural point of view, the carabiner should transfer the load between ropes without "breaking" and without deforming "too much", *i.e.*, the deformations must not jeopardize its performance. Therefore, an important objective for the mathematical modeling of this structure is to determine the magnitude of the load that can safely be transferred by the carabiner.

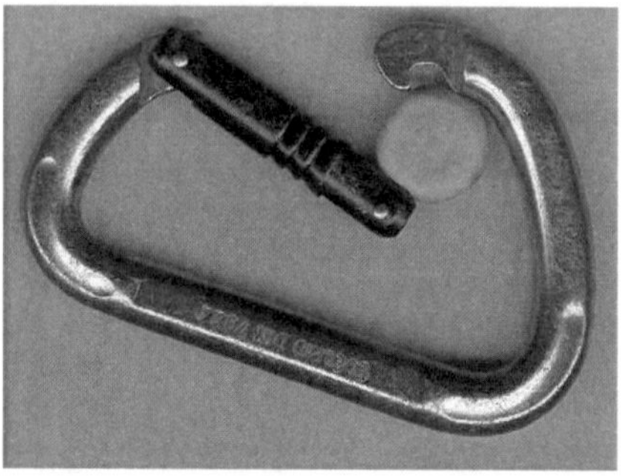

Fig. 1.2. Photograph of a carabiner

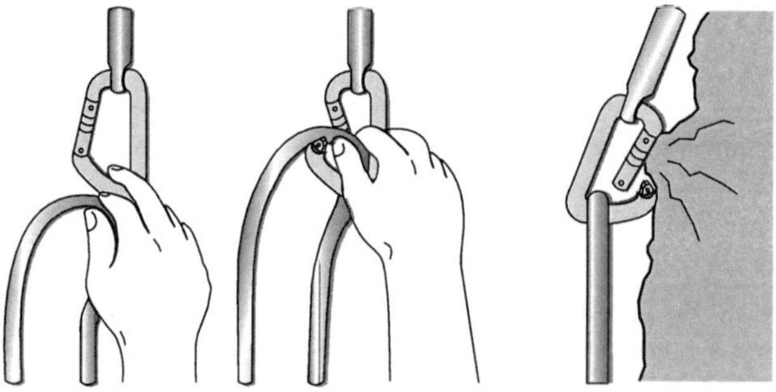

a) b)

Fig. 1.3. Schematic representation of the carabiner usage and of a situation that makes it to work open

Let us focus on the situation when the carabiner works open which could arise as schematically shown in Figure 1.3b. In order to discuss the modeling of the carabiner we need to use some structural mechanics terminology and concepts but do so in an introductory manner. We will revisit this problem in much more detail in Chapter 7.

The task at hand is to construct a mathematical model for the carabiner which allows us to predict the carabiner's structural response. Hence, there are some choices to be made that, although interconnected, can be organized as follows:

- How to represent the loads.
- How to represent the restrictions to motion (the displacement boundary conditions).
- How to represent the actual structure.

These are addressed below.

How to represent the loads

The load is transferred from the lower rope to the carabiner by the contact between the rope and the carabiner. A simplified way of modeling this action is by a concentrated load applied to the carabiner whose magnitude represents the load applied by the rope.

How to represent the restrictions to motion

We note that the motion is not clearly restricted at any specific point of the structure. Actually, the opposing ropes and the carabiner find an equilibrated position. Hence to restrict the motion we may consider the final equilibrated position. At one side the load is imposed as explained above and at the opposite side we prevent the motion. Of course, the reactions associated with restricting the motion equilibrate the applied load, see Figure 1.4.

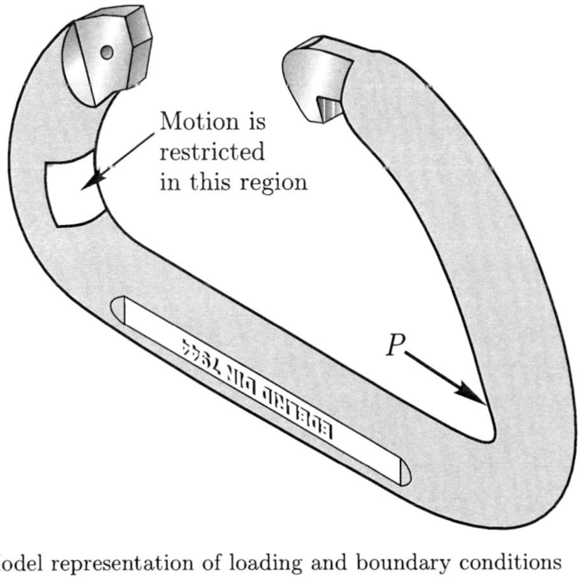

Fig. 1.4. Model representation of loading and boundary conditions

We note that since we are considering the carabiner to be open, Figure 1.4 shows only the part which is structurally relevant.

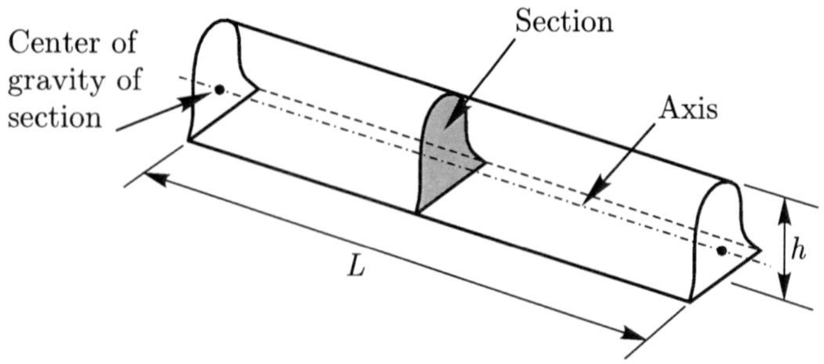

Fig. 1.5. A typical bar obtained by extruding a plane section through a straight line, normal to the plane of the section. The line, or axis, passes through the plane's center of gravity

How to represent the actual structure

In the first instance we may choose to represent the carabiner by an assemblage of straight bars. As we discuss in Chapter 4, the bar model corresponds to a structural theory in which the behavior of long solids is modeled. Figure 1.5 shows a typical bar. The solid of Figure 1.5 is called a bar when $L >> h$ (typically $\frac{1}{100} \leq h/L \leq \frac{1}{10}$). The bar model relates the loading applied on the bar to the displacements and internal forces/stresses of the bar. Of course, in the curved region, we need several straight bar elements to represent, approximately, the geometry. In Figure 1.6 we show a bar model of the carabiner. In this figure only the bar axis is represented and there is a straight bar element between each of two consecutive diamond symbols. The loading is also shown and the restriction to the movement is modeled by preventing the section of the bar at B to either displace or rotate. The solution gives us the displacements, rotations and stresses at each point of the model.

1.2.2 The case for hierarchical modeling

Suppose that we have limit values for the displacements/rotations and for the stresses, referred to as variables, below which we are sure that the carabiner considered in Section 1.2.1 would well function and not break. Then, the question that naturally arises is: "How accurate are the calculated values obtained by solving the model of Figure 1.6 for these variables?" Of course, we do not know the actual values of these variables in the actual physical problem and we need to answer this question based on mathematical modeling *alone.*

The reader might appreciate that several assumptions were made in the definition of the mathematical models. Besides imposing the loading and the displacement boundary conditions in a simplified manner, possibly the strongest assumption is to represent the carabiner as an assemblage of straight bars. Clearly, these assumptions will affect the accuracy of the predictions obtained for the variables when compared to the actual situation in a

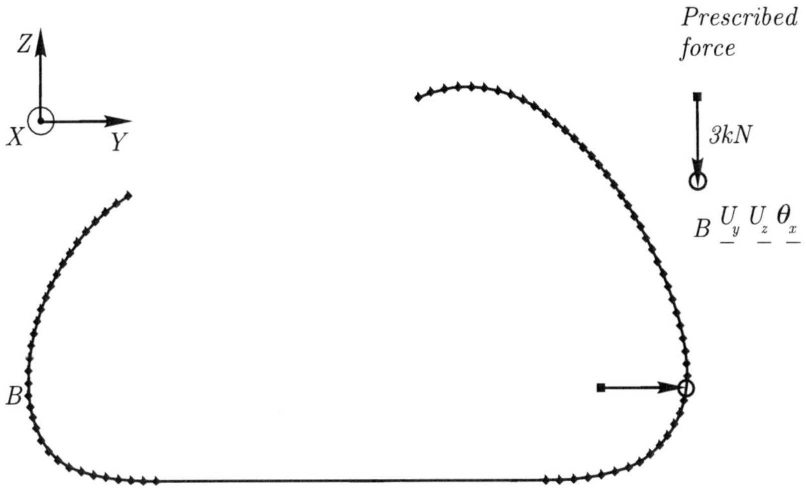

Fig. 1.6. Straight bar model in Y-Z plane. The material of the carabiner is aluminum with $E = 70$ GPa, $\nu = 0.33$

laboratory experiment (or in an actual rock climbing event). It is the purpose of hierarchical modeling to obtain good accuracy in the analysis results, and thus reliable analysis data for the design of the structure.

Let us introduce the hierarchical modeling process using the same demonstrative problem – the carabiner, see also Chapter 1 of Bathe, 1996, and Bathe, Lee and Bucalem, 1990.

1.2.3 Demonstrative hierarchical modeling example – a carabiner

The essence of hierarchical modeling is to set up and solve additional – and more accurate models – of the structure considered. For the carabiner, we can set up another more accurate mathematical model – a second model – by considering, for example, a curved bar structural theory. For such a theory the bar axis does not need to be straight and can be represented by a curve. This second model would be constructed by an assemblage of straight and curved bars. By representing the regions of the carabiner which are actually curved by curved bars we should arrive at a more accurate representation of the structural behavior of the carabiner. Of course, the curved bar structural theory is more complex and, hence, the model solution is more difficult to obtain.

The second model using curved bars is somewhat natural because there are curved parts in the carabiner. This second model is an example of the general and important fact that for a given physical problem there is not only one mathematical model to represent it. Indeed, there are always a number of mathematical models for the same physical problem. This idea is summarized in Figure 1.7, where it is shown that for a physical problem we can actually consider a sequence of mathematical models. The model number indicates an hierarchical order, *i.e.*, model 2 represents, through its assumptions, more phenomena of the physical problem than model 1. Therefore it leads to more complete predictions. This is always the case as we increase the index number of our model.

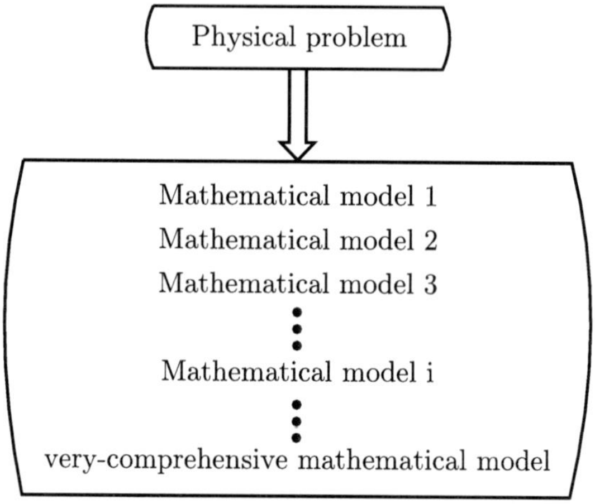

Fig. 1.7. Sequence of mathematical models for a physical problem

We can think – at least conceptually – of the highest order model as a "very-comprehensive mathematical model". This model would be the most precise and complete representation of the physical problem at hand. The very-comprehensive mathematical model is not to be actually solved but to provide a conceptual reference to which we compare the lower-order models.

For the carabiner problem, we have already described models 1 and 2. Let us consider, for illustrative purposes, a few higher-order models in the sequence.

If we evaluate the ratio of the radius of curvature to the thickness of the bars, we recognize that the curved bar model can only approximately represent the true 3-D behavior in the curved regions. Replacing the curved bar model by the 3-D elasticity model leads us to model 3.

A fourth model is motivated by evaluating the order of the displacements of the carabiner. Considering the material of which the carabiner is made (aluminum) and a typical design load (3 kN), deformed and undeformed configurations corresponding to the solution of the assemblage of straight bars are shown to scale in Figure 1.8, *i.e.*, without magnification. In this figure the solid line represents the undeformed bar axis, whereas the dotted line represents the same axis for the deformed configuration. We see that the

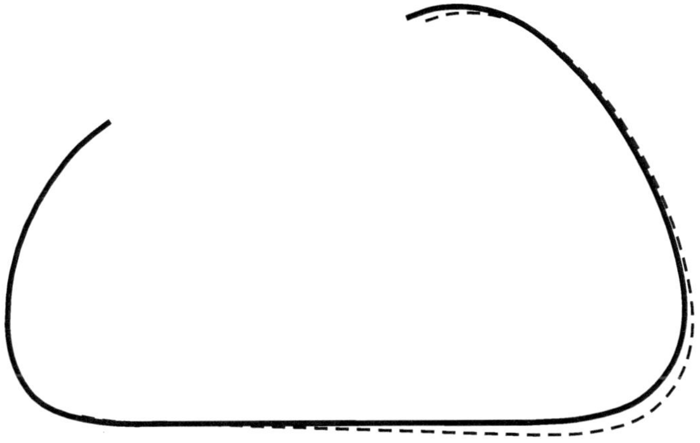

Fig. 1.8. Deformed (dashed line) and undeformed (solid line) configuration for the bar model

displacements are quite significant when compared to the dimensions of the structure. Hence, the modeling would more accurately represent the actual physical situation by considering that the geometry of the structure changes as the load is increased. This additional consideration would give rise to a geometrically nonlinear model (model 4).

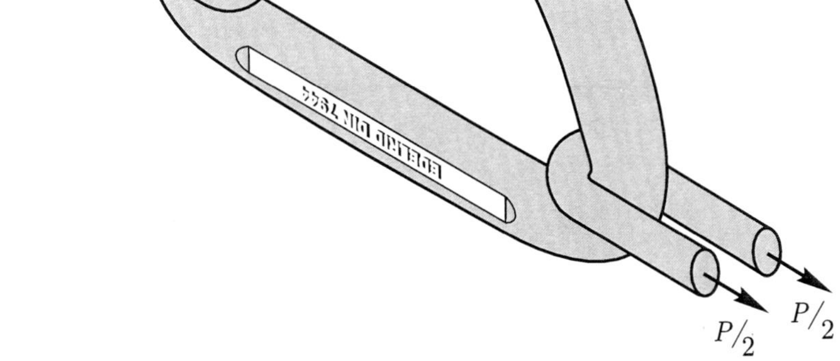

Fig. 1.9. Model of the carabiner with ropes included

We note that the modeling improvements discussed so far lead to a better description of the structural behavior of the carabiner when the models are all subjected to the loading and the displacement boundary conditions in Figure 1.4. Therefore, a fifth model could be established to improve the representation of the loads and boundary conditions. For example, we might consider the ropes as part of the model and represent the contact between the ropes and carabiner surfaces. The load would be introduced at the end of the ropes as schematically shown in Figure 1.9. This model is referred to as model 5 and its solution would be much more difficult (and therefore more expensive) to obtain than those of the former models. We summarize in Figure 1.10 the modeling process for the carabiner in which we show the sequence of models discussed so far.

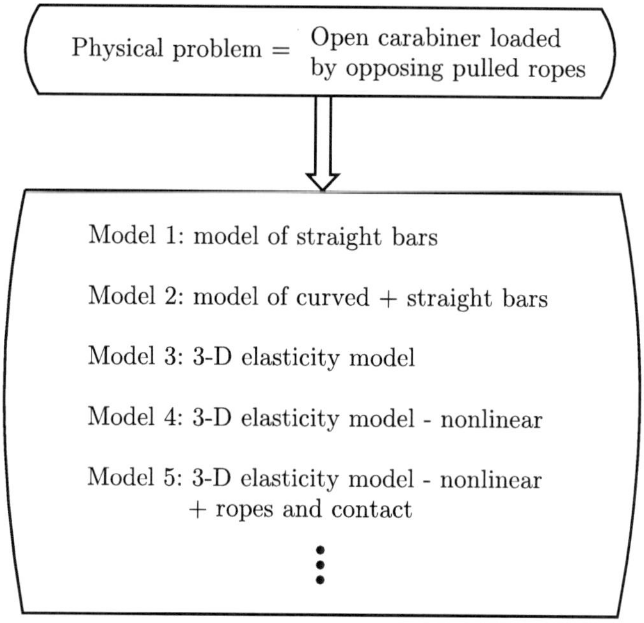

Fig. 1.10. Sequence of hierarchical models for the carabiner

Next we need to return to the question which we posed for model 1: How accurate are the predictions of the model for the variables of interest when compared to those of a laboratory test or a field event?

Suppose that we solve models 1 to 3 for the variables sought. In order to fix ideas, let us consider the vertical displacement at point A of the carabiner, as shown in Figure 1.11, as being representative of the displacement pattern of the structure. Let Δ be the required precision for δ. Figure 1.12 shows the solutions for the models.

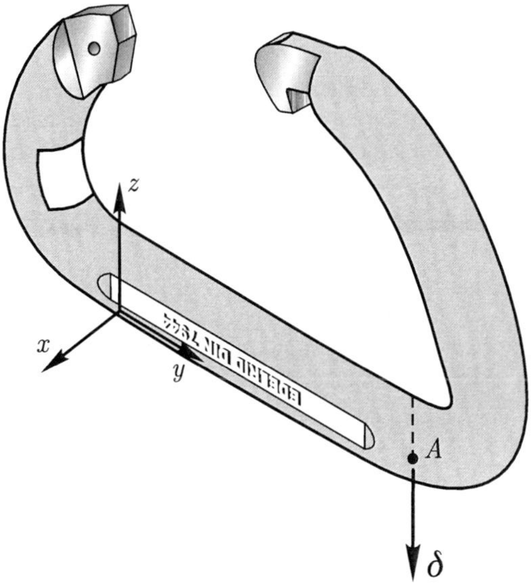

Fig. 1.11. Location of point A for which the vertical displacement is sought

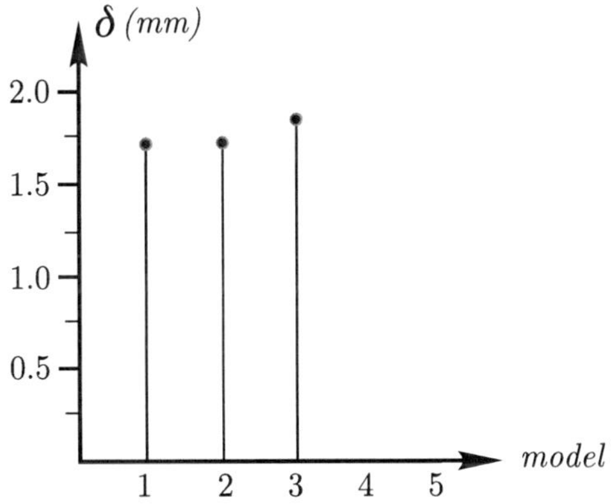

Fig. 1.12. Vertical displacements of point A in millimeters

A reliable model, say model i, for predicting the displacement δ would be a model for which the difference between its prediction δ_i and the displacement δ_a of the actual structure (measured in a laboratory experiment), would be less than Δ in absolute value, that is, $|\delta_i - \delta_a| < \Delta$.

However, since we are restricted to mathematical modeling and do not have the structure to experiment with in a laboratory, we do not have access to δ_a. Our best estimate for δ_a would be the solution of the very-comprehensive mathematical model, which we denote by δ_∞. Therefore, our criterion is $|\delta_i - \delta_\infty| < \Delta$. But also, we do not know δ_∞ since the very-comprehensive mathematical model will never be solved. However, the analyst knows all the phenomena that would be represented in the very-comprehensive mathematical model and that can influence this displacement. By comparison with the phenomena which are already represented in the mathematical model i, the analyst will have to assume whether or not $|\delta_i - \delta_\infty| < \Delta$. To make this decision properly is of major importance and requires knowledge of the structural problem and of the details of the mathematical models already solved as well as of those not yet solved. Of course, the solutions of the lower-order models in the sequence will already have given some insight into the behavior of the structure.

Assume that the analyst is not sure if a particular phenomenon, which is present in the very-comprehensive mathematical model and which is not taken into account in the currently considered model, might have an influence in the prediction of δ by a quantity greater than Δ. Then the analyst would have to conceive and include in the sequence a next model which should incorporate that particular phenomenon.

With the above considerations in mind, let us return to the data presented in Figure 1.12. We observe that the displacement of point A is changing (here increasing) as we proceed from model 1 to model 3. Some additional change in displacement will occur as the model is further refined (refer to models 4 and 5), and it is up to the analyst to decide, by experience or otherwise, if the results of a given model i are sufficiently accurate, that is, whether it can be assumed that $|\delta_i - \delta_\infty| < \Delta$.

Of course if the design requirement is a limit value, *i.e.*, for this example it could be $\delta < \delta_{Limit}$, the same reasoning is valid with the conditions $|\delta_i - \delta_\infty| < \Delta$ (as before) and additionally $\delta_{Limit} - \delta_i > \Delta$ which would allow to conclude that $\delta_\infty < \delta_{Limit}$[1]

From this discussion we can already draw some fundamental observations regarding the hierarchical modeling process:

- The process gives a logical and systematic framework to analyze engineering structures.
- The process relies on the understanding of the analyst of the problem at hand and of the mathematical models available. This understanding is revealed in the choice of the hierarchical models, the conceptual view of the very-comprehensive model and the interpretation of the results.
- The choice of the sequence of models is influenced by what variables are to be predicted and with which precision.

We can now further detail the diagram of Figure 1.7. The resulting diagram is shown in Figure 1.13. We implicitly assumed so far that we are always

[1] In this book we are always considering all the variables to be deterministic since our concern is with the modeling issues. Of course, additional considerations arise for a stochastic description of the variables. By the same argument we are supposing that an applicable safety factor is used when determining Δ or δ_{Limit}

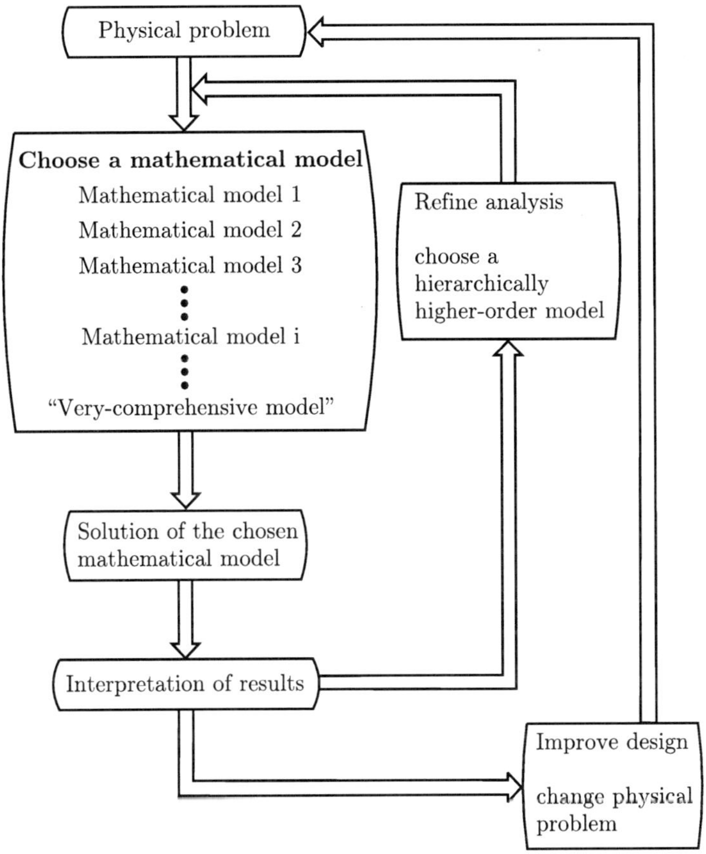

Fig. 1.13. The process of engineering design

able to solve the chosen mathematical model. We defer to later the discussion of how to solve the models when an analytical solution is not readily available.

Referring to Figure 1.13, we note that associated with the "interpretation of results" there is a decision to be made. At this point the analyst has to determine whether to assume that the modeling of this particular physical problem has "converged" or not. In other words, the analyst needs to decide if the variables to be predicted have been found with the required precision measured with respect to the very-comprehensive mathematical model ($|\delta_i - \delta_\infty| < \Delta$ in the terminology of the above discussion).

Every mathematical model in the sequence which gives the variables to be predicted with the required precision is termed a *reliable mathematical model.* As an engineer, when performing the hierarchical modeling procedure, we would like to find among all reliable models the one of lowest order which is therefore termed the *most effective mathematical model*. These two concepts are very important and will be used throughout the book.

In the diagram of Figure 1.13, we show that from the engineering point of view the modeling does not end when the hierarchical modeling process for a particular physical problem has converged. The predictions might suggest design improvements leading, for example, to geometric or material changes to the physical problem. Of course, such changes actually lead to a new physical problem and the hierarchical modeling should be restarted for the new physical problem. We would then restart the analysis process with more understanding based on the earlier hierarchical modeling, allowing us to take significant shortcuts in the choice of models.

1.3 Remarks on the hierarchical modeling process

The discussion presented so far could have been written, with small modifications, two hundred years ago. We should, therefore, point out what changes have occurred that make the hierarchical modeling process so important nowadays.

The key aspect is that due to developments in computational engineering during the last decades we can actually solve many more and much more complex mathematical models. Therefore, the relevance of using a structured approach towards analysis has dramatically increased and the hierarchical modeling provides this approach. In addition, of course, the fact that more complex mathematical models can now be solved has also spurred the development of new models, and regarding structural analysis, in particular, for the nonlinear analysis of solids and structures.

As it is well known, we are able to find analytical solutions only to very low-order models. Therefore, in the past, setting up a sequence of models would not result into significant benefits because the engineering decision of which model to choose was seldom based on whether such model was reliable or not. This decision had to be largely based on whether the next model in the sequence could actually be solved. On the other hand, in today's engineering environments very complex models can be solved using finite element methods. Hence the solution of higher-order models in structural and solid mechanics can now be achieved.

Of course, incredibly bold structures were built long time ago – even without any analysis at all. This was possible by introducing large safety factors, using the experience gained previously with similar structures and by relying on the physical intuition of great builders/designers. Clearly, the modern approach of analysis by no means diminishes the great earlier accomplishments in structural engineering. On the contrary, these accomplishments should stand out as a paradigm of what has been achieved without the finite element analysis tools that we have today. These tools enable us in a very exciting manner to now design even bolder, safer and more economical structures.

The finite element method is the predominant method to solve structural engineering problems. The study of the origin, history and development of the finite element method is *per se* rich and fascinating. We refrain ourselves from exploring in detail these items but we would like to only mention a few facts below, see Bathe, 1996, Bathe, 2009.

The finite element method, as used in engineering, was proposed in the 1950s based largely on intuition by *structural engineers* as an attempt to generalize matrix structural analysis from bars to continuum-like structures. At that time, with the development of digital computers, matrix structural analysis was successfully applied to solve relatively complex bar structures. Airplane wings – which were important structures to be analyzed – require, however, the representation of plate/shell like behavior. These needs led to the first finite element formulations for 2-D continua.

In the early 1970s, the tremendous value of the method to solve real engineering problems was widely realized when actual finite element computer programs became available to engineering analysts, and not just researchers. These tools made it possible that the method could be used in many industries and research centers to solve relatively complex problems of engineering interest. Various mathematical foundations were also identified, and from then onwards, the method developed at an exponential rate.

The availability of general purpose finite element computer programs brought a myriad of modeling options, which, associated with the ever increasing power and availability of computers, have made it possible to perform today on PCs analyses of very complex nature. Indeed, the widespread availability of powerful finite element resources has drastically altered the potential for analysis of very complex structures, Bathe 2007. The use of a conceptual framework for modeling, which was implied in the past but not explicitly detailed and used due to the scarcity of models which could be solved, is nowadays a requirement – although frequently overlooked.

Referring to the diagram of Figure 1.13 we could now replace the block "solution of the chosen mathematical model" by "solution of the chosen mathematical model by finite element procedures". In this way, considering the present software/hardware resources, we can proceed quite far in the sequence of mathematical models. For the carabiner problem we can easily find the solution of model 5 and beyond. However, we need to be aware of some new issues and concerns that arise, namely:

- Since powerful analysis resources are available, there is a tendency to immediately start the analysis using very complex models. This is actually bad engineering, since complex models usually cloud the basic structural behavior which is best revealed by low-order models. Also, the high-order models usually generate a large amount of results which are difficult and costly to interpret. On the other hand, it is quite different a situation to solve a complex high-order model after having solved lower-order models and considering that a higher-order model is necessary. Then the solution of the lower-order models has given enough understanding of the problem to recognize that for the response sought, and the required precision, the higher-order model is needed. We emphasize that we should not directly solve a high-order model only because the resources are available.

- The finite element solution is an approximation to the exact solution of the mathematical model considered. However, using *reliable finite element procedures*, we can obtain a finite element solution which is as close to the exact solution of the mathematical model as desired. Clearly, the use of reliable finite element procedures is imperative. The issues related to the reliability of finite element procedures will be discussed later in the book.
- The possibility of conceiving complex mathematical models and actually solving them is placing strong demands on the analyst. Besides knowing the formulations and the assumptions contained in the low-order models and being able to solve them analytically, the analyst needs to also be able to construct and solve applicable high-order models and be able to interpret the results obtained. Such demands for expertise are even higher when we consider coupled problems such as the solution of thermo-mechanical problems or fluid-structure interactions.

At this point a word of caution is in order. Of course, it is not reasonable to expect that every person working in structural analysis would have the knowledge and the experience required to tackle all high-order models. However, the above observations do apply for the engineering team as a whole if state-of-the-art modeling is to be used, and they give an idea of the challenges in this activity.

The analyst – the person or the team of people – who will perform the hierarchical modeling process should possess the required abilities to develop and solve hierarchical models. We emphasize that these requirements comprise:

- To know the candidate mathematical models involved: the analyst should be comfortable with the formulations, the assumptions contained in the models, the hypotheses and the basic behavior that can be predicted by the models.
- To know the related finite element procedures. The analyst should be able to set up appropriate finite element representations and guarantee that solutions have been obtained with the required accuracy measured with respect to the exact solution of the mathematical model.

These requirements highlight the basic aspects that we want to present in this book such that an analyst can harvest more of the full spectrum of mathematical models (and, hence, finite element analysis possibilities) currently available. We believe that the process of hierarchical modeling focused upon in this book is an art and the basis of modern analysis in engineering and the sciences; and we hope that this book will be valuable in the practice of this art – rich and full of challenges requiring knowledge and ample imagination.

1.4 Outline of book

We are now in a good position to outline the contents of the book. In the next chapter we discuss some fundamental concepts in structural mechanics which are applicable to all models discussed in the book. Then we apply these concepts to the simplest mathematical model – the one-dimensional truss/bar model and truss structures.

In Chapter 3 we motivate the need for two and three-dimensional theories and formulate the problem of 3-D linear elasticity.

Chapter 4 is dedicated to the formulation of mathematical models for structures. We proceed from linear plane elasticity to simple shells, discussing also bars and plates.

In Chapter 5 we discuss the equivalence between differential and variational formulations. The variational formulation is the basis of the finite element method. The emphasis is on a mechanical and physical approach through the introduction of the principle of virtual work for the mathematical models previously formulated in Chapters 3 and 4.

The finite element method is introduced in Chapter 6, first for the one-dimensional problem and then for two and three-dimensional analyses. The emphasis is placed on basic aspects and on the mechanical properties that the finite element solution satisfies. Effective finite elements which can be recommended for engineering practice are also presented. A brief discussion on locking, which was the major obstacle to arrive at reliable finite elements for structures, is also given.

In Chapter 7 we synthesize what we presented by applying the hierarchical modeling process to selected problems, using the mathematical models of Chapters 3 to 4 and the finite element procedures of Chapter 6.

Finally, in Chapter 8 we briefly address basic issues encountered in non-linear analyses.

All finite element solutions given in this book have been obtained using ADINA, 2008.

2. Fundamental steps in structural mechanics

It is easy to recognize that all structures in nature are three-dimensional. However, the reader has surely also encountered structures in simple forms such as an assemblage of geometrically slender members. As we discussed in the previous chapter, these simplified forms lead directly to models that are in accordance with the hierarchical modeling process. In fact, there are a number of structural mechanics mathematical models such as bars, beams, plates and shells, among others, that provide a convenient and efficient way to model structural behavior for design and analysis purposes. These mathematical models will be discussed in the following chapters due to their importance in the structural modeling process.

The objective of this chapter is to address some fundamental conditions that should be met whichever structural mathematical model is used. The discussion will be placed initially in a very general setting – the motion of a deformable body – and we will extract fundamental conditions which lead to "static equilibrium". The simplest structural mathematical model – the truss model – is then studied to exemplify, in a simple setting, the basic steps associated with the formulation and solution of structural mechanics problems. These steps are then systematized leading to the modern matrix approach of analysis.

2.1 General conditions

There exist some fundamental general conditions regarding the analysis of a structure which we present in this section.

2.1.1 Motion of a deformable three-dimensional body

Let us consider the motion of a three-dimensional body in 3-D space. The body can be idealized as a collection of particles which are assumed to have a mass density. At a given time, the set of positions occupied by the body material particles which defines a region in 3-D space is referred to as a configuration. In Figure 2.1 we show two configurations ${}^0V$ and ${}^tV$.

The configuration ${}^0V$ represents the configuration at the onset of the motion, *i.e.*, for time $t = 0$, which is also referred to as the undeformed

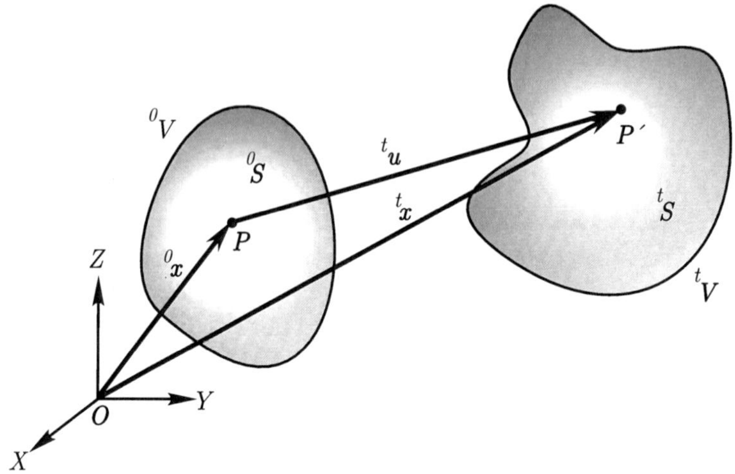

Fig. 2.1. Two configurations of a deformable body in a selected stationary Cartesian coordinate system

configuration, and ^tV stands for a generic deformed configuration[1], *i.e.*, for time t. We denote by ^0S and ^tS the boundary surfaces associated with ^0V and ^tV.

We adopt a Lagrangian description of the motion, *i.e.*, we "follow" the complete motion of the material particles, from time 0 to time t. The position of a particle at time t is given by[2] $^t\mathbf{x} = \mathbf{x}\left({}^0\mathbf{x},\, t\right)$ where $^0\mathbf{x}$ is the position vector of this particle at time $t = 0$.

The motion of the body is governed by the action of the "rest of the universe" onto the body. This action is represented by forces which are generically referred to as externally applied forces. These may partly be unknown, namely, when displacements of the body are prescribed, see below. There are two kinds of externally applied forces. There is the field of forces per unit of volume represented by $^t\mathbf{f}^B\left({}^t\mathbf{x},\, t\right)$, called body forces. The most common example of such a field is given by gravity acting on the body material particles. And there are externally applied forces on the surface of the body ^tS represented by a field of tractions – forces per unit of surface area – denoted by $^t\mathbf{f}^S\left({}^t\mathbf{x},\, t\right)$. These forces typically arise as a consequence of the contact of

[1] In the 20th century, continuum mechanics was also presented as a mathematical theory referred to as rational continuum mechanics, see Truesdell, 1977 and references therein. Although we recognize the importance of these works, we keep the mathematics in our presentation as simple as possible

[2] Since t is an argument to our function, we could have simply written $\mathbf{x} = \mathbf{x}\left({}^0\mathbf{x},\, t\right)$ to describe the particle position at time t. However, we choose to use the notation $^t\mathbf{x} = \mathbf{x}\left({}^0\mathbf{x},\, t\right)$ with the left superscript t to emphasize that we consider the configuration at the specific time t. This approach is also followed for many other quantities, see also Bathe, 1996

the body with other bodies, that is, the surrounding media, and included are here the effects of the restraints applied to part of the body's surface.

Let us define the resultant of all external forces applied to the body at time t by

$$ {}^t\mathbf{R} = \int_{{}^tV} {}^t\mathbf{f}^B \, d\,{}^tV + \int_{{}^tS} {}^t\mathbf{f}^S \, d\,{}^tS \tag{2.1} $$

and the resultant moment at time t about the system origin $\mathbf{O}$ by

$$ {}^t\mathbf{M_O} = \int_{{}^tV} {}^t\mathbf{x} \times {}^t\mathbf{f}^B \, d\,{}^tV + \int_{{}^tS} {}^t\mathbf{x} \times {}^t\mathbf{f}^S \, d\,{}^tS. \tag{2.2} $$

Here, ${}^t\mathbf{R}$, ${}^t\mathbf{M_O}$ are respectively called the external force and moment resultants.

The motion of the body is governed by two principles. The first one is the principle of linear momentum which states that

$$ {}^t\mathbf{R} = \frac{d}{dt} \int_{{}^tV} {}^t\rho \, {}^t\dot{\mathbf{x}} \, d\,{}^tV \tag{2.3} $$

where ${}^t\rho\,({}^t\mathbf{x},t)$ is the mass density function at time t and ${}^t\dot{\mathbf{x}} = d\,{}^t\mathbf{x}/dt$ is the material velocity at ${}^t\mathbf{x}$. The second principle, the principle of angular momentum, is given by

$$ {}^t\mathbf{M_O} = \frac{d}{dt} \int_{{}^tV} {}^t\mathbf{x} \times {}^t\rho \, {}^t\dot{\mathbf{x}} \, d\,{}^tV. \tag{2.4} $$

These two principles need to be satisfied, in any motion, in an inertial reference system[3].

The principle of linear momentum for the dynamics of a point mass is Newton's 2nd Law, *i.e.*, $\mathbf{R} = \frac{d}{dt}(\mathbf{p})$ where $\mathbf{p}$ is the linear momentum, $\mathbf{p} = m\mathbf{v}$, with m the point mass, $\mathbf{v}$ its velocity and $\mathbf{R}$ the resultant of the forces acting on the point mass. We note that the right-hand side of equation (2.3) gives the time derivative of the vectorial sum of the linear momenta of the material particles of the body.

The principle of angular momentum for a set of point masses is given by $\mathbf{M_O} = \frac{d}{dt} \sum_i \mathbf{x}_i \times m_i \mathbf{v}_i$ where $\mathbf{M_O}$ is the resultant moment of all forces acting on the point masses m_i about the origin $\mathbf{O}$ and $\mathbf{x}_i$ and $\mathbf{v}_i$ are the position and velocity of point mass i, respectively. Of course, the right-hand side of equation (2.4) represents the integrated effect for the mass particles in the continuum.

We emphasize that the fields of forces ${}^t\mathbf{f}^B$ and ${}^t\mathbf{f}^S$ represent all the influence of the "rest of the universe" on the motion of the body considered. Note

[3] For structural and solid mechanics applications a reference system which is either at rest or in rectilinear motion with constant velocity with respect to the planet Earth can be taken as inertial

that this general statement includes the very common situation in which the motion of a part of the body's surface is restrained. The physical devices which constrain the motion of the body's surface also belong to the "rest of the universe" and their effect on the body's motion also results into a field of surface tractions. These physical devices are generically referred to as restraints or supports.

Corresponding to these concepts, let us denote by ${}^tS_u$ that part of the body's surface ${}^tS$ which has its motion restrained and by ${}^tS_f$ the complementary part of the body's surface. Therefore, on ${}^tS_f$ there are the interactions with other bodies represented by surface tractions but no restraints. The restraints on ${}^tS_u$ give rise to surface tractions which are referred to as reactive surface tractions, or mostly, simply as reactions. This model situation is schematically summarized in Figure 2.2. The surface tractions on ${}^tS_f$ and the body forces ${}^t\mathbf{f}^B$ are the external loads.

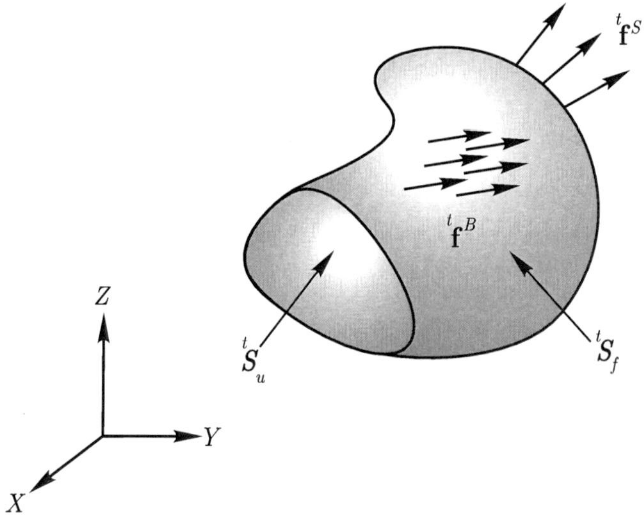

Fig. 2.2. Restrained body under external actions

In the above description we seem to have assumed that on ${}^tS_u$ all displacements (into the X, Y, and Z directions) are restrained. However, in three-dimensional analysis, the particles on the surface ${}^tS_u$ have three independent displacement degrees of freedom and only some of them may be restrained. For example, a particle may be prevented from moving into directions X and Y and free to move into direction Z. Hence, the definition of ${}^tS_u$ and ${}^tS_f$ given above should be generalized and we define ${}^tS_u$ and ${}^tS_f$ for each displacement degree of freedom. That is, we define ${}^tS_u$ and ${}^tS_f$ for the displacement degree of freedom into the direction X and also into the directions Y and Z. Therefore, a given material particle may belong to ${}^tS_f$ for some displacement degree of freedom and to ${}^tS_u$ for another degree of freedom.

2.1.2 Properly supported bodies

A solid[4] is *properly supported* if the supports prevent rigid body motions for any external loading. Hence, in this case the displacements of the material particles can only result from some straining of the material.

According to this definition, a properly supported rigid body can not display motion since rigid body motions are prevented and straining of the material can not occur. This concept can be also used to characterize a properly supported deformable body: indeed, assuming that this deformable body were rigid, if this rigid body can not display motion, then the deformable body is properly supported.

Considering a rigid body which is properly supported, we can conclude directly from equations (2.3) and (2.4) that ${}^t\mathbf{R} = \mathbf{0}$ and ${}^t\mathbf{M_O} = \mathbf{0}$ since the velocity field is always zero (${}^t\dot{\mathbf{x}} = \mathbf{0}$). Here, we note that while a constant velocity field also leads to ${}^t\mathbf{R} = \mathbf{0}$ and ${}^t\mathbf{M_O} = \mathbf{0}$, kinematic restraints that prevent rigid motions represent of course the sufficient condition for a rigid body to display no motion for any externally applied loading.

When assessing whether a deformable body is properly supported or not, we frequently investigate if the associated rigid body is properly supported because this is usually simpler.

In practice, we frequently find that a number of solids are connected by joints. In these cases the above concepts are also directly applicable, but each individual solid — considered just like the single solid above — must then be also properly supported by the rest of the assemblage. If this is not the case, despite the fact that global rigid motions of the whole assemblage are prevented by the supports, the assemblage of solids is said to contain one or more *internal mechanisms*. Corresponding to each internal mechanism there is an independent rigid motion that the solids can undergo while being still connected at the joints.

In all these cases when rigid motions are possible, we sometimes also simply say that the solid or the assemblage of solids is unstable (see Section 8.3 for a further discussion).

[4] The notion of a body encompasses both solids and fluids. In this book we are interested only in solids and suppose that the reader has an intuitive understanding of the behavior of deformable solids when contrasted with the behavior of a fluid. Frequently we use the term "body" with the implicit understanding that we are actually considering a solid

2.1.3 Internal actions

Consider next a generic part of a properly supported body in the configuration t with volume $\Delta^t V$ and total surface area $\Delta^t S$. Part of that surface area $\Delta^t S$ is the result of having sliced $\Delta^t V$ from the total body (refer to Figure 2.3). When subjected to external loads the deformable body develops internal forces. Let us represent these internal forces, which are given per unit of area, by ${}^t\mathbf{t}$. We postpone a more detailed discussion of these internal forces per unit of area – the stresses – until the next chapter. In Figure 2.3 a typical situation is summarized where the isolated part with all actions on it is shown. The figure showing the isolated part is referred to as a "free body diagram".

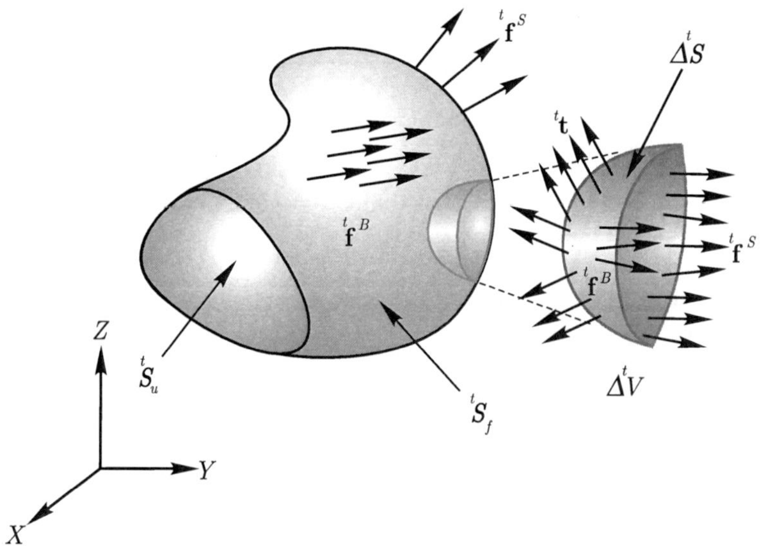

Fig. 2.3. Representation of a generic isolated part of the body showing the internal forces per unit of area: the "free body diagram"

Considering $\Delta^t V$, the principles of linear and angular momenta are, of course, directly applicable as long as we consider all the forces acting on $\Delta^t V$ and $\Delta^t S$ including the field ${}^t\mathbf{t}$. The forces ${}^t\mathbf{t}$ are now part of the "rest of the universe" acting on the body $\Delta^t V$ (and indeed can be thought of as forces ${}^t\mathbf{f}^S$). Hence, the above discussion is directly applicable: the body[5] considered is simply $\Delta^t V$ instead of ${}^tV$.

[5] Actually, we could have introduced the principles of linear and angular momenta for any part of the body, since all that matters to establish (2.3) and (2.4) is to represent all the actions of the "rest of the universe" on whatever body we consider: any body considered will always be a part of the universe

2.1.4 Assumptions for static analysis

Consider a properly supported body which is initially at rest and not subjected to external actions. Assume that the external loads are then applied very "slowly" from zero to their final values such that the induced accelerations ${}^t\ddot{\mathbf{x}}$ and velocities ${}^t\dot{\mathbf{x}}$ can be neglected, that is, the dynamic effects are negligible. These assumptions characterize *static analysis* and, then, equations (2.3) and (2.4) simplify to

$$ {}^t\mathbf{R} = \mathbf{0} \tag{2.5}$$

and

$$ {}^t\mathbf{M}_\mathbf{O} = \mathbf{0}. \tag{2.6}$$

Equations (2.5) and (2.6) are the principles of linear and angular momenta for static analysis. In this case, the variable t (used for "time") should be interpreted as a label used only to specify the external loading at time[6] t.

A useful concept is that of a system of forces in *static equilibrium.* A system of forces including body forces and surface tractions is said to be in static equilibrium in the configuration at time t if ${}^t\mathbf{R} = \mathbf{0}$ and ${}^t\mathbf{M}_\mathbf{O} = \mathbf{0}$. Hence in static analysis, the force system of all external actions on the body (due to all externally applied loads and all reactions) is to be always in static equilibrium (due to (2.5) and (2.6)).

Finally, we note that if (2.5) is satisfied and (2.6) is satisfied with respect to a point $\mathbf{O}$, then (2.6) is also satisfied when we select any other point $\mathbf{O}'$ instead of $\mathbf{O}$. In fact, since ${}^t\mathbf{R} = \mathbf{0}$ and

$$ {}^t\mathbf{M}_{\mathbf{O}'} = {}^t\mathbf{M}_\mathbf{O} - \mathbf{r}_{\mathbf{O}\mathbf{O}'} \times {}^t\mathbf{R}$$

where $\mathbf{r}_{\mathbf{O}\mathbf{O}'}$ is the vector from $\mathbf{O}$ to $\mathbf{O}'$, we obtain ${}^t\mathbf{M}_\mathbf{O} = {}^t\mathbf{M}_{\mathbf{O}'}$. Hence, the resultant moment about $\mathbf{O}'$ is equal to the moment about $\mathbf{O}$. This result implies that if a system of forces is in static equilibrium the moment resultant about any point is zero.

Of course (2.5) and (2.6) also apply to the generic part of the body $\Delta^t V$, but as pointed out above, ${}^t\mathbf{t}$ must then be included as actions from the "rest of the universe" onto this body part. Hence, in static analysis the condition is that the force system given by ${}^t\mathbf{f}^B$, ${}^t\mathbf{f}^S$ and ${}^t\mathbf{t}$ acting onto any part of the body is in static equilibrium. This fact will be used throughout the book.

2.1.5 Assumptions for a linear static analysis

In addition to considering static conditions (and not dynamic effects), most of this book is concerned with situations for which the material internal

[6] Of course, strictly, time is always present and can not be "switched off", and hence "t" is a convenient label to specify the loading

force-displacement relationships are linear and the loads are such that the body displacements are very small; indeed we can assume in the analysis that the displacements are infinitesimally small. As a result, when we use the linear and angular momentum principles given by equations (2.3) and (2.4), we assume that the deformed configurations are geometrically the same as (identical to) the undeformed configuration. This means that the equilibrium of the body is considered neglecting all body displacements – that is, as if the body did not displace.

With these assumptions

$$ {}^{t}\mathbf{R} = \int_{{}^{0}V} {}^{t}\mathbf{f}^{B}\, d^{0}V + \int_{{}^{0}S} {}^{t}\mathbf{f}^{S}\, d^{0}S \tag{2.7} $$

$$ {}^{t}\mathbf{M}_{\mathbf{O}} = \int_{{}^{0}V} {}^{0}\mathbf{x} \times {}^{t}\mathbf{f}^{B}\, d^{0}V + \int_{{}^{0}S} {}^{0}\mathbf{x} \times {}^{t}\mathbf{f}^{S}\, d^{0}S. \tag{2.8} $$

We note that all integrals are using the undeformed configuration.

In general, in linear static analysis we are only interested in the final configuration resulting from the applied loading. In such case, there are only two configurations of interest, the initial configuration – that prior to the application of the loads – and the deformed configuration reached due to the application of the loads. When we assume this situation, we drop the left superscript t associated with time.

2.1.6 Summary

Before we proceed to the next section, we would like to summarize the most important points discussed:

- The motion of a deformable body was characterized in a very general setting, *i.e.*, a body of arbitrary shape was considered undergoing arbitrary motions.
- The interaction of the body with the "rest of the universe" was characterized by fields of forces and displacement restraints. The fields of forces include forces due to the restraints.
- Two general principles, the principle of linear momentum and the principle of angular momentum were stated; they govern every motion of the body (and of any part thereof).
- The concept of a properly supported body was introduced. The concept is based on purely kinematic conditions and reflects the idea that a properly supported deformable body can experience displacements due only to the straining of its material fibers, *i.e.*, an associated rigid body displays no motion for any applied loads.
- The assumptions of linear static analysis were introduced.
- The concept and conditions for static equilibrium were given.

These general concepts and facts are applicable to any structural system. Indeed, they provide the theoretical framework upon which the mathematical models presented in this book will be built.

Until now we mentioned structures and structural systems without a precise definition of "a structure". In fact, we relied on the intuitive understanding of the reader. We can now better characterize a structure or structural system.

We define a structure or a structural system as an assemblage of three-dimensional deformable bodies. Although this definition encompasses all structures, the modeling of bodies as three-dimensional frequently does not lead to the most efficient mathematical models to predict their behavior. Fortunately, in many cases, deformable bodies possess specific characteristics that allow a more effective modeling. These characteristics are linked to the geometries of the bodies, the kind of external loading, the boundary conditions and the connections between bodies. The more effective modeling uses these characteristics to establish displacements and force flow assumptions which lead to the various mathematical models of structural mechanics. These mathematical models are the subject of the forthcoming chapters.

In the next section we study a simple structure – the truss structure. In a truss, the basic "bodies" that together make up the structure are slender bodies, the truss elements connected and only loaded at frictionless hinges. The ability to analyze a truss structure is of course of practical value but in presenting the general framework for truss analyses, we shall also introduce and explore the fundamental conditions which are always part of the formulation of every structural problem, namely: equilibrium, constitutive behavior and compatibility.

2.2 The analysis of truss structures – to exemplify general concepts of analysis

In order to convey the objectives outlined above, we start by characterizing a truss structure and then we apply the fundamental conditions introduced in Section 2.1 to a typical truss structure.

2.2.1 Model assumptions

We define a truss structure as an assemblage of slender prismatic solids of constant transverse cross-section which are called bars, and –

- The bars are connected to each other at frictionless pin joints (detailed later on).
- All external loading is applied as concentrated forces to the joints.
- The truss structure is only restrained at the joints.

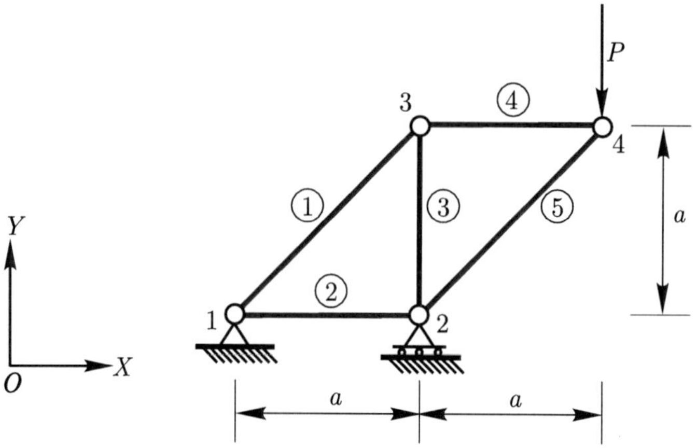

Fig. 2.4. A truss structure in the XY plane

In order to fix ideas let us consider the truss structural model in Figure 2.4. The structure is located in the XY plane and only loaded in that plane. Hence all actions take place in the XY plane (and we can refer to the structure as a "planar truss"). The truss bars are represented by straight lines which meet at the joints represented by the small circles. We shall refer to the joints as nodes; actually, more accurately, each node represents a joint. The supports are connected directly to nodes 1 and 2 and the external load is given by the concentrated force applied (also directly) to node 4.

In Figure 2.5 we show the detail of the connection represented by node 3. We consider not only the model representation used in Figure 2.4 but also a physical representation that gives the reader more insight into the pin-type joint behavior of a truss connection. Although the physical joint representation is still schematic, we can visualize the pin and how the bars are connected through the pin. The kinematics of the bars and the joint are assumed to be such that:

- The frictionless joint does not restrain the rotations of the bars.
- The lines representing the truss bars pass through the axes of the bars and the center of the joint.
- The bars displace with the joint.

2.2.2 Kinematic conditions for a properly supported truss

As we discussed in Section 2.1.2, we can identify whether an assemblage of solids and, hence, a structure is properly supported and does not have an internal mechanism by (first) assuming that the elements of the structure are rigid (where in a truss structure the joints are still assumed to be frictionless). If then the rigid structure and any part thereof can not undergo any motion,

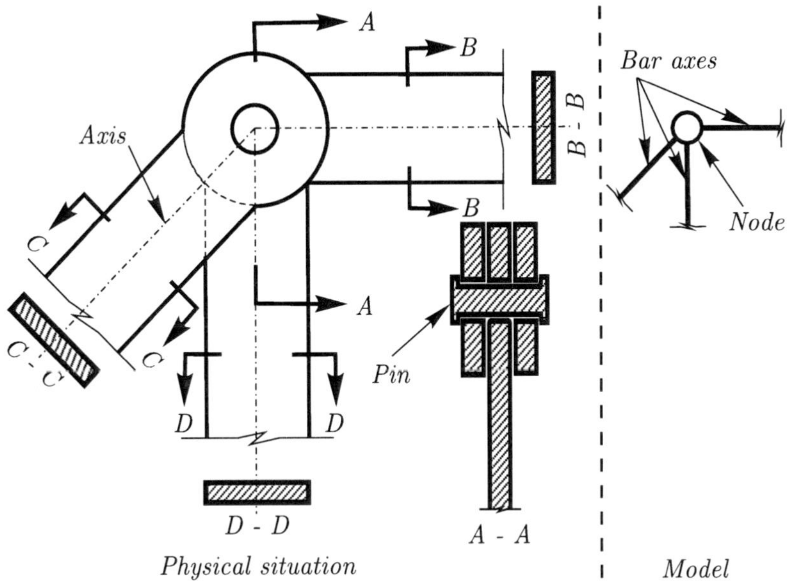

Fig. 2.5. Representation of a truss joint

the actual structure is properly supported and does not possess an internal mechanism.

Considering a truss structure, we should assume that each bar is rigid and verify whether a rigid body motion of the truss as a whole, or of any of its members, or of any of its parts is not possible.

Consider bars 1, 2 and 3 of the truss in Figure 2.4. They form a triangle and since each bar is assumed to be rigid, if bars 1, 2 and 3 were to move, they would have to do so as a rigid triangle. Namely, a triangle with sides of fixed lengths maintains its shape. For bars 3, 4 and 5 the same argument holds. Since bar 3 is common to both "rigid" triangles, the whole assemblage of bars, *i.e.*, bars 1 to 5 would behave as a rigid body.

Now, to examine if the assemblage of bars could have a rigid motion, we need to consider that the structure is supported. Hence, we can immediately conclude that the assemblage considered as rigid can not move since node 1 is fixed and a rotation about node 1 is prevented by the support at node 2.

Therefore, by kinematics alone, we conclude that the truss model of Figure 2.4 is properly supported and does not have an internal mechanism, and structural displacements can only be due to the straining of the bars. Although we consider here a very simple truss structure, this kinematics based approach can be applied to trusses of any complexity to arrive at a correct assessment of whether a truss structure is properly supported and does not have an internal mechanism. Note that this kind of analysis is independent of the external loading acting on the structure.

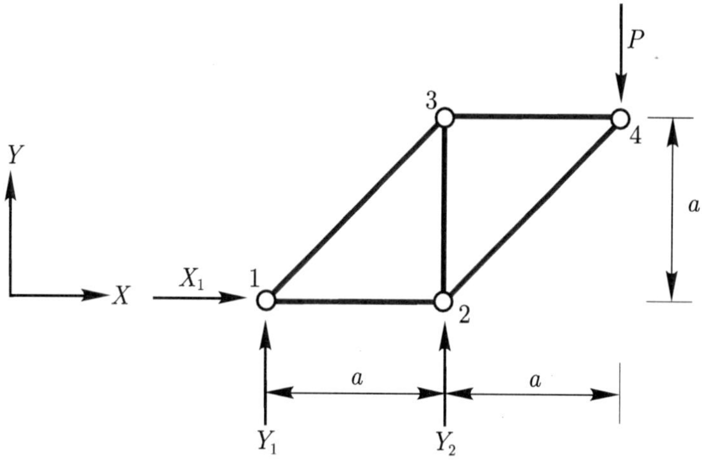

Fig. 2.6. Truss with reactions introduced explicitly

2.2.3 Equilibrium conditions for a truss model

Next, we detail how the "static equilibrium" condition can be applied to the truss structure to obtain the reactions at the supports and the internal forces of the truss bars.

Let us consider the equilibrium condition applied to the whole truss (this would correspond to taking $V \equiv \Delta V$ in the terminology for the solid body considered in Figure 2.3). Introducing the reactions as shown in Figure 2.6 and imposing the equilibrium conditions $\mathbf{R} = \mathbf{0}$ and $\mathbf{M_O} = \mathbf{0}$, we obtain

$$\begin{aligned} \sum F_X &= 0, \qquad X_1 = 0 \\ \sum F_Y &= 0, \qquad Y_1 + Y_2 - P = 0 \\ \sum M_1 &= 0, \qquad Y_2 \cdot a - P \cdot 2a = 0. \end{aligned}$$

In the above equations we are introducing the notation $\sum F_X$ and $\sum F_Y$ to represent the summation of all forces in the X and Y directions respectively, and $\sum M_1$ represents the summation of the moments of the external forces about node 1. We obtain

$$Y_2 = 2P \quad \text{and} \quad Y_1 = -P.$$

Next we impose the equilibrium condition to a generic bar of the truss (in Figure 2.3, this would correspond to taking ΔV as the bar in consideration), as shown in Figure 2.7. Here we also show the internal forces that could possibly arise. We note that no concentrated moment is introduced since we

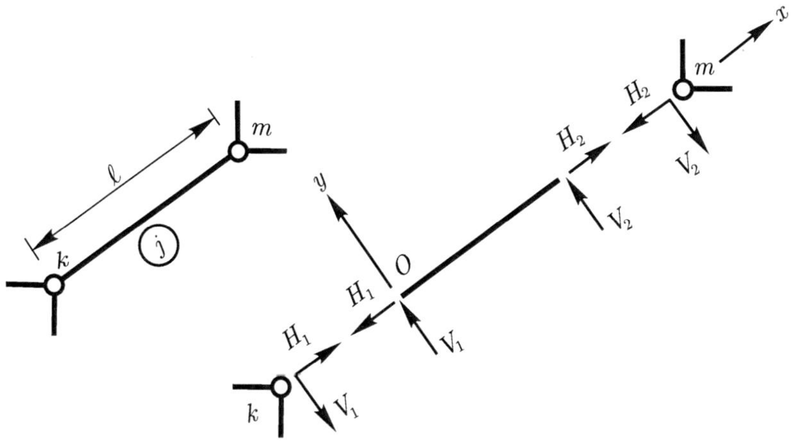

Fig. 2.7. Generic truss element, bar, of a truss. Shown are the magnitudes and directions of the forces acting onto the bar j and onto the nodes k and m (from the bar)

assume that the bars are free to rotate at the nodes. Hence, no such moment can arise.

Imposing $\mathbf{R} = \mathbf{0}$ and $\mathbf{M_O} = \mathbf{0}$ for the bar and using the local axis system shown in Figure 2.7, we arrive at

$$\sum F_x = 0, \qquad H_2 - H_1 = 0, \qquad H_1 = H_2$$

$$\sum F_y = 0, \qquad V_1 + V_2 - 0, \qquad V_1 = -V_2$$

$$\sum M_O = 0, \qquad V_2 \cdot \ell = 0, \qquad V_2 = 0$$

and hence $V_1 = 0$ also.

Therefore, each bar can only carry an axial force. We denote this force in bar j by N_j and a positive value is associated with tension. For the generic bar considered $N_j = H_1 = H_2$. This situation is shown in Figure 2.8.

If we next consider the equilibrium of the truss nodes, we can determine the forces in the truss structure of Figure 2.4.

In Figure 2.9 we show all nodes isolated from the rest of the truss structure. As we mentioned earlier, see Figure 2.3, *any* part of the structure must be in equilibrium and so must be each joint, that is each node. Hence, we can suppress the central portions of the truss bars and introduce the axial forces of the bars onto the remaining parts of the structure, namely the joints/nodes. Then, each node shown has to be in equilibrium and, in this example, we can directly solve for all bar forces.

In Figure 2.10, we also include the bars and indicate once more the condition that any part of the structure must be in equilibrium.

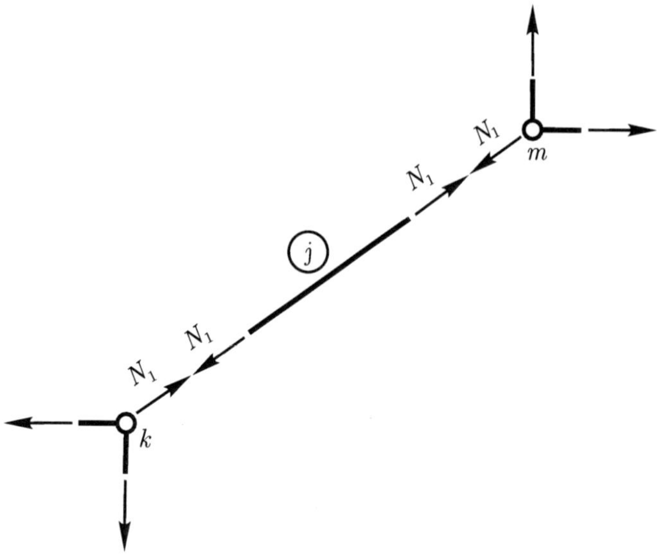

Fig. 2.8. Schematic representation of force in bar j and its action onto the end nodes

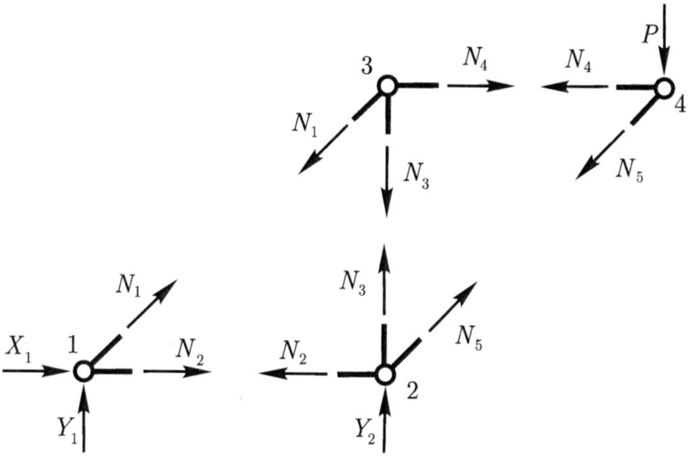

Fig. 2.9. Nodes of the truss structure considered as "free bodies"

Note that moment equilibrium is trivially satisfied for each joint since the lines of action of the forces pass through a point (the node). The condition $\mathbf{R} = \mathbf{0}$ implies for node 4

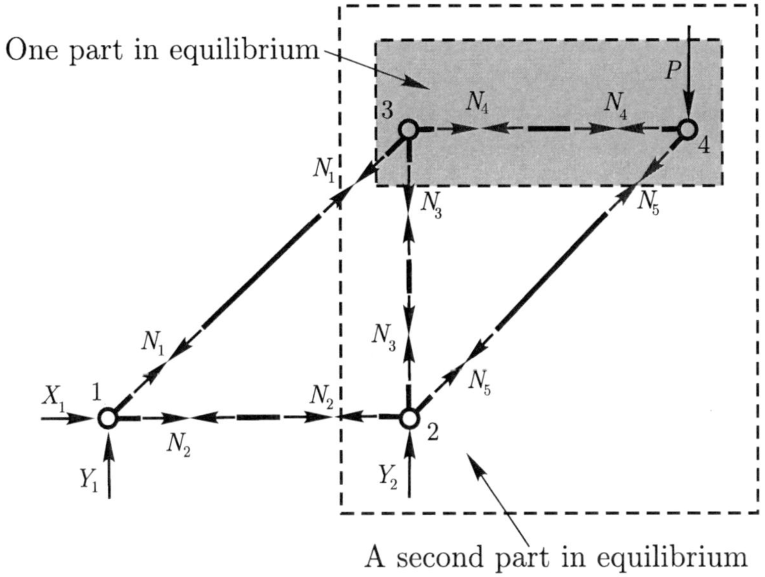

Fig. 2.10. Exploded view of bars and joints of the truss in Figure 2.4, and two typical parts that are in equilibrium

$$\sum F_X = 0, \qquad -N_4 - N_5\frac{\sqrt{2}}{2} = 0$$

$$\sum F_Y = 0, \qquad -P - N_5\frac{\sqrt{2}}{2} = 0$$

and hence

$$N_4 = P \quad \text{and} \quad N_5 = -P\sqrt{2}.$$

Consider next the equilibrium of node 3

$$\sum F_X = 0, \qquad N_4 - N_1\frac{\sqrt{2}}{2} = 0$$

$$\sum F_Y = 0, \qquad -N_3 - N_1\frac{\sqrt{2}}{2} = 0.$$

Using that $N_4 = P$ yields

$$N_1 = P\sqrt{2} \quad \text{and} \quad N_3 = -P.$$

We note that the equilibrium of bar 4 has already been implicitly taken into account. Next, let us impose the equilibrium of node 2

$$\sum F_X = 0, \qquad N_5\frac{\sqrt{2}}{2} - N_2 = 0$$

$$\sum F_Y = 0, \qquad Y_2 + N_5\frac{\sqrt{2}}{2} + N_3 = 0$$

which yields

$$N_2 = -P \quad \text{and} \quad Y_2 = 2P.$$

Finally, considering node 1

$$\sum F_X = 0, \qquad X_1 + N_1\frac{\sqrt{2}}{2} + N_2 = 0$$

$$\sum F_Y = 0, \qquad Y_1 + N_1\frac{\sqrt{2}}{2} = 0$$

which yields

$$X_1 = 0 \quad \text{and} \quad Y_1 = -P.$$

Note that the values of X_1, Y_1 and Y_2 are exactly the reactions already calculated by considering the global equilibrium of the complete structure (see Figure 2.6). This is the consequence of the important fact already mentioned but repeated now:

If equilibrium of each joint and each bar of a truss structure is satisfied, then also global equilibrium of any part the structure and hence of the complete structure is directly satisfied.

2.2.4 Constitutive behavior for a truss bar

Note that we did not consider so far the material of the truss bars; indeed the bars of the truss in Figure 2.4 could be of steel or wood, and the same forces would be transmitted. However, if – in general – we would like to evaluate the displacements of the nodes of any truss structure – an information of engineering interest – we need to characterize and quantify for each bar the relation between the internal force and the induced deformation. This is achieved by means of the constitutive behavior for each truss bar.

Consider a generic truss bar carrying the axial force N as shown in Figure 2.11.

The relevant quantity to characterize locally internal forces is the stress. For the one-dimensional truss bar the stress is constant over the section and normal to it. Therefore, it is given by $\tau = N/A$ as shown in Figure 2.12.

The material behavior of the truss bar is depicted in Figure 2.13. Here $\tau = E\varepsilon$ where ε is the strain

$$\varepsilon = \frac{\Delta\ell}{\ell}$$

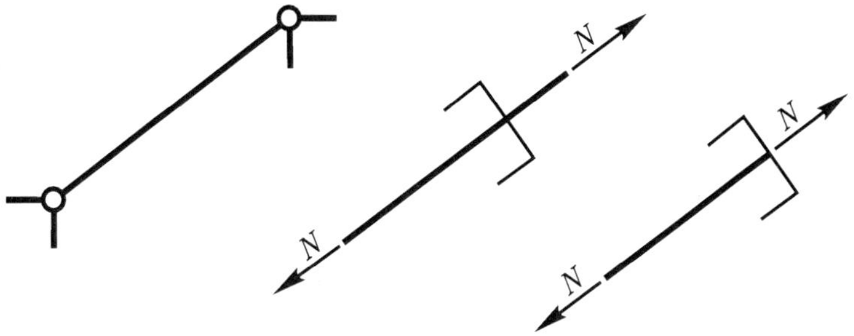

Fig. 2.11. Internal force in a generic truss bar; the force is the same at any section of the bar

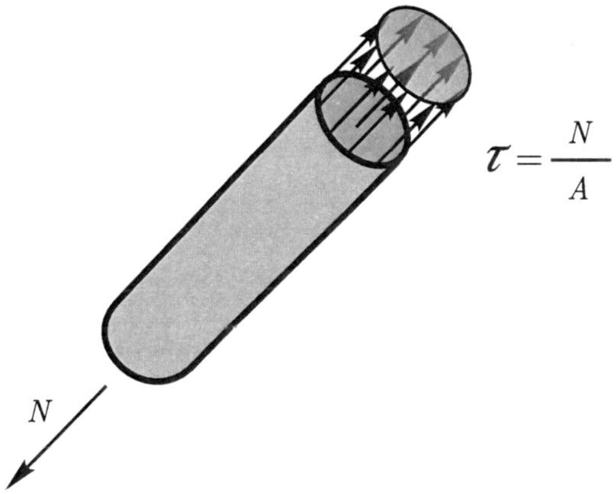

Fig. 2.12. Stress in a truss bar

and E is Young's modulus.

This stress-strain relationship is usually referred to as Hooke's law and materials with this stress-strain property are called "linear elastic". The "linear" refers to the fact that the stress is linearly proportional to the strain. The "elastic" means that the same (τ, ε) curve is followed for any loading or unloading causing an increase or a decrease in the stress/strain values. This property also means that for a given strain the stress value is unique and directly obtained from the (τ, ε) diagram.

Considering our truss model, if we assume that Hooke's law holds we can relate the axial force acting in a bar to the elongation of the bar, *i.e.*,

$$\tau = \frac{N}{A} = E\frac{\Delta\ell}{\ell} \quad \Rightarrow \quad N = \frac{EA}{\ell}\Delta\ell \quad \text{or} \quad \Delta\ell = N\frac{\ell}{EA}.$$

2.2.5 Compatibility conditions for a truss

So far we discussed the equilibrium requirements of a truss structure and the constitutive relation of the bars. Considering any truss structure, the

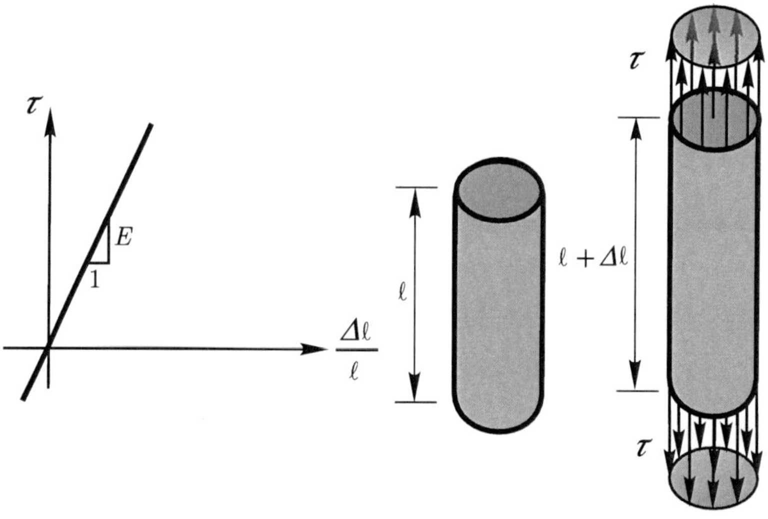

Fig. 2.13. Tension test data for a truss bar. Hooke's law

forces in the truss elements (bars) extend or shorten the bars and yet the bars must remain connected at the frictionless pins and some of the pins are restrained to move. The fact that the bars (the structural members) remain connected and the displacement boundary conditions need to be satisfied – for any externally applied loading – leads to the compatibility conditions: these conditions ensure that in any motion the structure "remains intact" (all elements stay connected) and the displacement boundary conditions are satisfied.

Considering again our truss structure in Figure 2.4, the change of length of each bar is given by

$$\Delta\ell_i = \frac{N_i}{E_i A_i}\ell_i. \tag{2.9}$$

Using the $\Delta\ell_i$, $i = 1, \cdots, 5$, we can now find the final positions of the nodes using the compatibility conditions: that the bars remain connected at the nodes and the structure satisfies the displacement boundary conditions.

We call this complete method of analysis the *elementary method* for analyzing truss structures: the determination of the internal forces of the bars and reactions as accomplished above and the evaluation of the nodal displacements using (2.9) and kinematics. The example below is an application of the elementary method.

Example 2.1

Use the elementary method to solve the truss problem of Figure 2.4.

Table 2.1. Data obtained for the truss in Figure 2.4 considering $a = 2m$, $E_i = 2.1 \times 10^{11} N/m^2$ (steel), $A_i = 1.439 \times 10^{-3} m^2$ for $i = 1, \cdots, 5$ and $P = 60kN$. Forces are given in Newtons and lengths in meters

Bar	N_i	$\Delta\ell_i = \frac{N_i}{E_i A_i}\ell_i$	$\varepsilon = \frac{\Delta\ell_i}{\ell_i}$	$\ell_i + \Delta\ell_i$
1	84853	7.942×10^{-4}	2.808×10^{-4}	2.8292
2	−60000	-3.971×10^{-4}	-1.986×10^{-4}	1.9996
3	−60000	-3.971×10^{-4}	-1.986×10^{-4}	1.9996
4	60000	3.971×10^{-4}	1.986×10^{-4}	2.0004
5	−84853	-7.942×10^{-4}	-2.808×10^{-4}	2.8276

Solution

For the evaluation of the forces of the bars and the reactions we refer to Section 2.2.3.

For the evaluation of the displacements, we summarize in Table 2.1 the data obtained by applying equation (2.9). This data is used for the calculation of the nodal positions of the deformed truss structure.

In Figure 2.14, we describe in four steps the determination of the deformed configuration of the truss. The change of length of the bars is magnified 300 times for visualization purposes. Note that the bars extend/shorten and freely rotate.

In Figure 2.14a, we show the final position of bar 2, which is obtained by introducing its shortening and taking into account the restraints. Hence nodes 1 and 2 are already in their final positions. We also show the stretching of bar 1.

In Figure 2.14b, we show the shortening of bar 3, which is shown in an intermediate position, considering the displacement of node 2 but not the rotation of bar 3. The final position of node 3 is also indicated, and it is obtained by the rotation of bars 1 and 3 around nodes 1 and 2, respectively.

In Figure 2.14c bars 1, 2 and 3 are in their final positions and bars 4 and 5 are shown in intermediate positions considering their extension/shortening and the displacements of nodes 2 and 3, but not the rotations of bars 4 and 5. We indicate how the final position of node 4 is obtained by the rotations of bars 4 and 5 around nodes 3 and 2, respectively.

Finally, in Figure 2.14d the deformed configuration of the complete truss structure is shown. With the steps detailed in Figure 2.14 and using the data of Table 2.1, it is possible to evaluate all nodal displacements of the truss.

□

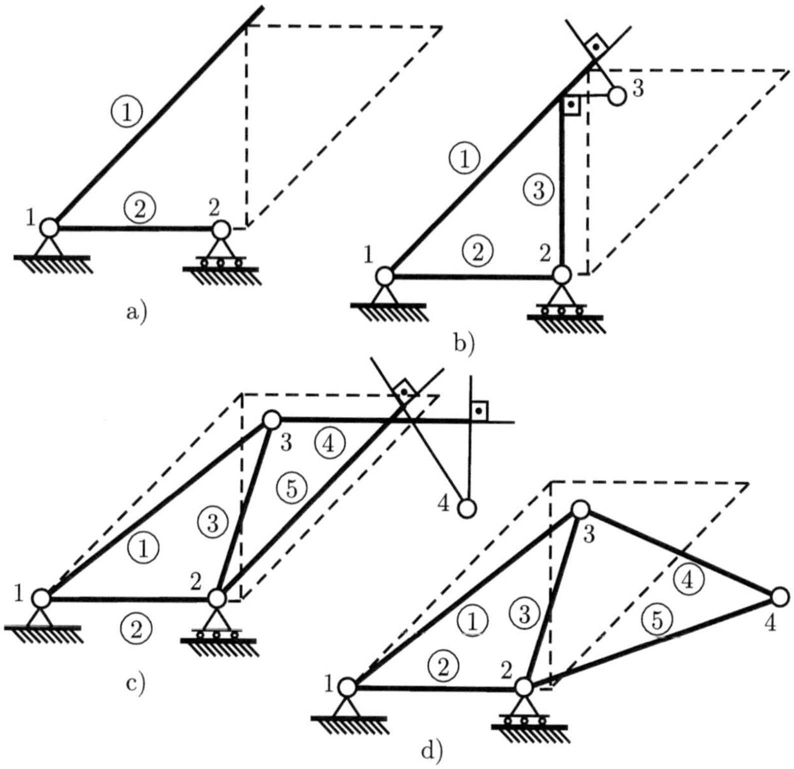

Fig. 2.14. Determination of the deformed configuration of a truss structure

Note that we used in Example 2.1 the important assumption of linear analysis: the displacements of the bars and their rotations are infinitesimally small.

In Figure 2.15a we schematically show a generic bar rotating about A for two conditions: large and infinitesimally small rotations. In Figure 2.15b, we detail the assumption of an infinitesimally small rotation. With θ assumed infinitesimally small, the displacement δ due to the rotation is assumed to take place at the 90 degree (right) angle to the bar. Also, the length of the deformed bar ℓ_d and the magnitude of the displacement δ are given by

$$\ell_d = \frac{\ell}{\cos\theta} \quad \text{and} \quad \delta = \ell_d \sin\theta.$$

which when θ is infinitesimally small (using $\cos\theta = 1$ and $\sin\theta = \theta$) leads to

$$\ell_d = \ell \quad \text{and} \quad \delta = \ell\theta.$$

Note that the bar does not change its length due to the rotation (and hence any force carried by the bar is not changed due to the rotation). For example,

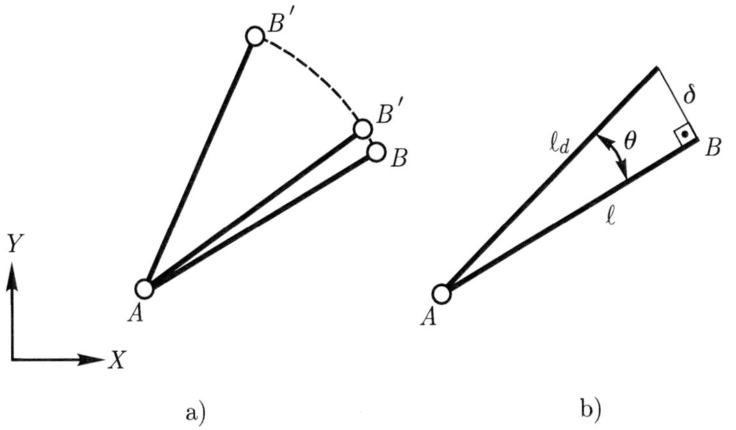

Fig. 2.15. Rotation of a generic truss bar. (a) Large rotation and (b) Infinitesimally small rotation

bars 1 and 3 when rotated about nodes 1 and 2, respectively, to meet at node 3 in the deformed configuration, see Figure 2.14b, do not change their lengths due to the rotations. We use this assumption throughout the book, except in Chapter 8.

Another important assumption due to considering that the displacements are infinitesimally small – already mentioned (see Section 2.1.5) – which we want to recall here once more is that the equilibrium conditions (for the bars, the joints, and any part of the truss) are established and satisfied in the original configuration of the structure. Hence, although the truss nodes and bars moved (see Figure 2.14 for the truss in Figure 2.4) the equilibrium conditions assume that these displacements are so small that they can be entirely neglected.

We finally note that as we use the linear model assumptions and solve a truss problem, as in Example 2.1, the calculated nodal displacements and bar rotations may not come out to be infinitesimally small. This fact is revealing that the solution of the linear model is only an approximation to the response of the actual physical problem – as the hierarchical modeling process emphasizes.

However, for actual engineering truss structures, the nodal displacements and bar rotations predicted by the linear model are mostly small as can be verified examining the numerical solution values and, hence, in most cases, the linear model is adequate for design purposes. If the linear model predicts large nodal displacements and bar rotations, then a nonlinear analysis may be necessary, see Chapter 8.

The objective of Example 2.1 was to present the elementary method for solving truss structures and to give insight into the use of constitutive relations and compatibility conditions to calculate the displacements of a truss structure. Of course, as the number of the bars increases – and there may be many bars in truss structures – the application of the above methodology becomes very cumbersome. We will see in Section 2.3, that the use of matrix methods leads to a much more efficient solution.

2.2.6 Statically determinate and indeterminate trusses

For the truss of Figure 2.4, the equilibrium conditions alone allowed us to determine the reactions and the forces in all bars. This kind of structure is termed statically determinate since the equilibrium conditions alone are sufficient to determine all bar forces and reactions.

However, this is not always the case. To understand when we can obtain the axial forces from equilibrium only and when not, we take a step back and consider two very simple truss structures.

Let us consider the truss structure shown in Figure 2.16a. Of course, this structure is properly supported.

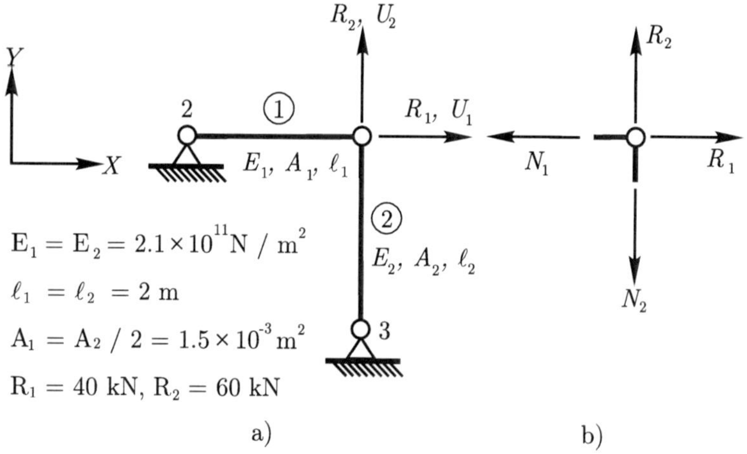

Fig. 2.16. a) Two-bar truss structure. R_1 and R_2 are concentrated applied loads and U_1 and U_2 are the node 1 displacements. Nodes and elements are numbered; b) Equilibrium of node 1 for the two bar truss

The structure can be solved by considering the equilibrium of node 1, as shown in Figure 2.16b, which leads to

$$\sum F_X = 0, \qquad R_1 - N_1 = 0 \qquad \Rightarrow \qquad N_1 = R_1$$

$$\sum F_Y = 0, \qquad R_2 - N_2 = 0 \qquad \Rightarrow \qquad N_2 = R_2.$$

The bar elongations are given by

$$\Delta\ell_1 = \frac{N_1}{E_1 A_1}\ell_1 = 2.53968\times 10^{-4} \text{ m}, \qquad \Delta\ell_2 = \frac{N_2}{E_2 A_2}\ell_2 = 1.90476\times 10^{-4} \text{ m}.$$

and

$$U_1 = \Delta\ell_1, \qquad U_2 = \Delta\ell_2.$$

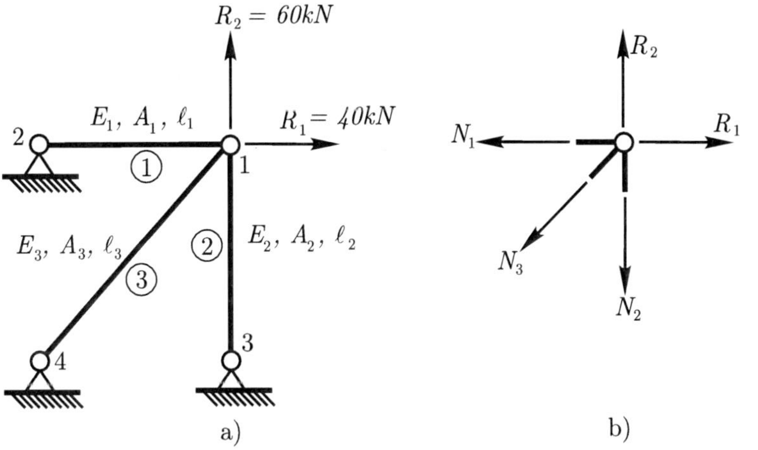

Fig. 2.17. a) Three-bar truss structure. The properties of bars 1 and 2 are as in Figure 2.16 and $E_3 = 2.1 \times 10^{11}$ N/m^2, $A_3 = 3A_1$; b) Equilibrium of node 1 for the three-bar truss

Let us now add to the structure of Figure 2.16 an inclined bar as shown in Figure 2.17a. Obviously, this new structure is still properly supported. The equilibrium of node 1, as shown in Figure 2.17b, now yields

$$\sum F_X = 0, \qquad -N_1 - N_3\frac{\sqrt{2}}{2} + R_1 = 0$$

$$\sum F_Y = 0, \qquad -N_2 - N_3\frac{\sqrt{2}}{2} + R_2 = 0$$

which leads to

$$N_1 + N_3\frac{\sqrt{2}}{2} = R_1, \qquad N_2 + N_3\frac{\sqrt{2}}{2} = R_2. \tag{2.10}$$

Therefore there are infinitely many values of N_1, N_2 and N_3 that satisfy the equilibrium conditions. However, if we consider the actual structure (the physical problem) – which admits as a mathematical model the truss model of Figure 2.17a – we would, of course, be able to experimentally measure unique forces in the truss bars for given loads R_1 and R_2. It is easy to conclude that the equilibrium conditions in equation (2.10) alone are not sufficient to determine the truss internal forces and the reactions. The truss of Figure 2.17a is a simple example of a statically indeterminate structure: for such structures we also need to consider the constitutive relations of the materials used and the compatibility conditions to solve for the internal forces.

Suppose we choose values for N_1, N_2 and N_3 which satisfy (2.10). These values would then satisfy the equilibrium conditions. Of course, these values could then be used to evaluate the bar elongations $\Delta\ell_1$, $\Delta\ell_2$ and $\Delta\ell_3$. However, in general these elongations will not lead to a valid deformed configuration, *i.e.*, the ends of bars 1, 2, and 3 would not connect to a single point, the supposedly new position of node 1. Therefore, the compatibility condition that the bars remain connected at node 1 would be violated.

In Figure 2.18a, we show bar elongations that lead to a kinematically admissible configuration, *i.e.*, a compatible deformed configuration. Of course, there is a relation that should be satisfied by $\Delta\ell_1$, $\Delta\ell_2$ and $\Delta\ell_3$, namely the compatibility condition. In Figure 2.18b, we show the region around node 1 magnified and we can write

$$\tan\alpha = \frac{\Delta\ell_1 - \Delta\ell_3\cos\alpha}{\Delta\ell_3\sin\alpha - \Delta\ell_2}$$

which for $\alpha = 45°$ leads to the compatibility condition

$$\Delta\ell_1 + \Delta\ell_2 = \Delta\ell_3\sqrt{2}. \tag{2.11}$$

Using the constitutive relations, equation (2.11) can be written in terms of the axial forces

$$\frac{N_1}{E_1A_1}\ell_1 + \frac{N_2}{E_2A_2}\ell_2 = \frac{N_3}{E_3A_3}\ell_3\sqrt{2}.$$

Introducing the data given in Figure 2.17a

$$N_1 + \frac{N_2}{2} - \frac{2N_3}{3} = 0. \tag{2.12}$$

Equations (2.10) and (2.12) contain the requirements of equilibrium, compatibility and material behavior. We can solve and obtain

$$N_1 = 11.34 \text{ kN}, \qquad N_2 = 31.34 \text{ kN}, \qquad N_3 = 40.52 \text{ kN}.$$

The nodal displacements are now given by

$$\begin{aligned} U_1 &= \Delta\ell_1 = \frac{N_1}{E_1A_1}\ell_1 = 7.203 \times 10^{-5} \text{ m} \\ U_2 &= \Delta\ell_2 = \frac{N_2}{E_2A_2}\ell_2 = 9.951 \times 10^{-5} \text{ m} \end{aligned}$$

Summarizing, we recognize that, since the three-bar truss structure is a statically indeterminate structure, we had to use the following conditions

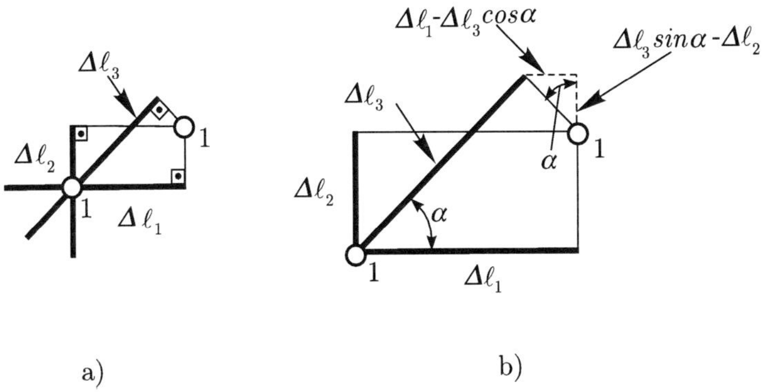

Fig. 2.18. Compatibility of displacements for node 1

- equilibrium
- compatibility
- constitutive

in order to solve for the internal forces in the structure. Once the internal forces have been calculated, the nodal displacement can be obtained as in the analysis of a statically determinate truss structure. These three conditions are the *fundamental conditions* that govern the behavior of every problem in structural mechanics.

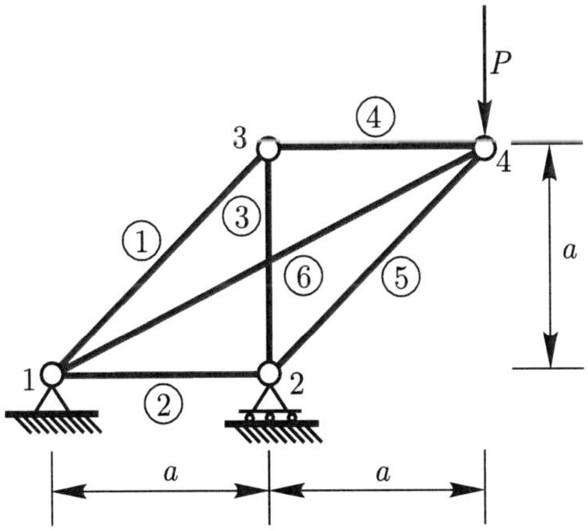

Fig. 2.19. New truss structure obtained by adding bar 6 to truss in Figure 2.4

In order to further elaborate on statically indeterminate structures, we show in Figure 2.19 the truss of Figure 2.4 with an extra bar linking nodes 1 and 4. The truss of Figure 2.19 is no longer statically determinate. Conceptually, the change from the statically determinate to the statically indeterminate structure is very similar to the change from the two-bar structure (Figure 2.16a) to the three-bar truss structure (Figure 2.17a). In fact, we can study the truss of Figure 2.19 as shown in Figure 2.20: the displacement of node 4 will depend on the value of N_6 and this bar force extends/shortens bar 6. The compatibility condition (that the bar 6 must fit into the distance between nodes 1 and 4 in the deformed geometry) can only be enforced by also using the constitutive relations of the bars.

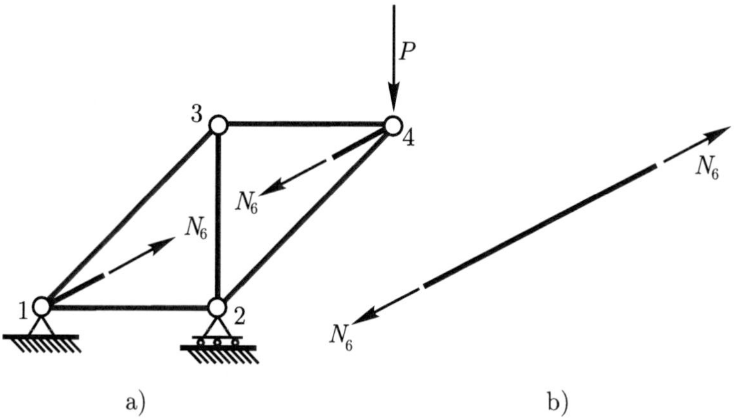

Fig. 2.20. Another representation of the previous truss

As for the three-bar truss in Figure 2.17, nodal equilibrium alone would not give the bar forces of the truss of Figure 2.19. We would again obtain a system of equations with one degree of indeterminacy.

This problem is solved later on in this chapter using the matrix method of analysis which provides a much more efficient solution procedure.

2.3 Matrix displacement method for trusses

In this section we introduce the matrix displacement method for planar truss structures. It is important to note that the concepts discussed in the context of trusses are also directly applicable to more complex structural analyses, like when considering frame structures. Therefore, the objective of this section is not only to present an efficient method for solving truss structures of arbitrary complexity but also to introduce the main concepts of matrix structural analysis.

We recall that the fundamental conditions of equilibrium, compatibility and constitutive behavior translate in the case of trusses, subjected to joint forces only, to:

- Every node should be in equilibrium considering the forces of the truss bars that connect to that node and, possibly, external forces applied directly to the node. Each bar is automatically in equilibrium as it only carries an axial (constant) force.
- The axial deformations of the truss bars must lead to a compatible deformation of the complete structure taking into account how the bars are linked to each other and to the supports. The joints (nodes) do not deform.

In the matrix formulation, the above conditions of equilibrium, constitutive behavior and compatibility are directly – and in a very elegant manner – enforced.

2.3.1 Truss bar stiffness matrix in its local system

We begin by establishing a relation between the end displacements and forces of a truss bar.

Let us use the convention given in Figure 2.21 for the end forces and displacements. The symbol $\sim$ over the quantities is used to show that a local coordinate system, aligned with the bar axis, is adopted. Then

$$\begin{aligned} \Delta\ell &= \tilde{u}_2 - \tilde{u}_1, \qquad N = EA\frac{\Delta\ell}{\ell} = \frac{EA}{\ell}(\tilde{u}_2 - \tilde{u}_1) \\ \tilde{f}_2 &= N, \qquad \tilde{f}_1 = -N \end{aligned}$$

with N positive when the bar is in tension.

Fig. 2.21. Local end-displacements and forces acting onto a truss bar

The equations above can be written in matrix form as follows

$$\begin{bmatrix} \frac{EA}{\ell} & -\frac{EA}{\ell} \\ -\frac{EA}{\ell} & \frac{EA}{\ell} \end{bmatrix} \begin{bmatrix} \tilde{u}_1 \\ \tilde{u}_2 \end{bmatrix} = \begin{bmatrix} \tilde{f}_1 \\ \tilde{f}_2 \end{bmatrix}.$$

Let us define

$$\begin{aligned} \tilde{\mathbf{u}} &= \begin{bmatrix} \tilde{u}_1 \\ \tilde{u}_2 \end{bmatrix}, \qquad \tilde{\mathbf{f}} = \begin{bmatrix} \tilde{f}_1 \\ \tilde{f}_2 \end{bmatrix} \\ \tilde{\mathbf{k}} &= \begin{bmatrix} \tilde{k}_{11} & \tilde{k}_{12} \\ \tilde{k}_{21} & \tilde{k}_{22} \end{bmatrix} = \begin{bmatrix} \frac{EA}{\ell} & -\frac{EA}{\ell} \\ -\frac{EA}{\ell} & \frac{EA}{\ell} \end{bmatrix} \end{aligned}$$

where $\tilde{\mathbf{u}}$ is referred to as the column matrix of element nodal point displacements, $\tilde{\mathbf{f}}$ as the column matrix of element nodal point forces (acting onto the bar) and $\tilde{\mathbf{k}}$ as the element stiffness matrix. All quantities are referred to the local coordinate system. The term "element" is representing a truss bar and is also used in later chapters to represent other structural members.

It is instructive to interpret the physical meaning of the coefficients in the stiffness matrix. For that purpose, let us impose a unit displacement at the left end and restrain the displacement to be zero at the right end, *i.e.*, $\tilde{u}_1 = 1$ and $\tilde{u}_2 = 0$. Then

$$\begin{bmatrix} \tilde{k}_{11} & \tilde{k}_{12} \\ \tilde{k}_{21} & \tilde{k}_{22} \end{bmatrix} \begin{bmatrix} 1 \\ 0 \end{bmatrix} = \begin{bmatrix} \tilde{f}_1 \\ \tilde{f}_2 \end{bmatrix}$$

leading to $\tilde{k}_{11} = \tilde{f}_1$ and $\tilde{k}_{21} = \tilde{f}_2$. In other words, the stiffness coefficient $\tilde{k}_{11} = \frac{EA}{\ell}$ is the force that must be applied in the displacement degree of freedom $\tilde{u}_1$ onto the bar to impose a unit displacement when $\tilde{u}_2 = 0$. The coefficient $\tilde{k}_{21} = -\frac{EA}{\ell}$ is the force (reaction) at the right end onto the bar, *i.e.*, the force of the restraint onto the bar. Of course, the interpretation of $\tilde{k}_{11}$ as a stiffness coefficient is now evident since it gives the magnitude of the force necessary to produce a unit displacement. An analogous interpretation can be given for $\tilde{k}_{12}$ and $\tilde{k}_{22}$ associated with imposing a unit displacement at the right end and fixing the left end. These results are summarized in Figure 2.22.

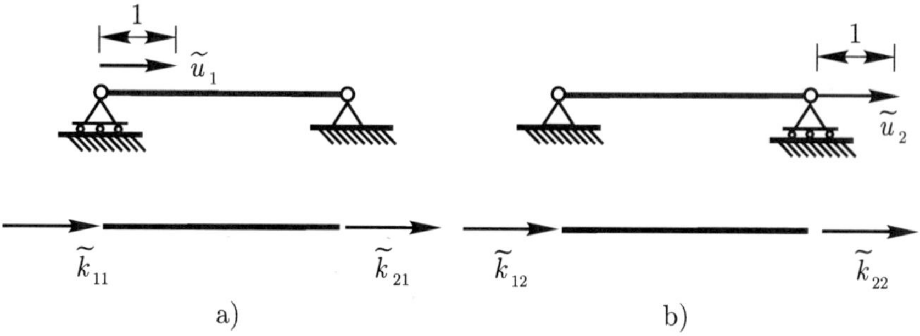

Fig. 2.22. Interpretation of the stiffness coefficients as forces applied *onto* the bar

Suppose now that we would like to solve the problem depicted in Figure 2.16 with the aid of the truss element stiffness matrix. This is a simple problem to demonstrate the matrix method of analysis.

Considering bar 1, we note that at its right end the node can displace not only along the axial direction but also along the transverse direction (the Y direction). In Figure 2.21, a nodal transverse displacement was not considered as a degree of freedom of the truss bar because there is no stiffness provided by the bar to such displacement. As a consequence, for example, if in Figure 2.22b there were no support at the right end, which prevents the transverse displacement, the bar could rigidly rotate about the left support. Of course, in the problem of Figure 2.16 the stiffness for the transverse displacement at the right end of bar 1 is provided by bar 2, which shares the node with bar 1, because for bar 2, such nodal displacement is along its axial direction.

Considering that a truss structure always consists of an assemblage of truss bars, we add, as shown in Figure 2.23, the transverse degrees of freedom for a generic truss element. Then the truss element stiffness matrix is given by

$$\tilde{\mathbf{k}} = \begin{bmatrix} \frac{EA}{\ell} & 0 & -\frac{EA}{\ell} & 0 \\ 0 & 0 & 0 & 0 \\ -\frac{EA}{\ell} & 0 & \frac{EA}{\ell} & 0 \\ 0 & 0 & 0 & 0 \end{bmatrix}. \tag{2.13}$$

Fig. 2.23. Degrees of freedom of a truss bar

The physical meaning of the second and fourth rows and columns being zero is that there is no stiffness associated with the degrees of freedom given by $\tilde{u}_2$ and $\tilde{u}_4$. In fact, the equation

$$\tilde{\mathbf{k}}\tilde{\mathbf{u}} = \tilde{\mathbf{f}} \tag{2.14}$$

tells that for any values of $\tilde{u}_2$ and $\tilde{u}_4$ (which would amount to vertical displacement and rotation of the bar) there is no induced internal bar force and there are no induced nodal forces (see also Figure 2.15 and the corresponding discussion).

Of course, the stiffness matrix given in equation (2.13) is useful as long as the truss bar is part of an assemblage and the stiffness for the transverse degrees of freedom $\tilde{u}_2$ and $\tilde{u}_4$ is provided by other bars.

2.3.2 Solution of a two-bar truss structure using the matrix method

Returning to the problem of Figure 2.16, if we obtain the displacements of node 1, which is the only node with free degrees of freedom, the problem is solved. Let U_1 and U_2 be such displacements as shown in Figure 2.24 (the same notation for these displacements has been used before). Note that we use capital letters to denote that these displacements are degrees of freedom defined for the whole structural assemblage. In this way, we distinguish such degrees of freedom from the individual bar degrees of freedom for which we use lower case letters. Note also that U_1 and U_2 (the structure degrees of freedom) are here referred to the global coordinate system X, Y.

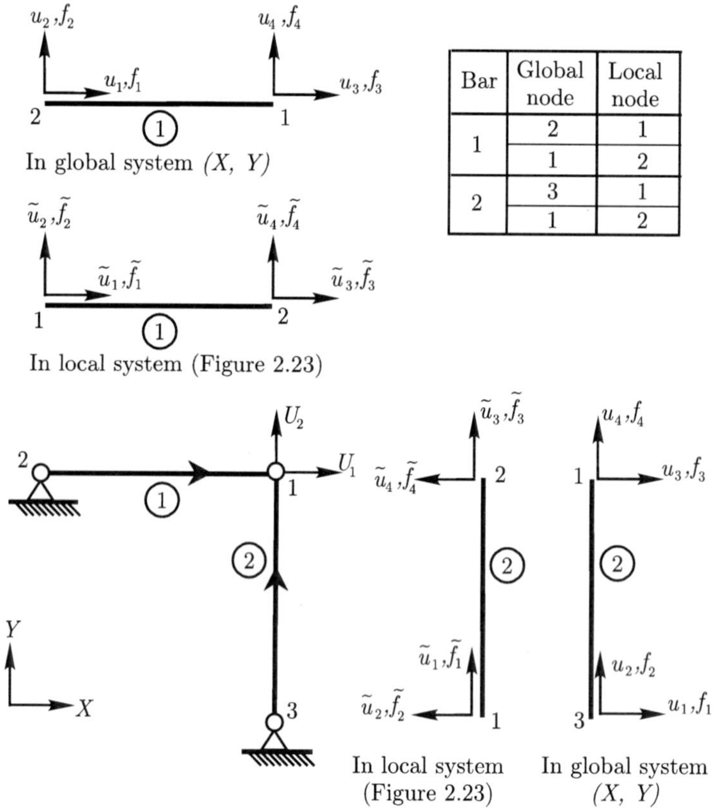

Bar	Global node	Local node
1	2	1
	1	2
2	3	1
	1	2

Fig. 2.24. Definitions for the two bar truss structure

In Figure 2.24, the degrees of freedom of bars 1 and 2 are also shown. The arrow on a bar axis defines the orientation of the bar and establishes a local (bar attached) numbering for the end nodes of the bar. The table included in Figure 2.24 shows, for each bar, the relation between the global node numbering (for the structure) and the local node numbering of the bar. We observe that the numbering of the displacements and forces of the bar starts always from the local node 1.

To solve the problem, we need to enforce equilibrium of the global node 1. We see that the end displacements and forces of the bars 1 and 2 (the quantities identified by a curl) at the global node 1 are not referred to a single coordinate system. Therefore, to facilitate the enforcement of nodal equilibrium, it is convenient to define the bar end displacements and forces in a common coordinate system which is chosen to be the global one. These nodal displacements and forces for bars 1 and 2 in the global system are also shown in Figure 2.24. The $\sim$ symbol over the lower case letters is dropped since we are considering a global system for these quantities.

The element stiffness matrices in the global system are, in general, different from those in the local system. Later on, in this section, we will derive a general expression which relates these stiffness matrices. However, for bars 1 and 2 in this problem we can easily obtain the global matrices as shown below. In fact, for bar 1 since the local and global systems are the same, we directly write

$$\mathbf{k}^{(1)} = \tilde{\mathbf{k}}^{(1)}$$

where again the $\mathbf{k}^{(1)}$ (without the $\sim$ symbol) indicates that we are using the global system, and therefore

$$\mathbf{k}^{(1)}\mathbf{u}^{(1)} = \mathbf{f}^{(1)}.$$

For bar 2, we have $\tilde{u}_1 = u_2$, $\tilde{u}_2 = -u_1$, $\tilde{u}_3 = u_4$ and $\tilde{u}_4 = -u_3$ with analogous relations for the forces. Using

$$\tilde{\mathbf{k}}^{(2)}\tilde{\mathbf{u}}^{(2)} = \tilde{\mathbf{f}}^{(2)}.$$

and the relations between the global and local quantities, we obtain

$$\begin{bmatrix} \frac{E_2A_2}{\ell_2} & 0 & -\frac{E_2A_2}{\ell_2} & 0 \\ 0 & 0 & 0 & 0 \\ -\frac{E_2A_2}{\ell_2} & 0 & \frac{E_2A_2}{\ell_2} & 0 \\ 0 & 0 & 0 & 0 \end{bmatrix} \begin{bmatrix} u_2 \\ -u_1 \\ u_4 \\ -u_3 \end{bmatrix} = \begin{bmatrix} f_2 \\ -f_1 \\ f_4 \\ -f_3 \end{bmatrix}. \tag{2.15}$$

Re-ordering the equations leads to

$$\begin{bmatrix} 0 & 0 & 0 & 0 \\ 0 & \frac{E_2A_2}{\ell_2} & 0 & -\frac{E_2A_2}{\ell_2} \\ 0 & 0 & 0 & 0 \\ 0 & -\frac{E_2A_2}{\ell_2} & 0 & \frac{E_2A_2}{\ell_2} \end{bmatrix} \begin{bmatrix} u_1 \\ u_2 \\ u_3 \\ u_4 \end{bmatrix} = \begin{bmatrix} f_1 \\ f_2 \\ f_3 \\ f_4 \end{bmatrix} \tag{2.16}$$

and, therefore, since $\mathbf{k}^{(2)}\mathbf{u}^{(2)} = \mathbf{f}^{(2)}$,

$$\mathbf{k}^{(2)} = \begin{bmatrix} 0 & 0 & 0 & 0 \\ 0 & \frac{E_2 A_2}{\ell_2} & 0 & -\frac{E_2 A_2}{\ell_2} \\ 0 & 0 & 0 & 0 \\ 0 & -\frac{E_2 A_2}{\ell_2} & 0 & \frac{E_2 A_2}{\ell_2} \end{bmatrix}.$$

It is important to note that the nodal forces are always defined in the same coordinate system as the nodal displacements and that the f_i are the forces acting *onto* the bar elements (just like the displacements are imposed *onto* the bar element).

We are now ready to enforce equilibrium at the global node 1. Referring to Figure 2.25, equilibrium in the X direction leads to

$$R_1 - \left(f_3^{(1)} + f_3^{(2)}\right) = 0 \qquad \Rightarrow \qquad R_1 = f_3^{(1)} + f_3^{(2)} \tag{2.17}$$

where we use the superscripts to identify the contributions from bar 1 and bar 2. However, the superscripts are used only when necessary. For example, in equation (2.15) the superscripts for u_i, f_i were not used since it is implicitly understood that we are working with bar 2. Equilibrium in the Y direction gives

$$R_2 - \left(f_4^{(1)} + f_4^{(2)}\right) = 0 \qquad \Rightarrow \qquad R_2 = f_4^{(1)} + f_4^{(2)}. \tag{2.18}$$

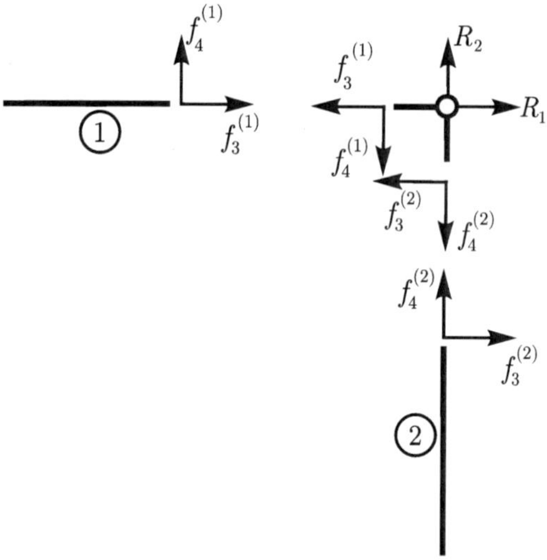

Fig. 2.25. Equilibrium of node 1, forces acting onto the bars and onto the node

Of course, since we have only an axial force in a truss bar

$$f_4^{(1)} = f_3^{(2)} = 0. \tag{2.19}$$

Introducing the stiffness relations and using equation (2.19), we can write (2.17) as

$$R_1 = k_{31}^{(1)} u_1^{(1)} + k_{33}^{(1)} u_3^{(1)}. \tag{2.20}$$

Considering that $u_1^{(1)} = 0$ (global node 2 is fixed) and the compatibility relation

$$U_1 = u_3^{(1)}$$

we arrive at

$$k_{33}^{(1)} U_1 = R_1. \tag{2.21}$$

Analogously, using (2.19) , equation (2.18) can be written as

$$R_2 = k_{42}^{(2)} u_2^{(2)} + k_{44}^{(2)} u_4^{(2)}. \tag{2.22}$$

Since node 3 is fixed $u_2^{(2)} = 0$ and using the compatibility relation

$$U_2 = u_4^{(2)}$$

we obtain

$$k_{44}^{(2)} U_2 = R_2. \tag{2.23}$$

Introducing the values of the stiffness coefficients, equations (2.21) and (2.23) can be written in matrix form as

$$\begin{bmatrix} \frac{E_1 A_1}{\ell_1} & 0 \\ 0 & \frac{E_2 A_2}{\ell_2} \end{bmatrix} \begin{bmatrix} U_1 \\ U_2 \end{bmatrix} = \begin{bmatrix} R_1 \\ R_2 \end{bmatrix}. \tag{2.24}$$

Let

$$\mathbf{U} = \begin{bmatrix} U_1 \\ U_2 \end{bmatrix}, \qquad \mathbf{R} = \begin{bmatrix} R_1 \\ R_2 \end{bmatrix} \tag{2.25}$$

where $\mathbf{U}$ is the column matrix of the free nodal degrees of freedom of the structure and $\mathbf{R}$ is the column matrix of the external nodal forces acting on the free degrees of freedom. We can write

$$\mathbf{KU} = \mathbf{R} \tag{2.26}$$

where $\mathbf{K}$, implicitly defined by (2.24) and (2.26), is the global stiffness matrix of the structure associated with the free degrees of freedom.

Note that the physical interpretations given earlier for stiffness coefficients also hold for those of the global **K** matrix. That is, the first column in **K** gives the external forces necessary to impose $U_1 = 1$ with $U_2 = 0$, and analogously the second column gives the external forces associated with $U_2 = 1$ with $U_1 = 0$. Due to the very simple nature of this problem, the stiffness matrix **K** could have been simply obtained in this way.

Solving (2.24) using the mechanical and geometrical properties of the bars leads to

$$U_1 = 2.53968 \times 10^{-4}\ \text{m}, \qquad U_2 = 1.90476 \times 10^{-4}\ \text{m} \tag{2.27}$$

which are the values obtained earlier.

Of course, since the bars are orthogonal there is no coupling between the vertical and horizontal displacements. This fact is reflected by the zero off-diagonal elements in the stiffness matrix.

Let us next consider the solution of the problem defined in Figure 2.17; that is, when an inclined bar is added to the structure. If we had the stiffness matrix of element 3 in the global coordinate system we could directly enter its contributions to the equilibrium of node 1. Therefore, we need to derive the stiffness matrix of a bar arbitrarily oriented in the global coordinate system.

2.3.3 Stiffness matrix of an arbitrarily oriented truss element

The degrees of freedom of an arbitrarily oriented truss element are summarized in Figure 2.26.

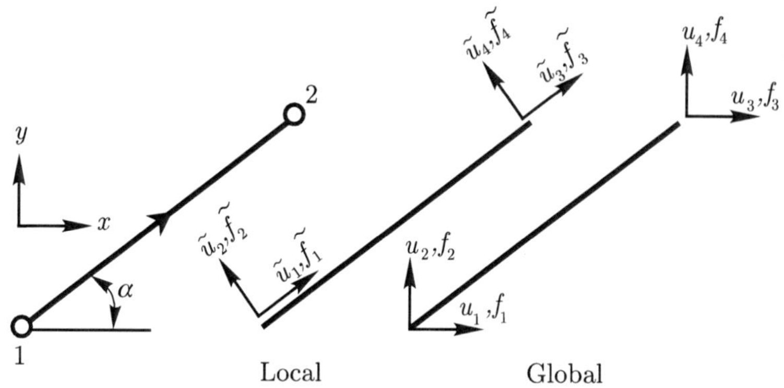

Fig. 2.26. Local and global degrees of freedom of an arbitrarily oriented truss element

We would like to obtain the matrix **k** such that

$$\mathbf{k}\mathbf{u} = \mathbf{f}$$

where $\mathbf{u}$ and $\mathbf{f}$ are as in (2.16). Of course, we have already derived $\tilde{\mathbf{k}}$ such that, see (2.13),

$$\tilde{\mathbf{k}}\tilde{\mathbf{u}} = \tilde{\mathbf{f}}. \tag{2.28}$$

Before deriving $\mathbf{k}$ based on a transformation matrix, let us show how $\mathbf{k}$ could be constructed column by column imposing unit displacements.

As an example, we obtain the first column by imposing a unit displacement $u_1 = 1$ and fixing the remaining degrees of freedom, *i.e.*, $u_2 = u_3 = u_4 = 0$. We know that under such conditions $k_{11} = f_1$, $k_{21} = f_2$, $k_{31} = f_3$ and $k_{41} = f_4$.

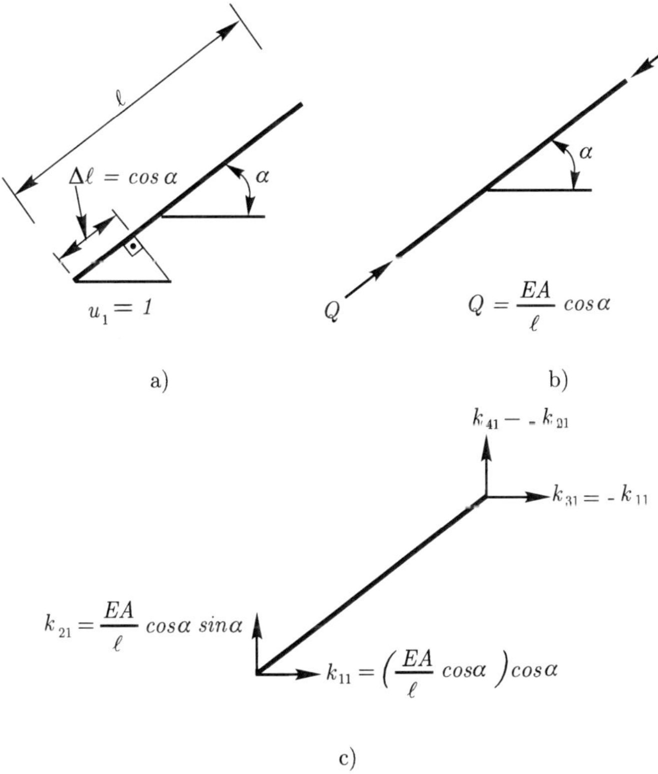

Fig. 2.27. Imposed horizontal unit displacement and corresponding forces. a) Imposed displacement and corresponding shortening of bar; b) Resulting force Q acting onto the bar; c) Nodal forces (stiffness coefficients) corresponding to (replacing) force Q

Referring to Figure 2.27, we have

$$\begin{aligned} \Delta\ell &= 1 \cdot \cos\alpha \\ Q &= EA\frac{\Delta\ell}{\ell} = \frac{EA}{\ell}\cos\alpha. \end{aligned}$$

Then, because the stiffness coefficients are forces into directions u_1, u_2, u_3 and u_4, and applied *onto* the element

$$\begin{aligned} k_{11} &= f_1 = Q\cos\alpha = \frac{EA}{\ell}\cos^2\alpha \\ k_{21} &= f_2 = Q\sin\alpha = \frac{EA}{\ell}\cos\alpha\sin\alpha \\ k_{31} &= f_3 = -Q\cos\alpha = -\frac{EA}{\ell}\cos^2\alpha \\ k_{41} &= f_4 = -Q\sin\alpha = -\frac{EA}{\ell}\cos\alpha\sin\alpha. \end{aligned} \tag{2.29}$$

Proceeding in an analogous way, we could construct the remaining columns. Of course, the bar axial force $N = -Q$.

However, a more effective procedure to obtain $\mathbf{k}$ is to use transformation matrices, where $\tilde{\mathbf{u}}$ and $\tilde{\mathbf{f}}$ in (2.28) are expressed in terms of $\mathbf{u}$ and $\mathbf{f}$, respectively.

The kinematic relation between the displacements at node 1 measured in the local $(\tilde{x}, \tilde{y})$ and global (x, y) systems is, see Figures 2.28 and 2.29,

$$\begin{aligned} \tilde{u}_1 &= u_1\cos\alpha + u_2\sin\alpha \\ \tilde{u}_2 &= -u_1\sin\alpha + u_2\cos\alpha. \end{aligned}$$

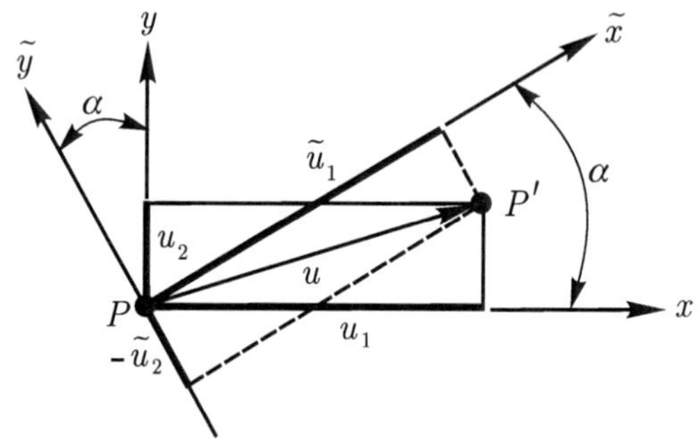

Fig. 2.28. Nodal point displacement vector $\mathbf{u}$ of local node 1 expressed in (x, y) and $(\tilde{x}, \tilde{y})$ coordinate systems

These relations can be written in matrix form

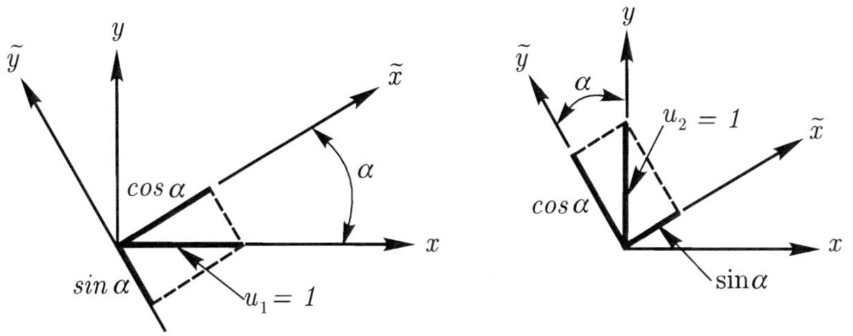

Fig. 2.29. Displacements $u_1 = 1$ and $u_2 = 1$ expressed in $(\tilde{x}, \tilde{y})$ coordinate system

$$\begin{bmatrix} \tilde{u}_1 \\ \tilde{u}_2 \end{bmatrix} = \begin{bmatrix} \cos\alpha & \sin\alpha \\ -\sin\alpha & \cos\alpha \end{bmatrix} \begin{bmatrix} u_1 \\ u_2 \end{bmatrix}.$$

For the displacements of the local node 2, the same kind of relationship holds

$$\begin{bmatrix} \tilde{u}_3 \\ \tilde{u}_4 \end{bmatrix} = \begin{bmatrix} \cos\alpha & \sin\alpha \\ -\sin\alpha & \cos\alpha \end{bmatrix} \begin{bmatrix} u_3 \\ u_4 \end{bmatrix}$$

which allows us to write

$$\tilde{\mathbf{u}} = \mathbf{T}\mathbf{u} \tag{2.30}$$

where

$$\mathbf{T} = \begin{bmatrix} \cos\alpha & \sin\alpha & 0 & 0 \\ -\sin\alpha & \cos\alpha & 0 & 0 \\ 0 & 0 & \cos\alpha & \sin\alpha \\ 0 & 0 & -\sin\alpha & \cos\alpha \end{bmatrix}.$$

Since we transformed vector components, the same relation holds for the forces

$$\tilde{\mathbf{f}} = \mathbf{T}\mathbf{f}. \tag{2.31}$$

It is easy to verify that $\mathbf{T}$ is an orthogonal matrix, *i.e.*,

$$\mathbf{T}^{-1} = \mathbf{T}^T, \qquad \mathbf{T}^T\mathbf{T} = \mathbf{I}.$$

Now substituting for $\tilde{\mathbf{u}}$ and $\tilde{\mathbf{f}}$ in (2.28) yields

$$\mathbf{T}\mathbf{f} = \tilde{\mathbf{k}}\mathbf{T}\mathbf{u}$$

and left multiplying both sides by $\mathbf{T}^T$

$$\mathbf{T}^T\mathbf{T}\mathbf{f} = \mathbf{T}^{\mathbf{T}}\tilde{\mathbf{k}}\mathbf{T}\mathbf{u}$$

we obtain

$$\mathbf{f} = \left(\mathbf{T}^{\mathbf{T}}\tilde{\mathbf{k}}\mathbf{T}\right)\mathbf{u}$$

yielding

$$\mathbf{k} = \mathbf{T}^{\mathbf{T}}\tilde{\mathbf{k}}\mathbf{T}. \tag{2.32}$$

Performing the matrix multiplications we arrive at

$$\mathbf{k} = \frac{EA}{\ell}\begin{bmatrix} \cos^2\alpha & \sin\alpha\cos\alpha & -\cos^2\alpha & -\sin\alpha\cos\alpha \\ \sin\alpha\cos\alpha & \sin^2\alpha & -\sin\alpha\cos\alpha & -\sin^2\alpha \\ -\cos^2\alpha & -\sin\alpha\cos\alpha & \cos^2\alpha & \sin\alpha\cos\alpha \\ -\sin\alpha\cos\alpha & -\sin^2\alpha & \sin\alpha\cos\alpha & \sin^2\alpha \end{bmatrix}. \tag{2.33}$$

Note that the first column in (2.33) corresponds to the results given in (2.29).

2.3.4 Solution of the three-bar truss structure using the matrix method

We can now efficiently solve the problem described in Figure 2.17. Using relation (2.32) we evaluate the stiffness matrix of element 3, $\mathbf{k}^{(3)}$, choosing node 4 as its initial node and impose the equilibrium of node 1. Equilibrium in the X direction gives

$$R_1 - \left(f_3^{(1)} + f_3^{(3)}\right) = 0 \tag{2.34}$$

where we used that $f_3^{(2)} = 0$. Equilibrium in the Y direction gives

$$R_2 - \left(f_4^{(2)} + f_4^{(3)}\right) = 0 \tag{2.35}$$

where we used that $f_4^{(1)} = 0$. Introducing the stiffness relations into (2.34) and (2.35) leads to

$$k_{33}^{(1)}u_3^{(1)} + k_{33}^{(3)}u_3^{(3)} + k_{34}^{(3)}u_4^{(3)} = R_1 \tag{2.36}$$

$$k_{44}^{(2)}u_4^{(2)} + k_{43}^{(3)}u_3^{(3)} + k_{44}^{(3)}u_4^{(3)} = R_2. \tag{2.37}$$

We note that in the above equations the stiffness coefficients $k_{34}^{(1)}$, $k_{43}^{(2)}$ are not included since, as bar 1 is horizontal and bar 2 is vertical, these coefficients are zero. Of course, for bar 3 these terms are not zero.

Introducing the compatibility relations

$$U_1 = u_3^{(1)} = u_3^{(3)}, \qquad U_2 = u_4^{(2)} = u_4^{(3)}$$

we can re-write equations (2.36) and (2.37) as

$$\begin{aligned} \left(k_{33}^{(1)} + k_{33}^{(3)}\right) U_1 + k_{34}^{(3)} U_2 &= R_1 \\ k_{43}^{(3)} U_1 + \left(k_{44}^{(2)} + k_{44}^{(3)}\right) U_2 &= R_2. \end{aligned}$$

The global stiffness coefficients are implicitly defined in the above equations and they are given by

$$\begin{aligned} K_{11} &= k_{33}^{(1)} + k_{33}^{(3)}, \qquad K_{12} = k_{34}^{(3)} \\ K_{22} &= k_{44}^{(2)} + k_{44}^{(3)}, \qquad K_{21} = k_{43}^{(3)}. \end{aligned} \tag{2.38}$$

Therefore the matrix equation for the structural assemblage is

$$\begin{bmatrix} K_{11} & K_{12} \\ K_{21} & K_{22} \end{bmatrix} \begin{bmatrix} U_1 \\ U_2 \end{bmatrix} = \begin{bmatrix} R_1 \\ R_2 \end{bmatrix} \tag{2.39}$$

and its numerical solution is given by

$$U_1 = 7.203 \times 10^{-5} \text{ m}, \qquad U_2 = 9.951 \times 10^{-5} \text{ m}.$$

We note that:

- The off-diagonal stiffness coefficients are now different from zero due to the inclined bar which couples the horizontal and vertical displacements U_1 and U_2.
- The structure stiffness matrix $\mathbf{K}$ is obtained by summing the appropriate stiffness coefficients of the bar elements. This assemblage process is a direct consequence of imposing the equilibrium and compatibility conditions at the nodes (joints).
- The equilibrium, compatibility and constitutive conditions for the bar elements are satisfied by use of the (correct) element stiffness matrices.
- The structure stiffness matrix is established for the bars in their original configuration, *i.e.*, the joint displacements which are caused by the applied loading do not enter $\mathbf{K}$.
- We have obtained U_1 and U_2 from (2.39) which completely characterize the solution of this statically indeterminate structure. Therefore the matrix method of solution gives directly the solution of statically indeterminate (and statically determinate) structures by enforcing all equilibrium, compatibility and constitutive conditions simultaneously.

- The displacements U_1 and U_2 have decreased due to adding the diagonal bar. This, of course, makes sense physically.

Let us further explore the three-bar truss structure problem. We formulated this problem only in terms of the free degrees of freedom U_1 and U_2. We could have also included the degrees of freedom at the supports which are shown in Figure 2.30.

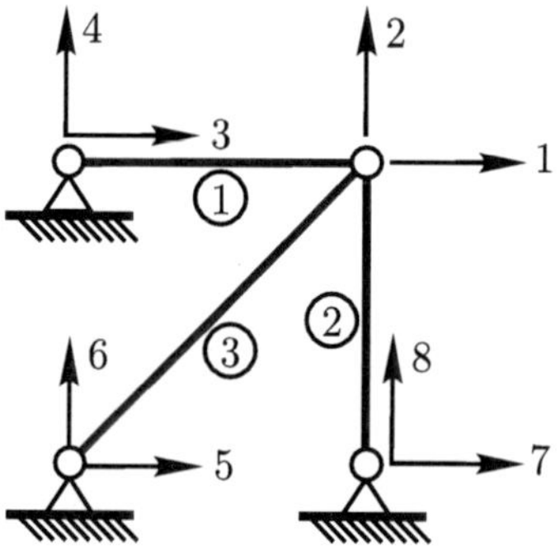

Fig. 2.30. Three-bar truss structure with all degrees of freedom shown

We observe that we have numbered all degrees of freedom. It is implied that for each degree of freedom i there is a displacement U_i and a force R_i. For the free degrees of freedom, the force is specified and the displacement is to be determined. For a restrained degree of freedom, the displacement is specified and the reaction force associated with such restraint is to be determined. The fixed conditions at the supports in this case imply, of course, that

$$U_3 = U_4 = U_5 = U_6 = U_7 = U_8 = 0.$$

Therefore, once the free degrees of freedom have been calculated, *i.e.*, when equation (2.39) has been solved, all the end bar displacements are known and all the bar end forces can be readily obtained by using the element stiffness matrices; that is, for bar (m)

$$\mathbf{f}^{(m)} = \mathbf{k}^{(m)}\mathbf{u}^{(m)}. \tag{2.40}$$

Consider bar 1, we have

$$U_1 = u_3^{(1)}, \qquad U_2 = u_4^{(1)}, \qquad U_3 = u_1^{(1)}, \qquad U_4 = u_2^{(1)}.$$

Equation (2.40) applied to bar 1 leads to

$$\mathbf{f}^{(1)} = \frac{E_1 A_1}{\ell_1} \begin{bmatrix} 1 & 0 & -1 & 0 \\ 0 & 0 & 0 & 0 \\ -1 & 0 & 1 & 0 \\ 0 & 0 & 0 & 0 \end{bmatrix} \begin{bmatrix} 0 \\ 0 \\ U_1 \\ U_2 \end{bmatrix}$$

which yields

$$\mathbf{f}^{(1)} = \begin{bmatrix} -11344 \\ 0 \\ 11344 \\ 0 \end{bmatrix}.$$

The reactions R_3 and R_4 can be evaluated using that node 2 is in equilibrium. Hence

$$R_3 = f_1^{(1)} = -11344 \text{ N}, \qquad R_4 = f_2^{(1)} = 0.$$

In an analogous manner, for bar 2

$$u_3^{(2)} = U_1, \qquad u_4^{(2)} = U_2, \qquad u_1^{(2)} = U_7, \qquad u_2^{(2)} = U_8.$$

$$\mathbf{f}^{(2)} = \frac{E_2 A_2}{\ell_2} \begin{bmatrix} 0 & 0 & 0 & 0 \\ 0 & 1 & 0 & -1 \\ 0 & 0 & 0 & 0 \\ 0 & -1 & 0 & 1 \end{bmatrix} \begin{bmatrix} 0 \\ 0 \\ U_1 \\ U_2 \end{bmatrix}, \qquad \mathbf{f}^{(2)} = \begin{bmatrix} 0 \\ -31344 \\ 0 \\ 31344 \end{bmatrix}.$$

Hence

$$R_7 = f_1^{(2)} = 0, \qquad R_8 = f_2^{(2)} = -31344 \text{ N}.$$

And for bar 3

$$u_3^{(3)} = U_1, \qquad u_4^{(3)} = U_2, \qquad u_1^{(3)} = U_5, \qquad u_2^{(3)} = U_6.$$

$$\mathbf{f}^{(3)} = \frac{E_3 A_3}{\ell_3} \begin{bmatrix} \frac{1}{2} & \frac{1}{2} & -\frac{1}{2} & -\frac{1}{2} \\ \frac{1}{2} & \frac{1}{2} & -\frac{1}{2} & -\frac{1}{2} \\ -\frac{1}{2} & -\frac{1}{2} & \frac{1}{2} & \frac{1}{2} \\ -\frac{1}{2} & -\frac{1}{2} & \frac{1}{2} & \frac{1}{2} \end{bmatrix} \begin{bmatrix} 0 \\ 0 \\ U_1 \\ U_2 \end{bmatrix}, \qquad \mathbf{f}^{(3)} = \begin{bmatrix} -28655 \\ -28655 \\ 28655 \\ 28655 \end{bmatrix}.$$

Hence

$$R_5 = f_1^{(3)} = -28655 \text{ N}, \qquad R_6 = f_2^{(3)} = -28655 \text{ N}.$$

Considering the three-bar truss structure, we observe that, in general, the displacements at the supports could be prescribed to have values different from zero. To consider this coupling explicitly, we evaluate the stiffness matrix corresponding to all degrees of freedom of the structure.

$$\left[\begin{array}{c|c|c|c|c|c|c|c}
k_{33}^{(1)} + k_{33}^{(3)} & k_{34}^{(3)} & k_{31}^{(1)} & 0 & k_{31}^{(3)} & k_{32}^{(3)} & 0 & 0 \\ \hline
k_{43}^{(3)} & k_{44}^{(2)} + k_{44}^{(3)} & 0 & 0 & k_{41}^{(3)} & k_{42}^{(3)} & 0 & k_{42}^{(2)} \\ \hline
k_{13}^{(1)} & 0 & k_{11}^{(1)} & 0 & 0 & 0 & 0 & 0 \\ \hline
0 & 0 & 0 & 0 & 0 & 0 & 0 & 0 \\ \hline
k_{13}^{(3)} & k_{14}^{(3)} & 0 & 0 & k_{11}^{(3)} & k_{12}^{(3)} & 0 & 0 \\ \hline
k_{23}^{(3)} & k_{24}^{(3)} & 0 & 0 & k_{21}^{(3)} & k_{22}^{(3)} & 0 & 0 \\ \hline
0 & 0 & 0 & 0 & 0 & 0 & 0 & 0 \\ \hline
0 & k_{24}^{(2)} & 0 & 0 & 0 & 0 & 0 & k_{22}^{(2)}
\end{array}\right]
\begin{bmatrix} U_1 \\ U_2 \\ U_3 \\ U_4 \\ U_5 \\ U_6 \\ U_7 \\ U_8 \end{bmatrix}
=
\begin{bmatrix} R_1 \\ R_2 \\ R_3 \\ R_4 \\ R_5 \\ R_6 \\ R_7 \\ R_8 \end{bmatrix}
\tag{2.41}$$

The additional stiffness coefficients – besides K_{11}, K_{12}, K_{21}, K_{22} which are already given in (2.38) – can be obtained by also considering the equilibrium and compatibility conditions of nodes 2 to 4, in the same way as given above for node 1. These considerations lead to adding the element stiffness matrices into the global structure stiffness matrix following the correspondence between the numbering of structure global and element local degrees of freedom. The non-zero contributions from each element can be identified in the matrix of (2.41).

The equations in (2.41) are then solved by specifying all known nodal displacements and the known externally applied nodal forces. The reactions can then be directly obtained by simply evaluating the left-hand side of (2.41).

We next develop this procedure in detail for general truss structures – and indeed for any other structural element assemblage.

2.3.5 Systematization of the matrix formulation for truss structures

We show in Figure 2.31 a part of a generic truss structure. In the matrix formulation of a truss problem, all bars, all nodes and all degrees of freedom are numbered, and the bar orientations are chosen. We select node g as a representative truss node to which we impose equilibrium.

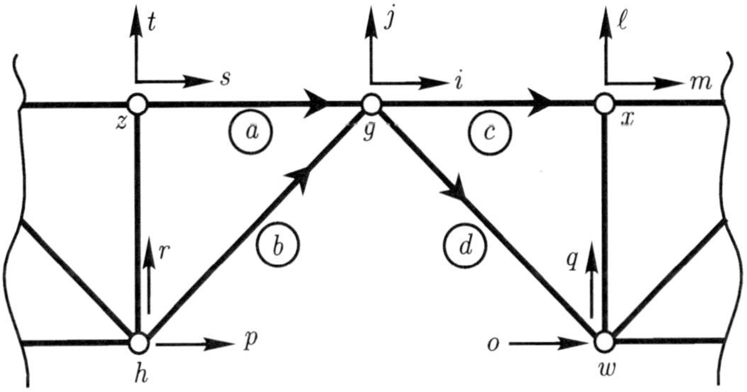

Fig. 2.31. Part of a truss structure

Referring to Figure 2.32, we can write for the node g

$$R_i = f_3^{(a)} + f_3^{(b)} + f_1^{(c)} + f_1^{(d)}, \qquad R_j = f_4^{(a)} + f_4^{(b)} + f_2^{(c)} + f_2^{(d)} \quad (2.42)$$

where we recall that the element end forces are applied onto the elements.

Hence, equations (2.42) reflect the fact that the external loads acting on a node are equilibrated by the sum of the bar end forces that connect to this node.

In order to facilitate the accounting of bar local and structure global numbering of degrees of freedom and the force summation process, we define for a generic bar (m) a $N \times 1$ column matrix $\mathbf{F}^{(m)}$ where N is the *total* number of degrees of freedom of the structure. The nodal forces of bar (m) are placed in $\mathbf{F}^{(m)}$ at the positions corresponding to the global numbering of the bar degrees of freedom. The remaining positions in $\mathbf{F}^{(m)}$ are each filled with 0. For example, for bar (b)

$$\mathbf{F}^{(b)^T} = [0 \cdots 0 \; \overset{i}{f_3^{(b)}} \; \overset{j}{f_4^{(b)}} \; 0 \cdots \overset{p}{f_1^{(b)}} \; 0 \cdots 0 \; \overset{r}{f_2^{(b)}} \; 0 \cdots 0]_{1\times N} \, .$$

With $\mathbf{F}^{(m)}$ given for every bar of the bar assemblage we can write the equilibrium equations for *every* degree of freedom as

$$\mathbf{R} = \sum_{m=1}^{n_e} \mathbf{F}^{(m)} \quad (2.43)$$

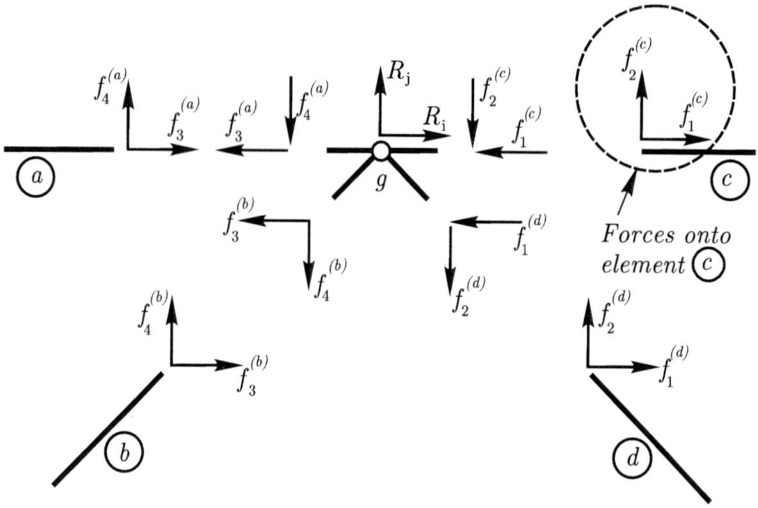

Fig. 2.32. Pictorial representation of equilibrium of node g

where $\mathbf{R}$ is the column matrix of external forces applied to the nodes of the truss corresponding to the global structural degrees of freedom and n_e is the total number of bar elements of the structure. Considering, for example, the degree of freedom i, we have

$$R_i = F_i^{(a)} + F_i^{(b)} + F_i^{(c)} + F_i^{(d)} \tag{2.44}$$

since (a), (b), (c) and (d) are the only bars which have end forces corresponding to the global degree of freedom i (that is, the ith entries of $\mathbf{F}^{(m)}$ are zero for all other bars ($m \neq a, b, c, d$)). Referring to Figure 2.32 we can write

$$F_i^{(a)} = f_3^{(a)}, \qquad F_i^{(b)} = f_3^{(b)}, \qquad F_i^{(c)} = f_1^{(c)}, \qquad F_i^{(d)} = f_1^{(d)}$$

and, hence, equation (2.44) is just the first equation of (2.42).

Now let us define a $N \times N$ matrix denoted by $\mathbf{K}^{(m)}$ such that

$$\mathbf{K}^{(m)}\mathbf{U} = \mathbf{F}^{(m)} \tag{2.45}$$

where $\mathbf{U}$ is the column matrix of the global displacement degrees of freedom and $\mathbf{K}^{(m)}$ has non-zero entries only at the positions associated with the nodal displacements of bar (m) when these nodal displacements are numbered according to the global ordering. Equation (2.45) contains, considering the relations between global and local numbering of degrees of freedom, the equations given by

$$\mathbf{k}^{(m)}\mathbf{u}^{(m)} = \mathbf{f}^{(m)} \tag{2.46}$$

and all remaining equations in (2.45) result into $0 = 0$ identities. Hence, all non-zero coefficients of $\mathbf{K}^{(m)}$ can be obtained from the coefficients of $\mathbf{k}^{(m)}$. For example, for bar (b), we have

$$K_{pp}^{(b)} = k_{11}^{(b)}, \quad K_{rr}^{(b)} = k_{22}^{(b)}, \quad K_{pr}^{(b)} = k_{12}^{(b)}, \quad K_{pi}^{(b)} = k_{13}^{(b)}, \quad K_{pj}^{(b)} = k_{14}^{(b)}$$

and so on.

Note that as we use equation (2.45) to reproduce (2.46), we are implicitly enforcing compatibility since the bar end displacements are taken to be the global displacements.

Now we are ready to present the following important derivation. Substituting (2.45) into (2.43) yields

$$\mathbf{R} = \left(\sum_{m=1}^{n_e} \mathbf{K}^{(m)} \right) \mathbf{U} \tag{2.47}$$

leading to

$$\mathbf{K}\mathbf{U} = \mathbf{R} \tag{2.48}$$

and, of course,

$$\mathbf{K} = \sum_{m=1}^{n_e} \mathbf{K}^{(m)} \tag{2.49}$$

is the stiffness matrix of the total structure obtained from the element stiffness matrices.

We observe that equation (2.48) represents the matrix formulation for a generic truss structure and this equation contains all three fundamental requirements:

- Equilibrium
- Compatibility
- Constitutive

"Equilibrium" because each truss element is always in equilibrium for any force it carries and (2.43) enforces the equilibrium of the nodes. "Compatibility" because the bars are connected to the joints which are undergoing unique displacements (some of which are imposed as displacement boundary conditions). "Constitutive" because the correct Young's modulus E is used for each element. Hence, once $\mathbf{U}$ has been calculated from (2.48), the truss problem has been solved.

The matrices $\mathbf{K}^{(m)}$, $\mathbf{F}^{(m)}$ were defined because they are very useful to present the above theoretical derivations in a rigorous and elegant manner. However, most entries of the matrices $\mathbf{K}^{(m)}$and $\mathbf{F}^{(m)}$ are zero and, in actual computations, we need to take advantage of this fact.

Let us briefly describe an efficient computational procedure to obtain $\mathbf{K}$. Since all non-zero entries of $\mathbf{K}^{(m)}$ are in $\mathbf{k}^{(m)}$, $\mathbf{K}$ can be obtained without constructing the $\mathbf{K}^{(m)}$. We define for every bar element of the truss the following row matrix

$$\mathbf{LM}^{(m)} = \begin{array}{c} \begin{matrix} u_1 & u_2 & u_3 & u_4 \end{matrix} \\ \begin{bmatrix} \ell & p & q & r \end{bmatrix} \end{array} \tag{2.50}$$

which is referred to as the element connectivity array. The number ℓ in the first entry is the number of the degree of freedom of the *structure* which corresponds to the displacement u_1 of bar (m). Analogously, p, q and r correspond to u_2, u_3 and u_4. For example, the element arrays for bars (a), (b), (c) and (d) defined in Figure 2.31 are

$$\mathbf{LM}^{(a)} = \begin{bmatrix} s & t & i & j \end{bmatrix}, \qquad \mathbf{LM}^{(b)} = \begin{bmatrix} p & r & i & j \end{bmatrix}$$

$$\mathbf{LM}^{(c)} = \begin{bmatrix} i & j & m & \ell \end{bmatrix}, \qquad \mathbf{LM}^{(d)} = \begin{bmatrix} i & j & o & q \end{bmatrix}.$$

The assemblage process implied by the summation sign in equation (2.49) can be effectively performed starting with an array of an empty $N \times N$ matrix (each entry in the matrix is initially zero) which eventually will contain $\mathbf{K}$. For every bar element in the structural assemblage, $m = 1, \cdots, n_e$, we then add the element stiffness matrix into this array. Considering bar (m) for which $\mathbf{LM}^{(m)}$ is given in (2.50) we add

$k_{11}^{(m)}$ to the entry $\ell\ell$ of the array

$k_{12}^{(m)}$ to the entry ℓp

$k_{13}^{(m)}$ to the entry ℓq

$k_{14}^{(m)}$ to the entry ℓr

$k_{22}^{(m)}$ to the entry pp

$k_{23}^{(m)}$ to the entry pq

$k_{24}^{(m)}$ to the entry pr

$k_{33}^{(m)}$ to the entry qq

$k_{34}^{(m)}$ to the entry qr

$k_{44}^{(m)}$ to the entry rr.

We note that since each bar stiffness matrix is symmetric (see equation (2.33)), the structure stiffness matrix $\mathbf{K}$ is also symmetric, see equation

(2.49). For this reason we construct in the assemblage procedure only the upper diagonal part of **K**.

To demonstrate the matrix procedure we consider the following examples.

Example 2.2

Formulate and solve the problem described in Figure 2.4 and considered in Example 2.1 using the matrix method.

Solution

We adopt the nodal and bar numbering already given in Figure 2.4 and define in Figure 2.33 the numbering of the degrees of freedom as well as the bar orientations.

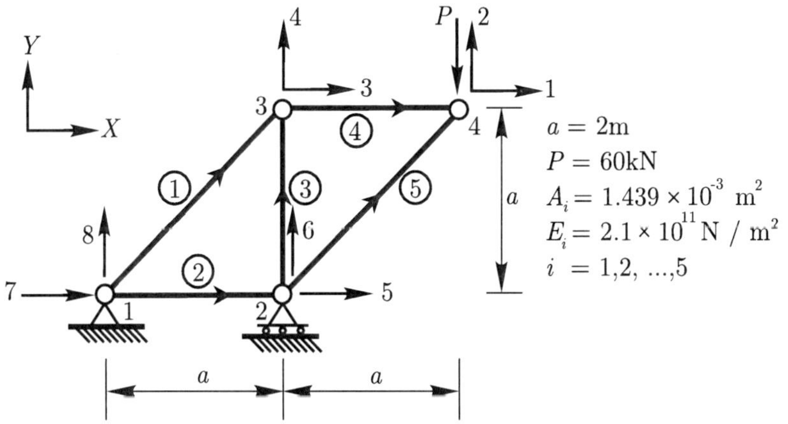

Fig. 2.33. Definitions for the matrix formulation of the problem in Figure 2.4

The next step is to obtain the stiffness matrices of the bar elements in the global coordinate system. We note that for bars 2 and 4, the local and global coordinate systems (of displacements and nodal forces) are the same. Therefore, we can write

$$\mathbf{k}^{(2)} = \mathbf{k}^{(4)} = \frac{EA}{a}\begin{bmatrix} 1 & 0 & -1 & 0 \\ 0 & 0 & 0 & 0 \\ -1 & 0 & 1 & 0 \\ 0 & 0 & 0 & 0 \end{bmatrix}.$$

The stiffness matrices of bars 1 and 5 are the same and can be obtained, corresponding to the global coordinate system, using equation (2.33) with $\alpha = 45°$ leading to

$$\mathbf{k}^{(1)} = \mathbf{k}^{(5)} = \frac{EA\sqrt{2}}{4a} \begin{bmatrix} 1 & 1 & -1 & -1 \\ 1 & 1 & -1 & -1 \\ -1 & -1 & 1 & 1 \\ -1 & -1 & 1 & 1 \end{bmatrix}$$

and for bar 3 we use again equation (2.33) with $\alpha = 90°$ arriving at

$$\mathbf{k}^{(3)} = \frac{EA}{a} \begin{bmatrix} 0 & 0 & 0 & 0 \\ 0 & 1 & 0 & -1 \\ 0 & 0 & 0 & 0 \\ 0 & -1 & 0 & 1 \end{bmatrix}.$$

To assemble the global stiffness matrix, we need the element connectivity arrays

$$\begin{aligned} \mathbf{LM}^{(1)} &= \begin{bmatrix} 7 & 8 & 3 & 4 \end{bmatrix}, \quad \mathbf{LM}^{(2)} = \begin{bmatrix} 7 & 8 & 5 & 6 \end{bmatrix} \\ \mathbf{LM}^{(3)} &= \begin{bmatrix} 5 & 6 & 3 & 4 \end{bmatrix}, \quad \mathbf{LM}^{(4)} = \begin{bmatrix} 3 & 4 & 1 & 2 \end{bmatrix} \\ \mathbf{LM}^{(5)} &= \begin{bmatrix} 5 & 6 & 1 & 2 \end{bmatrix}. \end{aligned}$$

Then performing the assemblage procedure leads to

$$\mathbf{K} = \begin{bmatrix} k_{33}^{(4)} + k_{33}^{(5)} & k_{34}^{(4)} + k_{34}^{(5)} & k_{13}^{(4)} & k_{23}^{(4)} & k_{13}^{(5)} & k_{23}^{(5)} & 0 & 0 \\ & k_{44}^{(4)} + k_{44}^{(5)} & k_{14}^{(4)} & k_{24}^{(4)} & k_{14}^{(5)} & k_{24}^{(5)} & 0 & 0 \\ & & k_{33}^{(1)} + k_{33}^{(3)} + k_{11}^{(4)} & k_{34}^{(1)} + k_{34}^{(3)} + k_{12}^{(4)} & k_{13}^{(3)} & k_{23}^{(3)} & k_{13}^{(1)} & k_{23}^{(1)} \\ & & & k_{44}^{(1)} + k_{44}^{(3)} + k_{22}^{(4)} & k_{14}^{(3)} & k_{24}^{(3)} & k_{14}^{(1)} & k_{24}^{(1)} \\ & & & & k_{33}^{(2)} + k_{11}^{(3)} + k_{11}^{(5)} & k_{34}^{(2)} + k_{12}^{(3)} + k_{12}^{(5)} & k_{13}^{(2)} & k_{23}^{(2)} \\ & \textit{symmetric} & & & & k_{44}^{(2)} + k_{22}^{(3)} + k_{22}^{(5)} & k_{14}^{(2)} & k_{24}^{(2)} \\ & & & & & & k_{11}^{(1)} + k_{11}^{(2)} & k_{12}^{(1)} + k_{12}^{(2)} \\ & & & & & & & k_{22}^{(1)} + k_{22}^{(2)} \end{bmatrix}$$

and introducing the numerical values we obtain

$$
\mathbf{K} = \frac{EA}{a}\begin{bmatrix}
\frac{\sqrt{2}}{4}+1 & \frac{\sqrt{2}}{4} & -1 & 0 & -\frac{\sqrt{2}}{4} & -\frac{\sqrt{2}}{4} & 0 & 0 \\
\frac{\sqrt{2}}{4} & \frac{\sqrt{2}}{4} & 0 & 0 & -\frac{\sqrt{2}}{4} & -\frac{\sqrt{2}}{4} & 0 & 0 \\
-1 & 0 & \frac{\sqrt{2}}{4}+1 & \frac{\sqrt{2}}{4} & 0 & 0 & -\frac{\sqrt{2}}{4} & -\frac{\sqrt{2}}{4} \\
0 & 0 & \frac{\sqrt{2}}{4} & \frac{\sqrt{2}}{4}+1 & 0 & -1 & -\frac{\sqrt{2}}{4} & -\frac{\sqrt{2}}{4} \\
-\frac{\sqrt{2}}{4} & -\frac{\sqrt{2}}{4} & 0 & 0 & \frac{\sqrt{2}}{4}+1 & \frac{\sqrt{2}}{4} & -1 & 0 \\
-\frac{\sqrt{2}}{4} & -\frac{\sqrt{2}}{4} & 0 & -1 & \frac{\sqrt{2}}{4} & \frac{\sqrt{2}}{4}+1 & 0 & 0 \\
0 & 0 & -\frac{\sqrt{2}}{4} & -\frac{\sqrt{2}}{4} & -1 & 0 & \frac{\sqrt{2}}{4}+1 & \frac{\sqrt{2}}{4} \\
0 & 0 & -\frac{\sqrt{2}}{4} & -\frac{\sqrt{2}}{4} & 0 & 0 & \frac{\sqrt{2}}{4} & \frac{\sqrt{2}}{4}
\end{bmatrix}.
$$

Hence the complete set of equilibrium equations with the applied nodal force P and imposed displacement restraints is

$$
\frac{EA}{a}\left[\begin{array}{ccccc|ccc}
\frac{\sqrt{2}}{4}+1 & \frac{\sqrt{2}}{4} & -1 & 0 & -\frac{\sqrt{2}}{4} & -\frac{\sqrt{2}}{4} & 0 & 0 \\
\frac{\sqrt{2}}{4} & \frac{\sqrt{2}}{4} & 0 & 0 & -\frac{\sqrt{2}}{4} & -\frac{\sqrt{2}}{4} & 0 & 0 \\
-1 & 0 & \frac{\sqrt{2}}{4}+1 & \frac{\sqrt{2}}{4} & 0 & 0 & -\frac{\sqrt{2}}{4} & -\frac{\sqrt{2}}{4} \\
0 & 0 & \frac{\sqrt{2}}{4} & \frac{\sqrt{2}}{4}+1 & 0 & -1 & -\frac{\sqrt{2}}{4} & -\frac{\sqrt{2}}{4} \\
-\frac{\sqrt{2}}{4} & -\frac{\sqrt{2}}{4} & 0 & 0 & \frac{\sqrt{2}}{4}+1 & \frac{\sqrt{2}}{4} & -1 & 0 \\
\hline
-\frac{\sqrt{2}}{4} & -\frac{\sqrt{2}}{4} & 0 & -1 & \frac{\sqrt{2}}{4} & \frac{\sqrt{2}}{4}+1 & 0 & 0 \\
0 & 0 & -\frac{\sqrt{2}}{4} & -\frac{\sqrt{2}}{4} & -1 & 0 & \frac{\sqrt{2}}{4}+1 & \frac{\sqrt{2}}{4} \\
0 & 0 & -\frac{\sqrt{2}}{4} & -\frac{\sqrt{2}}{4} & 0 & 0 & \frac{\sqrt{2}}{4} & \frac{\sqrt{2}}{4}
\end{array}\right]
\left[\begin{array}{c}
U_1 \\ U_2 \\ U_3 \\ U_4 \\ U_5 \\ \hline 0 \\ 0 \\ 0
\end{array}\right]
=
\left[\begin{array}{c}
0 \\ -P \\ 0 \\ 0 \\ 0 \\ \hline R_6 \\ R_7 \\ R_8
\end{array}\right]. \tag{2.51}
$$

Hence we can identify natural partitions for the load and displacement column matrices. The displacement partitioning is obtained according to whether the displacement degrees of freedom are free or restrained. Denoting by $\mathbf{U}_a$ the free displacement degrees of freedom and by $\mathbf{U}_b$ the restrained degrees of freedom, we can write

$$\mathbf{U}^T = \begin{bmatrix} \mathbf{U}_a^T & \mathbf{U}_b^T \end{bmatrix}$$

where for this particular case

$$\mathbf{U}_a^T = \begin{bmatrix} U_1 & U_2 & U_3 & U_4 & U_5 \end{bmatrix}$$

and

$$\mathbf{U}_b^T = \begin{bmatrix} U_6 & U_7 & U_8 \end{bmatrix}.$$

Analogously, for the load column matrix

$$\mathbf{R}^T = \begin{bmatrix} \mathbf{R}_a^T & \mathbf{R}_b^T \end{bmatrix}$$

where $\mathbf{R}_a$ collects the external loads for the free degrees of freedom and for this case is given by

$$\mathbf{R}_a^T = \begin{bmatrix} R_1 & R_2 & R_3 & R_4 & R_5 \end{bmatrix}$$

with $R_1 = R_3 = R_4 = R_5 = 0$ and $R_2 = -P$. The column matrix $\mathbf{R}_b$ collects the reactions and is given by

$$\mathbf{R}_b^T = \begin{bmatrix} R_6 & R_7 & R_8 \end{bmatrix}.$$

Furthermore, the partitions of the load and displacement column matrices also induce the following partitioning for the stiffness matrix

$$\begin{bmatrix} \mathbf{K}_{aa} & \mathbf{K}_{ab} \\ \mathbf{K}_{ba} & \mathbf{K}_{bb} \end{bmatrix} \begin{bmatrix} \mathbf{U}_a \\ \mathbf{U}_b \end{bmatrix} = \begin{bmatrix} \mathbf{R}_a \\ \mathbf{R}_b \end{bmatrix}. \tag{2.52}$$

Here $\mathbf{R}_a$ and $\mathbf{U}_b$ contain always known values whereas $\mathbf{U}_a$ and $\mathbf{R}_b$ contain always unknown values. In order to solve the system in (2.52) we use

$$\mathbf{K}_{aa}\mathbf{U}_a + \mathbf{K}_{ab}\mathbf{U}_b = \mathbf{R}_a \tag{2.53}$$

$$\mathbf{K}_{ba}\mathbf{U}_a + \mathbf{K}_{bb}\mathbf{U}_b = \mathbf{R}_b \tag{2.54}$$

and obtain

$$\mathbf{K}_{aa}\mathbf{U}_a = \mathbf{R}_a - \mathbf{K}_{ab}\mathbf{U}_b \tag{2.55}$$

which can be solved for $\mathbf{U}_a$. Having obtained $\mathbf{U}_a$, the reactions $\mathbf{R}_b$ can be evaluated from (2.54).

In this case, (2.55) reads

$$\frac{EA}{a}\begin{bmatrix} \frac{\sqrt{2}}{4}+1 & \frac{\sqrt{2}}{4} & -1 & 0 & -\frac{\sqrt{2}}{4} \\ \frac{\sqrt{2}}{4} & \frac{\sqrt{2}}{4} & 0 & 0 & -\frac{\sqrt{2}}{4} \\ -1 & 0 & \frac{\sqrt{2}}{4}+1 & \frac{\sqrt{2}}{4} & 0 \\ 0 & 0 & \frac{\sqrt{2}}{4} & \frac{\sqrt{2}}{4}+1 & 0 \\ -\frac{\sqrt{2}}{4} & -\frac{\sqrt{2}}{4} & 0 & 0 & \frac{\sqrt{2}}{4}+1 \end{bmatrix}\begin{bmatrix} U_1 \\ U_2 \\ U_3 \\ U_4 \\ U_5 \end{bmatrix} = \begin{bmatrix} 0 \\ -P \\ 0 \\ 0 \\ 0 \end{bmatrix}$$

since $\mathbf{U}_b = \mathbf{0}$. Introducing the numerical values for E, A, a and P and solving yields

$$\begin{aligned} U_1 &= 1.91737 \times 10^{-3} \text{ m}, \qquad U_2 = -3.43765 \times 10^{-3} \text{ m} \\ U_3 &= 1.52027 \times 10^{-3} \text{ m}, \qquad U_4 = -3.97101 \times 10^{-4} \text{ m} \\ U_5 &= -3.97101 \times 10^{-4} \text{ m}. \end{aligned}$$

The reactions are evaluated using (2.54)

$$\frac{EA}{a}\begin{bmatrix} -\frac{\sqrt{2}}{4} & -\frac{\sqrt{2}}{4} & 0 & -1 & \frac{\sqrt{2}}{4} \\ 0 & 0 & -\frac{\sqrt{2}}{4} & -\frac{\sqrt{2}}{4} & -1 \\ 0 & 0 & -\frac{\sqrt{2}}{4} & -\frac{\sqrt{2}}{4} & 0 \end{bmatrix}\begin{bmatrix} U_1 \\ U_2 \\ U_3 \\ U_4 \\ U_5 \end{bmatrix} = \begin{bmatrix} R_6 \\ R_7 \\ R_8 \end{bmatrix}$$

which leads to

$$R_6 = 120000 \text{ N}, \qquad R_7 = 0, \qquad R_8 = -60000 \text{ N}.$$

To complete the solution, we need to evaluate the internal forces in the truss bars. Considering bar 1 and using the nodal displacements, we obtain the end displacements of bar 1

$$\begin{aligned} u_1^{(1)} &= U_7 = 0, \qquad u_2^{(1)} = U_8 = 0 \\ u_3^{(1)} &= U_3 = 1.52027 \times 10^{-3} \text{ m}, \qquad u_4^{(1)} = U_4 = -3.97101 \times 10^{-4} \text{ m}. \end{aligned}$$

Therefore, the global nodal forces acting onto the bar are

$$\mathbf{f}^{(1)} = \mathbf{k}^{(1)}\mathbf{u}^{(1)} = \begin{bmatrix} -60000 \\ -60000 \\ 60000 \\ 60000 \end{bmatrix}$$

where the values are given in the units used throughout the solution (Newtons).
We could extract the axial force from $\mathbf{f}^{(1)}$ by projecting the components into the axial direction. Equivalently, we can evaluate $\tilde{\mathbf{f}}^{(1)}$ using equation (2.31) with $\alpha = 45°$, which leads to

$$\begin{bmatrix} \frac{\sqrt{2}}{2} & \frac{\sqrt{2}}{2} & 0 & 0 \\ -\frac{\sqrt{2}}{2} & \frac{\sqrt{2}}{2} & 0 & 0 \\ 0 & 0 & \frac{\sqrt{2}}{2} & \frac{\sqrt{2}}{2} \\ 0 & 0 & -\frac{\sqrt{2}}{2} & \frac{\sqrt{2}}{2} \end{bmatrix} \begin{bmatrix} -60000 \\ -60000 \\ 60000 \\ 60000 \end{bmatrix} = \begin{bmatrix} -84853 \\ 0 \\ 84853 \\ 0 \end{bmatrix}$$

and hence, of course,

$$N_1 = -\tilde{f}_1^{(1)} = \tilde{f}_3^{(1)} = 84853 \text{ N}.$$

In an analogous manner we can evaluate the remaining internal bar forces to obtain

$$N_2 = -60000 \text{ N}, \quad N_3 = -60000 \text{ N}, \quad N_4 = 60000 \text{ N}, \quad N_5 = -84853 \text{ N}.$$

□

Of course, the solution results given in Example 2.2 are those we obtained earlier (see Example 2.1).

It is instructive to compare the two solution methods. We solved the problem described in Figure 2.4, firstly, by using an elementary method in which the bar forces are determined by equilibrium of the joints and then the displacements are found and then, secondly, by the matrix method. We note that even in this case in which this elementary method allows to find the nodal forces by nodal equilibrium only – with good insight into the deformation of the structure – the determination of the nodal displacements is relatively complex and difficult to systematize. Hence, the matrix method has great appeal since the whole procedure can be easily systematized and implemented in computer programs leading to the solution of very large problems in a straight forward and fast manner.

Furthermore, the solution of statically indeterminate trusses requires no additional considerations.

Example 2.3

Consider the truss described in Figure 2.19, which is obtained by adding bar 6 to the truss structure of Figure 2.4. Assuming that bar 6 has the same Young's modulus and cross-sectional area as the other bars, modify the matrix formulation presented in Example 2.2 to solve this problem.

Solution

The only modification necessary is to add the contribution of bar 6 to the stiffness matrix of the structure. Considering that for bar 6 we have $\ell = a\sqrt{5}$ and $\alpha = arctg(1/2) = 26.565°$, its stiffness matrix is given by

$$\mathbf{k}^{(6)} = \frac{EA\sqrt{5}}{25a} \begin{bmatrix} 4 & 2 & -4 & -2 \\ 2 & 1 & -2 & -1 \\ -4 & -2 & 4 & 2 \\ -2 & -1 & 2 & 1 \end{bmatrix}$$

where we assume the bar orientation from node 1 to 4. The coefficients of the upper diagonal part of the global stiffness matrix which should be updated are

$$\begin{aligned} K_{11}^6 &= K_{11}^5 + k_{33}^{(6)}, & K_{12}^6 &= K_{12}^5 + k_{34}^{(6)} \\ K_{22}^6 &= K_{22}^5 + k_{44}^{(6)}, & K_{17}^6 &= 0 + k_{31}^{(6)} \\ K_{18}^6 &= 0 + k_{32}^{(6)}, & K_{27}^6 &= 0 + k_{41}^{(6)} \\ K_{28}^6 &= 0 + k_{42}^{(6)}, & K_{77}^6 &= K_{77}^5 + k_{11}^{(6)} \\ K_{78}^6 &= K_{78}^5 + k_{12}^{(6)}, & K_{88}^6 &= K_{88}^5 + k_{22}^{(6)} \end{aligned}$$

where we have used K_{ij}^5 and K_{ij}^6 to represent the stiffness coefficient K_{ij} of the truss structure with 5 and 6 bars respectively. The solution is then obtained, as for Example 2.2, by considering the updated stiffness matrix.

⊓

We emphasize that since equilibrium and compatibility are enforced simultaneously in the matrix method, there is no need to consider in the solution procedure whether the truss is a statically determinate or a statically indeterminate structure. Namely, adding bar 6 in Example 2.3, which makes the structure statically indeterminate, has very little impact on the complete solution effort. In fact, we only need to add the contribution of bar 6 to the structure stiffness matrix and there is no increase in the order of the system of linear algebraic equations to be solved. On the other hand, when we try to use the elementary method, the addition of bar 6 significantly increases the effort of solution because, since the structure becomes statically indeterminate, we can no longer determine the bar forces by nodal equilibrium only.

These observations reinforce the earlier conclusion that the matrix method is a very efficient method for the computerized analysis of complex truss structures.

2.3.6 Principle of superposition

We note that equation (2.48) establishes a linear relation between the applied forces and the resulting displacements. Of course, this linear relation is a direct consequence of the assumptions used in the formulation of the truss mathematical model, which all together result into a constant stiffness matrix $\mathbf{K}$ (that is independent of the nodal displacements). Such models are called "linear (elastic) models" and for such models the *principle of superposition* holds. Suppose that the total load $\mathbf{R}$ acting on a truss structure is decomposed into n_ℓ load sets given by $\mathbf{R}_i$, that is

$$\mathbf{R} = \sum_{i=1}^{n_\ell} \mathbf{R}_i.$$

If we solve for each load set

$$\mathbf{K}\mathbf{U}_i = \mathbf{R}_i$$

then the solution for the total load $\mathbf{R}$ is

$$\mathbf{U} = \sum_{i=1}^{n_\ell} \mathbf{U}_i.$$

Namely, we have

$$\mathbf{K}\mathbf{U} = \mathbf{K}\left(\sum_{i=1}^{n_\ell} \mathbf{U}_i\right) = \sum_{i=1}^{n_\ell} (\mathbf{K}\mathbf{U}_i) = \sum_{i=1}^{n_\ell} \mathbf{R}_i = \mathbf{R}.$$

and these relations hold true because $\mathbf{K}$ is constant.

The principle of superposition is valid for all linear mathematical models studied in this book. In practice, a structure may be analyzed for many different load cases (gravity, wind loading, snow loading, settlement of supports, etc.) and the analyst/designer needs to seek the worst valid combination of loads to identify the highest internal forces that the structure may possibly experience. Then, of course, for each load combination the principle of superposition is used to obtain the structural response.

To give a simple example, we mention that the three-bar structure described in Figure 2.17 could have been solved considering only R_1 and then R_2. Of course, the total response would be obtained by superimposing the two resulting solutions. However, the maximum force in bar 2 is reached when R_2 acts alone leading to $N_2 = 47.71$ kN (when R_1 acts alone $N_2 = -16.37$ kN).

Also, to obtain insight into the structural behavior, the principle of superposition is sometimes used to break up the structural response for complex loading, allowing the analyst/designer to examine the contribution of each load case to the total response.

2.3.7 Remarks about the structure stiffness matrix

We gave already a physical explanation of the stiffness coefficients of a bar element, see Figure 2.22. The same discussion also applies to the stiffness coefficients in $\mathbf{K}$. The element K_{ii} gives the force that should be applied to the degree of freedom i to impose a unit displacement at this same degree of freedom, when all other structural degrees of freedom are fixed. The stiffness coefficients K_{ji} for $j \neq i$ give the reaction forces associated with the degrees of freedom that have been fixed. Based on this property, the stiffness matrix can be constructed column by column. For example, referring to Example 2.2, the first column of the global stiffness matrix can be obtained by imposing a unit displacement at the degree of freedom 1, *i.e.*, for the horizontal displacement of node 4, and fixing all other degrees of freedom.

We can also interpret the contributions of the bar stiffness coefficients to the global stiffness matrix coefficients for such column. In fact, to impose $U_1 = 1$ we need to impose a unit displacement at the end sections of bars 4 and 5 which couple into node 4. Therefore

$$K_{11} = k_{33}^{(4)} + k_{33}^{(5)}$$

since $k_{33}^{(4)}$ and $k_{33}^{(5)}$ give the horizontal forces that should be imposed at the end sections of bars 4 and 5, respectively, for a unit end displacement when all the remaining bar degrees of freedom are fixed. By an analogous reasoning, the reactions at the fixed degrees of freedom can also be obtained as

$$\begin{aligned} K_{21} &= k_{43}^{(4)} + k_{43}^{(5)}, \qquad K_{31} = k_{13}^{(4)} \\ K_{41} &= k_{23}^{(4)}, \qquad K_{51} = k_{13}^{(5)}, \qquad K_{61} = k_{23}^{(5)}. \end{aligned}$$

In addition, $K_{71} = K_{81} = 0$ since there are no bars connecting nodes 1 and 4.

The above discussion also shows that the equation $\mathbf{KU} = \mathbf{R}$ represents a linear system of N algebraic equilibrium equations. The ith equation of the system given by

$$\sum_{j=1}^{N} K_{ij} U_j = R_i$$

reflects the equilibrium at the ith degree of freedom. In other words, $K_{ij}U_j$ gives the internal force contribution associated with the displacement U_j to the equilibrium at the ith degree of freedom.

2.3.8 Strain energy of a truss structure

In this section we introduce the strain energy concept for truss structures.

Elastic solids when subjected to external loading deform and store energy associated with the deformation like a spring which is called *strain energy.* We detail below this concept for a truss bar and then for truss structures.

The truss bar of linear elastic material shown in Figure 2.34a is subjected to a slowly increasing external load up to the value R. The final configuration is shown in Figure 2.34b.

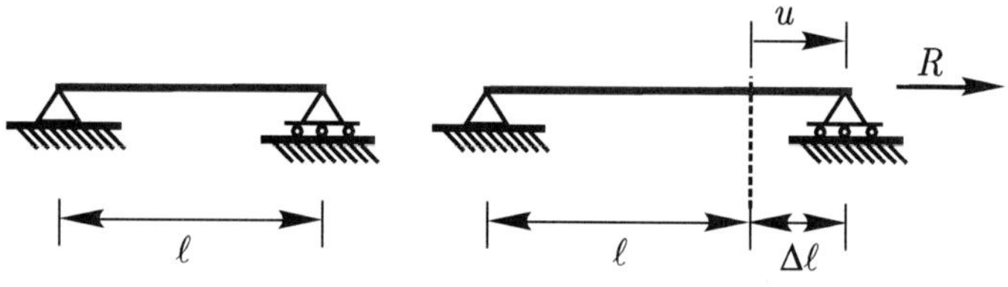

Fig. 2.34. Deformation of single bar structure

Let W_e be the external work done by the applied load. The differential external work dW_e associated with an induced differential displacement is given by the shaded area in Figure 2.35a, *i.e.*,

$$dW_e = R_u \, du_u$$

and, therefore,

$$W_e = \int_0^R R_u \, du_u = \frac{1}{2} Ru.$$

Let W_i be the internal work. The differential increment of internal work associated with an induced increment in strain is given by the shaded area in Figure 2.35b, multiplied by the differential volume element $dxdA$, *i.e.*,

$$dW_i = \tau_u d\varepsilon_u dx \, dA$$

and, therefore[7],

$$W_i = \int_V \left(\int_0^\tau \tau_u \, d\varepsilon_u \right) dxdA = \frac{1}{2} \tau\varepsilon A\ell.$$

Since

$$\tau = \frac{R}{A}, \qquad \varepsilon = \frac{\Delta\ell}{\ell} = \frac{u}{\ell}$$

[7] Note that although the bar will change its thickness (cross-sectional area) as indicated in Figure 2.35b, we are integrating over the original volume, in correspondence with the linear analysis assumptions (see Section 2.1.5)

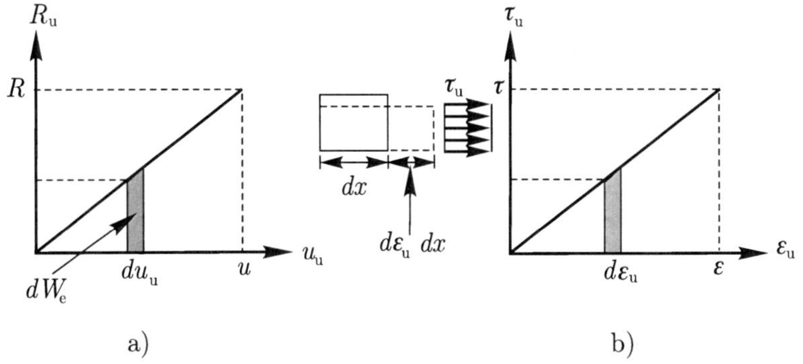

Fig. 2.35. a) Load displacement diagram for a truss bar; b) Stress-strain diagram for a generic point of a truss bar and side view of a differential volume element of the truss bar

we obviously have

$$W_i = W_e.$$

This is an important result that shows that the work done by the external force is equal to the work done by the internal forces/stresses.

Using Hooke's law, we can write

$$W_i = \frac{\varepsilon^2}{2} EA\ell.$$

Note that W_i depends only on the current strain and gives the work stored in the truss bar as elastic deformation. Therefore, we define

$$\mathcal{U}(\varepsilon) = W_i(\varepsilon) = \frac{\varepsilon^2}{2} EA\ell$$

as the strain energy of the truss bar. Of course, the strain energy per unit of volume is given by

$$W(\varepsilon) = \frac{E\varepsilon^2}{2} = \frac{1}{2}\tau\varepsilon.$$

It is usual to express the strain energy of a bar in terms of the axial force carried, then

$$\mathcal{U} = \frac{N^2\ell}{2EA}.$$

Since the strain energy is a scalar, we can evaluate the strain energy of an assemblage of bars by adding up the contribution of every bar. For a generic bar m of the assemblage, since $\mathbf{u}^{(m)}$ and $\mathbf{f}^{(m)}$ are end displacements and forces and the external work $W_e^{(m)}$ done by the forces $\mathbf{f}^{(m)}$ is equal to the bar strain energy $\mathcal{U}^{(m)}$, we have

$$\mathcal{U}^{(m)} = \frac{1}{2}\mathbf{u}^{(m)^T}\mathbf{f}^{(m)} = \frac{1}{2}\mathbf{u}^{(m)^T}\mathbf{k}^{(m)}\mathbf{u}^{(m)} = \frac{1}{2}\mathbf{U}^T\mathbf{K}^{(m)}\mathbf{U}.$$

Hence, for a complete truss structure the strain energy is

$$\mathcal{U} = \sum_{m=1}^{n_e}\mathcal{U}^{(m)} = \frac{1}{2}\mathbf{U}^T\left(\sum_{m=1}^{n_e}\mathbf{K}^{(m)}\right)\mathbf{U} = \frac{1}{2}\mathbf{U}^T\mathbf{K}\mathbf{U}.$$

Note that since $\mathbf{R} = \mathbf{K}\mathbf{U}$, we also have

$$\mathcal{U} = \frac{1}{2}\mathbf{U}^T\mathbf{R}.$$

2.3.9 Properly supported truss structures in the context of the matrix method

In Section 2.1.2 we discussed the concept of a properly supported deformable body and later we applied this concept to a truss structure. Recall that when a truss is properly supported and without an internal mechanism, any motion of its bars requires some bar to shorten or to extend.

Therefore, the strain energy associated with any motion of a properly supported truss structure, that is, corresponding to any non-trivial nodal displacements $\mathbf{U}$, will be positive

$$\mathcal{U}(\mathbf{U}) = \frac{1}{2}\mathbf{U}^T\mathbf{K}\mathbf{U} > 0 \qquad \text{for any } \mathbf{U} \neq \mathbf{0}. \tag{2.56}$$

Here "non-trivial $\mathbf{U}$" means $\mathbf{U} \neq \mathbf{0}$.

Mathematically, condition (2.56) defines $\mathbf{K}$ to be a positive definite matrix.

It is a mathematical property that a positive definite matrix has always an inverse, *i.e.*, it is not singular (Bathe, 1996). This leads to a very important result: for any given nodal load $\mathbf{R}$ acting on a truss structure which is properly supported and without an internal mechanism we can always find a unique nodal displacement $\mathbf{U}$ such that

$$\mathbf{K}\mathbf{U} = \mathbf{R}.$$

Now we would like to show that when a truss structure is not properly supported and/or has an internal mechanism the $\mathbf{K}$ matrix is singular and, therefore, there is no unique solution $\mathbf{U}$ for any $\mathbf{R}$. Before we do so, we note that:

- If the truss structure is not properly supported a global rigid motion of the complete structure is possible (see Figure 2.36).

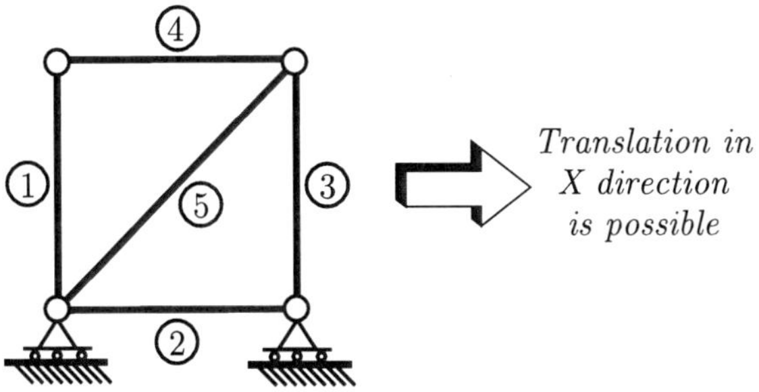

Fig. 2.36. Not properly supported truss structure: Rigid body motion of complete structure is possible

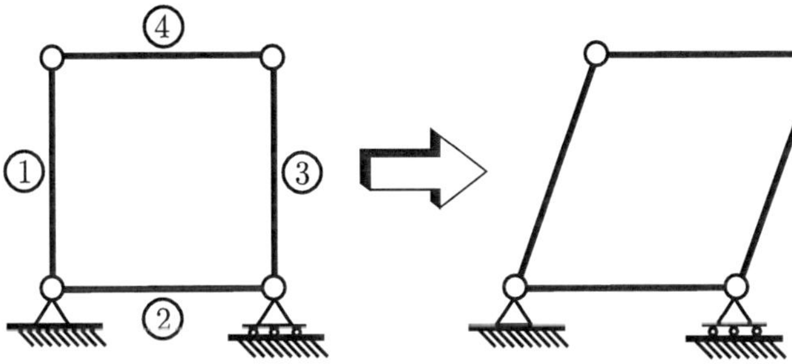

Fig. 2.37. Truss structure with an internal mechanism. Rigid body motion of some bars is possible, *i.e.*, there is a mechanism. Here bars 1, 3, 4 undergo a rigid body motion

- If the truss structure has an internal mechanism, a rigid motion of one or more of its parts without any motion of the remaining parts is possible (see Figure 2.37).

Note that the bars 1, 3 and 4 of the truss of Figure 2.37 display, individually, rigid motions but bar 2 displays no motion and hence does not strain. On the other hand, Figure 2.38 illustrates that for a properly supported truss structure without an internal mechanism parts of the structure may undergo rigid motion but then always cause straining in (some) other bars.

Consider a generic truss structure which either is not properly supported or has an internal mechanism (or both). Let us choose $\mathbf{U} \neq \mathbf{0}$ for which

$$\mathcal{U}(\mathbf{U}) = \frac{1}{2}\mathbf{U}^T\mathbf{K}\mathbf{U} = 0.$$

Of course, this choice of $\mathbf{U}$ is always possible since there is always a motion for which each bar (m) either displays a rigid motion or does not move. Therefore, the stiffness matrix is positive semidefinite and, hence, singular. Since the $\mathbf{K}$ matrix is singular we can not find a unique nodal displacement solution $\mathbf{U}$ for any given load $\mathbf{R}$.

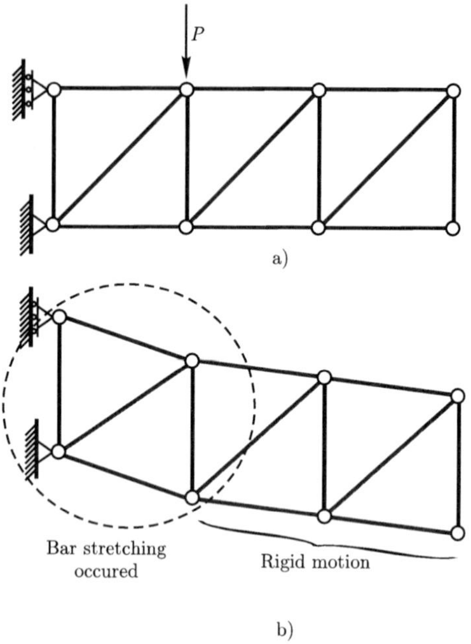

Fig. 2.38. a) Problem definition; b) Schematic and magnified deformed configuration

Summarizing, through the use of the strain energy concept, we arrived at two important results:

- For a truss structure which is properly supported and does not have an internal mechanism, given any loading **R**, there exists a unique nodal displacement solution **U**.
- If the truss structure is not properly supported and/or has an internal mechanism, then for any **R** there is no unique solution **U**.

Let us consider an example.

Example 2.4
Show using a purely kinematic approach that the stiffness matrix of a bar is singular whenever the bar can display rigid motions.

Solution
Let us consider a structure composed of a single truss bar as described in Figure 2.39.

Since there are no supports, this bar can display rigid body motions. In fact, there are three linearly independent rigid body motions: a translation along the X direction, a translation along the Y direction and a rigid body

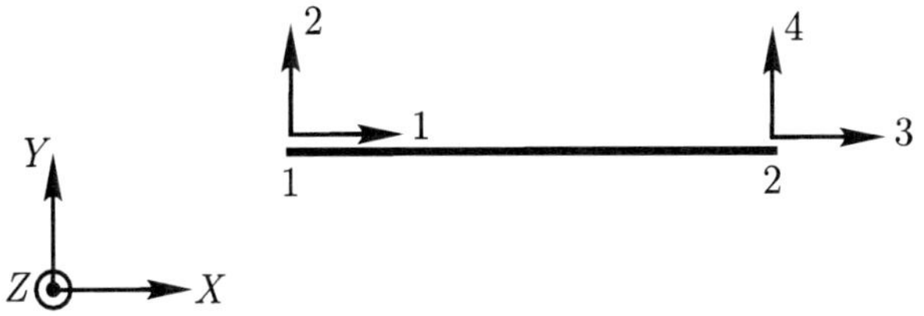

Fig. 2.39. One bar truss structure in X-Y plane

rotation in the XY plane. Any combination of these three rigid body motions also constitutes a rigid body motion.

Let us show that for each set of nodal displacements corresponding to a rigid body motion, we have

$$\mathbf{ku} = \mathbf{0}. \tag{2.57}$$

For a rigid body mode translation along the X direction

$$\mathbf{u}^T = \mathbf{u}_x^T = \begin{bmatrix} C_x & 0 & C_x & 0 \end{bmatrix}$$

where C_x is an arbitrary constant. Clearly (2.57) is satisfied when we use the stiffness matrix given by equation (2.13). An analogous result is found for a rigid body mode translation along the Y direction, which can be defined by

$$\mathbf{u}^T = \mathbf{u}_y^T = \begin{bmatrix} 0 & C_y & 0 & C_y \end{bmatrix}$$

where C_y is also an arbitrary constant.

Next we find the bar nodal displacements for an infinitesimally small rigid body rotation about the Z axis, as shown in Figure 2.40.

Referring to Figure 2.40 we can write

$$\begin{aligned} -u_1 &= r_1 d\varphi \sin\theta_1 \\ u_2 &= r_1 d\varphi \cos\theta_1 \end{aligned}$$

and since

$$\begin{aligned} X_1 &= r_1 \cos\theta_1 \\ Y_1 &= r_1 \sin\theta_1 \end{aligned}$$

we obtain

$$\begin{aligned} u_1 &= -d\varphi Y_1 \\ u_2 &= d\varphi X_1 \end{aligned}$$

Considering a similar derivation for node 2, we obtain

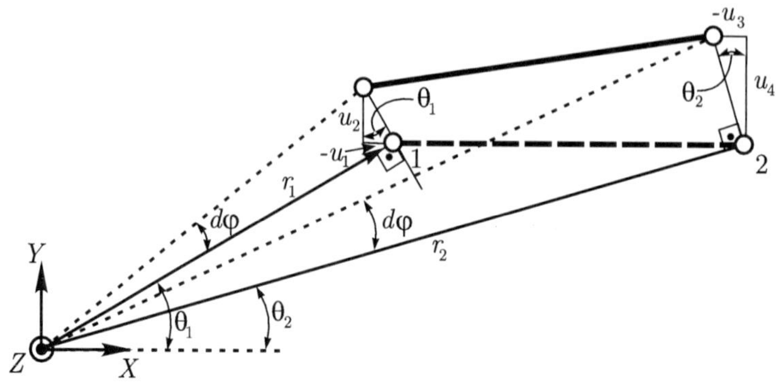

Fig. 2.40. Rigid rotation of the bar of an infinitesimally small angle $d\varphi$. Nodal coordinates are X_1, Y_1 and X_2, Y_2

$$\mathbf{u}_\varphi^T = \begin{bmatrix} -d\varphi Y_1 & d\varphi X_1 & -d\varphi Y_2 & d\varphi X_2 \end{bmatrix}.$$

Since $Y_1 = Y_2$, we also obtain (2.57). Furthermore, we note that for a displacement $\mathbf{u}$ which is any linear combination of $\mathbf{u}_x$, $\mathbf{u}_y$ and $\mathbf{u}_\varphi$, the relation (2.57) also holds.

Of course, when equation (2.57) holds, $\mathbf{k}$ is singular. Equation (2.57) also means that there are no bar end forces associated with a rigid motion.

We note that the choice of a bar which is aligned with the X axis does not imply lack of generality, since, if (2.57) holds for a given coordinate system, it is also satisfied for any other coordinate system.

Note that for the one bar structure of Figure 2.39 three restraints are necessary to kinematically suppress the three rigid body motions. For example, the restraints shown in Figure 2.22.

□

2.4 Modeling considerations for truss structures

While reading this chapter, the reader might have thought of the real physical truss structures that are part of the every day environment, for example, the structures that are frequently encountered in bridges and roofs. Possibly the joints of these "real" trusses are different from the joint described in Figure 2.5, in particular, there might not be an actual pin. In fact, in most truss structures, there are no pins and different joint options are used to connect the bars.

We need to resort to the hierarchical modeling approach to properly address this apparent inconsistency. The truss model with the pin joints is an adequate low-order mathematical model to represent the behavior of bar structures which are not pinned, as long as:

- The structure can not display either local or global rigid motions when its joints are considered as pin-type joints (the truss model is properly supported and does not have an internal mechanism).
- All external loads can be assumed to be introduced at the nodes as point forces only (no moments).
- The bars are long (typically the thickness/length ratio is smaller than 1/10) and have similar magnitudes of cross-sectional areas.
- The bar axes always intersect at unique geometric points (at the joints).

In such cases, the truss model provides a good description of the structure and, in general, it represents a reliable model for most analysis purposes.

The analysis and design of truss structures represent a very broad subject. While we did not discuss issues related to the design of engineering trusses, the analysis of truss structures is important in mechanics and, in our discussion, provided a convenient setting to introduce and explore the fundamental facts of structural mechanics.

In the next chapter we extend the ideas studied in this chapter to model 2-D and 3-D deformable solids.

3. The linear 3-D elasticity mathematical model

In Chapter 2 we examined some fundamental conditions that should be satisfied in the modeling of all deformable solids and structures. The study of truss structures provided an excellent setting to explore how these conditions can be used to formulate a mathematical model in structural mechanics, and how to apply modern procedures to solve this mathematical model. These solution procedures lead to the exact solution of the mathematical model (see examples in Section 2.2 and 2.3).

In engineering analysis, more complex mathematical models need in general to be considered and solved. Indeed, these mathematical models are so complex that exact solutions can mostly not be obtained. The objective in this chapter is to formulate the general mathematical model for three-dimensional (3-D) solids, but still assuming, as for the truss structures, infinitesimally small displacements and the linear constitutive relationship. The conditions to be satisfied, namely equilibrium, compatibility and the stress-strain relationship that we encountered in the analysis of truss structures are also the basic conditions to be satisfied when formulating the three-dimensional mathematical model of a solid. As for truss structures, these conditions need be satisfied for every differential element of the mathematical model – and now of the 3 D solid.

We shall use the discussion of the general three-dimensional mathematical model later in Chapter 4 to derive the various special models useful in engineering analysis. The solution of these models is then presented in the chapters thereafter.

Before embarking on the formulation of the mathematical model of 3-D solids, we motivate its need by means of simple problems which are related to the one-dimensional state of stress and strain of a truss bar, but which already display some 3-D behavior.

3.1 The analysis of a steel sheet problem

In Section 2.2 we considered one-dimensional stress and strain conditions in a truss bar. We now aim in this section to look deeper into the one-dimensional stress/strain conditions and to introduce in a physical manner new phenomena that indeed pertain to two- and three-dimensional conditions. The more formal physical and mathematical discussions of these phenomena are given in the sections to follow.

3.1.1 One-dimensional conditions

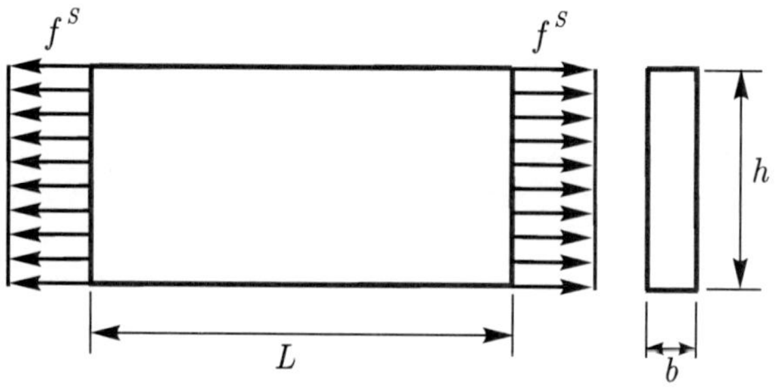

Fig. 3.1. Steel sheet subjected to self-equilibrated constant surface tractions f^S. There are no body forces

Consider the steel sheet shown in Figure 3.1. We note that the external forces are in static equilibrium, *i.e.*, $\mathbf{R} = \mathbf{0}$ and $\mathbf{M_O} = \mathbf{0}$. However, the steel sheet is not properly supported and can undergo rigid body motions. In order to suppress rigid body motions and not to interfere with the straining of the material, we introduce the supports shown in Figure 3.2.

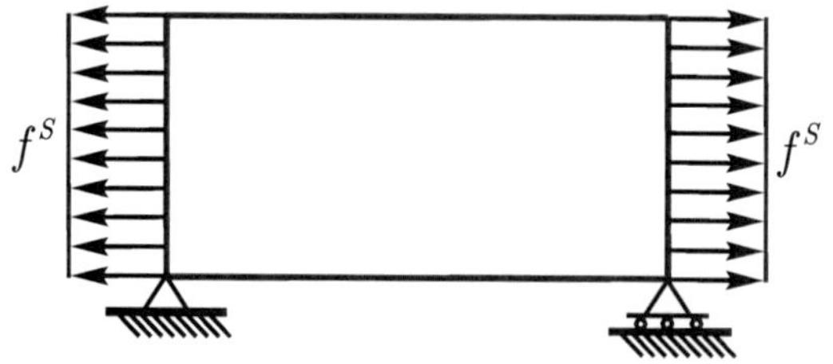

Fig. 3.2. Properly supported steel sheet

If we now cut the sheet by a plane orthogonal to its own plane and the x axis as shown in Figure 3.3a, and introduce the stress τ, see Figure 3.3b, we can, by equilibrium, evaluate the stress magnitude

$$\tau = f^S$$

where τ is constant along the transverse direction y and, of course, independent of the section position given by x (to show that τ is constant as a function of y, we could also cut the sheet with a plane orthogonal to the y-axis). Hence the state of stress is analogous to that in a truss bar. To find the deformed configuration of the sheet we need to consider the relation between the stresses and strains — the constitutive equation. In Section 2.2.4 we introduced Hooke's law for a one-dimensional state of stress and strain. This 1-D constitutive relation is applicable for the steel sheet considering the stress τ and the strain of the longitudinal fibers.

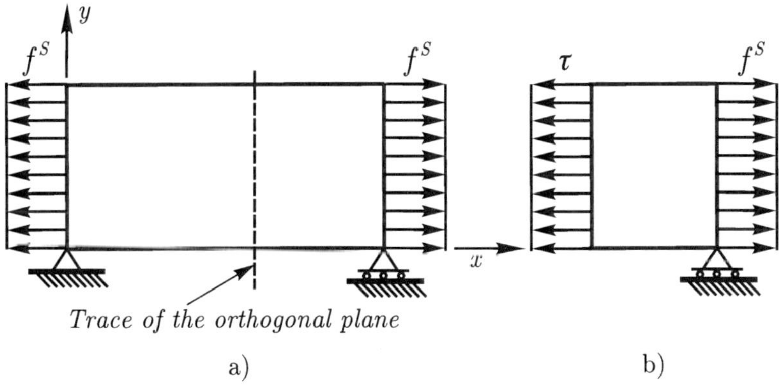

Fig. 3.3. a) Steel sheet cut by an orthogonal plane; b) Equilibrium of the extracted part

An important physical phenomenon in 3-D elasticity is what is referred to as "the Poisson effect". This effect corresponds to a contraction of the fibers that are orthogonal to the direction of the fibers being extended. The deformation of the steel sheet considering Poisson's effect is shown in Figure 3.4: the relative shortenings of the fibers in the transverse directions are proportional to the relative extension of the fibers in the longitudinal direction, that is, $\frac{\Delta h}{h} = \frac{\Delta b}{b} = -\nu \frac{\Delta L}{L}$. The material property constant ν is called Poisson's ratio.

Since in this section we want to concentrate on the longitudinal deformation of the sheet (1-D conditions), we assume, for now, that $\nu = 0$.

Considering Hooke's law

$$\tau = E\varepsilon$$

where

$$\varepsilon = \frac{\Delta L}{L}$$

is the strain of the longitudinal fibers. Hence

$$\tau = E\frac{\Delta L}{L}$$

and, we obtain

$$\Delta L = \tau\frac{L}{E} = f^S\frac{L}{E}.$$

The analogy to the truss model is evident since considering the cross-sectional area $A = hb$ and the axial force $N = \tau A$, we have

$$\Delta L = \tau\frac{L}{E} = \frac{\tau AL}{EA} = N\frac{L}{EA}.$$

As a side-note, this formula is in linear analysis valid for any ν.

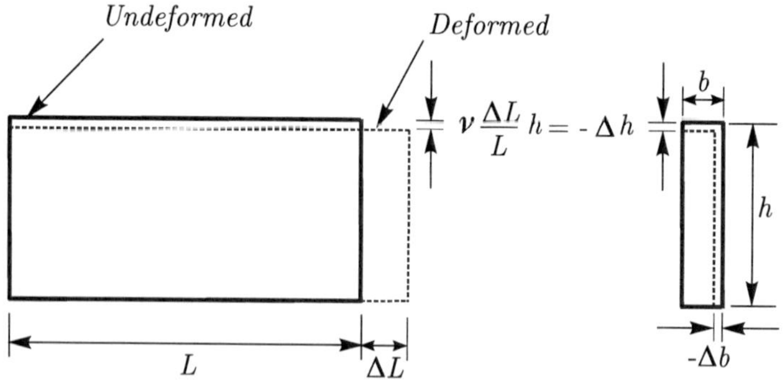

Fig. 3.4. Schematic and magnified deformation of steel sheet

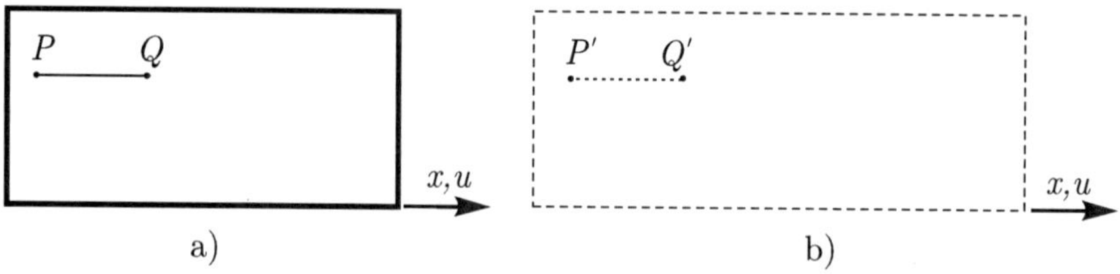

Fig. 3.5. Stretching of fiber **PQ**

Since the longitudinal stress is constant in the sheet the strain of any horizontal fiber such as **PQ**, see Figure 3.5a, is given by

$$\frac{\|\mathbf{P}'\mathbf{Q}'\| - \|\mathbf{PQ}\|}{\|\mathbf{PQ}\|} = \frac{\Delta L}{L} \tag{3.1}$$

where $\mathbf{P}'\mathbf{Q}'$ is the fiber in the deformed configuration, see Figure 3.5b, and $\|\mathbf{PQ}\|$ indicates the Euclidean norm of the vector defined by $\mathbf{PQ}$, *i.e.*, the vector length[1]. Figure 3.6 illustrates that both sides of (3.1) express the fiber extension measured with respect to its original length induced by the same constant stress field.

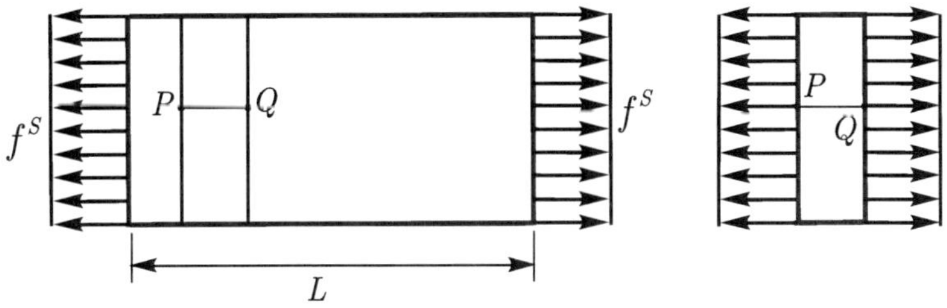

Fig. 3.6. Isolated part of the sheet corresponding to points **P** and **Q**

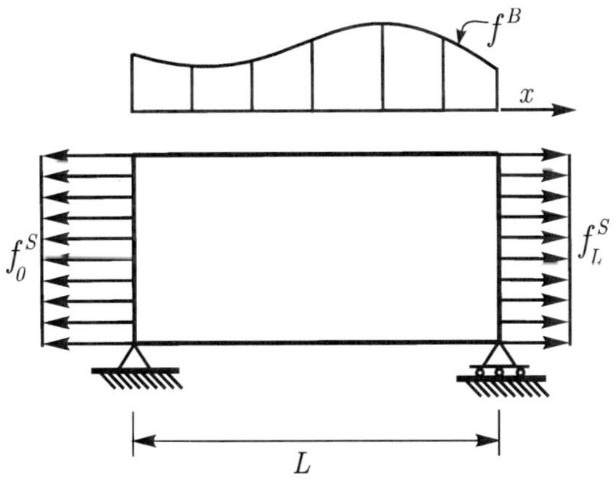

Fig. 3.7. Steel sheet problem considering a field of body forces. The field $\mathrm{f}^B(x)$ acts into the x direction, and $\mathrm{f}^B(x) > 0$ for all x

Next, let us increase the complexity of our loading by considering also a field of body forces acting into the x direction as shown in Figure 3.7. In this

[1] The Euclidian norm of a vector $\mathbf{a}$ can be evaluated by $\|\mathbf{a}\| = \sqrt{\sum_{i=1}^{3} (a_i)^2}$ where a_i are the components of $\mathbf{a}$ in an orthonormal basis (see *e.g.* Bathe, 1996)

figure, the function $f^B(x)$ gives the variation of the magnitude of f^B along the longitudinal direction. We assume that the surface tractions at $x = 0$ and $x = L$ considered together with the body forces are in equilibrium. Again, the stress at a generic section, given by x, can be evaluated by considering a cutting plane to obtain

$$\tau(x) = \frac{1}{A}\left[\int_x^L f^B(z)\, A\, dz + f_L^S A\right] \tag{3.2}$$

where $\tau(x)$ is the stress shown in Figure 3.8. We note that $\tau(x)$ is no longer constant with respect to x and since $f^B(x) > 0$ for all x, $\tau(x)$ decreases with increasing x.

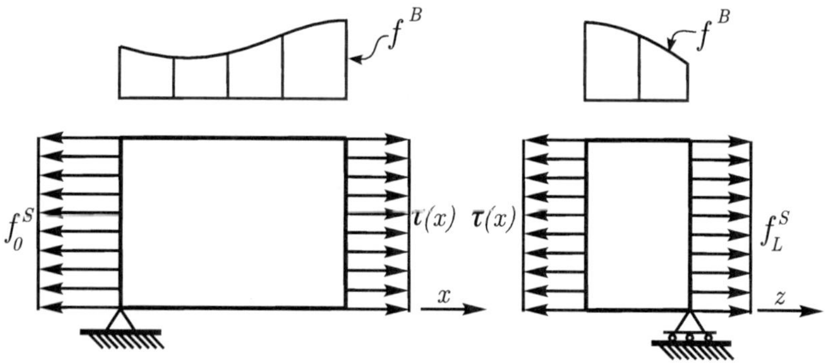

Fig. 3.8. Steel sheet with body forces cut at a generic section

In Figure 3.9 we show the deformation of two fibers: **PQ**, positioned at the same place as in Figure 3.5, and of another fiber originally of the same length as **PQ**, described by **MN**. Since the state of stress varies along the longitudinal direction we have that

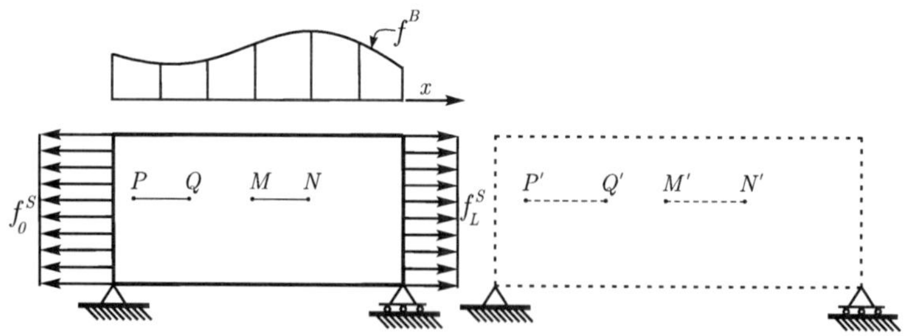

Fig. 3.9. Stretching of two horizontal fibers for problem with body forces

$$\frac{\|\mathbf{P}'\mathbf{Q}'\| - \|\mathbf{PQ}\|}{\|\mathbf{PQ}\|} > \frac{\|\mathbf{M}'\mathbf{N}'\| - \|\mathbf{MN}\|}{\|\mathbf{MN}\|}.$$

There are two pieces of information in the above expression. The first one is that the fiber strains are different since the stress level is different for the regions where the fibers are located. The second piece of information is that since the stress level is greater in the region of the fiber **PQ**, the fiber strain of **PQ** is greater than that of **MN**. Therefore (3.1) only holds for the case considered in Figure 3.2 because a special case is considered: a constant stress leading to an induced constant longitudinal strain over the domain. On the other hand, in the problem of Figure 3.7, the stress and the strain of the fibers vary along the length of the sheet.

Now, if we consider an intermediate point on fiber **PQ**, say $\mathbf{Q}_1$, as shown in Figure 3.10, by the above argument, we should have

$$\frac{\|\mathbf{P}'\mathbf{Q}_1'\| - \|\mathbf{PQ}_1\|}{\|\mathbf{PQ}_1\|} > \frac{\|\mathbf{P}'\mathbf{Q}'\| - \|\mathbf{PQ}\|}{\|\mathbf{PQ}\|}.$$

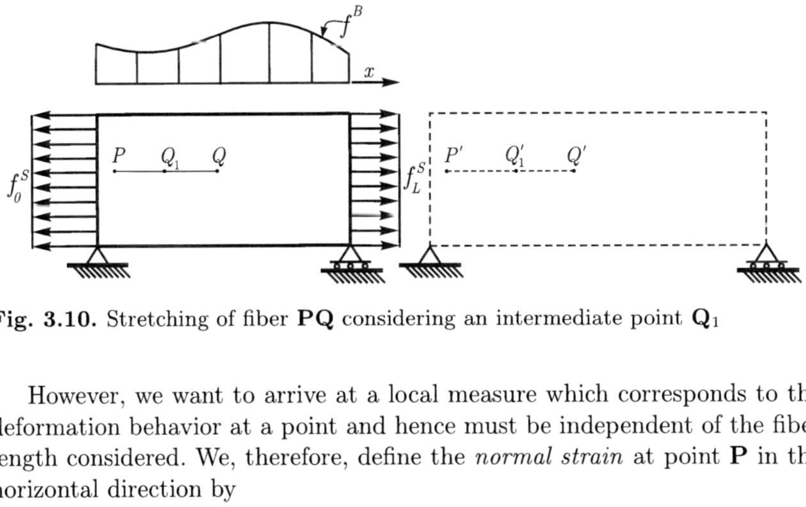

Fig. 3.10. Stretching of fiber **PQ** considering an intermediate point $\mathbf{Q}_1$

However, we want to arrive at a local measure which corresponds to the deformation behavior at a point and hence must be independent of the fiber length considered. We, therefore, define the *normal strain* at point **P** in the horizontal direction by

$$\varepsilon = \lim_{Q \to P} \frac{\|\mathbf{P}'\mathbf{Q}'\| - \|\mathbf{PQ}\|}{\|\mathbf{PQ}\|} \tag{3.3}$$

where point **Q** is always taken on a horizontal line through point **P**. The above quantity gives a measure of the straining of the horizontal fibers of infinitesimal length at point **P**. Hence, using $\|\mathbf{PQ}\| = ds$ and $\|\mathbf{P}'\mathbf{Q}'\| = ds'$, equation (3.3) can be equivalently written as

$$\varepsilon = \frac{ds' - ds}{ds}.$$

Note that as we use a fiber of infinitesimal length to evaluate the normal strain, the direction is given by the direction of the infinitesimal fiber. The terminology "normal" in "normal strain" is due to the fact that the normal stress, normal to a cutting plane, induces this strain[2].

Considering (3.3) it is clear that the change in length of the fiber **PQ** is only due to the displacements of the material particles in the sheet. We denote the displacement of a material particle of the sheet in the x direction by u. In Figure 3.11, we show explicitly the displacements of points **P** and **Q**, and hence we have

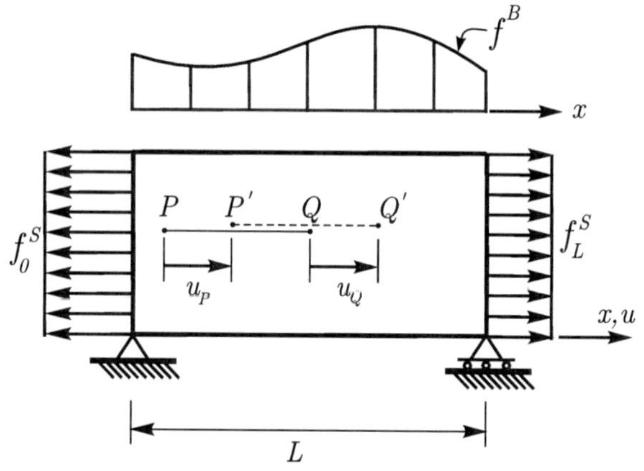

Fig. 3.11. Horizontal displacements of point P and Q

$$\|\mathbf{PQ}\| = x_Q - x_P$$

where x_Q and x_P are the x coordinate of the points **Q** and **P** respectively, and

$$\|\mathbf{PQ}\| + u_Q = u_P + \|\mathbf{P'Q'}\| .$$

The above relation is directly given by measuring the distance between points **P** and **Q′** in two alternative ways, see Figure 3.11. Then, we obtain

$$\lim_{Q \to P} \frac{\|\mathbf{P'Q'}\| - \|\mathbf{PQ}\|}{\|\mathbf{PQ}\|} = \lim_{x_Q \to x_P} \frac{u_Q - u_P}{x_Q - x_P} = \left. \frac{du}{dx} \right|_{x_P} . \tag{3.4}$$

The above limit corresponds to the definition of the derivative of $u(x)$ at x_P. Therefore, (3.4) gives the normal strain of an infinitesimal horizontal fiber with origin at a generic point x

[2] It is also used to distinguish the normal strain from the shear strain which is defined for a pair of fibers, see Section 3.2.2

$$\varepsilon(x) = \frac{du}{dx}.$$

Now we can find the displacements of the material particles in the sheet for the problem of Figure 3.10. Indeed

$$\frac{du}{dx} = \varepsilon(x) = \frac{\tau(x)}{E}$$

which can be integrated using $\tau(x)$ in (3.2) and leads to $u(x)$. While we assumed tensile conditions in the above discussion, this formula is general and also applicable in the case of compression.

3.1.2 Two Dimensional Conditions

Let us next consider the problem of Figure 3.2 with $\nu \neq 0$. Therefore, the deformation of the sheet is as given in Figures 3.4 and 3.12. If we proceed as before for the definition of a local strain measure for the fibers with origin in **P**, but vertical, we would arrive at

$$\varepsilon = \lim_{\mathbf{R}\to\mathbf{P}} \frac{\|\mathbf{P}'\mathbf{R}'\| - \|\mathbf{PR}\|}{\|\mathbf{PR}\|}.$$

This strain value is negative since the vertical fibers contract (due to Pois-

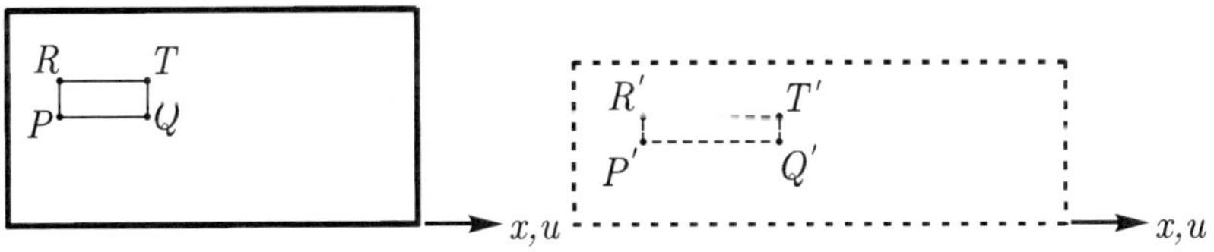

Fig. 3.12. Deformation of fibers **PQ** and **RT** for sheet in Figure 3.4

son's effect). If we take an inclined direction such as that given by **PT** we would again arrive at a different value for the normal strain given by

$$\varepsilon = \lim_{\mathbf{T}\to\mathbf{P}} \frac{\|\mathbf{P}'\mathbf{T}'\| - \|\mathbf{PT}\|}{\|\mathbf{PT}\|}.$$

Therefore, the normal strain depends not only on the fiber location but also on the fiber direction. Hence, we use the terminology "state of strain" at a point and will characterize this state in detail in Section 3.2.

Actually, this more general strain concept is required to solve more complex problems. Referring to Figure 3.2, we recall that the supports were only introduced for the purpose of suppressing rigid body motions. Consider the boundary conditions given in Figure 3.13, *i.e.*,

$$u(0,y) = 0, \; v(0,0) = 0.$$

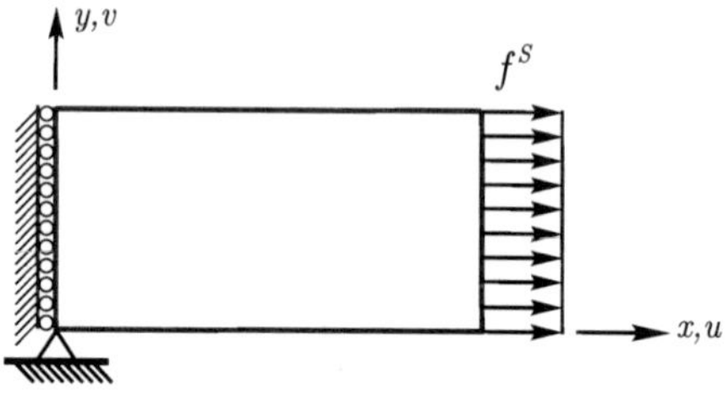

Fig. 3.13. Steel sheet of Figure 3.2 with modified boundary conditions

The quantity $v(0,0)$ indicates the displacement in the direction of y of the particle at point $(0,0)$. In general, the field of vertical displacements is denoted by $v(x,y)$. Since the deformation pattern for the problem of Figure 3.2 leads to $u(0,y) = 0$ the additional displacement restrictions introduced in Figure 3.13 do not interfere with the previous deformation. Hence, the solution to this problem can be determined solving for the stresses and strains considering 1-D conditions, as before, and then if $\nu > 0$ evaluating the contraction of the vertical fibers due to Poisson's effect.

Therefore, for any material (either a steel or a rubber sheet) we have the same stresses. Of course, the strain and the displacements depend on the material of the sheet.

The problem of Figure 3.13 represents an analogue, in the context of a 2-D problem, to the statically determinate truss structures (actually a single bar) in the sense that only the equilibrium condition is used to determine the stresses/internal forces in both cases.

Let us modify the boundary conditions of the problem of Figure 3.13 as shown in Figure 3.14a, *i.e.*, we now have the following boundary conditions

$$u(0,y) = 0, \; v(0,y) = 0. \tag{3.5}$$

We see that the shortening of the edge of the sheet given by $x = 0$ is now prevented and this is achieved by the reactions f_x^S and f_y^S at the clamped edge. We could think of solving this problem by starting from the problem of Figure 3.13 and looking for the field of forces f_y^S which should be applied at the edge given by $x = 0$ to obtain $v(0,y) = 0$. Figure 3.14b summarizes this approach.

The state of stress and strain for the problem of Figure 3.14 is no longer unidimensional and in addition to the longitudinal normal stresses, there are normal stresses in the transverse direction and also shear stresses.

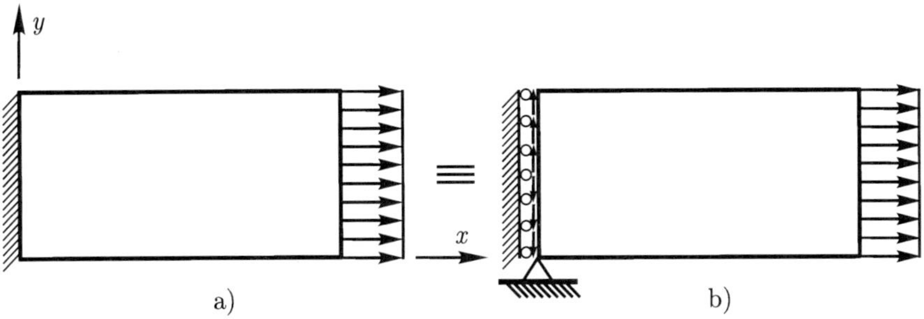

Fig. 3.14. Steel sheet with restrictions to the vertical displacement at the edge $x = 0$

In Figure 3.15, we introduce a notation to distinguish between these stresses, *i.e.*, τ_{xx} and τ_{yy} represent the normal stresses in the directions x and y respectively and τ_{xy} the shear stresses. We also show qualitatively their variations.

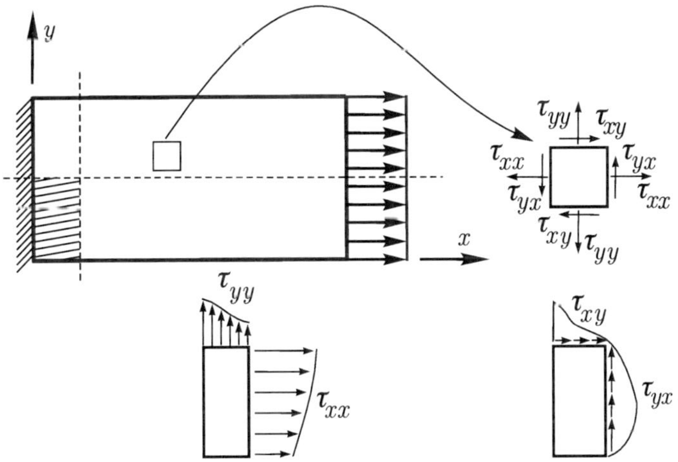

Fig. 3.15. Stresses schematically shown for internal faces defined by the cutting planes

In Figure 3.16, we present the data of a problem as described above. In Figures 3.17 and 3.18 we present the stress predictions for the lines shown in Figure 3.16. The solution was obtained by the finite element method with a mesh sufficiently refined to give a good quantitative description of the solution. We can appreciate that at line 2, far from the edge given by $x = 0$, the solution is very similar to that obtained for the problem of Figure 3.13 in which the supports at the edge prevent only horizontal motions, whereas at line 1, we see that the transverse normal stresses τ_{yy} and the shear stresses τ_{xy} are different from zero and that the normal stresses τ_{xx} are also not constant.

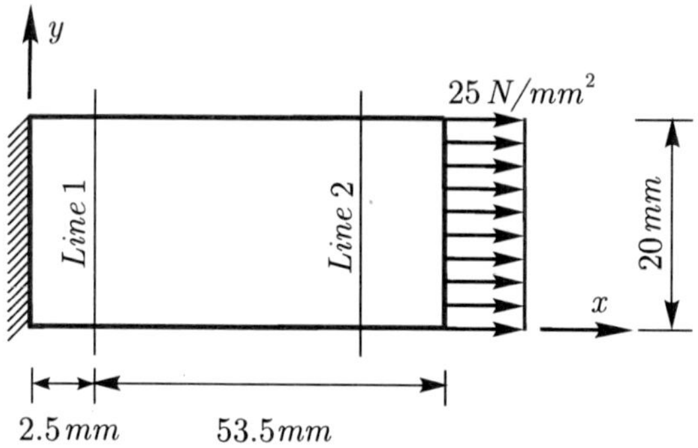

Fig. 3.16. A steel sheet problem. Thickness $t = 1\ mm, E = 70 \times 10^4\ N/mm^2, \nu = 0.25$

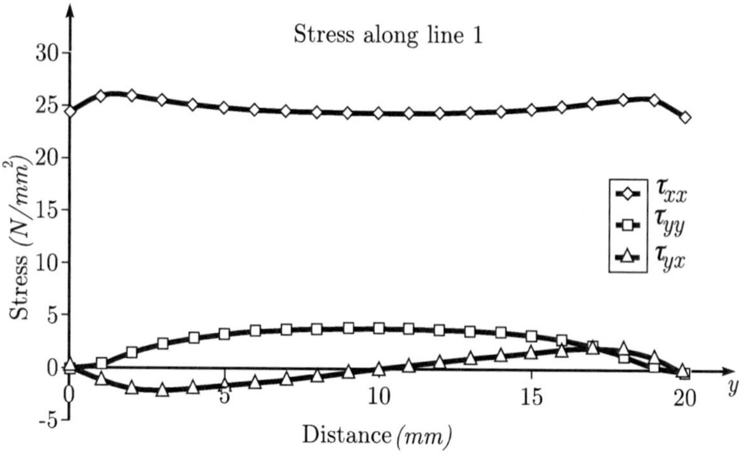

Fig. 3.17. Stress distributions at line 1

We note that this problem is not statically determinate, *i.e.*, we can not find the stress field without imposing also the constitutive equations and compatibility conditions.

When we considered statically indeterminate truss structures, we could find many sets of internal forces that satisfy equilibrium, but only one of these sets led to displacements at the nodes and at the supports that were compatible. The steel sheet is an example of an analogous behavior in the continuum

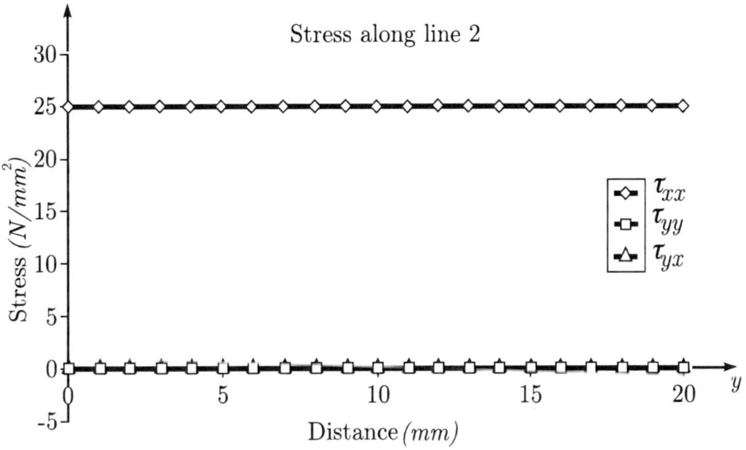

Fig. 3.18. Stress distributions at line 2; the τ_{yy} and τ_{yx} values are practically zero

case, since not all stress fields which satisfy equilibrium lead to fiber extensions/contractions, more generally to strains, that correspond to displacement fields with continuous displacements that satisfy the displacement boundary conditions. Therefore, even to solve the very "simple" case described in Figure 3.16, we need a multi-dimensional description of the strains, stresses and constitutive relations.

Motivated by the above discussion, we can now outline the contents of the remaining sections of this chapter. In Section 3.2 we present the study of the deformations considering first the motion of a three-dimensional body with no restriction on the magnitude of the displacements. Then, we examine the simplifications which are obtained when infinitesimally small displacements are assumed. In Section 3.3 we study the stresses, also in the context of a 3-D body, starting from the concepts already introduced in Chapter 2. In Section 3.4 the relations between the strains and stresses are discussed for a linear isotropic elastic behavior, *i.e.*, the constitutive equations are presented under such conditions. In Section 3.5 the complete formulation of the 3-D linear elasticity problem is presented. Finally, in Section 3.6 the 3-D elasticity solution of the torsion of a prismatic bar is discussed.

3.2 Deformations

In Section 3.1 we showed by means of a very simple example – the analysis of the steel sheet – that the strain of an infinitesimal fiber having origin at a point depends not only on the position of the point but also on the direction of the fiber.

In the analysis of the steel sheet, we also exemplified that the stress in a given direction induces a fiber strain in this direction and in the perpendicular directions as well. Actually, besides these fiber strains, the change in angle between fibers is also related to the stresses and, hence, should be included in the characterization of the complete state of strain at a point.

In order to establish a precise terminology for strains, we call a *normal strain* the strain of *a fiber* as introduced in Section 3.1 and a *shear strain* the change of angle between *pairs of orthogonal fibers.* The shear strain is sometimes referred to as angular distortion.

In this section we assume that the displacements of all material particles are given (from the undeformed to the deformed configuration of a solid) and calculate the normal and shear strains. Of course, the actual displacements can only be calculated once the complete mathematical model has been established.

3.2.1 Displacement field

Considering the motion of a deformable body, we described in Figure 2.1 generic configurations of the body, where ${}^{0}V$ stood for the initial configuration and ${}^{t}V$ for a generic configuration at time t. Since we are concerned with statics and, primarily, with linear theories, we can focus on only two configurations: the one prior to the application of external actions – the undeformed configuration – and the one after the introduction of the external actions – the deformed configuration. In order to simplify the notation, we redraw in Figure 3.19 the description presented in Figure 2.1. We choose to label the undeformed configuration as V and the deformed configuration as V_y. We also use x_1, x_2 and x_3 as coordinate axes.

In Figure 3.19, $\mathbf{P}$ is a material particle[3] of the solid in the undeformed configuration and $\mathbf{P}_y$ represents this particle in the deformed configuration. The position of $\mathbf{P}$ is characterized by the vector[4]

[3] We refer sometimes to a *material particle* as a *point*

[4] A *Euclidean vector* $\mathbf{v}$ can be defined as the oriented straight line segment that connects two points of the Euclidean three-dimensional space. Hence, the definition of a vector is independent of the choice of a coordinate system. However, to perform calculations with vectors we need to represent them in a coordinate system. For example, $\mathbf{v} = v_1\mathbf{e}_1 + v_2\mathbf{e}_2 + v_3\mathbf{e}_3$ where v_i are the components of the vector $\mathbf{v}$ in the coordinate system $(\mathbf{O}, \mathbf{e}_1, \mathbf{e}_2, \mathbf{e}_3)$. When $(\mathbf{e}_1, \mathbf{e}_2, \mathbf{e}_3)$ is an orthonormal basis the component v_i is obtained by the scalar product $v_i = \mathbf{v} \cdot \mathbf{e}_i$.

Sometimes we represent the vector $\mathbf{v}$ by $\mathbf{v} = \begin{bmatrix} v_1 \\ v_2 \\ v_3 \end{bmatrix}$. Actually, we are defining the vector by a column matrix which collects the components of $\mathbf{v}$ in the coordinate system $(\mathbf{O}, \mathbf{e}_1, \mathbf{e}_2, \mathbf{e}_3)$. Of course, in the coordinate system $(\mathbf{O}', \mathbf{e}_1', \mathbf{e}_2', \mathbf{e}_3')$ the components v_i' are different from v_i and, hence, another column matrix would correspond to the same vector $\mathbf{v}$. Notwithstanding, we adopt this convention to

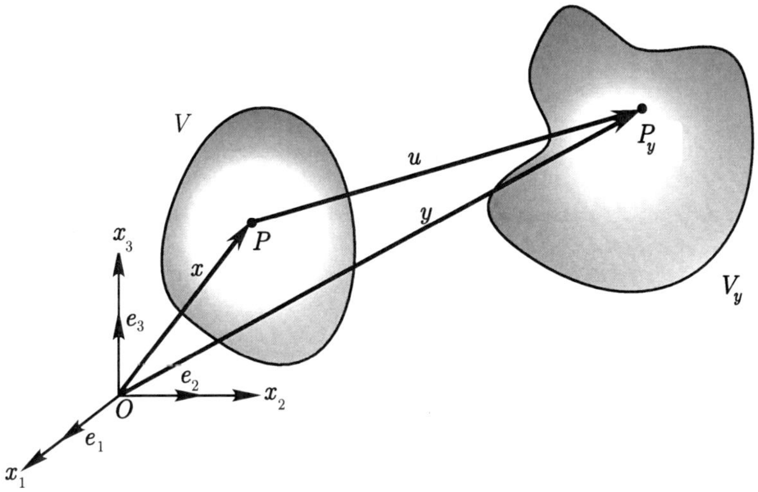

Fig. 3.19. Deformed and undeformed configurations in a stationary Cartesian coordinate system

$$\mathbf{x} = x_1\mathbf{e}_1 + x_2\mathbf{e}_2 + x_3\mathbf{e}_3 = \sum_{i=1}^{3} x_i\mathbf{e}_i$$

where the $\mathbf{e}_i$ are the base vectors of the coordinate system. The position of $\mathbf{P}_y$ is given by

$$\mathbf{y} = \sum_{i=1}^{3} y_i\mathbf{e}_i.$$

The displacement vector of the material particle is

$$\mathbf{u} = \mathbf{y} - \mathbf{x}.$$

The deformation of a solid can be fully described by defining the function which gives for every material particle in the undeformed configuration its position in the deformed configuration

$$\mathbf{y} = \mathbf{y}(\mathbf{x})$$

which in components can be written as

keep the notation as simple and as operational as possible with the implicit understanding that the coordinate system used is implied by the context. Note that the calculations are performed with matrix operations. For example, the scalar product (or dot product) between two vectors $\mathbf{v}$ and $\mathbf{w}$ is obtained by $\mathbf{v}^T\mathbf{w}$ (see Bathe, 1996)

$$
\begin{aligned}
y_1 &= y_1(x_1, x_2, x_3) \\
y_2 &= y_2(x_1, x_2, x_3) \\
y_3 &= y_3(x_1, x_2, x_3) \, .
\end{aligned}
$$

The displacement field can be also written as

$$
\begin{aligned}
u_1 &= u_1(x_1, x_2, x_3) \\
u_2 &= u_2(x_1, x_2, x_3) \\
u_3 &= u_3(x_1, x_2, x_3) \, .
\end{aligned}
$$

In order to allow the reader to become familiar with this notation, let us describe some very simple rigid plane motions in the example below.

Example 3.1

Consider a generic prismatic solid with transverse section parallel to the coordinate plane x_1x_2 as shown in Figure 3.20a. Characterize the displacement field for:

(i) a rigid body translation in the direction of x_1 of intensity Δ
(ii) a rigid body rotation of intensity φ with respect to the x_3 axis

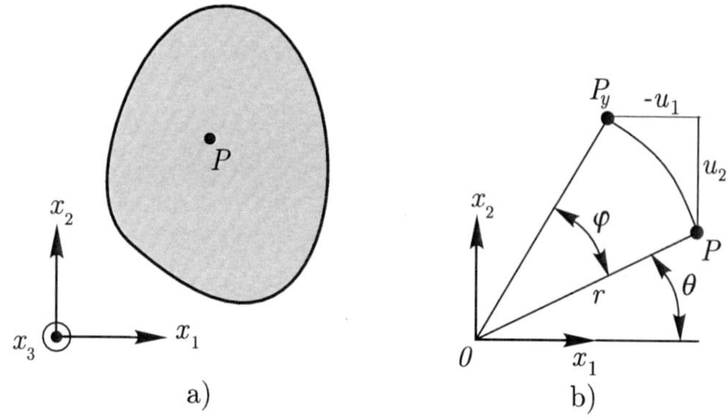

Fig. 3.20. a) Transverse section of a prismatic solid and a generic point **P**; b) Motion of a generic point in the section for an angle of rotation φ.

Solution

(i) It suffices to recognize that for every point of the solid denoting a material particle

$$\begin{aligned} u_1(x_1,x_2,x_3) &= \Delta \\ u_2(x_1,x_2,x_3) &= 0 \\ u_3(x_1,x_2,x_3) &= 0. \end{aligned}$$

(ii) In Figure 3.20b, **P** stands for a generic point denoting a material particle of coordinates x_1 and x_2 in the section of the solid and $\mathbf{P}_y$ represents its position in the deformed configuration, after the rigid body rotation. Note that point **P** moves on a circle of radius $r = \|\mathbf{OP}\|$. Therefore the displacements are given by

$$-u_1 = r\cos\theta - r\cos(\theta+\varphi), \qquad u_2 = r\sin(\theta+\varphi) - r\sin\theta$$

which leads to

$$u_1 = -r\cos\theta + r\cos\theta\cos\varphi - r\sin\theta\sin\varphi \tag{3.6}$$

$$u_2 = r\sin\theta\cos\varphi + r\sin\varphi\cos\theta - r\sin\theta. \tag{3.7}$$

Considering that

$$x_1 = r\cos\theta, \qquad x_2 = r\sin\theta \tag{3.8}$$

equations (3.6) and (3.7) can be re-written as

$$u_1(x_1,x_2,x_3) = x_1\cos\varphi - x_2\sin\varphi - x_1 \tag{3.9}$$

$$u_2(x_1,x_2,x_3) = x_1\sin\varphi + x_2\cos\varphi - x_2 \tag{3.10}$$

and, of course,

$$u_3(x_1,x_2,x_3) = 0.$$

□

3.2.2 Normal and shear strains

As discussed in Section 3.1, the normal strain depends both on the fiber location and its direction. In order to obtain a local measure at a point (independent of the fiber length), a limit process was pursued which means, in essence, to consider a fiber of infinitesimal length. This approach will be used to study deformations in 3-D and from now on, unless stated otherwise, a fiber is assumed to be of infinitesimal length.

Let us consider a deformation as described in Figure 3.19 and consider a fiber with origin at **P** given by the vector **dx**, as shown in Figure 3.21.

Due to the deformation, the fiber **dx** becomes **dy** in the deformed configuration, V_y, and we can write

$$\mathbf{dy} = \mathbf{dx} + \mathbf{u}(\mathbf{x}+\mathbf{dx}) - \mathbf{u}(\mathbf{x}) \tag{3.11}$$

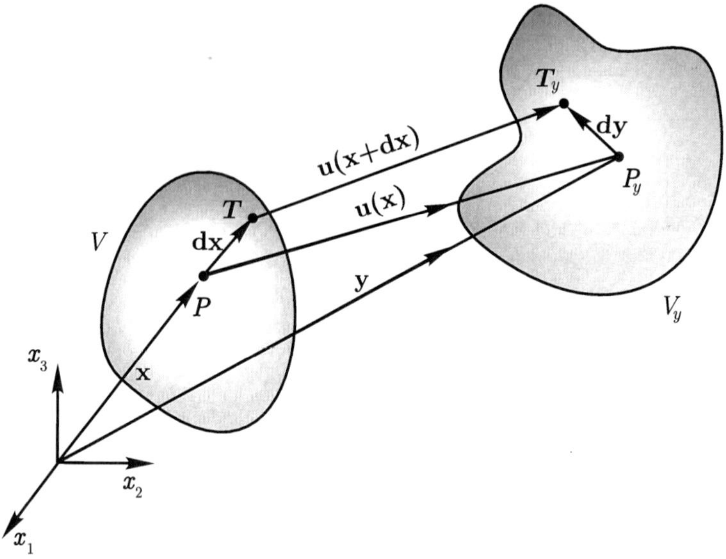

Fig. 3.21. Fibers in deformed and undeformed configurations

or in components

$$dy_i = dx_i + u_i\left(x_1 + dx_1, x_2 + dx_2, x_3 + dx_3\right) - u_i\left(x_1, x_2, x_3\right)$$

for $i = 1$, 2 and 3. Recall from multivariable calculus that

$$u_i\left(x_1 + dx_1, x_2 + dx_2, x_3 + dx_3\right) - u_i\left(x_1, x_2, x_3\right) = \frac{\partial u_i}{\partial x_1}dx_1 + \frac{\partial u_i}{\partial x_2}dx_2 + \frac{\partial u_i}{\partial x_3}dx_3. \tag{3.12}$$

Substituting (3.12) into (3.11) gives

$$dy_i = dx_i + \frac{\partial u_i}{\partial x_1}dx_1 + \frac{\partial u_i}{\partial x_2}dx_2 + \frac{\partial u_i}{\partial x_3}dx_3$$

and the range of indexes, such as i, is implicitly understood to vary from 1 to 3. We can write the above expression in matrix form

$$\begin{bmatrix} dy_1 \\ dy_2 \\ dy_3 \end{bmatrix} = \begin{bmatrix} dx_1 \\ dx_2 \\ dx_3 \end{bmatrix} + \begin{bmatrix} \frac{\partial u_1}{\partial x_1} & \frac{\partial u_1}{\partial x_2} & \frac{\partial u_1}{\partial x_3} \\ \frac{\partial u_2}{\partial x_1} & \frac{\partial u_2}{\partial x_2} & \frac{\partial u_2}{\partial x_3} \\ \frac{\partial u_3}{\partial x_1} & \frac{\partial u_3}{\partial x_2} & \frac{\partial u_3}{\partial x_3} \end{bmatrix} \begin{bmatrix} dx_1 \\ dx_2 \\ dx_3 \end{bmatrix} \tag{3.13}$$

or

$$\begin{bmatrix} dy_1 \\ dy_2 \\ dy_3 \end{bmatrix} = \begin{bmatrix} 1+\frac{\partial u_1}{\partial x_1} & \frac{\partial u_1}{\partial x_2} & \frac{\partial u_1}{\partial x_3} \\ \frac{\partial u_2}{\partial x_1} & 1+\frac{\partial u_2}{\partial x_2} & \frac{\partial u_2}{\partial x_3} \\ \frac{\partial u_3}{\partial x_1} & \frac{\partial u_3}{\partial x_2} & 1+\frac{\partial u_3}{\partial x_3} \end{bmatrix} \begin{bmatrix} dx_1 \\ dx_2 \\ dx_3 \end{bmatrix}. \tag{3.14}$$

Let us define the displacement gradient

$$\nabla \mathbf{u} = \begin{bmatrix} \frac{\partial u_1}{\partial x_1} & \frac{\partial u_1}{\partial x_2} & \frac{\partial u_1}{\partial x_3} \\ \frac{\partial u_2}{\partial x_1} & \frac{\partial u_2}{\partial x_2} & \frac{\partial u_2}{\partial x_3} \\ \frac{\partial u_3}{\partial x_1} & \frac{\partial u_3}{\partial x_2} & \frac{\partial u_3}{\partial x_3} \end{bmatrix} \tag{3.15}$$

and the deformation gradient[5]

$$\mathbf{X} = \begin{bmatrix} 1+\frac{\partial u_1}{\partial x_1} & \frac{\partial u_1}{\partial x_2} & \frac{\partial u_1}{\partial x_3} \\ \frac{\partial u_2}{\partial x_1} & 1+\frac{\partial u_2}{\partial x_2} & \frac{\partial u_2}{\partial x_3} \\ \frac{\partial u_3}{\partial x_1} & \frac{\partial u_3}{\partial x_2} & 1+\frac{\partial u_3}{\partial x_3} \end{bmatrix} = \mathbf{I} + \nabla \mathbf{u}. \tag{3.16}$$

Here, the identity matrix is denoted by $\mathbf{I}$. Equation (3.13) can be re-written as

$$\mathbf{dy} = \mathbf{dx} + \nabla \mathbf{u}\, \mathbf{dx}$$

and (3.14) as

$$\mathbf{dy} = (\mathbf{I} + \nabla \mathbf{u})\, \mathbf{dx} = \mathbf{X}\, \mathbf{dx}. \tag{3.17}$$

[5] Both the displacement gradient and the deformation gradient are linear transformations which relate vectors to vectors (see equation (3.17)). For instance, let $\mathbf{S}$ be a linear transformation. The linearity means that $\mathbf{S}(\alpha\mathbf{p}+\beta\mathbf{q}) = \alpha\mathbf{S}\mathbf{p}+\beta\mathbf{S}\mathbf{q}$ for any real numbers α, β and any vectors $\mathbf{p}$, $\mathbf{q}$. Considering a coordinate system ($\mathbf{O}$, $\mathbf{e}_1$, $\mathbf{e}_2$, $\mathbf{e}_3$), $\mathbf{w} = \mathbf{S}\mathbf{v} = \mathbf{S}\left(\sum_{i=1}^{3} v_i \mathbf{e}_i\right) = \sum_{i=1}^{3} v_i \mathbf{S}\mathbf{e}_i$. Now let $\mathbf{S}\mathbf{e}_i = \sum_{j=1}^{3} S_{ji}\mathbf{e}_j$ then $\mathbf{w} = \begin{bmatrix} w_1 \\ w_2 \\ w_3 \end{bmatrix} = \begin{bmatrix} S_{11} & S_{12} & S_{13} \\ S_{21} & S_{22} & S_{23} \\ S_{31} & S_{32} & S_{33} \end{bmatrix} \begin{bmatrix} v_1 \\ v_2 \\ v_3 \end{bmatrix} = \mathbf{S}\mathbf{v}$. In this book we call *tensor* a linear transformation which relates vectors to vectors (although a *tensor* is a more general concept). Note that a tensor can be represented in a basis by a 3×3 matrix. Of course, as for vectors, when we change basis the matrix components which represent the tensor will generally change. Nevertheless, we choose to represent a tensor by a 3×3 matrix and the basis is implied by the context (see Bathe, 1996 and Chapelle and Bathe, 2010*a*).

We also note that in many instances we will use matrices in their ordinary sense, that is, not as representing either a vector or a tensor. We mention that no special notation is used to distinguish between matrices when used in these distinct situations

Relation (3.17) is a very important result because it gives the change in any fiber due to the deformation. That is, the deformation gradient $\mathbf{X}$ relates a fiber $\mathbf{dx}$ in the undeformed configuration to the same fiber $\mathbf{dy}$ in the deformed configuration. Of course, $\mathbf{X}$ and $\nabla\mathbf{u}$ depend on the position (but not the direction) of the fiber in the undeformed configuration. However, if we choose a point, any fiber deformation at this point is obtained by the same $\mathbf{X}$ (or $\nabla\mathbf{u}$).

The relation (3.17) can now be used to calculate both the normal and shear strains.

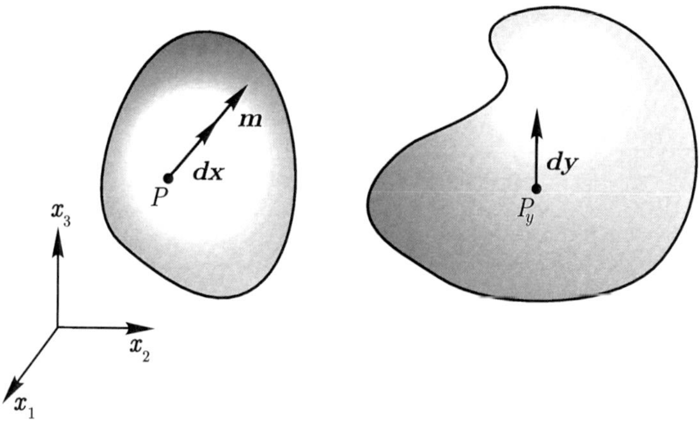

Fig. 3.22. Infinitesimal fibers in undeformed and deformed configurations

Let us first derive the expression that gives the normal strain of any fiber. Let $\mathbf{m}$ be the unit vector in the direction of $\mathbf{dx}$ as shown in Figure 3.22 and ds and ds_y be the lengths of the fibers $\mathbf{dx}$ and $\mathbf{dy}$, which are given by

$$\begin{aligned} ds &= \|\mathbf{dx}\| = \left((dx_1)^2 + (dx_2)^2 + (dx_3)^2\right)^{\frac{1}{2}} \\ ds_y &= \|\mathbf{dy}\| = \left((dy_1)^2 + (dy_2)^2 + (dy_3)^2\right)^{\frac{1}{2}} \end{aligned} \tag{3.18}$$

The normal strain[6] of the fiber $\mathbf{dx}$ is given by

$$\varepsilon_\ell = \frac{ds_y - ds}{ds} = \frac{ds_y}{ds} - 1.$$

It is usual to define the stretch of a fiber by

[6] For other strain measures, useful in nonlinear formulations, see *e.g.* Bathe, 1996

$$\lambda = \frac{ds_y}{ds}$$

i.e., the ratio between the deformed and undeformed lengths. Hence,

$$\varepsilon_\ell = \lambda - 1.$$

From equation (3.18)

$$ds_y = \|\mathbf{dy}\| = \sqrt{\mathbf{dy}^T\mathbf{dy}} = \sqrt{\mathbf{dx}^T\mathbf{X}^T\mathbf{X}\mathbf{dx}}.$$

which leads to

$$\varepsilon_\ell = \frac{ds_y}{ds} - 1 = \frac{\sqrt{\mathbf{dx}^T\mathbf{X}^T\mathbf{X}\mathbf{dx}}}{\sqrt{\mathbf{dx}^T\mathbf{dx}}} - 1$$

and since $\mathbf{dx} = ds\ \mathbf{m}$ we obtain

$$\varepsilon_\ell = \frac{ds\sqrt{\mathbf{m}^T\mathbf{X}^T\mathbf{X}\mathbf{m}}}{ds\sqrt{\mathbf{m}^T\mathbf{m}}} - 1$$

or equivalently

$$\varepsilon_\ell = \sqrt{\mathbf{m}^T\mathbf{X}^T\mathbf{X}\mathbf{m}} - 1 \tag{3.19}$$

since $\mathbf{m}$ is a unit vector.

The above expression gives the normal strain of a fiber, in the direction $\mathbf{m}$, *i.e.*, $\varepsilon_\ell(\mathbf{m})$. There is no restriction on the magnitude of the displacements involved in the deformation. However, when the displacements are assumed to be infinitesimally small, we can introduce simplifications (this is the case in linear elasticity).

Using equation (3.16) we can write

$$\begin{aligned} \mathbf{X}^T\mathbf{X} &= (\mathbf{I} + \nabla\mathbf{u})^T(\mathbf{I} + \nabla\mathbf{u}) \\ &= \mathbf{I} + \nabla\mathbf{u} + \nabla\mathbf{u}^T + \nabla\mathbf{u}^T\nabla\mathbf{u} \end{aligned} \tag{3.20}$$

Note that the components of $\nabla\mathbf{u}^T\nabla\mathbf{u}$ involve sums of terms of the form $\frac{\partial u_i}{\partial x_k}\frac{\partial u_j}{\partial x_k}$ (refer to relation (3.15)). When the displacements are considered infinitesimally small, the terms $\frac{\partial u_i}{\partial x_k}$ are also considered infinitesimally small and therefore we can neglect $\frac{\partial u_i}{\partial x_k}\frac{\partial u_j}{\partial x_k}$ with respect to $\frac{\partial u_i}{\partial x_k}$. Hence we can neglect $\nabla\mathbf{u}^T\nabla\mathbf{u}$ with respect to $\nabla\mathbf{u}$ and use

$$\mathbf{X}^T\mathbf{X} = \mathbf{I} + \nabla\mathbf{u} + \nabla\mathbf{u}^T. \tag{3.21}$$

and can define the infinitesimal strain tensor $\mathbf{E}$ by

$$\mathbf{E} = \frac{1}{2}\left(\nabla\mathbf{u} + \nabla\mathbf{u}^T\right). \tag{3.22}$$

The components of $\mathbf{E}$ are

$$\mathbf{E} = \begin{bmatrix} \frac{\partial u_1}{\partial x_1} & \frac{1}{2}\left(\frac{\partial u_1}{\partial x_2} + \frac{\partial u_2}{\partial x_1}\right) & \frac{1}{2}\left(\frac{\partial u_1}{\partial x_3} + \frac{\partial u_3}{\partial x_1}\right) \\ \frac{1}{2}\left(\frac{\partial u_1}{\partial x_2} + \frac{\partial u_2}{\partial x_1}\right) & \frac{\partial u_2}{\partial x_2} & \frac{1}{2}\left(\frac{\partial u_2}{\partial x_3} + \frac{\partial u_3}{\partial x_2}\right) \\ \frac{1}{2}\left(\frac{\partial u_1}{\partial x_3} + \frac{\partial u_3}{\partial x_1}\right) & \frac{1}{2}\left(\frac{\partial u_2}{\partial x_3} + \frac{\partial u_3}{\partial x_2}\right) & \frac{\partial u_3}{\partial x_3} \end{bmatrix}. \quad (3.23)$$

The strain tensor $\mathbf{E}$ is symmetric, $\mathbf{E} = \mathbf{E}^T$, whereas $\mathbf{X}$ is in general not symmetric. Hence assuming infinitesimally small displacements we have

$$\mathbf{X}^T\mathbf{X} = \mathbf{I}+2\mathbf{E} \quad (3.24)$$

and equation (3.19) can be re-written as

$$\varepsilon_\ell = \sqrt{\mathbf{m}^T(\mathbf{I}+2\mathbf{E})\mathbf{m}} - 1 = (1 + 2\mathbf{m}^T\mathbf{E}\mathbf{m})^{\frac{1}{2}} - 1. \quad (3.25)$$

Using (3.23) we can verify that

$$\mathbf{m}^T\mathbf{E}\mathbf{m} = \sum_{i=1}^{3}\sum_{j=1}^{3} \frac{\partial u_i}{\partial x_j} m_i m_j. \quad (3.26)$$

Of course, $\mathbf{m}^T\mathbf{E}\mathbf{m}$ is a real number. We next recall the useful mathematical identity

$$(1+\delta)^s = 1 + s\delta + (\text{higher-order terms in } \delta) \quad (3.27)$$

for δ and s real numbers and δ small. Substituting $\delta = 2\mathbf{m}^T\mathbf{E}\mathbf{m}$ and $s = \frac{1}{2}$, we can write

$$\left(1 + 2\mathbf{m}^T\mathbf{E}\mathbf{m}\right)^{\frac{1}{2}} = 1 + \frac{1}{2}\, 2\mathbf{m}^T\mathbf{E}\mathbf{m} + \left(\text{higher-order terms in } \mathbf{m}^T\mathbf{E}\mathbf{m}\right). \quad (3.28)$$

If we neglect higher-order terms in $\frac{\partial u_i}{\partial x_j}$, we arrive at

$$\left(1 + 2\mathbf{m}^T\mathbf{E}\mathbf{m}\right)^{1/2} = 1 + \mathbf{m}^T\mathbf{E}\mathbf{m} \quad (3.29)$$

which is exact for infinitesimally small displacements. Introducing (3.29) into (3.25), we finally obtain for infinitesimally small displacements

$$\varepsilon_\ell(\mathbf{m}) = \mathbf{m}^T\mathbf{E}\mathbf{m}. \quad (3.30)$$

Therefore in analyses assuming infinitesimally small displacements, the normal strain ε_ℓ of any fiber with origin at point $\mathbf{P}$ and direction $\mathbf{m}$ can be

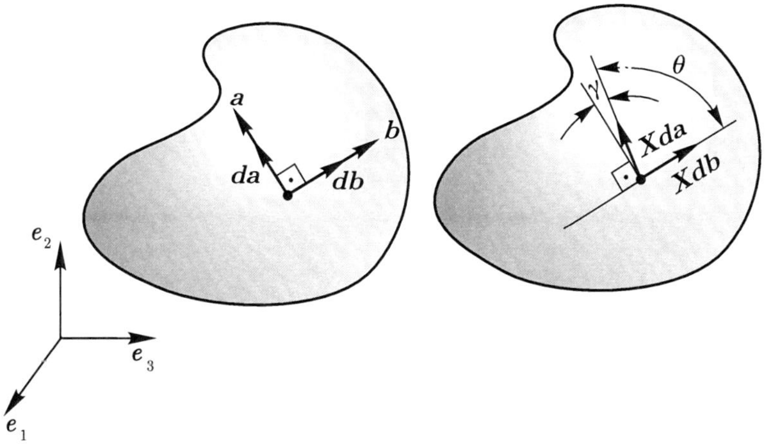

Fig. 3.23. Shear strain γ for fibers **da** and **db**

determined by the strain tensor **E** at this point, and is given by equation (3.30).

As we mentioned earlier, the change in angle between fibers due to the deformation needs also to be evaluated.

Hence, let us consider two orthogonal fibers **da** and **db** in the undeformed configuration whose directions are given by the unit vectors **a** and **b** as indicated in Figure 3.23.

In general, these fibers are no longer orthogonal in the deformed configuration. Let θ be the angle between the fibers in the deformed configuration. We define the shear strain (angular distortion) for this pair of fibers by the angle $\gamma = \frac{\pi}{2} - \theta$. Hence, the shear strain measures the deviation from the orthogonality and γ is positive when the angle between fibers, originally equal to $\pi/2$ has decreased. To obtain γ, using the definition of the scalar product, we write

$$\mathbf{X}\,\mathbf{da} \cdot \mathbf{X}\,\mathbf{db} = \|\mathbf{X}\,\mathbf{da}\|\,\|\mathbf{X}\,\mathbf{db}\| \cos\theta$$

and using that $\mathbf{da} = da\,\mathbf{a}$ and $\mathbf{db} = db\,\mathbf{b}$, we obtain

$$\mathbf{Xa} \cdot \mathbf{Xb} = \|\mathbf{Xa}\|\,\|\mathbf{Xb}\| \cos\theta.$$

Then also

$$\mathbf{Xa} \cdot \mathbf{Xb} = \mathbf{a}^T\,\mathbf{X}^T\mathbf{Xb}, \quad \|\mathbf{Xa}\| = \sqrt{\mathbf{a}^T\,\mathbf{X}^T\mathbf{Xa}}, \quad \|\mathbf{Xb}\| = \sqrt{\mathbf{b}^T\,\mathbf{X}^T\mathbf{Xb}}$$

leading to

$$\sin\gamma = \frac{\mathbf{a}^T\,\mathbf{X}^T\mathbf{Xb}}{\sqrt{\mathbf{a}^T\,\mathbf{X}^T\mathbf{Xa}}\sqrt{\mathbf{b}^T\,\mathbf{X}^T\mathbf{Xb}}} \tag{3.31}$$

where we used $\cos\theta = \sin\gamma$. From (3.19)

$$1+\varepsilon_\ell(\mathbf{a}) = \sqrt{\mathbf{a}^T\mathbf{X}^T\mathbf{X}\mathbf{a}}, \qquad 1+\varepsilon_\ell(\mathbf{b}) = \sqrt{\mathbf{b}^T\mathbf{X}^T\mathbf{X}\mathbf{b}}$$

where $\varepsilon_\ell(\mathbf{a})$ and $\varepsilon_\ell(\mathbf{b})$ are the normal strains of the fibers in the directions of $\mathbf{a}$ and $\mathbf{b}$. Therefore (3.31) can be written as

$$\sin\gamma = \frac{\mathbf{a}^T\mathbf{X}^T\mathbf{X}\mathbf{b}}{(1+\varepsilon_\ell(\mathbf{a}))(1+\varepsilon_\ell(\mathbf{b}))}. \tag{3.32}$$

Both expressions (3.31) and (3.32) give the shear strain without any approximation no matter how large the displacements might be. However, considering infinitesimally small displacements and using (3.24), the numerator of expression (3.31) can be written as

$$\mathbf{b}^T(\mathbf{I}+2\mathbf{E})\,\mathbf{a} = 2\mathbf{b}^T\mathbf{E}\mathbf{a} + \mathbf{b}^\mathbf{T}\mathbf{a} = 2\mathbf{b}^T\mathbf{E}\mathbf{a}$$

since $\mathbf{a}$ and $\mathbf{b}$ are orthogonal to each other. Using (3.26) and (3.30) we can see that ε_ℓ will always be given by a sum of terms of the type $\frac{\partial u_i}{\partial x_j}$. Hence, we can neglect ε_ℓ with respect to 1 in the denominator of (3.32). Also, using $\sin\gamma = \gamma$ since we are considering infinitesimally small displacements, we obtain

$$\gamma = 2\mathbf{b}^T\mathbf{E}\mathbf{a}. \tag{3.33}$$

Hence, in summary, the infinitesimal strain tensor $\mathbf{E}$ gives, through the very simple expressions (3.30) and (3.33), the normal and shear strains. Writing

$$\mathbf{E} = \begin{bmatrix} E_{11} & E_{12} & E_{13} \\ E_{21} & E_{22} & E_{23} \\ E_{31} & E_{32} & E_{33} \end{bmatrix}$$

the normal strain of a fiber in the direction of $\mathbf{e}_1$ is given by

$$\varepsilon_\ell = \mathbf{e}_1^T\mathbf{E}\mathbf{e}_1$$

or

$$\varepsilon_\ell = \begin{bmatrix} 1 & 0 & 0 \end{bmatrix} \begin{bmatrix} E_{11} & E_{12} & E_{13} \\ E_{21} & E_{22} & E_{23} \\ E_{31} & E_{32} & E_{33} \end{bmatrix} \begin{bmatrix} 1 \\ 0 \\ 0 \end{bmatrix}$$

which leads to

Table 3.1. Normal and shear strains for finite and infinitesimally small displacements

	Finite displacements	Infinitesimally small displacements
Normal strain of fiber with direction **m**	$\varepsilon_\ell = \sqrt{\mathbf{m}^T\mathbf{X}^T\mathbf{X}\mathbf{m}} - 1$	$\varepsilon_\ell = \mathbf{m}^T\mathbf{E}\mathbf{m}$
Shear strain of a pair of orthogonal fibers with direction **a** and **b**	$\sin\gamma = \frac{\mathbf{a}^T\mathbf{X}^T\mathbf{X}\mathbf{b}}{(1+\varepsilon_\ell(\mathbf{a}))(1+\varepsilon_\ell(\mathbf{b}))}$	$\gamma = 2\mathbf{b}^T\mathbf{E}\mathbf{a}$

$$\varepsilon_\ell(\mathbf{e}_1) = E_{11}$$

i.e., E_{11} gives the normal strain of a fiber in the direction of $\mathbf{e}_1$. Analogously,

$$\varepsilon_\ell(\mathbf{e}_2) = E_{22}, \qquad \varepsilon_\ell(\mathbf{e}_3) = E_{33}.$$

The shear strain of two fibers whose directions are $\mathbf{e}_1$ and $\mathbf{e}_2$ is given by

$$\begin{aligned} \gamma &= 2\mathbf{e}_1^T\mathbf{E}\mathbf{e}_2 \\ &= 2\begin{bmatrix} 1 & 0 & 0 \end{bmatrix}\begin{bmatrix} E_{11} & E_{12} & E_{13} \\ E_{21} & E_{22} & E_{23} \\ E_{31} & E_{32} & E_{33} \end{bmatrix}\begin{bmatrix} 0 \\ 1 \\ 0 \end{bmatrix} \end{aligned}$$

which leads to

$$\gamma(\mathbf{e}_1, \mathbf{e}_2) = 2E_{12}$$

and allows for the physical interpretation of the component E_{12} (see Example 3.2). We note that $\gamma(\mathbf{e}_1, \mathbf{e}_2) = 2E_{12} = 2E_{21} = \gamma(\mathbf{e}_2, \mathbf{e}_1)$ due to the symmetry of $\mathbf{E}$. Analogously

$$\gamma(\mathbf{e}_1, \mathbf{e}_3) = 2E_{13} = 2E_{31} = \gamma(\mathbf{e}_3, \mathbf{e}_1), \gamma(\mathbf{e}_2, \mathbf{e}_3) = 2E_{23} = 2E_{32} = \gamma(\mathbf{e}_3, \mathbf{e}_2).$$

In Table 3.1 we summarize the equations that give the normal and shear strains considering finite and infinitesimally small displacements. Note that the term "finite" (as opposed to "infinitesimal") is used to characterize displacements of arbitrary magnitude (which of course includes the infinitesimally small displacement case). The terminology "large displacements" and "small displacements" is also used to describe these conditions.

There is one important point to recognize. While we have six strain components in **E**, see (3.23), these are clearly not linearly independent since they are all calculated from the three independent displacement fields u, v and w. However, as long as the strain components are calculated from the (continuous) displacements as discussed above, the strains will be compatible. Throughout the presentations in this book we shall assume continuous displacements for all problems considered, and hence strain compatibility reduces to calculating the strain components appropriately from the applicable strain-displacement relationships.

Example 3.2

Consider a block which has a generic section parallel to the x_1x_2 plane as shown in Figure 3.24a. The deformed configuration to be studied is defined by

$$\begin{aligned} y_1 &= x_1 + \tan\beta\, x_2 \\ y_2 &= x_2 \\ y_3 &= x_3 \end{aligned}$$

with β large as shown in Figure 3.24b.

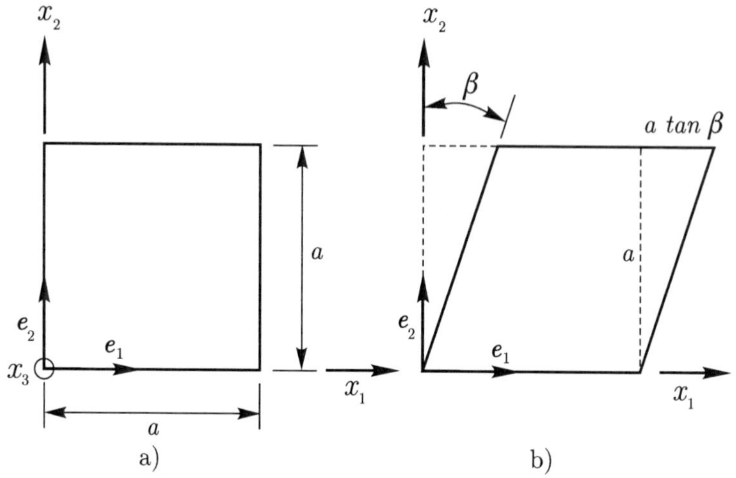

Fig. 3.24. Section of block and deformation

(i) Calculate the displacement field and the normal strains of the fibers with directions $\mathbf{e}_1$, $\mathbf{e}_2$, $\mathbf{m}_1 = \frac{\sqrt{2}}{2}\mathbf{e}_1 + \frac{\sqrt{2}}{2}\mathbf{e}_2$ and $\mathbf{m}_2 = -\frac{\sqrt{2}}{2}\mathbf{e}_1 + \frac{\sqrt{2}}{2}\mathbf{e}_2$.

(ii) Calculate the shear strains of the pairs of fibers with directions $\mathbf{e}_1$, $\mathbf{e}_2$ and $\mathbf{m}_1$, $\mathbf{m}_2$.

(iii) Repeat items (i) and (ii) assuming that β is small. Obtain the results first directly using small displacement theory, and then show that these same results are also obtained from (i) and (ii).

Solution

(i) The displacement field is given by

$$u_i = y_i - x_i$$

which leads to

$$\begin{aligned} u_1 &= y_1 - x_1 = \tan\beta\, x_2 \\ u_2 &= y_2 - x_2 = 0 \\ u_3 &= y_3 - x_3 = 0. \end{aligned}$$

Considering large displacements, we need to evaluate the deformation gradient

$$\mathbf{X} = \begin{bmatrix} 1 & \tan\beta & 0 \\ 0 & 1 & 0 \\ 0 & 0 & 1 \end{bmatrix}$$

where we note that throughout the block the elements in $\mathbf{X}$ are constant (independent of x_1, x_2, x_3). Then

$$\mathbf{X}^T\mathbf{X} = \begin{bmatrix} 1 & \tan\beta & 0 \\ \tan\beta & 1 + (\tan\beta)^2 & 0 \\ 0 & 0 & 1 \end{bmatrix}.$$

Now using the left column in Table 3.1

$$\varepsilon_\ell(\mathbf{e}_1) = \sqrt{\mathbf{e}_1^T\mathbf{X}^T\mathbf{X}\mathbf{e}_1} - 1 = 0 \tag{3.34}$$

$$\varepsilon_\ell(\mathbf{e}_2) = \sqrt{\mathbf{e}_2^T\mathbf{X}^T\mathbf{X}\mathbf{e}_2} - 1 = \sqrt{1 + (\tan\beta)^2} - 1 \tag{3.35}$$

$$\varepsilon_\ell(\mathbf{m}_1) = \sqrt{\mathbf{m}_1^T\mathbf{X}^T\mathbf{X}\mathbf{m}_1} - 1 = \sqrt{1 + \tan\beta + \frac{(\tan\beta)^2}{2}} - 1 \tag{3.36}$$

$$\varepsilon_\ell(\mathbf{m}_2) = \sqrt{\mathbf{m}_2^T\mathbf{X}^T\mathbf{X}\mathbf{m}_2} - 1 = \sqrt{1 - \tan\beta + \frac{(\tan\beta)^2}{2}} - 1. \tag{3.37}$$

We can interpret these results with the aid of Figure 3.25, where we can see that the fibers in the directions given by AB and AC extended, whereas the fibers along the diagonal DB shortened. Of course, the fibers parallel to AD did not change their lengths.

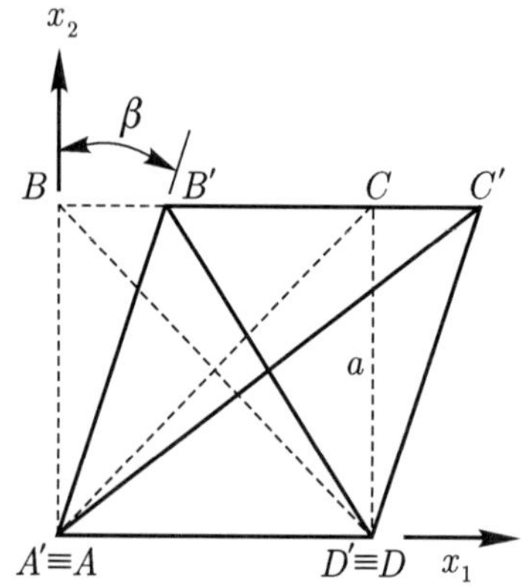

Fig. 3.25. Deformed and undeformed configurations with diagonals represented

(ii) The shear strains can be evaluated using again Table 3.1

$$\sin\gamma\left(\mathbf{e}_1,\mathbf{e}_2\right) = \frac{\mathbf{e}_1^T\mathbf{X}^T\mathbf{X}\mathbf{e}_2}{\left(1+\varepsilon_\ell(\mathbf{e}_1)\right)\left(1+\varepsilon_\ell(\mathbf{e}_2)\right)} = \frac{\tan\beta}{\sqrt{1+(\tan\beta)^2}} = \sin\beta \quad (3.38)$$

leading to $\gamma = \beta$, and

$$\begin{aligned} \sin\gamma\left(\mathbf{m}_1,\mathbf{m}_2\right) &= \frac{\mathbf{m}_1^T\mathbf{X}^T\mathbf{X}\mathbf{m}_2}{\left(1+\varepsilon_\ell(\mathbf{m}_1)\right)\left(1+\varepsilon_\ell(\mathbf{m}_2)\right)} \\ &= \frac{(\tan\beta)^2}{2\sqrt{1+\tan\beta+\frac{(\tan\beta)^2}{2}}\sqrt{1-\tan\beta+\frac{(\tan\beta)^2}{2}}}. \quad (3.39) \end{aligned}$$

(iii) When the displacements are infinitesimally small, we can evaluate the normal and shear strains using the infinitesimal strain tensor whose components are given by

$$E_{ij} = \frac{1}{2}\left(\frac{\partial u_i}{\partial x_j} + \frac{\partial u_j}{\partial x_i}\right).$$

Then,

$$E_{11} = \frac{\partial u_1}{\partial x_1} = 0, \qquad E_{22} = \frac{\partial u_2}{\partial x_2} = 0, \qquad E_{33} = \frac{\partial u_3}{\partial x_3} = 0$$

and since for infinitesimally small displacements $\tan\beta = \beta$

$$\begin{aligned} E_{12} &= E_{21} = \frac{1}{2}\left(\frac{\partial u_1}{\partial x_2} + \frac{\partial u_2}{\partial x_1}\right) = \frac{\beta}{2} \\ E_{13} &= E_{31} = \frac{1}{2}\left(\frac{\partial u_1}{\partial x_3} + \frac{\partial u_3}{\partial x_1}\right) = 0 \\ E_{23} &= E_{32} = \frac{1}{2}\left(\frac{\partial u_2}{\partial x_3} + \frac{\partial u_3}{\partial x_2}\right) = 0 \end{aligned}$$

leading to

$$\mathbf{E} = \begin{bmatrix} 0 & \beta/2 & 0 \\ \beta/2 & 0 & 0 \\ 0 & 0 & 0 \end{bmatrix}$$

and, hence, we obtain

$$\varepsilon_\ell(\mathbf{e}_1) = E_{11} = 0, \qquad \varepsilon_\ell(\mathbf{e}_2) = E_{22} = 0, \qquad \gamma(\mathbf{e}_1, \mathbf{e}_2) = 2E_{12} = \beta.$$

The normal strains of the fibers in the directions of $\mathbf{m}_1$ and $\mathbf{m}_2$ can be evaluated by

$$\begin{aligned} \varepsilon_\ell(\mathbf{m}_1) &= \mathbf{m}_1^T \mathbf{E} \mathbf{m}_1 \\ &= \begin{bmatrix} \frac{\sqrt{2}}{2} & \frac{\sqrt{2}}{2} & 0 \end{bmatrix} \begin{bmatrix} 0 & \beta/2 & 0 \\ \beta/2 & 0 & 0 \\ 0 & 0 & 0 \end{bmatrix} \begin{bmatrix} \frac{\sqrt{2}}{2} \\ \frac{\sqrt{2}}{2} \\ 0 \end{bmatrix} = \frac{\beta}{2} \end{aligned}$$

$$\begin{aligned} \varepsilon_\ell(\mathbf{m}_2) &= \mathbf{m}_2^T \mathbf{E} \mathbf{m}_2 \\ &= \begin{bmatrix} -\frac{\sqrt{2}}{2} & \frac{\sqrt{2}}{2} & 0 \end{bmatrix} \begin{bmatrix} 0 & \beta/2 & 0 \\ \beta/2 & 0 & 0 \\ 0 & 0 & 0 \end{bmatrix} \begin{bmatrix} -\frac{\sqrt{2}}{2} \\ \frac{\sqrt{2}}{2} \\ 0 \end{bmatrix} = -\frac{\beta}{2}. \end{aligned}$$

The shear strain of the fibers with directions $\mathbf{m}_1$, $\mathbf{m}_2$ is given by

$$\begin{aligned}\gamma(\mathbf{m}_1,\mathbf{m}_2) &= 2\mathbf{m}_1^T\mathbf{E}\mathbf{m}_2 \\ &= 2\begin{bmatrix} \frac{\sqrt{2}}{2} & \frac{\sqrt{2}}{2} & 0 \end{bmatrix}\begin{bmatrix} 0 & \beta/2 & 0 \\ \beta/2 & 0 & 0 \\ 0 & 0 & 0 \end{bmatrix}\begin{bmatrix} -\frac{\sqrt{2}}{2} \\ \frac{\sqrt{2}}{2} \\ 0 \end{bmatrix} = 0.\end{aligned}$$

In order to show that we obtain the above results from the values calculated in (i) and (ii), we need to consider β infinitesimally small in the expressions (3.35) to (3.39). Using the mathematical identity (3.27) we obtain

$$\varepsilon_\ell(\mathbf{e}_2) = 1 + \frac{1}{2}(\tan\beta)^2 + \left(\text{higher order terms in } (\tan\beta)^2\right) - 1$$

leading to

$$\varepsilon_\ell(\mathbf{e}_2) = 0. \tag{3.40}$$

Similarly

$$\varepsilon_\ell(\mathbf{m}_1) = 1 + \frac{1}{2}\left(\tan\beta + \frac{(\tan\beta)^2}{2}\right) + \left(\text{higher order terms in } \left(\tan\beta + \frac{(\tan\beta)^2}{2}\right)\right) - 1$$

leading to

$$\varepsilon_\ell(\mathbf{m}_1) = \frac{1}{2}\tan\beta. \tag{3.41}$$

Analogously, we obtain

$$\varepsilon_\ell(\mathbf{m}_2) = -\frac{1}{2}\tan\beta. \tag{3.42}$$

Note that to obtain these relations we neglected the terms $(\tan\beta)^2$ and of higher–order. And now using that for infinitesimally small displacements $\tan\beta = \beta$ gives

$$\varepsilon_\ell(\mathbf{e}_2) = 0, \quad \varepsilon_\ell(\mathbf{m}_1) = \frac{\beta}{2} \quad \text{and} \quad \varepsilon_\ell(\mathbf{m}_2) = -\frac{\beta}{2}.$$

Introducing the same approximations for (3.38) and (3.39), we directly obtain

$$\gamma(\mathbf{e}_1,\mathbf{e}_2) = \beta$$

and

$$\gamma(\mathbf{m}_1,\mathbf{m}_2) = 0.$$

Referring to Figure 3.25, we note that fibers that are parallel to the diagonals AC and DB, hence orthogonal in the undeformed configuration, remain orthogonal for infinitesimally small displacements.

□

3.2.3 Finite and infinitesimal rigid deformations

The kinematic description presented so far is of course very general and therefore includes a rigid body motion/deformation of a solid.

We start by studying a rigid body rotation which is always referred to an axis – the axis of rotation which is shown in Figure 3.26a. The rotation can

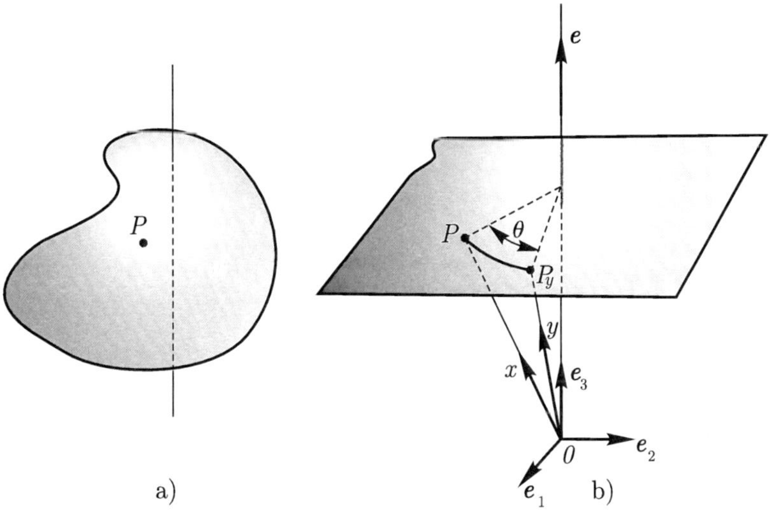

Fig. 3.26. Kinematic description of a rigid rotation

be kinematically characterized by describing the motion of a generic point **P** of the solid as shown in Figure 3.26b. We see that point **P** moves on the plane which is orthogonal to the axis of rotation and passes through **P**, and on the circle centered on the axis of rotation. The angle θ defines the magnitude of the rotation. Hence, a rotation is fully characterized by the vector $\theta = \theta\mathbf{e}$, where $\mathbf{e}$ is a unit vector in the direction of the axis of rotation.

Let us choose a reference system such that $\mathbf{e}_3 = \mathbf{e}$ as shown in Figure 3.26b. Referring to Example 3.1, the deformation corresponding to the rotation described above is given by

$$\begin{bmatrix} y_1 \\ y_2 \\ y_3 \end{bmatrix} = \begin{bmatrix} \cos\theta & -\sin\theta & 0 \\ \sin\theta & \cos\theta & 0 \\ 0 & 0 & 1 \end{bmatrix} \begin{bmatrix} x_1 \\ x_2 \\ x_3 \end{bmatrix} \tag{3.43}$$

$$\mathbf{y} = \mathbf{Q}\mathbf{x} \tag{3.44}$$

where $\mathbf{Q}$ is an orthogonal tensor. Orthogonal tensors are defined by either one of the following equivalent statements:

$$\|\mathbf{Q}\mathbf{w}\| = \|\mathbf{w}\| \quad \text{for every vector } \mathbf{w} \tag{3.45}$$

$$\mathbf{Q}\mathbf{Q}^T = \mathbf{Q}^T\mathbf{Q} = \mathbf{I} \quad \Rightarrow \quad \mathbf{Q}^{-1} = \mathbf{Q}^T. \tag{3.46}$$

From (3.46), we can write $\det(\mathbf{Q}\mathbf{Q}^T) = \det\mathbf{Q}\,\det\mathbf{Q}^T = (\det\mathbf{Q})^2 = 1$. Hence $\det\mathbf{Q} = \pm 1$.

It can be shown that every orthogonal tensor with positive determinant ($\det\mathbf{Q} = 1$) describes a rotation through equation (3.44). The example below is an illustration.

Example 3.3

Consider a tensor $\mathbf{Q}$ given in the basis $(\mathbf{e}_1, \mathbf{e}_2, \mathbf{e}_3)$ by

$$\mathbf{Q} = \begin{bmatrix} \frac{4}{9} + \frac{5\sqrt{3}}{18} & \frac{5}{18} - \frac{2\sqrt{3}}{9} & \frac{5}{9} - \frac{\sqrt{3}}{9} \\ \frac{11}{18} - \frac{2\sqrt{3}}{9} & \frac{4}{9} + \frac{5\sqrt{3}}{18} & -\frac{1}{9} - \frac{\sqrt{3}}{9} \\ -\frac{1}{9} - \frac{\sqrt{3}}{9} & \frac{5}{9} - \frac{\sqrt{3}}{9} & \frac{1}{9} + \frac{4\sqrt{3}}{9} \end{bmatrix}.$$

Verify that $\mathbf{Q}$ is an orthogonal tensor. Then, obtain the axis and the magnitude of the rotation given by $\mathbf{Q}$.

Solution

It suffices to verify that $\mathbf{Q}^T\mathbf{Q} = \mathbf{I}$ to conclude that $\mathbf{Q}$ is an orthogonal tensor.

Let $\mathbf{e}$ be a unit vector in the direction of the axis of rotation defined by $\mathbf{Q}$. Then

$$\mathbf{Q}\mathbf{e} = \mathbf{e}.$$

We can determine $\mathbf{e}$ solving

$$(\mathbf{Q} - \mathbf{I})\,\mathbf{e} = \mathbf{0} \tag{3.47}$$

to obtain

$$\mathbf{e} = \begin{bmatrix} \frac{2}{3} \\ \frac{2}{3} \\ \frac{1}{3} \end{bmatrix}.$$

Let us define a unit vector $\mathbf{g}$ which is orthogonal to the axis of rotation using

$$\mathbf{g} = \frac{\mathbf{e} \times \mathbf{e}_2}{\|\mathbf{e} \times \mathbf{e}_2\|} = \begin{bmatrix} -\frac{\sqrt{5}}{5} \\ 0 \\ \frac{2\sqrt{5}}{5} \end{bmatrix}.$$

Note that

$$\mathbf{f} = \mathbf{Q}\mathbf{g} = \begin{bmatrix} -\frac{\sqrt{15}}{10} + \frac{2\sqrt{5}}{15} \\ -\frac{\sqrt{5}}{6} \\ \frac{\sqrt{15}}{5} + \frac{\sqrt{5}}{15} \end{bmatrix}$$

gives a unit vector which is rotated with respect to $\mathbf{g}$ by the angle of rotation θ. Then

$$\mathbf{g} \cdot \mathbf{f} = \|\mathbf{g}\| \, \|\mathbf{f}\| \cos\theta = \frac{\sqrt{3}}{2}$$

and since

$$(\mathbf{g} \times \mathbf{f}) \cdot \mathbf{e} = \frac{1}{2}$$

we have

$$\theta = \frac{\pi}{6}.$$

□

Now let us show that the deformation given by

$$\mathbf{y}(\mathbf{x}) = \mathbf{y}_0 + \mathbf{Q}(\mathbf{x} - \mathbf{x}_0) \tag{3.48}$$

is rigid for any given vectors $\mathbf{y}_0$ and $\mathbf{x}_0$, and any orthogonal tensor $\mathbf{Q}$. As detailed later (see Figure 3.27), $\mathbf{y}_0$ gives the translation part of the rigid deformation while $\mathbf{Q}(\mathbf{x} - \mathbf{x}_0)$ gives the rotation part and $\mathbf{x}_0$ is a point on the axis of rotation.

The fundamental property of a rigid deformation is that the distance between any two points of the solid does not change with the deformation. Let us verify that this property holds for any deformation defined by (3.48). Let $\mathbf{x}_1$ and $\mathbf{x}_2$ be the position vectors of two arbitrarily selected points of the solid. Then, their positions in the deformed configuration are given by

$$\mathbf{y}_1 = \mathbf{y}_0 + \mathbf{Q}(\mathbf{x}_1 - \mathbf{x}_0) \tag{3.49}$$

$$\mathbf{y}_2 = \mathbf{y}_0 + \mathbf{Q}(\mathbf{x}_2 - \mathbf{x}_0). \tag{3.50}$$

Subtracting (3.49) from (3.50), we obtain

$$\mathbf{y}_2 - \mathbf{y}_1 = \mathbf{Q}\mathbf{x}_2 - \mathbf{Q}\mathbf{x}_1 = \mathbf{Q}\left(\mathbf{x}_2 - \mathbf{x}_1\right). \tag{3.51}$$

Taking norms in both sides of (3.51) and using (3.45), we can write

$$\|\mathbf{y}_2 - \mathbf{y}_1\| = \|\mathbf{Q}\left(\mathbf{x}_2 - \mathbf{x}_1\right)\| = \|\mathbf{x}_2 - \mathbf{x}_1\|$$

which shows that the distance between any two points of the solid is maintained and, indeed, (3.48) defines a rigid deformation or a *finite rigid deformation* since the magnitude of the rotation may be arbitrarily large.

Of course, we can select $\mathbf{x}_1$ and $\mathbf{x}_2$ close enough such that $\mathbf{dx} = \mathbf{x}_2 - \mathbf{x}_1$ defines a fiber. Then, (3.51) leads to

$$\mathbf{dy} = \mathbf{Q}\mathbf{dx}$$

showing that the deformation gradient of a rigid deformation given by (3.48) is $\mathbf{Q}$, that is,

$$\mathbf{X} = \mathbf{Q} \tag{3.52}$$

and due to (3.16)

$$\nabla\mathbf{u} = \mathbf{Q} - \mathbf{I}. \tag{3.53}$$

Note that $\mathbf{X}$ and $\nabla\mathbf{u}$ as given in equations (3.52) and (3.53) are the same for every point of the solid.

Since

$$\|\mathbf{dy}\| = \|\mathbf{Q}\mathbf{dx}\| = \|\mathbf{dx}\| \tag{3.54}$$

the normal strain in any direction is zero. Also, there is no shear strain between any pair of orthogonal fibers. Indeed, using (3.32)

$$\sin\gamma = \frac{\mathbf{a}^T\mathbf{Q}^T\mathbf{Q}\mathbf{b}}{(1+\varepsilon_\ell(\mathbf{a}))(1+\varepsilon_\ell(\mathbf{b}))} = \mathbf{a}^T\mathbf{b} = 0. \tag{3.55}$$

The results given by (3.54) and (3.55) actually confirm our expectation regarding strains for a rigid deformation.

Defining $\mathbf{u}_0 = \mathbf{y}_0 - \mathbf{x}_0$, we obtain the displacement field for a rigid deformation

$$\mathbf{u} = \mathbf{y}(\mathbf{x}) - \mathbf{x} = \mathbf{u}_0 + \mathbf{Q}(\mathbf{x} - \mathbf{x}_0) - (\mathbf{x} - \mathbf{x}_0) = \mathbf{u}_0 + (\mathbf{Q} - \mathbf{I})(\mathbf{x} - \mathbf{x}_0) \tag{3.56}$$

and using (3.53), we obtain

$$\mathbf{u} = \mathbf{u}_0 + \nabla\mathbf{u}(\mathbf{x} - \mathbf{x}_0). \tag{3.57}$$

We emphasize that $\nabla\mathbf{u}$ in (3.57) is the displacement gradient for the finite rigid deformation given by (3.53).

In order to motivate the definition of an infinitesimal rigid deformation we consider the following derivations.

Using (3.20) and (3.52), we obtain

$$\mathbf{X}^T\mathbf{X} = \mathbf{Q}^T\mathbf{Q} = \mathbf{I} = \mathbf{I} + \nabla\mathbf{u} + \nabla\mathbf{u}^T + \nabla\mathbf{u}^T\nabla\mathbf{u}$$

hence,

$$\frac{\nabla \mathbf{u} + \nabla \mathbf{u}^T}{2} = -\frac{1}{2}\nabla \mathbf{u}^T \nabla \mathbf{u}$$

and the infinitesimal strain tensor for the rigid deformation is given by (see (3.22))

$$\mathbf{E} = \frac{1}{2}(\nabla \mathbf{u} + \nabla \mathbf{u}^T) = -\frac{1}{2}\nabla \mathbf{u}^T \nabla \mathbf{u} \tag{3.58}$$

which is not exactly zero. This fact motivates the definition of an *infinitesimal rigid deformation* as the deformation which gives $\mathbf{E} = \mathbf{0}$.

Consider

$$\mathbf{u}(\mathbf{x}) = \mathbf{u}_0 + \mathbf{W}(\mathbf{x} - \mathbf{x}_0) \tag{3.59}$$

where $\mathbf{W}$ is a *skew tensor*, that is, a tensor for which $\mathbf{W}^T = -\mathbf{W}$. Then the displacement gradient for the $\mathbf{u}(\mathbf{x})$ defined in (3.59) is

$$\nabla \mathbf{u} = \mathbf{W}$$

and

$$\mathbf{E} = \frac{1}{2}\left(\nabla \mathbf{u} + \nabla \mathbf{u}^T\right) = \frac{1}{2}\left(\mathbf{W} + \mathbf{W}^T\right) = \mathbf{0}. \tag{3.60}$$

Equation (3.60) shows that every displacement field defined by (3.59) with $\mathbf{W}$ skew leads to $\mathbf{E} = \mathbf{0}$, *i.e.*, defines an infinitesimal rigid deformation.

Since for every skew tensor $\mathbf{W}$ there is a unique vector $\mathbf{w}$, called the axial vector, such that

$$\mathbf{W}\mathbf{a} = \mathbf{w} \times \mathbf{a} \quad \text{for every vector } \mathbf{a}$$

and vice-versa (see Crandall, Dahl and Lardner, 1978), we can re-write (3.59) as

$$\mathbf{u}(\mathbf{x}) = \mathbf{u}_0 + \mathbf{w} \times (\mathbf{x} - \mathbf{x}_0). \tag{3.61}$$

Now, let us interpret the finite and infinitesimal rigid deformations. In Figure 3.27a, we detail the displacement of a generic point for a finite rigid deformation. The unit vector $\mathbf{e}$ gives the direction of the axis of rotation associated with the orthogonal tensor $\mathbf{Q}$ and $\mathbf{x}_0$ is a point on the axis of rotation. According to equation (3.56) there are two contributions for the displacement. The first, $\mathbf{Q}(\mathbf{x} - \mathbf{x}_0) - (\mathbf{x} - \mathbf{x}_0)$, is due to the rotation and the second, $\mathbf{u}_0$, is due to the translation. In Figure 3.27b, we consider the displacement of a generic point $\mathbf{x}$ for an infinitesimal rigid deformation. The vector $\mathbf{w}$ is the axial vector associated with $\mathbf{W}$ and according to (3.61) the term $\mathbf{w} \times (\mathbf{x} - \mathbf{x}_0)$ gives the contribution of the rotation and $\mathbf{u}_0$ of the translation.

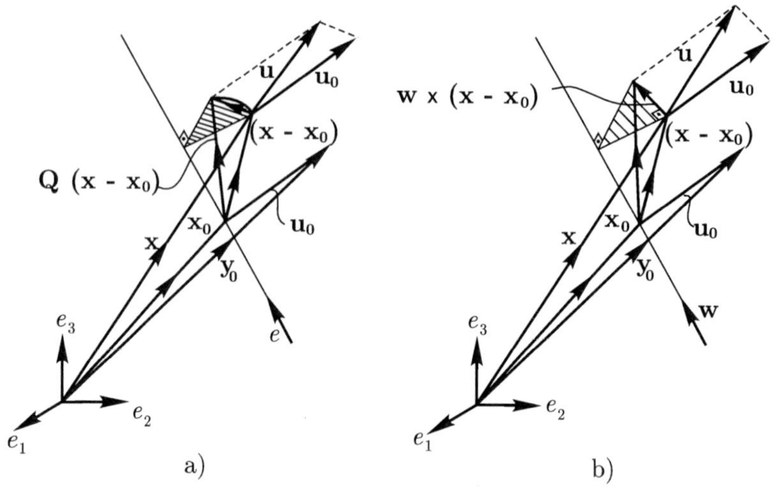

Fig. 3.27. Displacement of a point for finite and infinitesimal rigid deformations

Comparing both rigid deformations, we note that the difference lies on how the rotation is described. While for the finite rotation the point displaces along the circumference centered on the axis of rotation, for the infinitesimal rotation the point displaces over the straight line which makes a right angle with the radius of this circumference, that is, over the tangent to this circumference. We recall that this distinction in describing the rotation was also addressed in Section 2.2.5 for the rotation of a truss bar.

Example 3.4
Consider the rigid body rotation described in Example 3.1 (ii). Let the angle of rotation be infinitesimally small, that is, $d\varphi$ instead of φ. Calculate the displacement field associated with this infinitesimal rigid deformation. Then, verify that $\mathbf{E} = \mathbf{0}$.

Solution
In Figure 3.28a we repeat Figure 3.20b which describes the finite rotation of a generic point of the solid and in Figure 3.28b we show the analogous situation for an infinitesimal rotation. Since the rotation is infinitesimal, we should take the displacements over the tangent to the circumference which has center in O and radius $r = \|\mathbf{OP}\|$. Therefore

$$\begin{aligned} -u_1 &= d\varphi\, r \sin\theta \\ u_2 &= d\varphi\, r \cos\theta. \end{aligned}$$

Using (3.8) , we obtain

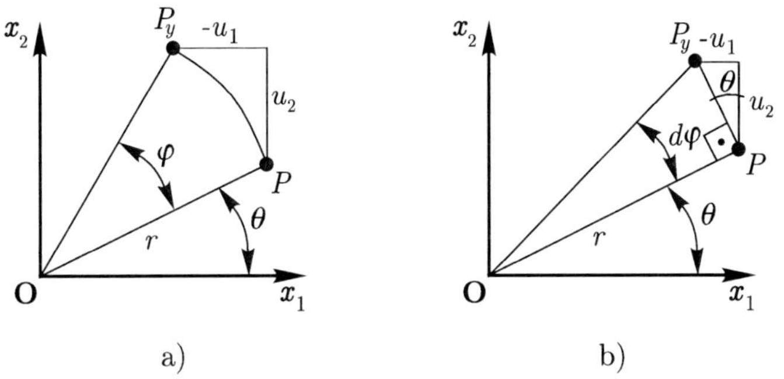

Fig. 3.28. Motion of a generic point in the section. a) Finite rotation φ; b) Infinitesimal rotation $d\varphi$

$$
\begin{aligned}
u_1 &= -d\varphi\, x_2 \\
u_2 &= d\varphi\, x_1
\end{aligned}
$$

Finally

$$
\begin{aligned}
u_1\left(x_1, x_2, x_3\right) &= -d\varphi\, x_2 \\
u_2\left(x_1, x_2, x_3\right) &= d\varphi\, x_1 \\
u_3\left(x_1, x_2, x_3\right) &= 0.
\end{aligned}
$$

Using the displacement field above and (3.23) we obtain $\mathbf{E} = \mathbf{0}$.

□

3.2.4 Technical or engineering notation for the strains

In the engineering literature it is very common to use what is called technical or engineering notation in which the coordinate axes are denoted by x, y, and z, *i.e.*, $x \equiv x_1$, $y \equiv x_2$ and $z \equiv x_3$ and the displacements by u, v, and w, *i.e.*, $u \equiv u_1$, $v \equiv u_2$ and $w \equiv u_3$.

In this notation the components of the strain tensor are given and denoted by

$$
\begin{aligned}
\varepsilon_{xx} &= \varepsilon_x = \frac{\partial u}{\partial x} \\
\varepsilon_{yy} &= \varepsilon_y = \frac{\partial v}{\partial y}
\end{aligned}
$$

$$\begin{aligned}
\varepsilon_{zz} &= \varepsilon_z = \frac{\partial w}{\partial z} \\
\varepsilon_{xy} &= \varepsilon_{yx} = \frac{1}{2}\left(\frac{\partial u}{\partial y} + \frac{\partial v}{\partial x}\right) \\
\varepsilon_{yz} &= \varepsilon_{zy} = \frac{1}{2}\left(\frac{\partial v}{\partial z} + \frac{\partial w}{\partial y}\right) \\
\varepsilon_{zx} &= \varepsilon_{xz} = \frac{1}{2}\left(\frac{\partial w}{\partial x} + \frac{\partial u}{\partial z}\right).
\end{aligned}$$

These relations are called strain-displacement relations. We can also write the shear strain for the pairs of fibers parallel to the coordinate axes by

$$\gamma_{xy} = 2\varepsilon_{xy}, \qquad \gamma_{yz} = 2\varepsilon_{yz}, \qquad \gamma_{zx} = 2\varepsilon_{zx}. \tag{3.62}$$

Note that these shear strains are twice the corresponding tensorial components and we call $\gamma_{xy}, \gamma_{yz}, \gamma_{zx}$ *the engineering shear strains* and $\varepsilon_{xy}, \varepsilon_{yz}, \varepsilon_{zx}$ the *tensorial shear strains.*

3.2.5 Deformation in the vicinity of a point

In order to obtain more insight into the deformation, let us examine the deformation in a small region around a point – its vicinity. Consider a deformation with infinitesimally small displacements and $w = 0$, $u = u(x, y)$ and $v = v(x, y)$. Under such conditions the deformation observed in any plane parallel to the plane xy is identical, and the displacement gradient and strain tensor are given by

$$\nabla\mathbf{u} = \begin{bmatrix} \frac{\partial u}{\partial x} & \frac{\partial u}{\partial y} & 0 \\ \frac{\partial v}{\partial x} & \frac{\partial v}{\partial y} & 0 \\ 0 & 0 & 0 \end{bmatrix}, \quad \mathbf{E} = \begin{bmatrix} \frac{\partial u}{\partial x} & \frac{1}{2}\left(\frac{\partial u}{\partial y} + \frac{\partial v}{\partial x}\right) & 0 \\ \frac{1}{2}\left(\frac{\partial u}{\partial y} + \frac{\partial v}{\partial x}\right) & \frac{\partial v}{\partial y} & 0 \\ 0 & 0 & 0 \end{bmatrix}.$$

Assume that at point $\mathbf{P}$ in Figure 3.29 $u = 0$, $v = 0$, $\frac{\partial u}{\partial x} = 0$ and $\frac{\partial v}{\partial y} = 0$, *i.e.*, the normal strains in the directions of x and y are zero. In Figure 3.30a, we show the deformed and undeformed configurations for the selected part. We note that the geometrical interpretation of the engineering shear strain $\gamma_{xy} = \frac{\partial u}{\partial y} + \frac{\partial v}{\partial x}$ is evident. Now, in Figure 3.30b we show a special case in which $\frac{\partial v}{\partial x} = -\frac{\partial u}{\partial y}$. Then, of course, $\gamma_{xy} = 0$ and we have a rigid body rotation of the part of intensity $\alpha = \frac{\partial v}{\partial x} = -\frac{\partial u}{\partial y}$. We note that despite the fact we have a rigid rotation, the components of the displacement gradient are not zero since $\frac{\partial u}{\partial y} \neq 0$ and $\frac{\partial v}{\partial x} \neq 0$.

Considering the deformation of Figure 3.30a, we can also write

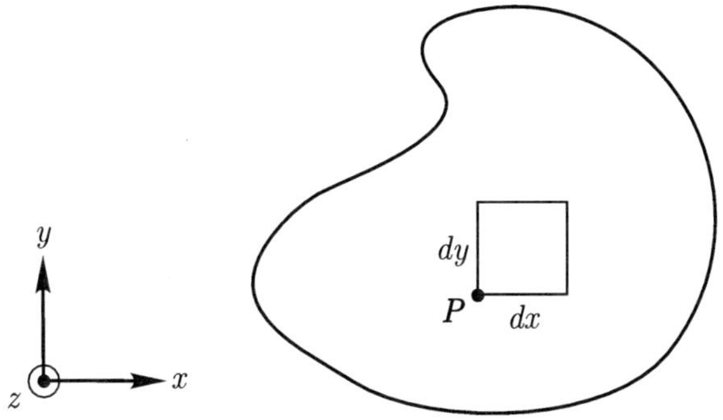

Fig. 3.29. Schematic representation of a generic square part defining the region of interest

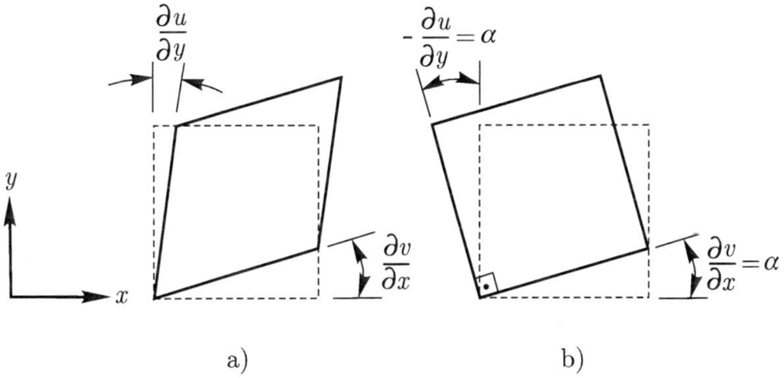

Fig. 3.30. a) Deformation of the selected part; b) Geometrical interpretation of an infinitesimal rigid rotation

$$\frac{\partial u}{\partial y} = \frac{1}{2}\left(\frac{\partial u}{\partial y} + \frac{\partial v}{\partial x}\right) + \frac{1}{2}\left(\frac{\partial u}{\partial y} - \frac{\partial v}{\partial x}\right) \tag{3.63}$$

$$\frac{\partial v}{\partial x} = \frac{1}{2}\left(\frac{\partial u}{\partial y} + \frac{\partial v}{\partial x}\right) - \frac{1}{2}\left(\frac{\partial u}{\partial y} - \frac{\partial v}{\partial x}\right) \tag{3.64}$$

The left hand sides of equations (3.63) and (3.64) give components of the displacement gradient and the right hand sides represent one half of the engineering shear strain plus a term which can be interpreted as a rigid body rotation. The geometrical interpretation of equations (3.63) and (3.64) is given in Figure 3.31[7].

Although in the above discussion some assumptions were made to simplify the visualization of the deformation, the interpretation given is quite general. Let us define

[7] In Section 3.4 we will relate γ_{xy} by a material constant to a shear stress. Note that the rotation in (3.63) and (3.64) does not cause a stress

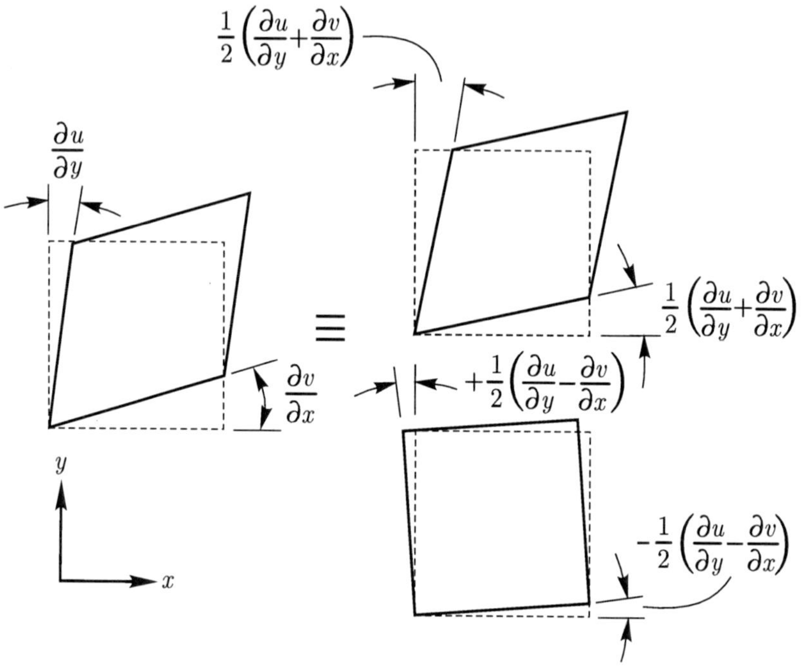

Fig. 3.31. Geometrical interpretation of the decomposition of the displacement gradient. The inherent rotation does not cause a stress (see Section 3.4)

$$\mathbf{W} = \nabla\mathbf{u} - \mathbf{E}$$
$$\mathbf{W} = \frac{1}{2}\left(\nabla\mathbf{u} - \nabla\mathbf{u}^T\right)$$

or

$$\mathbf{W} = \begin{bmatrix} 0 & \frac{1}{2}\left(\frac{\partial u}{\partial y} - \frac{\partial v}{\partial x}\right) & \frac{1}{2}\left(\frac{\partial u}{\partial z} - \frac{\partial w}{\partial x}\right) \\ -\frac{1}{2}\left(\frac{\partial u}{\partial y} - \frac{\partial v}{\partial x}\right) & 0 & \frac{1}{2}\left(\frac{\partial v}{\partial z} - \frac{\partial w}{\partial y}\right) \\ -\frac{1}{2}\left(\frac{\partial u}{\partial z} - \frac{\partial w}{\partial x}\right) & -\frac{1}{2}\left(\frac{\partial v}{\partial z} - \frac{\partial w}{\partial y}\right) & 0 \end{bmatrix}.$$

Of course,

$$\nabla\mathbf{u} = \mathbf{E} + \mathbf{W}. \tag{3.65}$$

We can also write

$$\mathbf{u}(\mathbf{x} + \mathbf{dx}) - \mathbf{u}(\mathbf{x}) = \nabla\mathbf{u}\mathbf{dx} = (\mathbf{E} + \mathbf{W})\,\mathbf{dx}. \tag{3.66}$$

Since $\mathbf{W}$ is skew, $\mathbf{W}\mathbf{dx}$ represents the displacement due to an infinitesimal rigid rotation (refer to equation (3.59)). Therefore, (3.66) allows the interpretation that, locally, the increment in displacements has a contribution of a rigid rotation plus that of the straining of the material fibers.

Equation (3.65) is referred to as the additive decomposition of the displacement gradient $\nabla \mathbf{u}$ into its symmetric part – the infinitesimal strain tensor – and into its skew part $\mathbf{W}$ – the infinitesimal rotation tensor.

Example 3.5

Consider the cylinder shown in Figure 3.32. Suppose that the transverse sections rotate without deformation in the plane yz around the cylinder's axis by an angle $\theta(x)$ with the constant rate of rotation $\frac{d\theta}{dx} = \alpha$. Calculate, assuming infinitesimal displacement conditions:

(i) The displacement field.
(ii) The strain tensor within the cylinder.

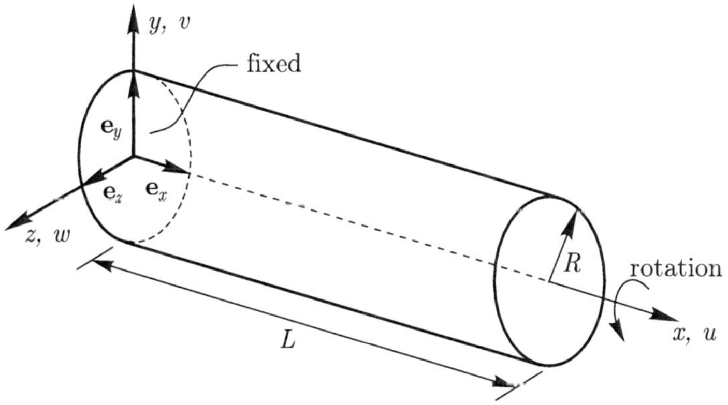

Fig. 3.32. Cylinder under study

Solution

(i) We obtain by integration

$$\theta(x) = \alpha x + C.$$

Since the rotation at $x = 0$ is prevented

$$\theta(0) = 0 \Rightarrow C = 0$$

and therefore the rotation of a generic section is given by

$$\theta(x) = \alpha x.$$

Since the section rotations are infinitesimal, we can use directly the results derived in Example 3.4. Therefore, considering $x_1 \equiv y$, $x_2 \equiv z$, $x_3 \equiv x$, we obtain

$$\begin{aligned} u &= 0 \\ v &= -\theta z = -\alpha x z \\ w &= \theta y = \alpha x y. \end{aligned}$$

(ii) The strain components are

$$\begin{aligned} \varepsilon_{xx} &= \frac{\partial u}{\partial x} = 0, \quad \varepsilon_{yy} = \frac{\partial v}{\partial y} = 0, \quad \varepsilon_{zz} = \frac{\partial w}{\partial z} = 0 \\ \varepsilon_{xy} &= \frac{1}{2}\left(\frac{\partial u}{\partial y} + \frac{\partial v}{\partial x}\right) = -\frac{1}{2}\alpha z \\ \varepsilon_{xz} &= \frac{1}{2}\left(\frac{\partial u}{\partial z} + \frac{\partial w}{\partial x}\right) = \frac{1}{2}\alpha y \\ \varepsilon_{yz} &= \frac{1}{2}\left(\frac{\partial v}{\partial z} + \frac{\partial w}{\partial y}\right) = \frac{1}{2}(-\alpha x + \alpha x) = 0. \end{aligned}$$

Therefore

$$\mathbf{E} = \begin{bmatrix} 0 & -\frac{1}{2}\alpha z & \frac{1}{2}\alpha y \\ -\frac{1}{2}\alpha z & 0 & 0 \\ \frac{1}{2}\alpha y & 0 & 0 \end{bmatrix}$$

and the engineering shear strains are

$$\begin{aligned} \gamma_{xy} &= -\alpha z \\ \gamma_{xz} &= \alpha y \\ \gamma_{yz} &= 0. \end{aligned}$$

In Figure 3.33 a geometrical interpretation of γ_{xz} is given. Referring to Figure 3.33 we can calculate γ_{xz} for a point of coordinates x, $y = R$, $z = 0$ as the ratio

$$\gamma_{xz} = \frac{d\theta R}{dx} = \alpha R$$

which is in accordance with the derived expression.

□

3.3 Stresses

In Section 2.1.3 we introduced the concept of stress, see Figure 2.3. In this figure, a field of forces per unit area – the field of stresses – is acting on the internal surface of the part $\Delta^t V$.

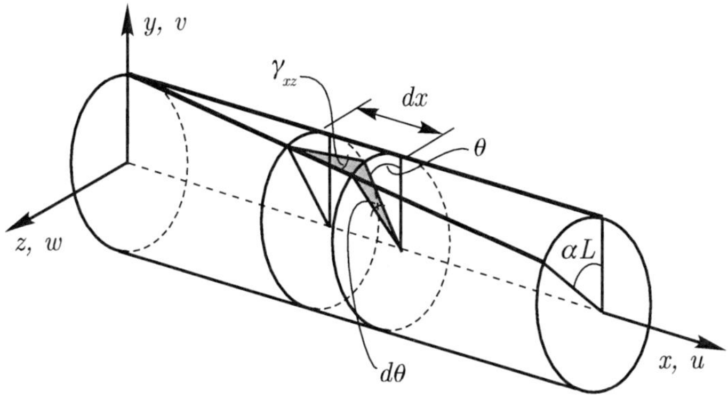

Fig. 3.33. Deformation of the cylinder

In Section 2.1.4 we arrived at the conditions for a solid, subjected to external forces, to be in static equilibrium. These conditions are given by ${}^t\mathbf{R} = \mathbf{0}$ and ${}^t\mathbf{M_O} = \mathbf{0}$ where ${}^t\mathbf{R}$ is the resultant of all externally applied forces and ${}^t\mathbf{M_O}$ is the moment of these forces about the system origin[8]. We also pointed out that the static equilibrium condition should hold for any part $\Delta^t V$ extracted from the body when, of course, we consider the stresses and the external actions on $\Delta^t V$.

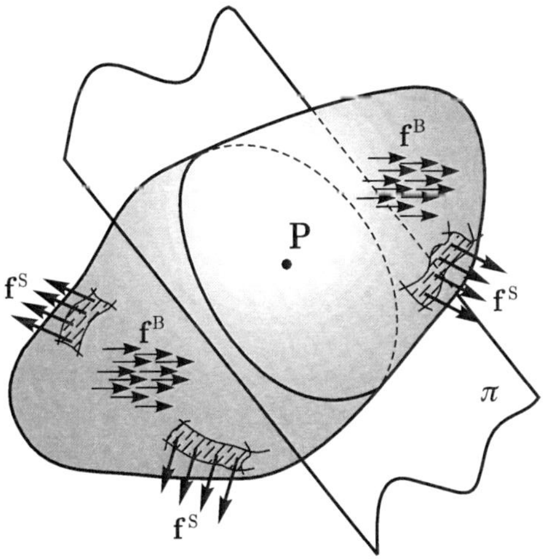

Fig. 3.34. Solid in equilibrium

[8] Actually the moment has to be zero about any point since ${}^t\mathbf{R} = \mathbf{0}$ as discussed in Chapter 2

The objective of this section is to examine in detail the concept of stress at a point, the stress field and the relations which link the stress field to the external field of forces acting on the solid.

We start in the next section by introducing in a classical manner the concept of stress.

3.3.1 Classical concept of stress

Let V_y in Figure 3.19 represent the deformed configuration of a solid which is in static equilibrium subjected to the field of body forces $\mathbf{f}^B$ and surface tractions $\mathbf{f}^S$. We suppose that $\mathbf{f}^S$ is defined over the entire external surface of the solid, therefore it includes possible reactions associated with motion restraints. Since we consider only this deformed configuration in the following discussion, we use simply V to denote V_y and this situation is summarized in Figure 3.34. Let $\mathbf{P}$ be a point in the interior of the solid and let π be a plane that passes through $\mathbf{P}$ and which sections the solid into two parts, ΔV_I and ΔV_{II}. These two parts are shown separately in Figure 3.35.

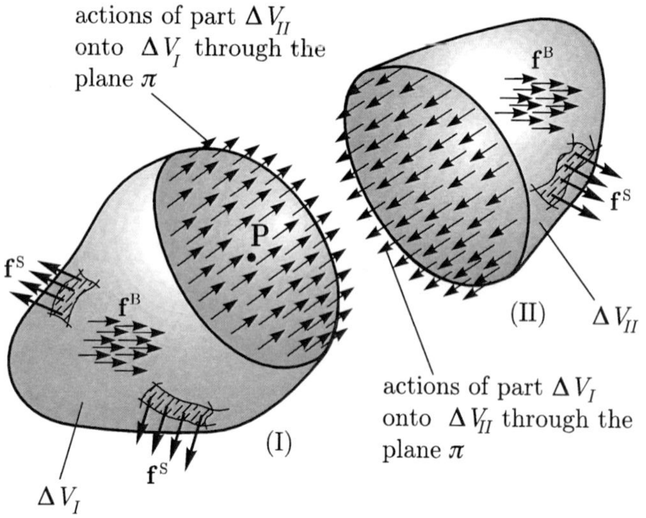

Fig. 3.35. Two parts of solid V sectioned by plane π (part rotated merely for better visualization)

The fields of internal forces that act on ΔV_I and ΔV_{II} on that plane are also shown. We note that the force actions of part ΔV_{II} onto part ΔV_I are such that ΔV_I considered alone is in static equilibrium, and similarly for part ΔV_{II}.

Let ΔS be a surface region around $\mathbf{P}$ as shown in Figure 3.36 and let $\Delta\mathbf{F}$ be the resultant of the internal forces acting on ΔS by the action of part

ΔV_{II} onto ΔV_I. We can define

$$\mathbf{t}_m = \frac{\Delta \mathbf{F}}{\Delta S}$$

which is an average force per unit of surface area. This vector quantity $\mathbf{t}_m$ is called the average stress at $\mathbf{P}$ considering the plane π and the area ΔS.

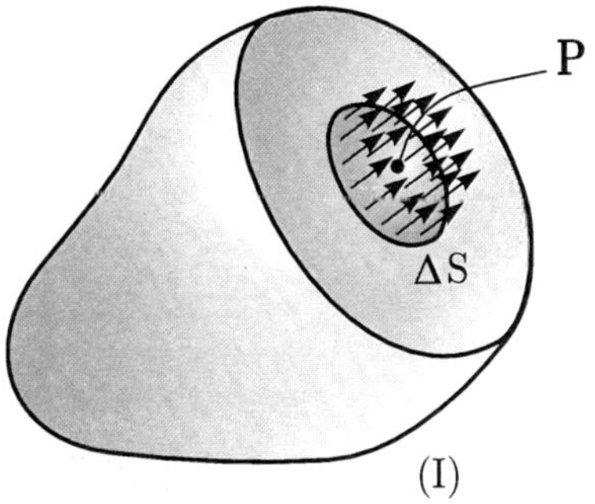

Fig. 3.36. Forces transmitted through ΔS

Since we want to obtain a local measure of the force transmitted per unit of area, we define the stress $\mathbf{t}$ at $\mathbf{P}$ on the plane π as

$$\mathbf{t} = \lim_{\Delta S \to 0} \frac{\Delta \mathbf{F}}{\Delta S} = \lim_{\Delta S \to 0} \mathbf{t}_m .$$

Since the plane π is well defined by its normal unit vector $\mathbf{n}$ we have

$$\mathbf{t} = \mathbf{t}\left(\mathbf{P}, \mathbf{n}\right). \tag{3.67}$$

Here $\mathbf{n}$ is taken to point outward from the face of the solid on which the stress is acting. Hence, the stress at the same point $\mathbf{P}$ and considering the same plane π but representing the action of part ΔV_I onto ΔV_{II} is given by

$$\mathbf{t} = \mathbf{t}\left(\mathbf{P}, -\mathbf{n}\right).$$

It is usual to decompose the stress into two vector components: $\mathbf{t}_n$ normal to the plane, *i.e.*, in the direction $\mathbf{n}$, and $\mathbf{t}_s$ tangential to the plane, *i.e.*, normal to $\mathbf{n}$, see Figure 3.37. Therefore

$$\begin{aligned} \mathbf{t} &= \mathbf{t}_n + \mathbf{t}_s \\ \mathbf{t}_n &= \left(\mathbf{t} \cdot \mathbf{n}\right)\mathbf{n} = t_n \mathbf{n}, \qquad \mathbf{t}_s = \mathbf{t} - \mathbf{t}_n . \end{aligned}$$

Then $\mathbf{t}_n$ is called the normal stress and $\mathbf{t}_s$ the shear stress. This decomposition is not merely formal, since the physical effects associated with the action of these components are very different. Indeed, the normal stress when $t_n > 0$

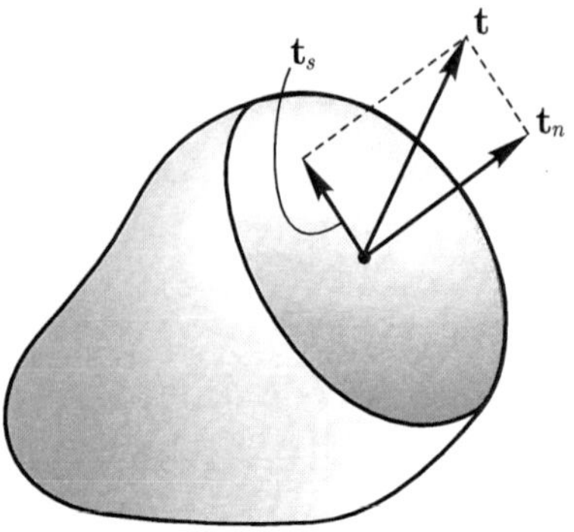

Fig. 3.37. Decomposition of the stress

(tensile stress) induces extension of the fiber along the normal direction, and when $t_n < 0$ (compressive stress) induces shortening of the same fiber. Also, as detailed later, the shear stress induces shear strain.

In the discussion above, we have considered the plane π as a generic plane to pass through $\mathbf{P}$. Therefore for each plane defined by a generic normal $\mathbf{n}$ there is a different value of $\mathbf{t}$ representing the action of part ΔV_{II} on ΔV_I (of course, as $\mathbf{n}$ changes ΔV_I and ΔV_{II} also change). This fact was acknowledged in equation (3.67) as we explicitly indicated the dependency on the plane through the normal $\mathbf{n}$. As there are infinitely many planes that pass through $\mathbf{P}$ (each time characterized by the corresponding $\mathbf{n}$), it is usual to refer to a "state of stress" at a point.

3.3.2 Characterization of the state of stress at a point

Our next objective is to better characterize the state of stress at a point.

First consequence of equilibrium

Let us first establish a relation between the stresses $\mathbf{t}(\mathbf{P},\mathbf{n})$ and $\mathbf{t}(\mathbf{P},-\mathbf{n})$, *i.e.*, they are defined for the same point and for the same plane but act on different parts of the sectioned solid, see Figure 3.34. Consider as a part ΔV of the solid of Figure 3.34, the parallelepiped shown in Figure 3.38. It has two faces parallel to the plane π which also sections the parallelepiped into two equal halves. This parallelepiped is also shown in Figure 3.39. Its thickness is $(\delta)^q$, $q \geq 2$, with δ infinitesimally small.

Equilibrium requires

$$\mathbf{t}(\mathbf{P},\mathbf{n})\,\delta^2 + \mathbf{t}(\mathbf{P},-\mathbf{n})\,\delta^2 + \mathbf{t}(\mathbf{P},\mathbf{n}_1)\,\delta^{q+1} + \mathbf{t}(\mathbf{P},-\mathbf{n}_1)\,\delta^{q+1} +$$
$$\mathbf{t}(\mathbf{P},\mathbf{n}_2)\,\delta^{q+1} + \mathbf{t}(\mathbf{P},-\mathbf{n}_2)\,\delta^{q+1} + \mathbf{f}^B(\mathbf{P})\,\delta^{q+2} = \mathbf{0}$$

Since δ is infinitesimal, the stress is constant over each face and therefore

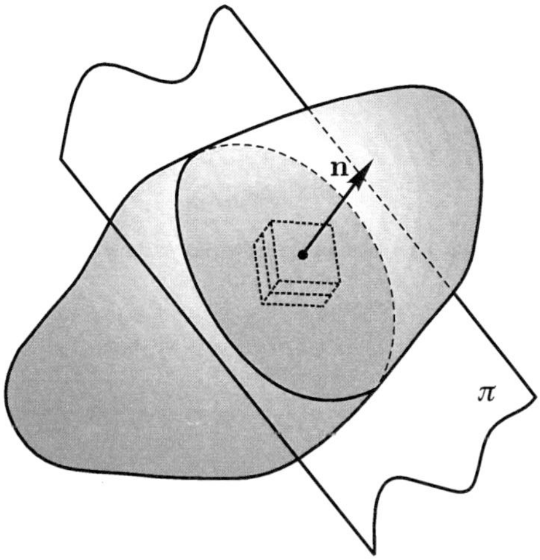

Fig. 3.38. Selection of a parallelepiped part around P

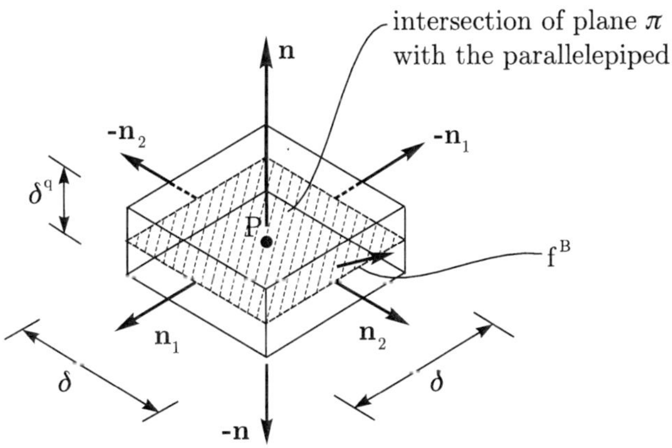

Fig. 3.39. Definition of the parallelepiped part, $\mathbf{f}^B$ is the body force per unit of volume

can be evaluated at $\mathbf{P}$. Neglecting terms of higher-order in δ, we obtain

$$\mathbf{t}(\mathbf{P}, \mathbf{n}) = -\mathbf{t}(\mathbf{P}, -\mathbf{n}). \tag{3.68}$$

Of course, taking the height to be δ^q and the base to be δ was an appropriate choice and certainly allowed since every part of the solid should be in equilibrium.

The relation (3.68) simply states that the stress acting on a plane but on the two different parts of the sectioned solid is of same magnitude and direction but of opposite orientation.

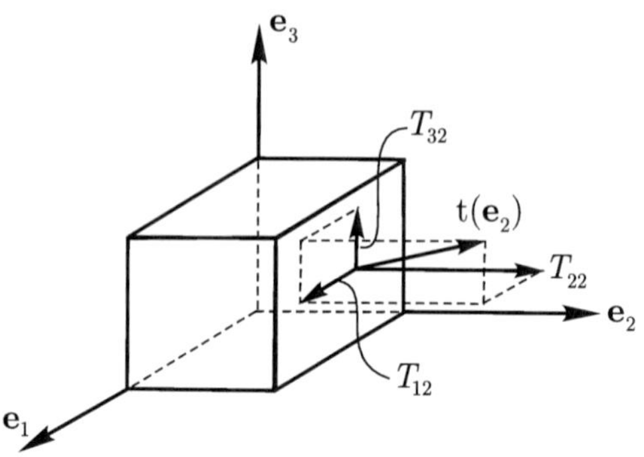

Fig. 3.40. Decomposition of the stress on the plane with unit normal $\mathbf{e}_2$

Stress components

Consider the situation in Figure 3.40 where we isolated a parallelepiped with infinitesimally small sides from the solid to examine the stress acting on a plane with normal $\mathbf{e}_2$. Let T_{22} be the component of $\mathbf{t}(\mathbf{e}_2)$ in the direction of $\mathbf{e}_2$ and T_{12} and T_{32} be the components in the directions $\mathbf{e}_1$ and $\mathbf{e}_3$ respectively, *i.e.*,

$$\mathbf{t}(\mathbf{e}_2) = T_{12}\mathbf{e}_1 + T_{22}\mathbf{e}_2 + T_{32}\mathbf{e}_3 = \sum_{i=1}^{3} T_{i2}\mathbf{e}_i. \tag{3.69}$$

where $T_{i2} = \mathbf{t}(\mathbf{e}_2) \cdot \mathbf{e}_i$.

In Figure 3.41 we still represent the same part but now we indicate explicitly the stresses and their components, defined as above, also for planes whose normals are $\mathbf{e}_1$ and $\mathbf{e}_3$. We can write

$$\mathbf{t}(\mathbf{e}_j) = \sum_{i=1}^{3} T_{ij}\mathbf{e}_i. \tag{3.70}$$

The stress acting at the same point as $\mathbf{t}(\mathbf{e}_2)$, but considering a plane defined by $-\mathbf{e}_2$, is also represented in Figure 3.41. This stress represents the action of part (I) onto part (II) whereas $\mathbf{t}(\mathbf{e}_2)$ represents the action of part (II) onto part (I) (note that part (II) was displaced vertically only to permit visualization). Using (3.68) and (3.69) we have that

$$\mathbf{t}(-\mathbf{e}_2) = -\mathbf{t}(\mathbf{e}_2) = -T_{12}\mathbf{e}_1 - T_{22}\mathbf{e}_2 - T_{32}\mathbf{e}_3$$

and the orientation indicated in Figure 3.41 are those corresponding to positive values of T_{i2}. Here T_{ii} represents the normal stress to the plane given by $\mathbf{e}_i$ and -$\mathbf{e}_i$ and a positive value indicates tension, T_{ij}, $i \neq j$, represents a shear stress component in the direction of $\mathbf{e}_i$ for the plane defined by the

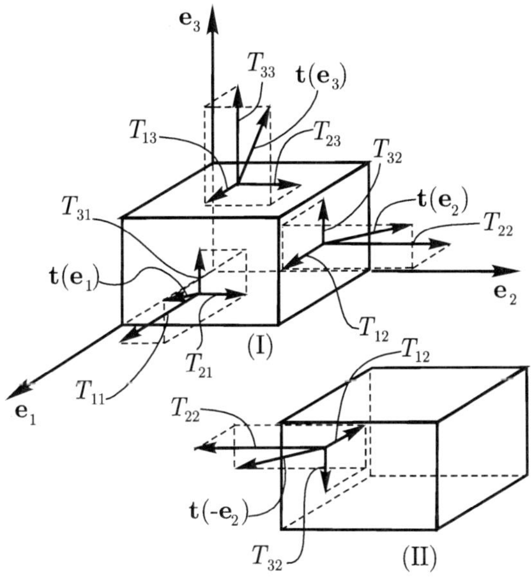

Fig. 3.41. Stress components in planes parallel to coordinate planes

normal $\mathbf{e}_j$ and a positive value indicates that the orientation is the same as $\mathbf{e}_i$. However, when the plane is defined by $-\mathbf{e}_j$, then T_{ij} with a positive value indicates that the orientation is opposite to that given by $\mathbf{e}_i$.

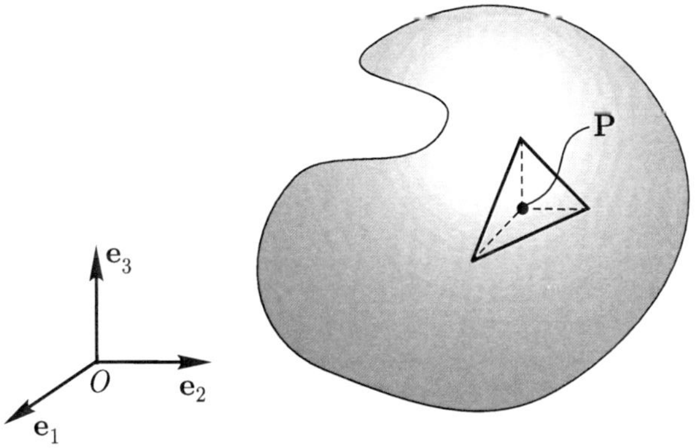

Fig. 3.42. Tetrahedral part to be isolated from the solid

Second consequence of equilibrium

We are now ready to demonstrate an important result which helps to further characterize the state of stress at a point. Consider a solid in static equilibrium and a part of the solid, the tetrahedron, shown in Figure 3.42. The tetrahedron has one vertex denoted by **P**, and as shown three edges are parallel to the coordinate axes. The edges are assumed to be of infinitesimal length. We can isolate the tetrahedron from the solid, and the action of the rest of the solid onto the tetrahedron is represented by the stresses acting on its faces.

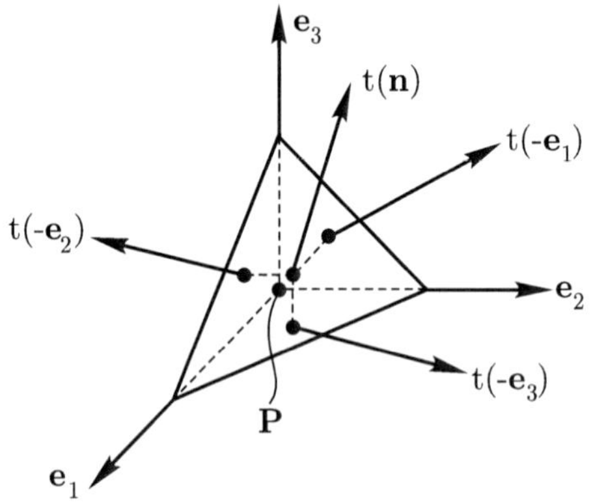

Fig. 3.43. Stress on tetrahedral faces, **n** is the unit outward normal vector on the face S

Therefore, in Figure 3.43, we are representing by $\mathbf{t}\,(-\mathbf{e}_i)$ the stress on the face with exterior normal $-\mathbf{e}_i$ and area S_i $(i = 1, 2, 3)$. The stress acting on the inclined face, which has normal $\mathbf{n}$ and area S, is given by $\mathbf{t}\,(\mathbf{n})$. The stresses are taken to be constant over each face since the edges are of infinitesimal lengths and are given by $\mathbf{t}\,(\mathbf{n})$ and $\mathbf{t}\,(-\mathbf{e}_i)$ $i = 1, 2, 3$ (representing $\mathbf{t}\,(\mathbf{P}, \mathbf{n})$, $\mathbf{t}\,(\mathbf{P}, -\mathbf{e}_i)$ respectively). Similarly, we do not include the spatial variation of the stress from **P** to the inclined face at the distance h from **P** since this variation is given by an infinitesimal of higher order and its contribution would drop out in the derivation not affecting the final result.

Some geometric properties of the tetrahedron are shown in Figure 3.44. Note that $n_i = \mathbf{n} \cdot \mathbf{e}_i = \|\mathbf{n}\| \, \|\mathbf{e}_i\| \cos\alpha_i = \cos\alpha_i$, where α_i is the angle between $\mathbf{n}$ and $\mathbf{e}_i$; $\cos\alpha_1$, $\cos\alpha_2$ and $\cos\alpha_3$ are usually called the direction cosines of $\mathbf{n}$. Therefore the volume of the tetrahedron can be evaluated either taking S as the base and h as the height or S_i as the base and $\frac{h}{n_i}$ as the height leading to

$$V = \frac{1}{3} S h = \frac{1}{3} S_i \left(\frac{h}{n_i} \right) \tag{3.71}$$

from which

$$S_i = n_i S. \tag{3.72}$$

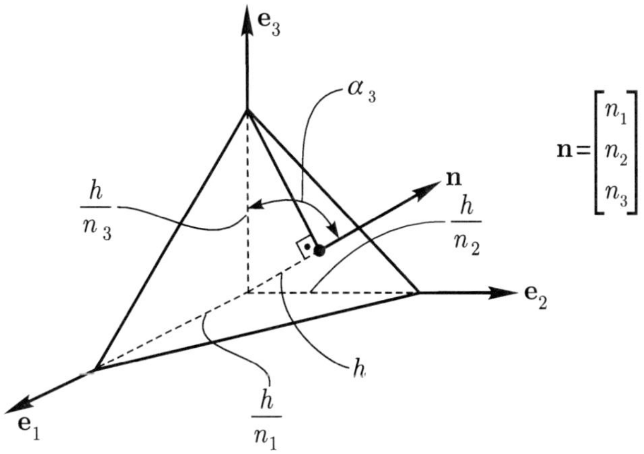

Fig. 3.44. Geometric properties of the tetrahedron of volume V

The tetrahedron must be in equilibrium and, therefore, the resultant of all forces acting on the tetrahedron should be zero, *i.e.*,

$$\mathbf{R} = \mathbf{t}(\mathbf{n})\, S + \mathbf{t}(-\mathbf{e}_1)\, S_1 + \mathbf{t}(-\mathbf{e}_2)\, S_2 + \mathbf{t}(-\mathbf{e}_3)\, S_3 + \mathbf{f}^B V = \mathbf{0} \tag{3.73}$$

where $\mathbf{f}^B$ is the externally applied body force for points inside the tetrahedron which can also be taken as constant, since the edges are of infinitesimal lengths. Dividing (3.73) by S and using (3.71) and (3.72) gives

$$\mathbf{t}(\mathbf{n}) + \mathbf{t}(-\mathbf{e}_1)\, n_1 + \mathbf{t}(-\mathbf{e}_2)\, n_2 + \mathbf{t}(-\mathbf{e}_3)\, n_3 + \frac{1}{3}\mathbf{f}^B h = \mathbf{0}. \tag{3.74}$$

Since h is also infinitesimal (the edges are infinitesimal) we can neglect the term $\frac{1}{3}\mathbf{f}^B h$ with respect to the others terms in (3.74). Using (3.68) we obtain

$$\mathbf{t}(\mathbf{n}) = \mathbf{t}(\mathbf{e}_1)\, n_1 + \mathbf{t}(\mathbf{e}_2)\, n_2 + \mathbf{t}(\mathbf{e}_3)\, n_3. \tag{3.75}$$

The fundamental result is that the stress on the inclined face, defined by the normal $\mathbf{n}$, can be obtained from the stresses on the three planes parallel to the coordinate planes.

Substituting (3.70) into (3.75) results into

$$\mathbf{t}(\mathbf{n}) = \left(\sum_{i=1}^{3} T_{i1}\mathbf{e}_i\right) n_1 + \left(\sum_{i=1}^{3} T_{i2}\mathbf{e}_i\right) n_2 + \left(\sum_{i=1}^{3} T_{i3}\mathbf{e}_i\right) n_3. \tag{3.76}$$

Since $\mathbf{t}(\mathbf{n}) = \sum_{i=1}^{3} t_i \mathbf{e}_i$, we obtain

$$\begin{aligned} t_1 &= T_{11}n_1 + T_{12}n_2 + T_{13}n_3 \\ t_2 &= T_{21}n_1 + T_{22}n_2 + T_{23}n_3 \\ t_3 &= T_{31}n_1 + T_{32}n_2 + T_{33}n_3 \end{aligned}$$

or in matrix form

$$\begin{bmatrix} t_1 \\ t_2 \\ t_3 \end{bmatrix} = \begin{bmatrix} T_{11} & T_{12} & T_{13} \\ T_{21} & T_{22} & T_{23} \\ T_{31} & T_{32} & T_{33} \end{bmatrix} \begin{bmatrix} n_1 \\ n_2 \\ n_3 \end{bmatrix}. \tag{3.77}$$

This relation defines implicitly the *Cauchy stress tensor* $\mathbf{T}$ given by

$$\mathbf{T} = \begin{bmatrix} T_{11} & T_{12} & T_{13} \\ T_{21} & T_{22} & T_{23} \\ T_{31} & T_{32} & T_{33} \end{bmatrix}.$$

We note that the Cauchy stress components listed in $\mathbf{T}$ correspond to the base vectors $\mathbf{e}_1$, $\mathbf{e}_2$ and $\mathbf{e}_3$. Equation (3.77) shows that for any point in the solid we can use $\mathbf{T}$ to calculate the stresses acting on any plane through the point, that is we have

$$\mathbf{t}(\mathbf{n}) = \mathbf{T}\mathbf{n}. \tag{3.78}$$

Third consequence of equilibrium

For the tetrahedron to be in equilibrium it is also necessary to have that the moment of the forces acting on it about any point is equal to zero.

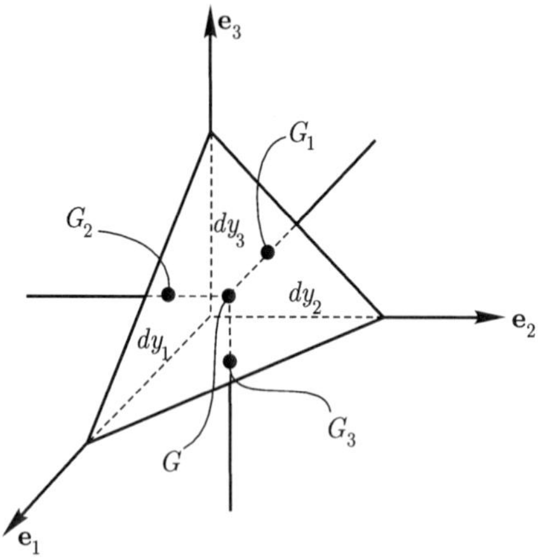

Fig. 3.45. Geometric property that holds for the centers of gravity of tetrahedron faces

We note that since the stresses can be taken as constant over each face of the tetrahedron, the resultant of the stresses on each face should be applied to the center of gravity of the face. Let G_i be the center of gravity of the face S_i and G the center of gravity of the inclined face. It is a geometric property that straight lines parallel to the coordinate axes through the points G_i meet at the center of gravity of the inclined face, as shown in Figure 3.45.

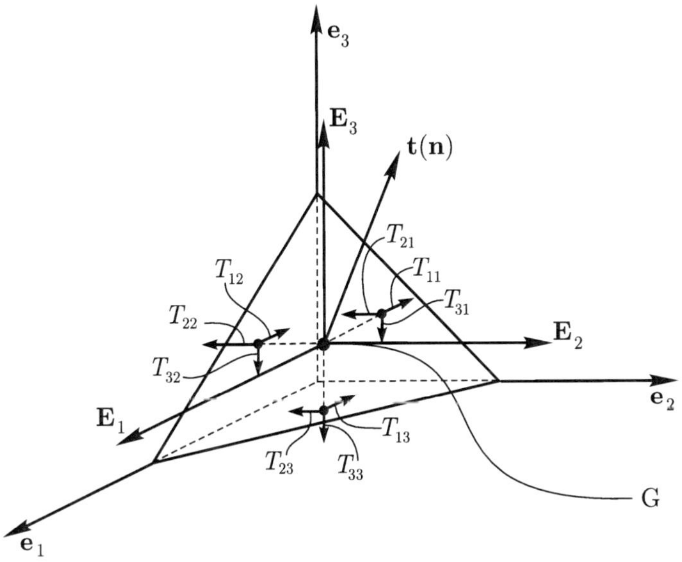

Fig. 3.46. Quantities used for moment balance

Referring to Figure 3.46 and imposing moment equilibrium we obtain

$$\sum M_{E_1} = 0 = T_{32} S_2 \frac{dy_2}{3} - T_{23} S_3 \frac{dy_3}{3}$$

therefore

$$T_{32} \left(\frac{dy_1 dy_3}{2} \right) \frac{dy_2}{3} = T_{23} \left(\frac{dy_1 dy_2}{2} \right) \frac{dy_3}{3}$$

which leads to

$$T_{23} = T_{32}.$$

Note that the moment produced by the body forces was not considered since it corresponds to an infinitesimal quantity of higher–order.

Also

$$\sum M_{E_2} = 0 = -T_{31} S_1 \frac{dy_1}{3} + T_{13} S_3 \frac{dy_3}{3}$$

$$T_{31}\left(\frac{dy_2dy_3}{2}\right)\frac{dy_1}{3} = T_{13}\left(\frac{dy_1dy_2}{2}\right)\frac{dy_3}{3}$$

leading to

$$T_{31} = T_{13}$$

and

$$\sum M_{E_3} = 0 = T_{21}S_1\frac{dy_1}{3} - T_{12}S_2\frac{dy_2}{3}$$

$$T_{21}\left(\frac{dy_2dy_3}{2}\right)\frac{dy_1}{3} = T_{12}\left(\frac{dy_1dy_3}{2}\right)\frac{dy_2}{3}$$

resulting into

$$T_{21} = T_{12}.$$

Hence, based on the moment equilibrium condition we showed that the Cauchy stress tensor $\mathbf{T}$ is symmetric.

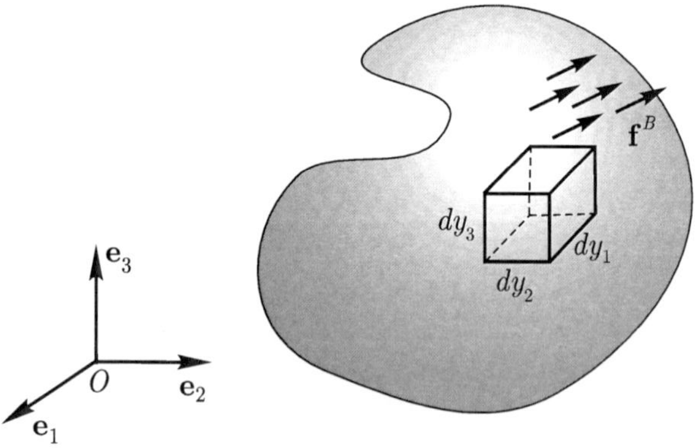

Fig. 3.47. Generic parallelepiped isolated from the solid

3.3.3 Differential equilibrium equations

So far we did not consider any spatial variations in the stresses – which however, of course, exist in almost all analyses. As we mentioned, if the spatial variation of the stresses were included in all above derivations, they would result into higher–order infinitesimal contributions and the final results would be the same as those given. However, we now want to study how every differential element of sides dy_1, dy_2, dy_3 (see Figure 3.47) is in equilibrium when body forces are present and then need to include the spatial variation of the stresses.

Let us isolate the parallelepiped in Figure 3.47 as a free body and consider the stresses on its faces and the body forces in its volume. We show the stresses in three separate figures, 3.48, 3.49 and 3.50 merely to facilitate the visualization; each figure shows the stresses on opposing parallel faces.

Imposing equilibrium in the $\mathbf{e}_1$ direction

$$\begin{aligned}
&-T_{11}dy_2dy_3 + T_{11}dy_2dy_3 + \frac{\partial T_{11}}{\partial y_1}dy_1dy_2dy_3 \\
&-T_{12}dy_1dy_3 + T_{12}dy_1dy_3 + \frac{\partial T_{12}}{\partial y_2}dy_2dy_1dy_3 \\
&-T_{13}dy_1dy_2 + T_{13}dy_1dy_2 + \frac{\partial T_{13}}{\partial y_3}dy_3dy_1dy_2 \\
&+f_1^B dy_1dy_2dy_3 = 0
\end{aligned} \tag{3.79}$$

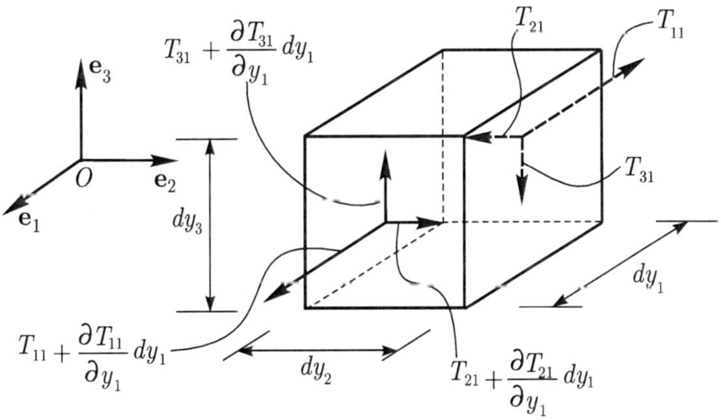

Fig. 3.48. Components of stress in planes dy_1 apart

Since the edges are infinitesimal we may assume the stresses to be constant on each face and the body force $\mathbf{f}_i^B = f_1^B\mathbf{e}_1 + f_2^B\mathbf{e}_2 + f_3^B\mathbf{e}_3$ to be constant inside the parallelepiped. If spatial variations for these quantities were included, they would result in infinitesimal contributions of higher-order and not affect the final equations[9]. Simplifying (3.79) and dividing it by the parallelepiped's volume $dy_1dy_2dy_3$ we obtain

[9] The general approach here, and in Section 3.3.2, is to include only those variations that lead to infinitesimal quantities of low-order and need to be included to extract the final result

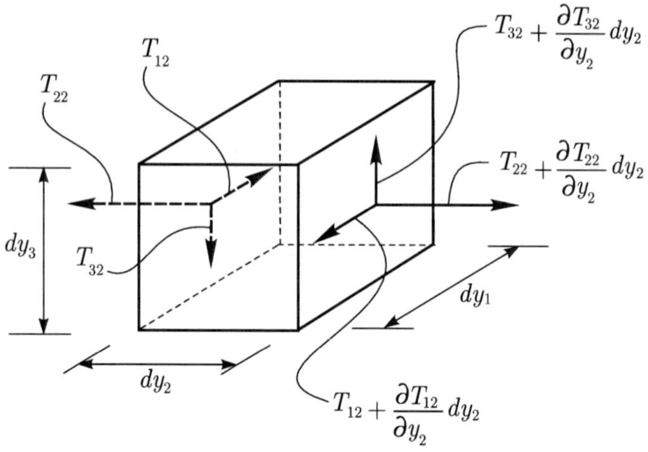

Fig. 3.49. Components of stress in planes dy_2 apart

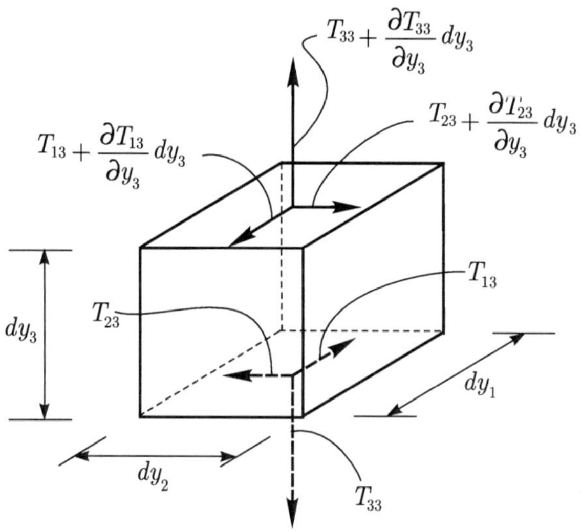

Fig. 3.50. Components of stress in planes dy_3 apart

$$\frac{\partial T_{11}}{\partial y_1} + \frac{\partial T_{12}}{\partial y_2} + \frac{\partial T_{13}}{\partial y_3} + f_1^B = 0 \tag{3.80}$$

Analogously, imposing equilibrium in the directions $\mathbf{e}_2$ and $\mathbf{e}_3$ gives

$$\frac{\partial T_{21}}{\partial y_1} + \frac{\partial T_{22}}{\partial y_2} + \frac{\partial T_{23}}{\partial y_3} + f_2^B = 0 \tag{3.81}$$

$$\frac{\partial T_{31}}{\partial y_1} + \frac{\partial T_{32}}{\partial y_2} + \frac{\partial T_{33}}{\partial y_3} + f_3^B = 0. \tag{3.82}$$

The equations (3.80) to (3.82) are referred to as the differential equilibrium equations, and these need to be satisfied throughout the body.

3.3.4 Principal stresses

Since the stress $\mathbf{t}$ at a point depends on the plane it is acting given by the unit normal vector $\mathbf{n}$, see Figure 3.43 and equation (3.78), natural questions are:

- On which plane do we have the maximum and minimum normal stresses?
- What values do the maximum and minimum normal stresses have?
- On which plane do we have the maximum and minimum shear stresses?
- What values do the maximum and minimum shear stresses have?
- Are there planes on which the shear stress is zero, and if yes, on which planes?

Mathematically, since the stress vector is given by equation (3.78), this last question is answered by solving for t_n and $\mathbf{n}$, the equation[10]

$$\mathbf{T}\mathbf{n} = t_n \mathbf{n}. \tag{3.83}$$

Namely, only when this equation holds, are the shear stresses zero. Equation (3.83) can be re-written as

$$(\mathbf{T} - t_n \mathbf{I})\, \mathbf{n} = \mathbf{0} \tag{3.84}$$

or as

$$\begin{bmatrix} T_{11}-\lambda & T_{12} & T_{13} \\ T_{21} & T_{22}-\lambda & T_{23} \\ T_{31} & T_{32} & T_{33}-\lambda \end{bmatrix} \begin{bmatrix} n_1 \\ n_2 \\ n_3 \end{bmatrix} = \begin{bmatrix} 0 \\ 0 \\ 0 \end{bmatrix}. \tag{3.85}$$

In equation (3.85) we have replaced t_n by λ.

A trivial solution of (3.85) is $n_1 = n_2 = n_3 = 0$, but since $\mathbf{n}$ should be a unit vector, *i.e.*, $\|\mathbf{n}\| = 1$, the trivial solution is of no value. The non-trivial solutions of (3.85) are obtained by enforcing that the coefficient matrix be singular, that is, we enforce

$$\det \begin{bmatrix} T_{11}-\lambda & T_{12} & T_{13} \\ T_{21} & T_{22}-\lambda & T_{23} \\ T_{31} & T_{32} & T_{33}-\lambda \end{bmatrix} = 0 \tag{3.86}$$

which leads to an equation for λ

---

[10] The relation (3.83) is an eigenvalue problem (see *e.g.* Bathe, 1996)

$$\lambda^3 - I_1\lambda^2 + I_2\lambda - I_3 = 0 \tag{3.87}$$

where

$$\begin{aligned} I_1 &= T_{11} + T_{22} + T_{33} \\ I_2 &= \begin{vmatrix} T_{11} & T_{12} \\ T_{21} & T_{22} \end{vmatrix} + \begin{vmatrix} T_{22} & T_{23} \\ T_{32} & T_{33} \end{vmatrix} + \begin{vmatrix} T_{11} & T_{13} \\ T_{31} & T_{33} \end{vmatrix} \\ I_3 &= \begin{vmatrix} T_{11} & T_{12} & T_{13} \\ T_{21} & T_{22} & T_{23} \\ T_{31} & T_{32} & T_{33} \end{vmatrix}. \end{aligned}$$

The scalar quantities I_1, I_2 and I_3 are called the stress invariants since they are independent of the reference system used.

The roots of equation (3.87) are the eigenvalues and are denoted as λ_i, $i = 1, 2$ and 3. For each λ_i we can find an eigenvector $\mathbf{n}_i$ by solving[11] (3.85) with $\lambda = \lambda_i$. It can be proven that, since $\mathbf{T}$ is symmetric, the solutions of (3.87) are always real values and three situations can arise (see Bathe, 1996):

(i) The three values of λ which satisfy (3.87) are distinct, *i.e.*, $\lambda_1 \neq \lambda_2 \neq \lambda_3$. Then the vectors $\mathbf{n}_1$, $\mathbf{n}_2$ and $\mathbf{n}_3$ are automatically orthogonal. In fact, considering $\mathbf{n}_1$ and $\mathbf{n}_2$, for example, we have

$$\begin{aligned} \mathbf{T}\mathbf{n}_1 &= \lambda_1\mathbf{n}_1 \\ \mathbf{T}\mathbf{n}_2 &= \lambda_2\mathbf{n}_2 \end{aligned}$$

and pre multiplying by $\mathbf{n}_2^T$ and $\mathbf{n}_1^T$, respectively, we obtain

$$\mathbf{n}_2^T\mathbf{T}\mathbf{n}_1 = \lambda_1\mathbf{n}_2^T\mathbf{n}_1 \tag{3.88}$$

$$\mathbf{n}_1^T\mathbf{T}\mathbf{n}_2 = \lambda_2\mathbf{n}_1^T\mathbf{n}_2. \tag{3.89}$$

Since $\mathbf{T}$ is symmetric, we have

$$\mathbf{n}_1^T\mathbf{T}\mathbf{n}_2 = \mathbf{n}_2^T\mathbf{T}\mathbf{n}_1 \tag{3.90}$$

and obtain

$$(\lambda_1 - \lambda_2)\,\mathbf{n}_1^T\mathbf{n}_2 = 0. \tag{3.91}$$

Since $\lambda_1 \neq \lambda_2$, $\mathbf{n}_1^T\mathbf{n}_2 = 0$ and hence $\mathbf{n}_1$ and $\mathbf{n}_2$ are orthogonal vectors.

(ii) Assume $\lambda_1 = \lambda_2$, $\lambda_3 \neq \lambda_1$. Then there is a unique $\mathbf{n}_3$ corresponding to λ_3. Also, it can be proven that any unit vector in a plane orthogonal to $\mathbf{n}_3$ is a vector that satisfies (3.85) for $\lambda = \lambda_1 = \lambda_2$. Hence we can choose two such vectors, orthogonal to each other, and assign them to correspond to λ_1 and λ_2.

[11] In this solution we enforce $\|\mathbf{n}_i\|$ to be equal to 1

(iii) Assume $\lambda_1 = \lambda_2 = \lambda_3$. In this case any unit vector $\mathbf{n}$ satisfies (3.85) for $\lambda = \lambda_1 = \lambda_2 = \lambda_3$ and we can choose three orthogonal vectors and assign them to correspond to λ_1, λ_2 and λ_3.

From these properties, it is easy to see that we can always find three orthogonal unit vectors, $\mathbf{n}_1$, $\mathbf{n}_2$ and $\mathbf{n}_3$ which satisfy (3.85) and which are associated with the solutions of (3.87), *i.e.*, λ_1, λ_2 and λ_3. We order λ_1, λ_2 and λ_3 according to their algebraic values

$$\lambda_1 \geq \lambda_2 \geq \lambda_3.$$

If instead of using the orthonormal unit vectors $\mathbf{e}_i$, we now use the orthonormal vectors $\mathbf{n}_i$ to define our reference system[12], then the stress tensor in such a system reads

$$\mathbf{T} = \begin{bmatrix} \lambda_1 & 0 & 0 \\ 0 & \lambda_2 & 0 \\ 0 & 0 & \lambda_3 \end{bmatrix}. \tag{3.92}$$

Of course, the stress tensor given in (3.92), tells that there are no shear stresses on the planes with normals $\mathbf{n}_1$, $\mathbf{n}_2$ and $\mathbf{n}_3$.

Let us now address the question of determining the maximum and minimum values of the normal stresses. Recall that the normal stress acting on a plane with normal $\mathbf{n}$ is given by

$$t_n = \mathbf{n} \cdot \mathbf{t} = \mathbf{n} \cdot \mathbf{T}\mathbf{n} = \mathbf{n}^T\mathbf{T}\mathbf{n}. \tag{3.93}$$

To find the maximum and minimum values of t_n that can be reached by varying $\mathbf{n}$, it is effective to evaluate t_n as a function of the normal $\mathbf{n}$ using the reference system defined by $\mathbf{n}_1$, $\mathbf{n}_2$ and $\mathbf{n}_3$, *i.e.*,

$$t_n(\mathbf{n}) = \mathbf{n}^T\mathbf{T}\mathbf{n} = \begin{bmatrix} n_1 & n_2 & n_3 \end{bmatrix} \begin{bmatrix} \lambda_1 & 0 & 0 \\ 0 & \lambda_2 & 0 \\ 0 & 0 & \lambda_3 \end{bmatrix} \begin{bmatrix} n_1 \\ n_2 \\ n_3 \end{bmatrix}.$$

Hence

$$t_n = \lambda_1 (n_1)^2 + \lambda_2 (n_2)^2 + \lambda_3 (n_3)^2 \tag{3.94}$$

Since $\mathbf{n}$ is a unit vector

$$(n_1)^2 + (n_2)^2 + (n_3)^2 = 1$$

[12] Orthogonal vectors $\mathbf{n}_i$ and $\mathbf{n}_j$ are also orthonormal if $\mathbf{n}_i^T\mathbf{n}_j = \delta_{ij}$ (δ_{ij}, the Kronecker delta, $= 1$ for $i = j$ and $= 0$ for $i \neq j$)

and substituting in (3.94)

$$(n_1)^2 = 1 - (n_2)^2 - (n_3)^2$$

gives

$$t_n = \lambda_1 \left(1 - (n_2)^2 - (n_3)^2\right) + \lambda_2 (n_2)^2 + \lambda_3 (n_3)^2$$

or

$$t_n = \lambda_1 + (\lambda_2 - \lambda_1)(n_2)^2 + (\lambda_3 - \lambda_1)(n_3)^2 . \tag{3.95}$$

Since $\lambda_1 \geq \lambda_2 \geq \lambda_3$ we can conclude that

$$t_n \leq \lambda_1$$

and hence the maximum value of t_n that can be reached by varying the plane at the point (varying $\mathbf{n}$) is λ_1 and, of course, is reached for the plane with normal $\mathbf{n}_1$. Analogously, substituting $(n_3)^2 = 1 - (n_1)^2 - (n_2)^2$ into (3.94) leads to

$$t_n = \lambda_3 + (\lambda_1 - \lambda_3)(n_1)^2 + (\lambda_1 - \lambda_2)(n_2)^2$$

which implies that

$$t_n \geq \lambda_3$$

and hence λ_3 is the minimum value[13] of t_n and, of course, is reached for the plane with normal $\mathbf{n}_3$.

Since λ_1 and λ_3 are the maximum and minimum stress values that can be reached, we call λ_1, λ_2 and λ_3 principal values and denote these principal stresses as τ_1, τ_2 and τ_3 respectively, and the associated normals $\mathbf{n}_1$, $\mathbf{n}_2$, $\mathbf{n}_3$ the principal stress directions.

Example 3.6

Consider the cylinder as shown in Figure 3.32. The solid cylinder is subjected to self-equilibrating torsional moments $\mathbf{M}_t$ at the end sections, *i.e.*, $\mathbf{M}_t = M_t \mathbf{e}_x$ at $x = L$ and $\mathbf{M}_t = -M_t \mathbf{e}_x$ at $x = 0$. The stresses at a generic cross-section are given in Figure 3.51 (see Section 3.6). Therefore, at a generic cross-section there is only a shear stress distribution and the maximum value τ is related to the twisting moment by $\tau = \frac{M_t}{I_t} R$ where $I_t = \frac{\pi R^4}{2}$. Let us consider the point $\mathbf{P}$ with coordinates $y = R$, $z = 0$, $0 \leq x \leq L$. Evaluate the principal stresses and the principal stress directions.

[13] The expression $t_n(\mathbf{n}) = \mathbf{n}^T \mathbf{T} \mathbf{n}$ is actually a Rayleigh quotient, and we have in general for the Rayleigh quotient $\rho(\mathbf{v}) = \mathbf{v}^T \mathbf{A} \mathbf{v}$, with $\mathbf{v}^T \mathbf{v} = 1$ and $\mathbf{A}$ a symmetric matrix of order n, that $\lambda_n \leq \rho(\mathbf{v}) \leq \lambda_1$, where λ_n and λ_1 are the smallest and largest eigenvalues of $\mathbf{A}$, see Bathe, 1996 for a proof

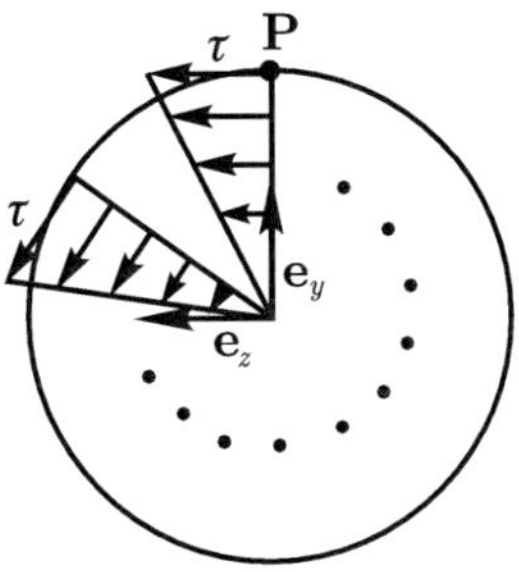

Fig. 3.51. Shear stress for a cross-section of a cylinder in torsion

Solution

The stress tensor at point **P** is given by

$$\mathbf{T} = \begin{bmatrix} 0 & 0 & \tau \\ 0 & 0 & 0 \\ \tau & 0 & 0 \end{bmatrix}$$

and the principal stresses can be obtained solving

$$\det(\mathbf{T}-\lambda\mathbf{I}) = \det \begin{bmatrix} -\lambda & 0 & \tau \\ 0 & -\lambda & 0 \\ \tau & 0 & -\lambda \end{bmatrix} = 0$$

which leads to

$$-\lambda\left(\lambda^2 - \tau^2\right) = 0.$$

The roots of this equation are $\lambda_1 = \tau$, $\lambda_2 = 0$ and $\lambda_3 = -\tau$. Hence, the principal stresses are

$$\tau_1 = \tau,\ \tau_2 = 0,\ \tau_3 = -\tau.$$

To obtain the principal stress directions, we need to solve

$$(\mathbf{T}-\tau_i\mathbf{I})\mathbf{n}_i = \mathbf{0} \qquad \text{with} \qquad \|\mathbf{n}_i\| = 1$$

for $i = 1$ to 3. For the first principal stress, $\tau_1 = \tau$ we have

$$\begin{bmatrix} -\tau & 0 & \tau \\ 0 & -\tau & 0 \\ \tau & 0 & -\tau \end{bmatrix} \begin{bmatrix} n_1 \\ n_2 \\ n_3 \end{bmatrix} = \begin{bmatrix} 0 \\ 0 \\ 0 \end{bmatrix} \tag{3.96}$$

or

$$\begin{cases} -\tau n_1 + \tau n_3 = 0 \\ -\tau n_2 = 0 \\ \tau n_1 - \tau n_3 = 0 \end{cases} .$$

Of course, the first and third equations in the above system are the same. Considering the first and the second equations, we obtain

$$n_1 = n_3, \qquad n_2 = 0$$

and imposing that

$$n_1^2 + n_2^2 + n_3^2 = 1 \tag{3.97}$$

we obtain

$$2n_1^2 = 1 \quad \Rightarrow \quad n_1 = \pm\frac{\sqrt{2}}{2}.$$

Choosing $n_1 = \frac{\sqrt{2}}{2}$ we have

$$\mathbf{n}_1 = \frac{\sqrt{2}}{2}\mathbf{e}_x + \frac{\sqrt{2}}{2}\mathbf{e}_z.$$

Note that if we had used $n_1 = -\frac{\sqrt{2}}{2}$, then we would have simply obtained $-\mathbf{n}_1$ as the solution. Considering τ_2 we have

$$\begin{bmatrix} 0 & 0 & \tau \\ 0 & 0 & 0 \\ \tau & 0 & 0 \end{bmatrix} \begin{bmatrix} n_1 \\ n_2 \\ n_3 \end{bmatrix} = \begin{bmatrix} 0 \\ 0 \\ 0 \end{bmatrix}$$

or

$$\begin{cases} \tau n_3 = 0 \\ 0 = 0 \\ \tau n_1 = 0 \end{cases}$$

which yields

$$n_1 = 0, \qquad n_3 = 0$$

and equation (3.97) gives

$$n_2 = \pm 1.$$

Therefore we can select

$$\mathbf{n}_2 = \mathbf{e}_y.$$

Finally, for τ_3 we have

$$\begin{bmatrix} \tau & 0 & \tau \\ 0 & \tau & 0 \\ \tau & 0 & \tau \end{bmatrix} \begin{bmatrix} n_1 \\ n_2 \\ n_3 \end{bmatrix} = \begin{bmatrix} 0 \\ 0 \\ 0 \end{bmatrix}$$

or

$$\begin{cases} \tau n_1 + \tau n_3 = 0 \\ \tau n_2 = 0 \\ \tau n_1 + \tau n_3 = 0 \end{cases}$$

which gives

$$n_1 = -n_3, \qquad n_2 = 0$$

and substituting in (3.97)

$$n_1 = \pm\frac{\sqrt{2}}{2}$$

Selecting $n_1 = -\frac{\sqrt{2}}{2}$ yields $n_3 = \frac{\sqrt{2}}{2}$ and we have in summary the following orthonormal vectors defining the principal directions, *i.e.*,

$$\begin{aligned} \mathbf{n}_1 &= \frac{\sqrt{2}}{2}\mathbf{e}_x + \frac{\sqrt{2}}{2}\mathbf{e}_z \\ \mathbf{n}_2 &= \mathbf{e}_y \\ \mathbf{n}_3 &= -\frac{\sqrt{2}}{2}\mathbf{e}_x + \frac{\sqrt{2}}{2}\mathbf{e}_z. \end{aligned}$$

These can be used as new base vectors of a new reference coordinate system, and in this system,

$$\mathbf{T} = \begin{bmatrix} \tau & 0 & 0 \\ 0 & 0 & 0 \\ 0 & 0 & -\tau \end{bmatrix}.$$

Of course, since τ_1 and τ_3 are the maximum and minimum normal stresses, we have

$$\max(t_n) = \tau$$
$$\min(t_n) = -\tau.$$

□

Mohr's circles

In order to obtain further insight into the state of stress at a point, we study below the stresses for planes which contain one of the principal directions. For example, let us consider a generic plane π which contains $\mathbf{n}_3$. This situation is summarized in Figure 3.52.

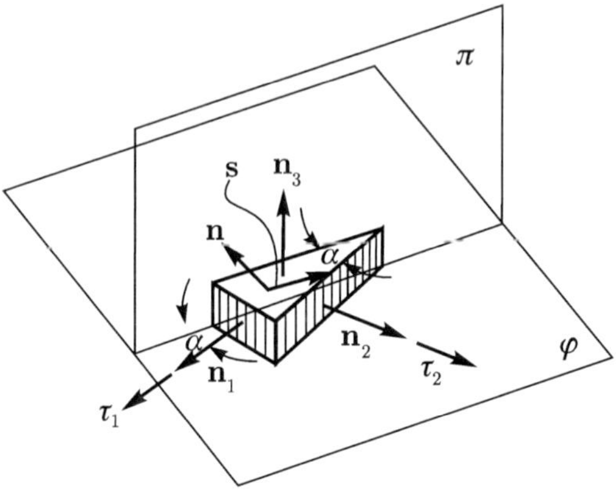

Fig. 3.52. Definition of a generic plane π which contains $\mathbf{n}_3$; the angle α is varying; the stress state in the plane π is analyzed

Since $\mathbf{n}_1$ and $\mathbf{n}_2$ are orthogonal to $\mathbf{n}_3$, they lie in a plane orthogonal to π which we call φ. Let α be the angle between π and $\mathbf{n}_1$ measured in the clockwise sense from $\mathbf{n}_1$, see Figure 3.52. The stress vector acting on π can be decomposed in the normal and tangential directions, *i.e.*,

$$\mathbf{t} = \mathbf{t}_n + \mathbf{t}_s = t_n \mathbf{n} + t_s \mathbf{s}$$

where $\mathbf{n}$ is the normal to π and we let $\mathbf{s}$ be a unit vector in the direction of $\mathbf{t}_s$. The sense of $\mathbf{s}$ is chosen such that when t_s is positive it tends to rotate the prism of triangular base shown in Figure 3.52 in the clockwise direction. The unit vectors $\mathbf{n}$ and $\mathbf{s}$ and hence $\mathbf{t}$ lie in plane φ and the stress on the plane π is

$$\mathbf{t} = \mathbf{T}\mathbf{n} = \begin{bmatrix} \tau_1 & 0 & 0 \\ 0 & \tau_2 & 0 \\ 0 & 0 & \tau_3 \end{bmatrix} \begin{bmatrix} -\sin\alpha \\ -\cos\alpha \\ 0 \end{bmatrix} = \begin{bmatrix} -\tau_1 \sin\alpha \\ -\tau_2 \cos\alpha \\ 0 \end{bmatrix}$$

where we used the reference system defined by $\mathbf{n}_1$, $\mathbf{n}_2$ and $\mathbf{n}_3$. Here α is an angle that we consider to vary. We can evaluate t_n using

$$t_n = \mathbf{n}^T\mathbf{T}\mathbf{n} = \begin{bmatrix} -\sin\alpha & -\cos\alpha & 0 \end{bmatrix} \begin{bmatrix} \tau_1 & 0 & 0 \\ 0 & \tau_2 & 0 \\ 0 & 0 & \tau_3 \end{bmatrix} \begin{bmatrix} -\sin\alpha \\ -\cos\alpha \\ 0 \end{bmatrix}$$

which leads to

$$t_n = \tau_1 \sin^2\alpha + \tau_2 \cos^2\alpha$$

and using the identities $\sin^2\alpha + \cos^2\alpha = 1$ and $\cos^2\alpha = \frac{1+\cos 2\alpha}{2}$, we obtain

$$t_n = \frac{1}{2}(\tau_1 + \tau_2) - \frac{1}{2}(\tau_1 - \tau_2)\cos 2\alpha. \tag{3.98}$$

We can also evaluate t_s

$$t_s = = \mathbf{s}^T\mathbf{T}\mathbf{n} = \begin{bmatrix} -\cos\alpha & \sin\alpha & 0 \end{bmatrix} \begin{bmatrix} \tau_1 & 0 & 0 \\ 0 & \tau_2 & 0 \\ 0 & 0 & \tau_3 \end{bmatrix} \begin{bmatrix} -\sin\alpha \\ -\cos\alpha \\ 0 \end{bmatrix}$$

which yields

$$t_s = (\tau_1 - \tau_2)\cos\alpha \sin\alpha = \frac{1}{2}(\tau_1 - \tau_2)\sin 2\alpha. \tag{3.99}$$

Equations (3.98) and (3.99) allow a graphical representation of the pair (t_n, t_s) as shown in Figure 3.53.

This representation is referred to as Mohr's Circle and represents the variation of stress in a plane orthogonal to a principal stress direction (this principal stress might be zero, as in a plane stress analysis, see Section 4.1.2).

If we consider the variation of the stress in planes that contain the principal directions $\mathbf{n}_1$ and $\mathbf{n}_2$, we obtain the additional results summarized in Figure 3.54. It is possible to show that the pairs (t_n, t_s) for planes which do not contain any of the principal directions are in the dashed region of Figure 3.54. This result permits to determine the maximum absolute value of the shear stress, *i.e.*,

$$t_{s_{\max}} = \max(\|t_s\|) = \frac{(\tau_1 - \tau_3)}{2}.$$

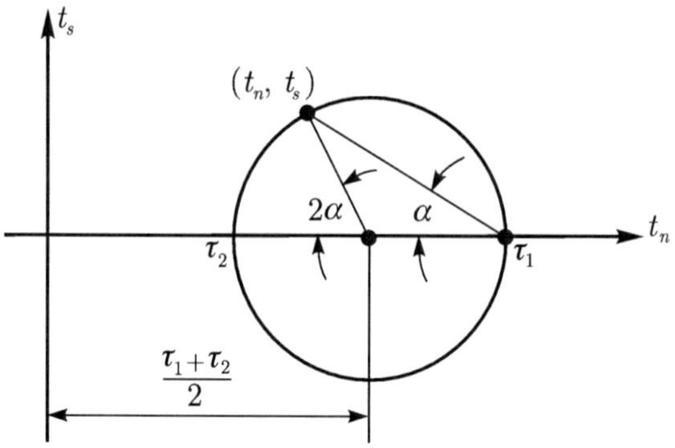

Fig. 3.53. Mohr's circle graphical representation for pairs (t_n, t_s)

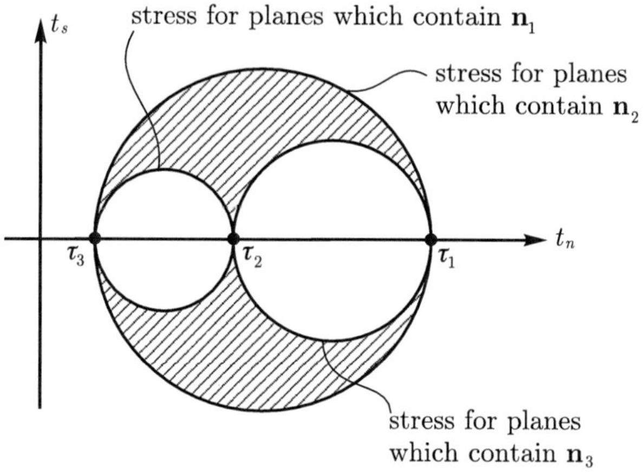

Fig. 3.54. Mohr's Circles for planes which contain one of the principal stress directions. Dashed region represents allowed pairs (t_n, t_s)

3.3.5 Principal strains

The results derived in the study of the stress at a point can directly be used to also study the strains at a point and evaluate the principal strain values.

As we mentioned, an eigenvalue problem was considered, see (3.83) with $\mathbf{T}$ a symmetric tensor. The infinitesimal strain tensor $\mathbf{E}$ is also symmetric and hence we can consider the problem of finding the eigenvalues and eigenvectors of

$$\mathbf{E}\mathbf{x} = \lambda\mathbf{x}.$$

The eigenvalues are represented by

$$\varepsilon_1 \geq \varepsilon_2 \geq \varepsilon_3$$

and are called the principal strains. The eigenvectors which can be denoted also by $\mathbf{h}_1$, $\mathbf{h}_2$ and $\mathbf{h}_3$ define the principal strain directions. In the reference system defined by the principal strain directions, $\mathbf{E}$ is given by

$$\mathbf{E} = \begin{bmatrix} \varepsilon_1 & 0 & 0 \\ 0 & \varepsilon_2 & 0 \\ 0 & 0 & \varepsilon_3 \end{bmatrix}.$$

We note that the off–diagonal terms are zero hence there are no shear strains between fibers aligned with the principal strain directions. We also recall that the strain of a fiber whose direction is defined by the unit vector $\mathbf{m}$ is given by

$$\varepsilon(\mathbf{m}) = \mathbf{m}^T \mathbf{E} \mathbf{m}.$$

Therefore, referring to (3.93) we can conclude that ε_1 is the maximum normal strain and ε_3 is the minimum normal strain and these occur for fibers in the directions of $\mathbf{h}_1$ and $\mathbf{h}_3$ respectively.

3.3.6 Infinitesimally small displacements

The discussion presented so far in Section 3.3 is valid for arbitrarily large displacements. In fact, the concept of stress was introduced for a generic deformed configuration of the solid. However, as discussed in Section 2.1.5, when we consider infinitesimally small displacements, the principles of linear and angular momenta are written for the undeformed configuration. Then, the conditions of equilibrium apply for the undeformed configuration, that is, as if the solid had not displaced.

Hence, when we consider infinitesimally small displacements all the discussion presented in Section 3.3 is valid considering the undeformed configuration. In particular, the differential equilibrium equations now read

$$\begin{aligned}
\frac{\partial T_{11}}{\partial x_1} + \frac{\partial T_{12}}{\partial x_2} + \frac{\partial T_{13}}{\partial x_3} + f_1^B &= 0 \\
\frac{\partial T_{21}}{\partial x_1} + \frac{\partial T_{22}}{\partial x_2} + \frac{\partial T_{23}}{\partial x_3} + f_2^B &= 0 \\
\frac{\partial T_{31}}{\partial x_1} + \frac{\partial T_{32}}{\partial x_2} + \frac{\partial T_{33}}{\partial x_3} + f_3^B &= 0.
\end{aligned}$$

The formulation of the mathematical model for three-dimensional solids is greatly simplified for infinitesimally small displacement conditions. One of the reasons is that equilibrium is enforced in the undeformed configuration which is known.

3.3.7 Technical or engineering notation for the stresses

In the technical or engineering notation, which is commonly used in the engineering literature, the coordinate axes are denoted by x, y and z, *i.e.*, $x \equiv x_1$, $y \equiv x_2$ and $z \equiv x_3$ and the stress components are represented by τ_{xx}, τ_{yy}, τ_{zz}, $\tau_{xy} = \tau_{yx}$, $\tau_{xz} = \tau_{zx}$ and $\tau_{yz} = \tau_{zy}$.

3.4 Constitutive equations

In this section, we introduce the characterization of the material of deformable solids. Recall that, both, the study of the deformations in Section 3.2 and the study of the stresses in Section 3.3 were carried out without having specified a material behavior for the solid. But in the analysis of the truss structures (see Section 2.2.4) and the analysis of the steel sheet (Section 3.1) we used the fact, justified by experimental observations, that the stresses induce strains.

For the one-dimensional stress conditions in the bars, the truss structures could be completely analyzed using Hooke's law relating the one-dimensional normal stress to the one-dimensional normal strain.

For the analysis of the steel sheet, we introduced the Poisson's ratio effect and the problem could be solved only for very simple conditions. Actually, the observed interaction between the strains and the stresses prompted to undertake a 3-D study of the deformations and stresses.

When we consider a three-dimensional solid undergoing arbitrarily large displacements, the constitutive equation is defined as the relation that gives the Cauchy stress at a point in a given configuration as a function of the complete history of deformation, that is, from the initial configuration of the solid up to the current configuration (see Bathe, 1996 and Kojic and Bathe, 2005). Such a general constitutive equation can be used to describe not only usual structural engineering materials but also very unusual ones.

The concept explored for the 1-D problem, that for an elastic material the normal stress depends only on the current state of deformation given by the current normal strain, can be generalized to define a constitutive relation for elastic materials in 3-D conditions. Considering three-dimensional solids undergoing arbitrarily large displacements, we can define a constitutive relation for elastic behavior by

$$\mathbf{T} = \mathcal{F}(\mathbf{X}) \qquad \text{for every point of the solid}$$

that is, the Cauchy stress at a point in the current configuration depends on the deformation gradient $\mathbf{X}$ at that point through a function $\mathcal{F}$.

However, our present objective is to introduce the elastic behavior considering infinitesimally small displacements. This is achieved by relating, for a generic point in the solid, the state of stress in the undeformed configuration to the state of strain for infinitesimally small displacement conditions.

Our discussion is restricted to linear elastic materials, *i.e.*, to those materials for which the deformation is immediate (with loading) and reversible, and moreover the relation between the stresses and strains is linear. We also assume that the material is isotropic, *i.e.*, it behaves in the same manner in all directions, and homogenous, *i.e.*, its behavior does not change from point to point. Under these conditions the relation between the stresses and strains can be represented by the generalized Hooke's law which we detail below.

3.4.1 Hooke's law for three-dimensional isotropic material conditions

Consider a parallelepiped of differential volume $dx_1 dx_2 dx_3$ subjected to a uniform one-dimensional state of stress as shown in Figure 3.55a. Since this is a one-dimensional state of normal stress we can use Hooke's law

$$\varepsilon = \frac{t_n}{E}$$

where E is the Young modulus. In the two directions orthogonal to the di-

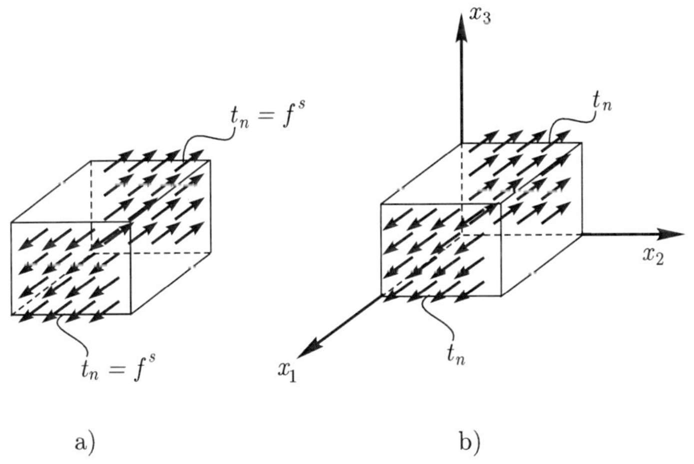

Fig. 3.55. a) Surface forces inducing an uniform state of stress; b) Placement of the parallelepiped with respect to the coordinate axes

rection of the applied stress a contraction takes place due to Poisson's effect given by $-\nu\frac{t_n}{E}$, where ν is the Poisson ratio. If we place the parallelepiped with respect to the coordinate axes shown in Figure 3.55b the Cauchy stress tensor would be

$$\mathbf{T} = \begin{bmatrix} t_n & 0 & 0 \\ 0 & 0 & 0 \\ 0 & 0 & 0 \end{bmatrix}$$

i.e.,

$$T_{11} = t_n$$

and the infinitesimal strain tensor would be given by

$$\mathbf{E} = \begin{bmatrix} \frac{t_n}{E} & 0 & 0 \\ 0 & -\nu\frac{t_n}{E} & 0 \\ 0 & 0 & -\nu\frac{t_n}{E} \end{bmatrix}$$

or

$$E_{11} = \frac{t_n}{E} = \frac{T_{11}}{E}$$

and

$$E_{22} = E_{33} = -\nu\frac{t_n}{E} = -\nu\frac{T_{11}}{E} = -\nu E_{11}.$$

If we consider analogous situations for the directions $\mathbf{e}_2$ and $\mathbf{e}_3$, and use superposition, we can write

$$E_{11} = \frac{T_{11}}{E} - \frac{\nu}{E}(T_{22} + T_{33}) \tag{3.100}$$

$$E_{22} = \frac{T_{22}}{E} - \frac{\nu}{E}(T_{11} + T_{33}) \tag{3.101}$$

$$E_{33} = \frac{T_{33}}{E} - \frac{\nu}{E}(T_{11} + T_{22}) \tag{3.102}$$

where T_{22} and T_{33} are normal stresses for the planes with normals $\mathbf{e}_2$ and $\mathbf{e}_3$, respectively. We note that an isotropic behavior was implicitly assumed since the same behavior was assumed for the three directions (expressed by E and ν). For material with non-isotropic behavior, the Young's modulus and Poisson's ratios depend on the material direction considered, see Crandall, Dahl and Lardner, 1978 and Jones, 1975. We will focus in this book on isotropic material behaviors.

Now let us suppose that shearing tractions shown in Figure 3.56a are applied to the surface of the parallelepiped. The stress tensor is given by

$$\mathbf{T} = \begin{bmatrix} 0 & t_s & 0 \\ t_s & 0 & 0 \\ 0 & 0 & 0 \end{bmatrix}$$

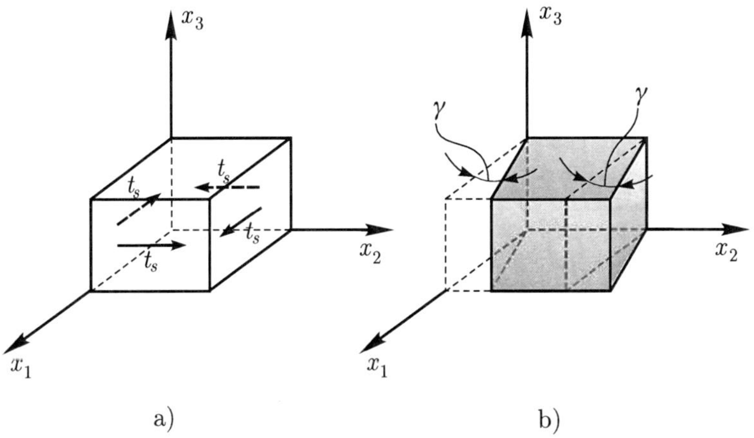

Fig. 3.56. a) Uniform shearing state of stress; b) Deformation associated with uniform shearing state of stress

It is an experimentally supported observation that the deformation induced by this state of stress is that given in Figure 3.56b[14], *i.e.*, a shear strain is induced and the strain tensor is given by

$$\mathbf{E} = \begin{bmatrix} 0 & \frac{\gamma}{2} & 0 \\ \frac{\gamma}{2} & 0 & 0 \\ 0 & 0 & 0 \end{bmatrix}$$

as detailed in Example 3.2.

The relation between the shear stress and the shear strain is

$$\gamma = \frac{t_s}{G} \tag{3.103}$$

where G is a material constant called the shear modulus. Relation (3.103) expressed in components reads

$$2E_{12} = \frac{T_{12}}{G} \tag{3.104}$$

and considering analogous uniform shearing stress states corresponding to T_{13} and T_{23}, we have

$$2E_{13} = \frac{T_{13}}{G} \tag{3.105}$$

[14] Of course, an infinitesimal rigid rotation could be added to the deformation which would not affect the resulting $\mathbf{E}$ (see Section 3.2.5)

$$2E_{23} = \frac{T_{23}}{G}. \tag{3.106}$$

Collecting equations (3.100) to (3.102) and (3.104) to (3.106), we have the (so-called) *Generalized Hooke's law* which relates the stress components to the strain components

$$\begin{aligned}
E_{11} &= \frac{T_{11}}{E} - \frac{\nu}{E}(T_{22} + T_{33}) \\
E_{22} &= \frac{T_{22}}{E} - \frac{\nu}{E}(T_{11} + T_{33}) \\
E_{33} &= \frac{T_{33}}{E} - \frac{\nu}{E}(T_{11} + T_{22}) \\
E_{12} &= \frac{T_{12}}{2G} \quad \text{or} \quad \gamma_{12} = \frac{T_{12}}{G} \\
E_{13} &= \frac{T_{13}}{2G} \quad \text{or} \quad \gamma_{13} = \frac{T_{13}}{G} \\
E_{23} &= \frac{T_{23}}{2G} \quad \text{or} \quad \gamma_{23} = \frac{T_{23}}{G}.
\end{aligned}$$

When the solid is subjected to changes in its temperature field, thermal strains are induced. Let α_{th} be the thermal expansion coefficient and

$$\Delta\theta(\mathbf{x}) = \theta_f - \theta_i$$

be the field of temperature changes, where θ_i and θ_f are the initial and final temperature fields respectively. The induced thermal strain field is given by

$$\mathbf{E}^{th} = \begin{bmatrix} \alpha_{th}\Delta\theta & 0 & 0 \\ 0 & \alpha_{th}\Delta\theta & 0 \\ 0 & 0 & \alpha_{th}\Delta\theta \end{bmatrix}.$$

Then, the Generalized Hooke's law reads

$$\begin{aligned}
E_{11} &= \frac{T_{11}}{E} - \frac{\nu}{E}(T_{22} + T_{33}) + \alpha_{th}\Delta\theta \\
E_{22} &= \frac{T_{22}}{E} - \frac{\nu}{E}(T_{11} + T_{33}) + \alpha_{th}\Delta\theta \\
E_{33} &= \frac{T_{33}}{E} - \frac{\nu}{E}(T_{11} + T_{22}) + \alpha_{th}\Delta\theta \\
E_{12} &= \frac{T_{12}}{2G} \quad \text{or} \quad \gamma_{12} = \frac{T_{12}}{G}
\end{aligned}$$

$$E_{13} = \frac{T_{13}}{2G} \quad \text{or} \quad \gamma_{13} = \frac{T_{13}}{G}$$

$$E_{23} = \frac{T_{23}}{2G} \quad \text{or} \quad \gamma_{23} = \frac{T_{23}}{G}.$$

Of course, the strains above are the total strains, that is, the strains due to the stress field – the mechanical strains – plus those due to the temperature changes – the thermal strains.

3.4.2 Relation between G and E, ν

The following analysis shows that there is a relation between the shear modulus G, Young's modulus E and Poisson's ratio ν.

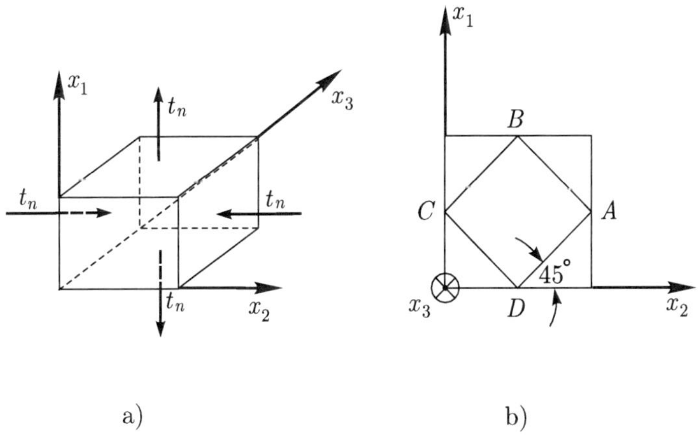

Fig. 3.57. a) Uniform state of stress in a cube; b) Section of the cube by a cutting plane parallel to x_1x_2. The square $ABCD$ represents the trace of the parallelepiped on the cutting plane

Consider the uniform state of stress in the cube shown in Figure 3.57a. The stress tensor is given by

$$\mathbf{T} = \begin{bmatrix} t_n & 0 & 0 \\ 0 & -t_n & 0 \\ 0 & 0 & 0 \end{bmatrix}. \tag{3.107}$$

Let us select a parallelepiped in the interior of the cube of Figure 3.57a as shown in Figure 3.57b.

The stress acting on the face defined by AB can be evaluated by

$$\mathbf{t} = \mathbf{T}\mathbf{n} = \begin{bmatrix} t_n & 0 & 0 \\ 0 & -t_n & 0 \\ 0 & 0 & 0 \end{bmatrix} \begin{bmatrix} \frac{\sqrt{2}}{2} \\ \frac{\sqrt{2}}{2} \\ 0 \end{bmatrix} = t_n \begin{bmatrix} \frac{\sqrt{2}}{2} \\ -\frac{\sqrt{2}}{2} \\ 0 \end{bmatrix}.$$

This stress is tangential to the parallelepiped's face since

$$\mathbf{n}^T\mathbf{T}\mathbf{n} = \begin{bmatrix} \frac{\sqrt{2}}{2} & \frac{\sqrt{2}}{2} & 0 \end{bmatrix} t_n \begin{bmatrix} \frac{\sqrt{2}}{2} \\ -\frac{\sqrt{2}}{2} \\ 0 \end{bmatrix} = 0$$

and its direction is given by the unit vector $\mathbf{s}$, $\mathbf{s}^T = \begin{bmatrix} \frac{\sqrt{2}}{2} & -\frac{\sqrt{2}}{2} & 0 \end{bmatrix}$. It is therefore a shear stress given by

$$\mathbf{t}_s = t_s\mathbf{s}$$

with $t_s = t_n$.

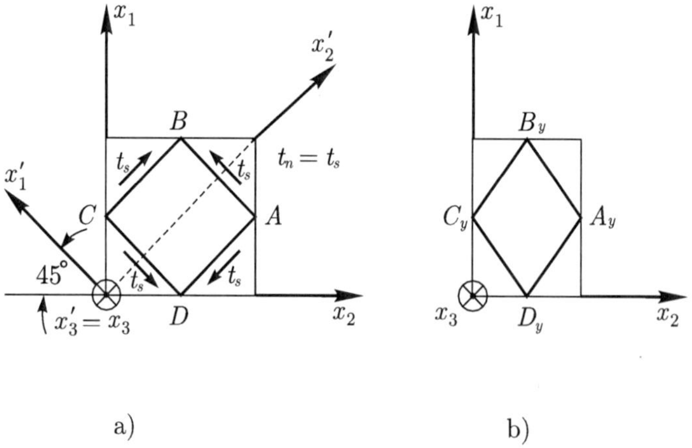

Fig. 3.58. a) Stress acting on the parallelepiped faces; b) Induced deformation

Analogous derivations for the other three faces of the parallelepiped lead to the situation shown in Figure 3.58a. If we choose a new set of axes to represent the stress tensor denoted by x_1', x_2' and x_3' with base vector ($\mathbf{e}_1'$, $\mathbf{e}_2'$, $\mathbf{e}_3'$) (see Figure 3.58a), we have

$$\mathbf{T}_{e'} = \begin{bmatrix} 0 & t_s & 0 \\ t_s & 0 & 0 \\ 0 & 0 & 0 \end{bmatrix}$$

and by the generalized Hooke's law, we obtain

$$\mathbf{E}_{e'} = \begin{bmatrix} 0 & \frac{t_s}{2G} & 0 \\ \frac{t_s}{2G} & 0 & 0 \\ 0 & 0 & 0 \end{bmatrix}.$$

The notation $\mathbf{T}_{e'}$ $\mathbf{E}_{e'}$ is being used in the above expressions to emphasize that we are referring the stresses/strains to the reference system defined by $\mathbf{e}'$, that is, $(\mathbf{e}'_1, \mathbf{e}'_2, \mathbf{e}'_3)$.

The strain tensor for the reference system $(\mathbf{e}_1, \mathbf{e}_2, \mathbf{e}_3)$ denoted by $\mathbf{e}$ can also be evaluated by Hooke's law. From (3.107)

$$\begin{aligned} E_{11} &= \frac{T_{11}}{E} - \nu\,(T_{22} + T_{33}) = \frac{t_n}{E} - \frac{\nu}{E}\,(-t_n) = \frac{t_n\,(1+\nu)}{E} \\ E_{22} &= \frac{T_{22}}{E} - \nu\,(T_{11} + T_{33}) = -\frac{t_n}{E} - \frac{\nu}{E} t_n = -\frac{t_n\,(1+\nu)}{E} \\ E_{33} &= \frac{T_{33}}{E} - \nu\,(T_{11} + T_{22}) = -\frac{\nu}{E}\,(t_n - t_n) = 0 \\ E_{12} &= E_{13} = E_{23} = 0 \end{aligned}$$

leading to

$$\mathbf{E}_e = \begin{bmatrix} \frac{t_n(1+\nu)}{E} & 0 & 0 \\ 0 & -\frac{t_n(1+\nu)}{E} & 0 \\ 0 & 0 & 0 \end{bmatrix}. \tag{3.108}$$

We note that the normal strain of a fiber in the direction of $\mathbf{e}_1$ is $\varepsilon_\ell\,(\mathbf{e}_1) = \frac{t_n(1+\nu)}{E}$ (refer to (3.108)). This strain can also be calculated by using the strain tensor $\mathbf{E}_{e'}$. The unit vector $\mathbf{e}_1$ can be written as $\frac{\sqrt{2}}{2}\mathbf{e}'_1 + \frac{\sqrt{2}}{2}\mathbf{e}'_2$ and therefore the strain in this direction can also be evaluated by

$$\begin{aligned} \varepsilon_\ell(\frac{\sqrt{2}}{2}\mathbf{e}'_1 + \frac{\sqrt{2}}{2}\mathbf{e}'_2) &= \begin{bmatrix} \frac{\sqrt{2}}{2} & \frac{\sqrt{2}}{2} & 0 \end{bmatrix} \mathbf{E}_{e'} \begin{bmatrix} \frac{\sqrt{2}}{2} \\ \frac{\sqrt{2}}{2} \\ 0 \end{bmatrix} \\ &= \begin{bmatrix} \frac{\sqrt{2}}{2} & \frac{\sqrt{2}}{2} & 0 \end{bmatrix} \begin{bmatrix} 0 & \frac{t_s}{2G} & 0 \\ \frac{t_s}{2G} & 0 & 0 \\ 0 & 0 & 0 \end{bmatrix} \begin{bmatrix} \frac{\sqrt{2}}{2} \\ \frac{\sqrt{2}}{2} \\ 0 \end{bmatrix} \\ &= \frac{t_s}{2G}. \end{aligned}$$

Equating the values, we obtain

$$\frac{t_s}{2G} = \frac{t_n}{E}(1+\nu)$$

since $t_s = t_n$, we arrive at

$$G = \frac{E}{2(1+\nu)}. \tag{3.109}$$

Hence, of the three elastic constants E, ν and G only two are independent.

In Figure 3.58b we show schematically the deformed configuration of the section. The fiber considered above is given by DB in the undeformed configuration and $D_y B_y$ in the deformed configuration. We note that the only way this fiber displays the same deformation when interpreted as being in the cube subjected to normal stresses and in the parallelepiped subjected to pure shear is when relation (3.109) holds.

3.4.3 Generalized Hooke's law for an isotropic material in matrix notation

We may define column matrices to collect the independent components of the stress and strain tensors denoting them by τ and ε respectively

$$\begin{aligned} \tau^T &= \begin{bmatrix} \tau_{11} & \tau_{22} & \tau_{33} & \tau_{12} & \tau_{13} & \tau_{23} \end{bmatrix} \\ \varepsilon^T &= \begin{bmatrix} \varepsilon_{11} & \varepsilon_{22} & \varepsilon_{33} & \gamma_{12} & \gamma_{13} & \gamma_{23} \end{bmatrix} \end{aligned}$$

where $\tau_{ij} = T_{ij}$, $\varepsilon_{ii} = E_{ii}$ for $i, j = 1, 2, 3$ and $\gamma_{12} = 2E_{12}$, $\gamma_{13} = 2E_{13}$, $\gamma_{23} = 2E_{23}$. Considering the above definitions the generalized Hooke's law reads

$$\varepsilon = \mathbf{D}\tau \tag{3.110}$$

where

$$\mathbf{D} = \frac{1}{E}\begin{bmatrix} 1 & -\nu & -\nu & 0 & 0 & 0 \\ -\nu & 1 & -\nu & 0 & 0 & 0 \\ -\nu & -\nu & 1 & 0 & 0 & 0 \\ 0 & 0 & 0 & 2(1+\nu) & 0 & 0 \\ 0 & 0 & 0 & 0 & 2(1+\nu) & 0 \\ 0 & 0 & 0 & 0 & 0 & 2(1+\nu) \end{bmatrix}. \tag{3.111}$$

We also have

$$\tau = \mathbf{D}^{-1}\varepsilon = \mathbf{C}\varepsilon \tag{3.112}$$

where

$$\mathbf{C} = \frac{E\,(1-\nu)}{(1+\nu)\,(1-2\nu)} \begin{bmatrix} 1 & \frac{\nu}{1-\nu} & \frac{\nu}{1-\nu} & 0 & 0 & 0 \\ \frac{\nu}{1-\nu} & 1 & \frac{\nu}{1-\nu} & 0 & 0 & 0 \\ \frac{\nu}{1-\nu} & \frac{\nu}{1-\nu} & 1 & 0 & 0 & 0 \\ 0 & 0 & 0 & \frac{1-2\nu}{2(1-\nu)} & 0 & 0 \\ 0 & 0 & 0 & 0 & \frac{1-2\nu}{2(1-\nu)} & 0 \\ 0 & 0 & 0 & 0 & 0 & \frac{1-2\nu}{2(1-\nu)} \end{bmatrix}. \tag{3.113}$$

We note that the Hooke's law written in the form of (3.112) leads directly to the expected result that an infinitesimally small rigid motion produces no stresses, since this motion results in zero strains.

We can also conclude either from (3.110) or (3.112) that for an isotropic elastic material the principal directions of the stress and strain tensors coincide, since zero shear stresses lead to zero shear strains and vice-versa and, hence, the principal stresses and strains are also related by (3.110) and (3.112).

3.5 Formulation of the linear elasticity problem

In Chapter 2 we examined, in the context of truss structures, the *Three Fundamental Conditions* to be satisfied in the formulation of a structural mechanics problem:

- Equilibrium.
- Compatibility.
- Constitutive equations.

In Section 3.1, by means of a simple problem – the analysis of the steel sheet – we motivated the need for expressing these conditions for a three-dimensional deformable body.

Compatibility was addressed in Section 3.2 in which the deformations of a 3-D body were studied, strain measures were defined and the relations between the strains and displacements were derived.

In Section 3.3, the state of stress at a point was characterized by means of the Cauchy stress tensor and the equations of differential equilibrium were established.

In Section 3.4, the relations between the strains and stresses were detailed for an elastic, isotropic and homogeneous material. Now we are ready to formulate the elasticity problem.

Consider a deformable solid subjected to the action of a field of body forces $\mathbf{f}^B$ defined for every point $\mathbf{x}$ within the solid. On part of its surface, S_u, prescribed displacements $\hat{\mathbf{u}}$ are imposed and on the complementary part of the surface, S_f, with unit normal $\mathbf{n}$, prescribed surface tractions $\mathbf{f}^S$ are imposed[15]. We note that $S_u \cup S_f = S$ which represents the surface of the body and $S_u \cap S_f = \emptyset$. This situation is schematically represented in Figure 3.59 where the undeformed configuration V is shown. The linear elasticity problem can be posed as follows.

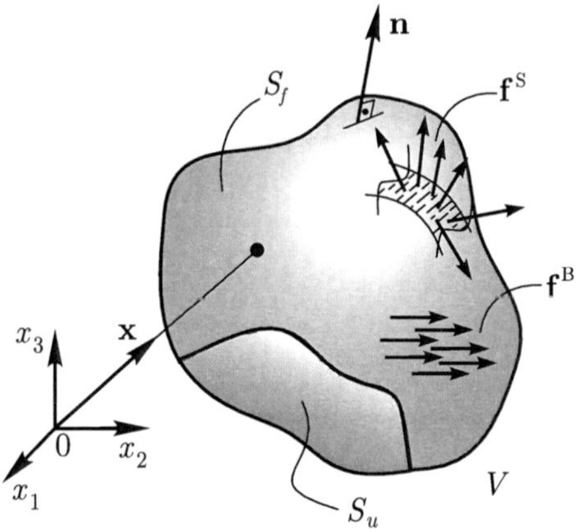

Fig. 3.59. Schematic representation of actions and boundary conditions for linear elasticity problem

Formulation of the 3-D elasticity problem

Find the displacement field $\mathbf{u}(\mathbf{x})$, the stress field $\tau(\mathbf{x})$ and the strain field $\varepsilon(\mathbf{x})$ such that for every point $\mathbf{x}$ within V

$$\sum_{j=1}^{3} \frac{\partial \tau_{ij}}{\partial x_j} + f_i^B = 0 \qquad i = 1, 2, 3 \tag{3.114}$$

[15] There are situations for which at a point on the surface the displacements are restricted in some directions and the surface tractions are prescribed in the remaining directions. Then S_u and S_f should be defined for each component of surface displacement and surface traction (see Section 2.1.1)

$$\varepsilon_{ij} = \frac{1}{2}\left(\frac{\partial u_i}{\partial x_j} + \frac{\partial u_j}{\partial x_i}\right) \qquad i,j = 1,2,3 \tag{3.115}$$

$$\tau = \mathbf{C}\varepsilon \tag{3.116}$$

with the boundary conditions for points $\mathbf{x}$ on S_f

$$\sum_{j=1}^{3} \tau_{ij} n_j = f_i^S, \qquad i = 1,2,3, \tag{3.117}$$

where the n_i are the components of $\mathbf{n}$, and for the points $\mathbf{x}$ on S_u

$$\mathbf{u} = \hat{\mathbf{u}} \tag{3.118}$$

or

$$u_i = \hat{u}_i, \qquad i = 1,2,3.$$

The equations (3.114) to (3.118) are field equations in the volume V and on the surfaces S_f and S_u, as applicable. The *Equilibrium Condition* corresponds to satisfying the differential equilibrium in (3.114) and (3.117); the *Compatibility Condition* corresponds to finding continuous displacements $\mathbf{u}(\mathbf{x})$ that satisfy (3.118) and from which the strains are calculated as given in (3.115); and the *Constitutive Condition* is given in (3.116). Therefore all the requirements for the solution of the 3-D problem are contained in the above equations and the formulation is complete.

Note that when the solid is also subjected to known temperature changes, we should use in the above formulation the generalized Hooke's law which includes the thermal strains as given in Section 3.4.1.

The search for analytical solutions of the above problem when specialized to specific geometries, loading and boundary conditions has challenged mathematicians and engineers alike for a number of centuries. A great deal of solutions have been derived – albeit frequently based on simplifying assumptions – and they are available in the literature. These solutions represent a very important resource for understanding the behavior of elastic solids subjected to external actions.

In the past, prior to the availability of computer resources and numerical methods for the solution of (3.114) to (3.118), these analytical solutions to special problems, derived and idealized from the full 3-D problem, were the only solutions available for analysis. While, at present, full 3-D problems with arbitrary boundary conditions can be solved using the finite element method, the solution of the full 3-D problem is in many cases not an effective way to perform engineering analysis.

The 3-D elasticity model is of great importance, since it is our highest order hierarchical model assuming linear elastic behavior. Therefore, it provides a conceptual reference model from which other specific and frequently more effective models can be derived. The solutions of these models can then be compared with the solution of the full 3-D model giving good insight into the modeling of physical problems – all within the objective of using hierarchical mathematical modeling in engineering analysis.

Before we close this chapter, we present two illustrative cases of analytical solutions of the 3-D elasticity model which are important for structural analysis: the pure bending of a prismatic bar of rectangular cross-section (given in Example 3.7) and the uniform torsion of a prismatic bar (given in Section 3.6).

Example 3.7

Consider the undeformed configuration of a solid as shown in Figure 3.60.

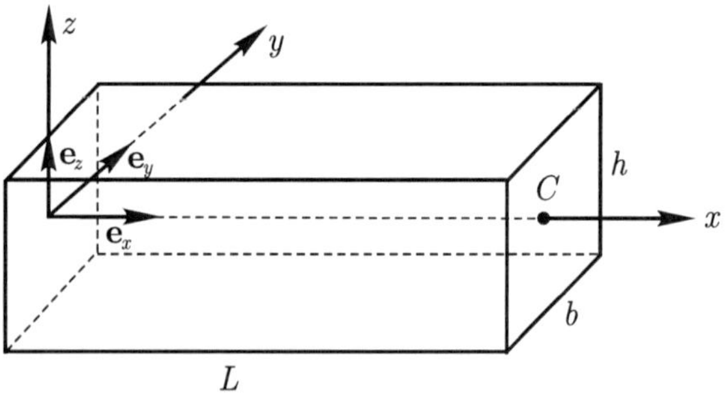

Fig. 3.60. Definition of the solid under study

Let the displacement field be given by

$$u = -\frac{M}{EI}xz \tag{3.119}$$

$$v = \nu\frac{M}{EI}yz \tag{3.120}$$

$$w = \frac{M}{2EI}\left[x^2 + \nu\left(z^2 - y^2\right)\right]. \tag{3.121}$$

where M is a positive constant, E and ν are Young's modulus and Poisson's ratio of the material and I is the moment of inertia about the y axis, $I = \frac{bh^3}{12}$ (see Section 4.2.2).

(i) Find the stress field associated with the given displacement field.

(ii) Identify the problem for which the given displacement field represents the exact solution of the 3-D elasticity model.

Solution

(i) The strains are given as

$$\begin{aligned}
\varepsilon_{xx} &= \frac{\partial u}{\partial x} = -\frac{M}{EI}z \\
\varepsilon_{yy} &= \frac{\partial v}{\partial y} = \nu\frac{M}{EI}z \\
\varepsilon_{zz} &= \frac{\partial w}{\partial z} = \nu\frac{M}{EI}z \\
\varepsilon_{xy} &= \frac{1}{2}\left(\frac{\partial u}{\partial y} + \frac{\partial v}{\partial x}\right) = 0 \\
\varepsilon_{xz} &= \frac{1}{2}\left(\frac{\partial u}{\partial z} + \frac{\partial w}{\partial x}\right) = \frac{1}{2}\left(-\frac{M}{EI}x + \frac{M}{EI}x\right) = 0 \\
\varepsilon_{yz} &= \frac{1}{2}\left(\frac{\partial v}{\partial z} + \frac{\partial w}{\partial y}\right) = \frac{1}{2}\left(\nu\frac{M}{EI}y - \nu\frac{M}{EI}y\right) = 0.
\end{aligned}$$

The stresses can be obtained using the generalized Hooke's law (see equation (3.112))

$$\tau_{xx} = \frac{E(1-\nu)}{(1+\nu)(1-2\nu)}\left[\varepsilon_{xx} + \frac{\nu}{1-\nu}\left(\varepsilon_{yy} + \varepsilon_{zz}\right)\right].$$

Noting that for our problem

$$\varepsilon_{yy} = \varepsilon_{zz} = -\nu\varepsilon_{xx} \tag{3.122}$$

we obtain

$$\tau_{xx} = E\varepsilon_{xx} = -\frac{M}{I}z.$$

We also have

$$\begin{aligned}
\tau_{yy} &= \frac{E(1-\nu)}{(1+\nu)(1-2\nu)}\left[\varepsilon_{yy} + \frac{\nu}{1-\nu}\left(\varepsilon_{xx} + \varepsilon_{zz}\right)\right] \\
\tau_{zz} &= \frac{E(1-\nu)}{(1+\nu)(1-2\nu)}\left[\varepsilon_{zz} + \frac{\nu}{1-\nu}\left(\varepsilon_{xx} + \varepsilon_{yy}\right)\right]
\end{aligned}$$

and considering (3.122) we obtain

$$\tau_{yy} = \tau_{zz} = 0.$$

Of course, since $\gamma_{xy} = \gamma_{xz} = \gamma_{yz} = 0$, we have

$$\tau_{xy} = \tau_{xz} = \tau_{yz} = 0.$$

This completes part (i) of the problem.

(ii) To identify the 3-D problem actually solved, we need to determine which fields of body forces and surface tractions are in equilibrium with this stress field. Considering the differential equilibrium equations and introducing this stress field, we have

$$\begin{aligned}
\frac{\partial \tau_{xx}}{\partial x} + \frac{\partial \tau_{xy}}{\partial y} + \frac{\partial \tau_{xz}}{\partial z} + f_x^B &= 0 \Rightarrow f_x^B = 0 \\
\frac{\partial \tau_{xy}}{\partial x} + \frac{\partial \tau_{yy}}{\partial y} + \frac{\partial \tau_{yz}}{\partial z} + f_y^B &= 0 \Rightarrow f_y^B = 0 \\
\frac{\partial \tau_{xz}}{\partial x} + \frac{\partial \tau_{yz}}{\partial y} + \frac{\partial \tau_{zz}}{\partial z} + f_z^B &= 0 \Rightarrow f_z^B = 0.
\end{aligned}$$

Therefore, there should be no body forces. Since the only nonzero stress component is τ_{xx}, the surface tractions should be zero on all four lateral surfaces. Further, for the surface defined by $x = L$

$$\mathbf{Tn} = \mathbf{f}^S \Rightarrow \tau_{xx}\mathbf{e}_x = \mathbf{f}^S \Rightarrow \mathbf{f}^S = -\frac{M}{I}z\mathbf{e}_x$$

and for $x = 0$

$$\mathbf{Tn} = \mathbf{f}^S \Rightarrow -\tau_{xx}\mathbf{e}_x = \mathbf{f}^S \Rightarrow \mathbf{f}^S = \frac{M}{I}z\mathbf{e}_x.$$

If we reduce the surface tractions at the section given by $x = L$ to its center of gravity we obtain for the force resultant

$$\mathbf{R} = \int_A \mathbf{f}^S \, dA = \int_A -\frac{M}{I}z\mathbf{e}_x \, dA = 0$$

and for the moment resultant with respect to the center of the section, represented by $\mathbf{C}$

$$\begin{aligned}
\mathbf{M_C} &= \int_A (y\mathbf{e}_y + z\mathbf{e}_z) \times \mathbf{f}^S \, dA \\
&= \int_A \left(\frac{M}{I}yz\mathbf{e}_z - \frac{M}{I}z^2\mathbf{e}_y\right) dA = -\frac{M}{I}\int_A z^2 dA \, \mathbf{e}_y = -M \, \mathbf{e}_y.
\end{aligned}$$

Hence the applied surface tractions at $x = L$ correspond to a bending moment, see Figure 3.61. Analogously, for the end section defined by $x = 0$ we obtain $\mathbf{R} = \mathbf{0}$ and $M\mathbf{e}_y$ leading to a self-equilibrated force system.

Note also that the displacement field given in (3.119) – (3.121) satisfies $u = 0$ for $x = 0$, and $u = v = w = 0$ for $x = y = z = 0$.

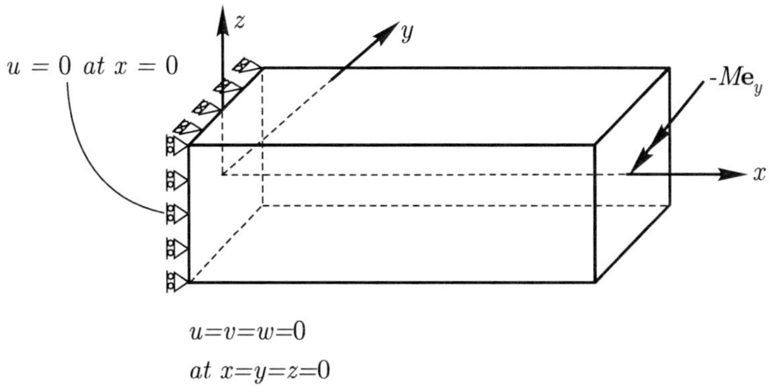

Fig. 3.61. Problem summary. Body forces are zero

We can conclude that the displacements given in (3.119) to (3.121), the derived strains and stresses correspond to the exact solution of the cantilever beam subjected to a pure bending moment at the tip, as summarized in Figure 3.61, as long as the bending moment at $x = L$ is introduced by the surface traction field $\mathbf{f}^S = -\frac{M}{I} z \mathbf{e}_x$ and the displacement restrictions at the "built-in section" $x = 0$ are as shown in Figure 3.61.

In Figure 3.62, we show the deformed and undeformed configuration of the solid. We see that a line parallel to the y axis in the undeformed configuration is deformed into an arc, whose curvature is opposite to the curvature of the deformed axis. This transverse curvature is known as *anticlastic curvature* and it is due to the Poisson effect.

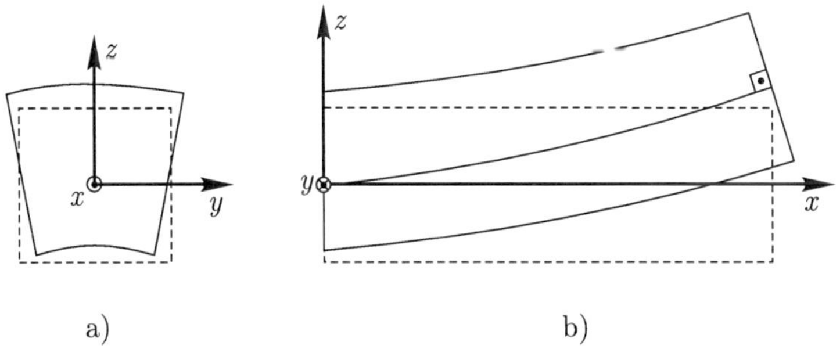

Fig. 3.62. Deformations for selected planes. a) Plane given by $x = 0$; b) Plane given by $y = 0$. The magnitude of the displacements is chosen for visualization purposes; the shown displacements are much larger than those for which the linear model would be adequate

□

3.6 Torsion of a prismatic bar

We present below the exact solution of a 3-D elasticity problem – the Saint Venant torsion solution for a prismatic bar – which is of considerable practical importance. It is also a classical example of the so-called semi-inverse method for deriving solutions. In this method, some assumptions on the functional form of the displacements are made, either motivated by experimental observations or by intuition. These displacement assumptions are the starting point towards obtaining a solution.

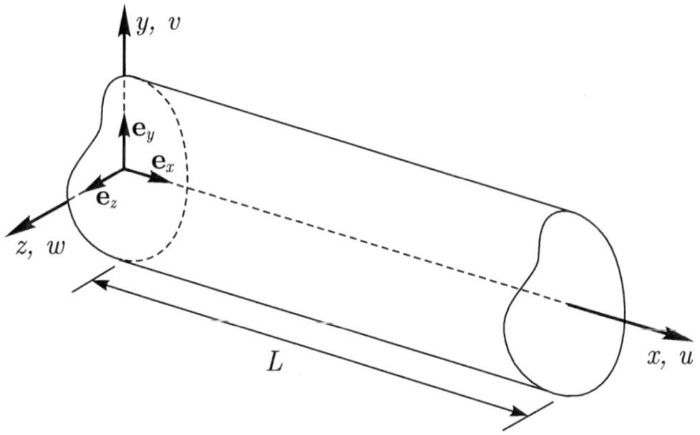

Fig. 3.63. Generic prismatic bar which will be subjected to torsion. Solid section with no holes

Consider a prismatic bar with a generic cross-section as shown in Figure 3.63. Suppose that the bar is subjected to self-equilibrated torsional moments at the end sections, *i.e.*, $\mathbf{M}_t = M_t \mathbf{e}_x$ at $x = L$ and $-\mathbf{M}_t$ at $x = 0$. It is an experimental observation that for a bar subjected to such loading the cross-sections rotate as rigid bodies in their own plane (see Example 3.5 for the in-plane displacements). However, these sections do not remain plane, they display some warping. Motivated by these observations, we will seek a solution of the 3-D elasticity mathematical model of this problem using the following displacement assumptions

$$u = \theta' \psi(y, z) \tag{3.123}$$

$$v = -\theta' x z \tag{3.124}$$

$$w = \theta' x y \tag{3.125}$$

where θ' is the rate of rotation of the cross-sections with respect to the x axis which is assumed to be constant. Hence, denoting by $\theta(x)$ the angle of rotation of a generic section, we have

$$\theta' = \frac{d\theta}{dx} = \text{constant}.$$

Referring to Example 3.5, v and w as given by equations (3.124) and (3.125) are the displacements associated with the section rotations given by $\theta(x) = \theta' x$. Note that we are implicitly assuming that there is no section rotation at $x = 0$. The functional form of u reflects the warping of the section. In fact, $\psi(y, z)$ gives the spatial variation of the out of section displacements and is called the warping function. Its precise form should be determined based on additional conditions as will be seen shortly. We also note that the intensity of warping is assumed to be proportional to the rate of rotation of the sections, θ'.

Starting from the displacement assumptions given in equations (3.123) to (3.125), we derive the associated strain field

$$\begin{aligned}
\varepsilon_{xx} &= \frac{\partial u}{\partial x} = 0, \quad \varepsilon_{yy} = \frac{\partial v}{\partial y} = 0, \quad \varepsilon_{zz} = \frac{\partial w}{\partial z} = 0 \\
\gamma_{xy} &= \frac{\partial u}{\partial y} + \frac{\partial v}{\partial x} = \theta' \frac{\partial \psi}{\partial y} - \theta' z = \theta' \left(\frac{\partial \psi}{\partial y} - z \right) \\
\gamma_{xz} &= \frac{\partial u}{\partial z} + \frac{\partial w}{\partial x} = \theta' \frac{\partial \psi}{\partial z} + \theta' y = \theta' \left(\frac{\partial \psi}{\partial z} + y \right) \\
\gamma_{yz} &= \frac{\partial v}{\partial z} + \frac{\partial w}{\partial y} = -\theta' x + \theta' x = 0.
\end{aligned}$$

The stresses can be obtained using the generalized Hooke's law. Since we are considering a homogenous and isotropic material we have

$$\begin{aligned}
\tau_{xx} &= \frac{E(1-\nu)}{(1+\nu)(1-2\nu)} \left[\varepsilon_{xx} + \frac{\nu}{1-\nu} (\varepsilon_{yy} + \varepsilon_{zz}) \right] \\
\tau_{yy} &= \frac{E(1-\nu)}{(1+\nu)(1-2\nu)} \left[\varepsilon_{yy} + \frac{\nu}{1-\nu} (\varepsilon_{xx} + \varepsilon_{zz}) \right] \\
\tau_{zz} &= \frac{E(1-\nu)}{(1+\nu)(1-2\nu)} \left[\varepsilon_{zz} + \frac{\nu}{1-\nu} (\varepsilon_{xx} + \varepsilon_{yy}) \right]
\end{aligned}$$

which lead to

$$\tau_{xx} = \tau_{yy} = \tau_{zz} = 0 \tag{3.126}$$

since $\varepsilon_{xx} = \varepsilon_{yy} = \varepsilon_{zz} = 0$. For the shear stresses we have

$$\tau_{xy} = G\gamma_{xy} = G\theta' \left(\frac{\partial \psi}{\partial y} - z \right) \tag{3.127}$$

$$\tau_{xz} = G\gamma_{xz} = G\theta' \left(\frac{\partial \psi}{\partial z} + y \right) \tag{3.128}$$

$$\tau_{yz} = G\gamma_{yz} = 0. \tag{3.129}$$

Consider now the equilibrium conditions. Assuming that we have no body forces, the differential equilibrium relations read

$$\frac{\partial \tau_{xx}}{\partial x} + \frac{\partial \tau_{xy}}{\partial y} + \frac{\partial \tau_{xz}}{\partial z} = 0 \tag{3.130}$$

$$\frac{\partial \tau_{xy}}{\partial x} + \frac{\partial \tau_{yy}}{\partial y} + \frac{\partial \tau_{yz}}{\partial z} = 0 \tag{3.131}$$

$$\frac{\partial \tau_{xz}}{\partial x} + \frac{\partial \tau_{yz}}{\partial y} + \frac{\partial \tau_{zz}}{\partial z} = 0. \tag{3.132}$$

Referring to the stresses given in equations (3.126) to (3.129), we verify that the equilibrium conditions (3.131) and (3.132) are identically satisfied and equation (3.130) leads to

$$\frac{\partial^2 \psi}{\partial y^2} + \frac{\partial^2 \psi}{\partial z^2} = 0 \tag{3.133}$$

which is the condition on the warping function that guarantees equilibrium.

Next, we consider the boundary conditions. Since the lateral surfaces are free from any surface tractions, we should have

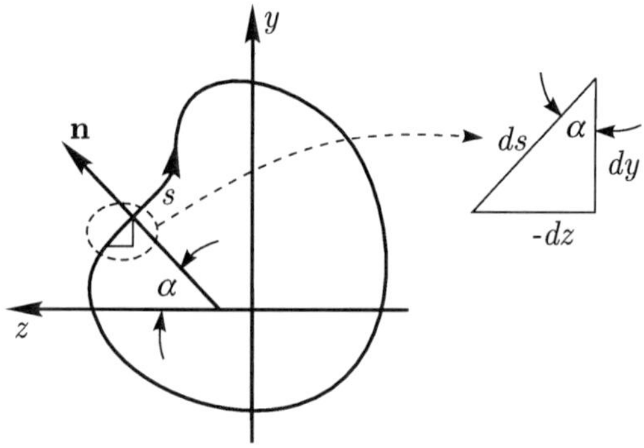

Fig. 3.64. Generic cross section

$$\mathbf{Tn} = \begin{bmatrix} 0 & \tau_{xy} & \tau_{xz} \\ \tau_{xy} & 0 & 0 \\ \tau_{xz} & 0 & 0 \end{bmatrix} \begin{bmatrix} 0 \\ n_y \\ n_z \end{bmatrix} = \begin{bmatrix} 0 \\ 0 \\ 0 \end{bmatrix}$$

or

$$\tau_{xy} n_y + \tau_{xz} n_z = 0. \tag{3.134}$$

Let s be the arc length coordinate along the section boundary as shown in Figure 3.64. Of course, for a given cross-section, the section boundary is characterized by $y = y(s)$ and $z = z(s)$. Defining α as the angle that the surface outward's unit normal makes with the z axis, we can write

$$n_y = \sin\alpha, \ n_z = \cos\alpha$$

$$ds\cos\alpha = dy, \quad ds\sin\alpha = -dz$$

and introducing these relations into equation (3.134)

$$\tau_{xz}\frac{dy}{ds} - \tau_{xy}\frac{dz}{ds} = 0. \tag{3.135}$$

This equation can be written in terms of the warping function

$$G\theta'\left(\frac{\partial\psi}{\partial z} + y\right)\frac{dy}{ds} - G\theta'\left(\frac{\partial\psi}{\partial y} - z\right)\frac{dz}{ds} = 0$$

which finally leads to

$$\left(\frac{\partial\psi}{\partial z} + y\right)\frac{dy}{ds} - \left(\frac{\partial\psi}{\partial y} - z\right)\frac{dz}{ds} = 0. \tag{3.136}$$

This condition is to be satisfied for every point on the section boundary.

The field equation (3.133) subjected to the boundary condition (3.136) should determine the warping function $\psi(y, z)$. We note that the boundary condition (3.136) depends only on the geometry of the cross-section. Therefore, given a cross-section, the warping function can be determined and depends only on the shape of the cross-section.

Although the warping function can be determined as described above, Prandtl (see Timoshenko and Goodier, 1970) has introduced a stress function $\phi(y, z)$ to propose an alternative, somewhat simpler, formulation of the torsion problem. In order to introduce the stress function, let us consider the equilibrium equation (3.130) and re-write it as

$$\frac{\partial\tau_{xy}}{\partial y} = -\frac{\partial\tau_{xz}}{\partial z}. \tag{3.137}$$

Therefore, we can define a function $\phi(y, z)$ such that

$$\frac{\partial\phi}{\partial z} = \tau_{xy} \quad \text{and} \quad \frac{\partial\phi}{\partial y} = -\tau_{xz} \tag{3.138}$$

and (3.137) is automatically satisfied.

Using (3.127) and (3.128) we also have

$$\frac{\partial\phi}{\partial z} = G\theta'\left(\frac{\partial\psi}{\partial y} - z\right) \tag{3.139}$$

$$\frac{\partial\phi}{\partial y} = -G\theta'\left(\frac{\partial\psi}{\partial z} + y\right). \tag{3.140}$$

To eliminate ψ and obtain the governing differential equation in terms of ϕ we take derivatives of equations (3.139) and (3.140) with respect to z and y respectively, and then add to obtain

$$\frac{\partial^2 \phi}{\partial y^2} + \frac{\partial^2 \phi}{\partial z^2} = -2G\theta'. \tag{3.141}$$

Expressing the boundary condition (3.135) in terms of ϕ, we obtain

$$\frac{\partial \phi}{\partial y}\frac{dy}{ds} + \frac{\partial \phi}{\partial z}\frac{dz}{ds} = 0$$

which leads to

$$\frac{d\phi}{ds} = 0 \;\Rightarrow\; \phi \equiv \text{constant over the boundary.}$$

Since the stresses are obtained from ϕ by taking derivatives we can choose

$$\phi = 0 \text{ on the boundary of the cross-section.} \tag{3.142}$$

Let us summarize what we have obtained so far.

The formulation based on the warping function can be written as:

Find $\psi(y, z)$ defined on the cross-section domain such that

$$\begin{cases} \frac{\partial^2 \psi}{\partial y^2} + \frac{\partial^2 \psi}{\partial z^2} = 0 & \text{for every point in the cross-section domain} \\ \left(\frac{\partial \psi}{\partial z} + y\right)\frac{dy}{ds} - \left(\frac{\partial \psi}{\partial y} - z\right)\frac{dz}{ds} = 0 & \text{for every point on the cross-section boundary.} \end{cases}$$

Having determined $\psi(y, z)$, the displacements can be obtained from equations (3.123) – (3.125) and the stresses from (3.126) – (3.129). We note that θ' is still to be found as will be discussed shortly. The formulation based on the stress function can be summarized as:

Find $\phi(y, z)$ defined on the cross-section domain such that

$$\begin{cases} \frac{\partial^2 \phi}{\partial y^2} + \frac{\partial^2 \phi}{\partial z^2} = -2G\theta' & \text{for every point in the cross-section domain} \\ \phi = 0 & \text{for every point on the cross-section boundary.} \end{cases}$$

Having determined $\phi(y, z)$, we can obtain $\psi(y, z)$ by integration of equations (3.139)–(3.140) and hence the remaining quantities can be found as described above. The same observation regarding θ' applies.

To complete the formulation of the torsion problem and to obtain θ' we consider the tractions on the boundary surfaces at $x = L$ and $x = 0$. Specifically

$$\mathbf{f}^S = \mathbf{T}\mathbf{e}_x \text{ at } x = L$$

or

$$\begin{bmatrix} f_x^S \\ f_y^S \\ f_z^S \end{bmatrix} = \begin{bmatrix} 0 & \tau_{xy} & \tau_{xz} \\ \tau_{xy} & 0 & 0 \\ \tau_{xz} & 0 & 0 \end{bmatrix} \begin{bmatrix} 1 \\ 0 \\ 0 \end{bmatrix} = \begin{bmatrix} 0 \\ \tau_{xy} \\ \tau_{xz} \end{bmatrix}$$

where τ_{xy} and τ_{xz} are given by equation (3.127) and (3.128) or alternatively by (3.138). These stress components are shown in Figure 3.65.

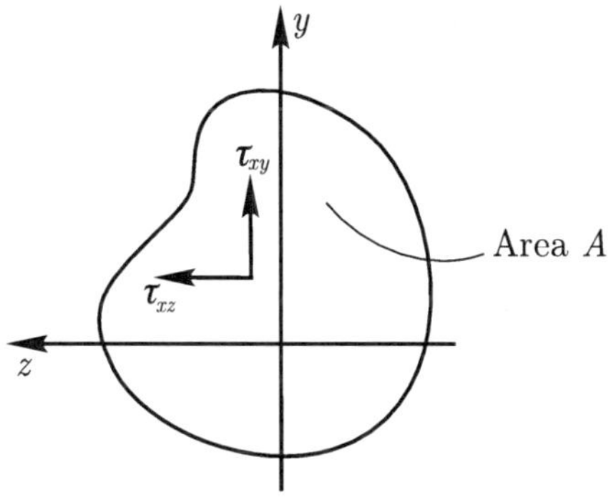

Fig. 3.65. Stress components at section $x = L$

Let us evaluate the stress resultants at the end section $x = L$. The normal force N and the bending moments $\mathbf{M}_y = M_y \mathbf{e}_y$ and $\mathbf{M}_z = M_z \mathbf{e}_z$ are trivially found to be zero since

$$N = \int_A \tau_{xx}\, dA = 0, \quad M_y = \int_A \tau_{xx}\, z\, dA = 0, \quad M_z = \int_A -\tau_{xx}\, y\, dA = 0$$

when we take into account that $\tau_{xx} = 0$. The shear forces $\mathbf{V}_y = V_y \mathbf{e}_y$ and $\mathbf{V}_z = V_z \mathbf{e}_z$ are given by

$$V_y = \int_A \tau_{xy}\, dA, \quad V_z = \int_A \tau_{xz}\, dA$$

which can be shown to be zero by expressing τ_{xy} and τ_{xz} in terms of ϕ (see equation (3.138)) and performing the above integrations.

Therefore, we can conclude that the stresses at the end section $x = L$ are mechanically equivalent to a torsional moment M_t only, since the resultants $N = M_y = M_z = V_y = V_z = 0$ and we have

$$\begin{aligned} M_t &= \int_A (\tau_{xz} y - \tau_{xy} z)\, dA \\ &= \int_A \tau_{xz}\, y\, dA - \int_A \tau_{xy}\, z\, dA. \end{aligned} \tag{3.143}$$

Performing the integrations it can be shown that

$$M_t = 2\int_A \phi\, dA \tag{3.144}$$

which relates the stress function to the torsional moment applied. We can also relate the torsional moment to the warping function. Starting from (3.143) and introducing (3.127) and (3.128), we obtain

$$\begin{aligned} M_t &= G\theta' \int_A \left[\left(\frac{\partial \psi}{\partial z} + y\right) y - \left(\frac{\partial \psi}{\partial y} - z\right) z\right] dA \\ &= G\theta' \int_A \left[\left(\frac{\partial \psi}{\partial z} y - \frac{\partial \psi}{\partial y} z\right) + \left(y^2 + z^2\right)\right] dA. \end{aligned}$$

Recalling that the warping function ψ depends only on the geometry of the cross-section, we can define the torsional moment of inertia of the cross-section

$$I_t = \int_A \left[\left(\frac{\partial \psi}{\partial z} y - \frac{\partial \psi}{\partial y} z\right) + \left(y^2 + z^2\right)\right] dA. \tag{3.145}$$

and hence

$$\frac{M_t}{GI_t} = \theta' \tag{3.146}$$

establishing the link between the torsional moment and θ'. Analogous derivations hold for the end section given by $x = 0$.

The formulation of the torsion problem is now complete and its solutions is an exact solution to the 3–D elasticity mathematical model as long as M_t is applied by means of the surface tractions $\mathbf{f}^S = \mathbf{T}\mathbf{e}_x$ at $x = L$ and $\mathbf{f}^S = \mathbf{T}\left(-\mathbf{e}_x\right)$ at $x = 0$.

As an example, we present below the solution of the torsion problem for an elliptical cross-section.

Example 3.8

Consider a bar with the elliptical cross-section in Figure 3.66. The bar is subjected to self–equilibrated torsional moments $\mathbf{M}_t = M_t\mathbf{e}_x$ at $x = L$ and $-\mathbf{M}_t$ at $x = 0$. Solve for the stress and displacement fields.

Solution

We consider the solution in terms of the stress function ϕ. Since the boundary equation is

$$\frac{y^2}{a^2} + \frac{z^2}{b^2} - 1 = 0$$

we can investigate solutions of the form

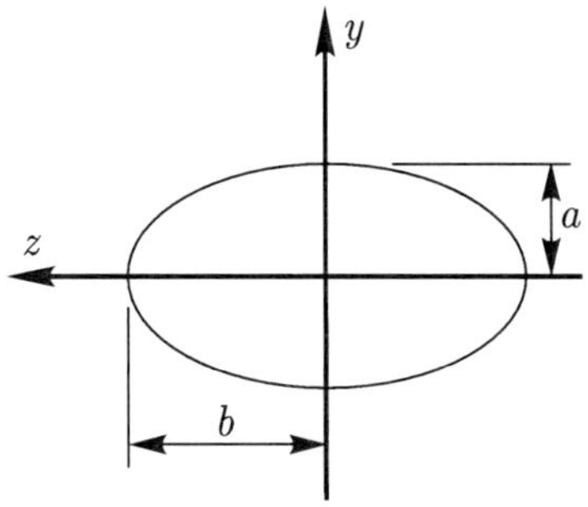

Fig. 3.66. Definitions for elliptic cross section

$$\phi = C\left(\frac{y^2}{a^2} + \frac{z^2}{b^2} - 1\right) \tag{3.147}$$

where C is a real constant. Of course, (3.147) satisfies the condition $\phi = 0$ at the boundary and substituting (3.147) in (3.141) yields

$$\frac{2C}{a^2} + \frac{2C}{b^2} = -2G\theta'$$

which leads to

$$C = -\frac{a^2b^2G\theta'}{a^2+b^2}.$$

Therefore

$$\phi = -\frac{a^2b^2}{a^2+b^2}G\theta'\left(\frac{y^2}{a^2} + \frac{z^2}{b^2} - 1\right)$$

and the stresses

$$\tau_{xy} = \frac{\partial\phi}{\partial z} = -\frac{2a^2}{a^2+b^2}G\theta' z, \quad \tau_{xz} = -\frac{\partial\phi}{\partial y} = \frac{2b^2}{a^2+b^2}G\theta' y.$$

Using relations (3.139) and (3.140) we obtain

$$\left(\frac{\partial\psi}{\partial y} - z\right) = -\frac{2a^2}{a^2+b^2}z, \quad \left(\frac{\partial\psi}{\partial z} + y\right) = \frac{2b^2}{a^2+b^2}y \tag{3.148}$$

which lead to $\psi(y,z)$ by integration. In fact, integrating (3.148) with respect to y and z respectively leads to

$$\psi = yz - \frac{2a^2}{a^2+b^2}yz + f(z), \quad \psi = -yz + \frac{2b^2}{a^2+b^2}yz + f(y).$$

Hence

$$\psi = \frac{b^2 - a^2}{a^2 + b^2} yz + K$$

where K is a real constant. Assuming that $u = 0$ for $y = z = 0$ yields

$$\psi(y, z) = \frac{b^2 - a^2}{a^2 + b^2} yz.$$

Having the warping function $\psi(y, z)$, we can obtain I_t from (3.145) which gives

$$I_t = \frac{\pi a^3 b^3}{a^2 + b^2}$$

and we can relate M_t to θ', *i.e.*,

$$\theta' = \frac{M_t}{GI_t} = \frac{M_t}{G} \frac{a^2 + b^2}{\pi a^3 b^3}. \tag{3.149}$$

Alternatively we could have used

$$M_t = 2 \int_A \phi dA$$

leading also to the result given in (3.149). The displacements are

$$\begin{aligned} u &= \theta' \psi(y, z) = \frac{M_t}{G} \frac{b^2 - a^2}{\pi a^3 b^3} yz \\ v &= -\theta' xz = -\frac{M_t}{G} \frac{a^2 + b^2}{\pi a^3 b^3} xz \\ w &= \theta' xy = \frac{M_t}{G} \frac{a^2 + b^2}{\pi a^3 b^3} xy \end{aligned}$$

and the stresses can be written in terms of M_t by

$$\tau_{xy} = -\frac{2M_t a^2}{\pi a^3 b^3} z, \quad \tau_{xz} = \frac{2M_t b^2}{\pi a^3 b^3} y.$$

The stress variation along the selected lines is schematically represented in Figure 3.67a and the out-of-plane displacements $u(y, z)$ – the warping displacements – are shown in Figure 3.67b. We note that the above solution includes the classical torsion solution of a prismatic bar of circular cross-section when $a = b = r$, r being the radius of the circular cross-section. In this case, there is no warping of the cross-section and since the solution derived above is an exact solution of the 3-D elasticity problem, the classical solution of a circular prismatic bar is also exact.

□

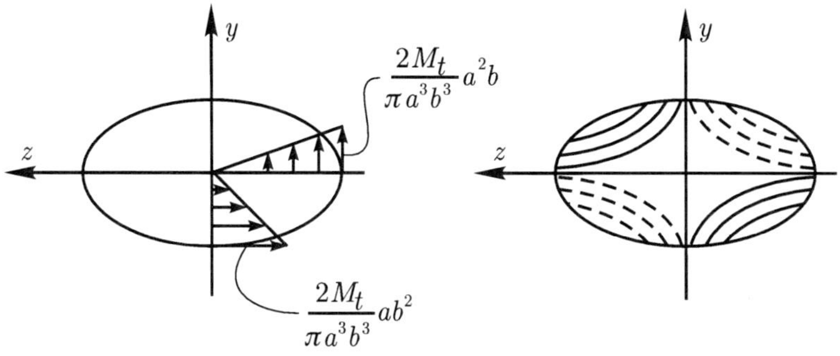

a) b)

Fig. 3.67. a) Some shear stresses for the solution of a prismatic bar of elliptic cross section; b) Lines connecting points of equal values of warping displacements. The solid lines represent positive displacements and the dashed lines represent negative displacements

The membrane analogy

A valuable physical interpretation of the solution reached in (3.141) is obtained by the Prandtl membrane analogy. Namely, a membrane prestressed with uniform tension T and subjected to transverse pressure p undergoes the transverse displacement $u(y,z)$ calculated from (see *e.g.* Bathe, 1996)

$$\frac{\partial^2 u}{\partial y^2} + \frac{\partial^2 u}{\partial z^2} = -\frac{p}{T} \tag{3.150}$$

with $u = 0$ on the boundary. Hence from (3.141), we obtain

$$\phi = c\,u(y,z) \text{ with } c = \frac{2G\theta' T}{p}. \tag{3.151}$$

Therefore, the transverse displacement of the membrane is proportional to ϕ with c the constant of proportionality.

There are two properties linked to the fact that the shear stresses are obtained as derivatives of ϕ which are of interest. In Figure 3.68a, we show a generic cross-section and a curve for which ϕ is constant. It can be shown (Timoshenko and Goodier, 1970) that

$$\mathbf{t}_s = t_s \mathbf{s}$$

that is, $\mathbf{t}_s$ is always tangent to the curve of constant ϕ since $\mathbf{s}$ is the tangent unit vector to this curve, and that

$$t_s = -\frac{d\phi}{dn} \tag{3.152}$$

that is, the derivative of ϕ with respect to the normal direction to the curve of constant ϕ gives the magnitude of the shear stress t_s. In Figure 3.68b a typical distribution is shown. Due to the membrane analogy, a curve which gives equal values of ϕ corresponds to a level curve of the deformed membrane and hence we can qualitatively anticipate the shear stress distributions.

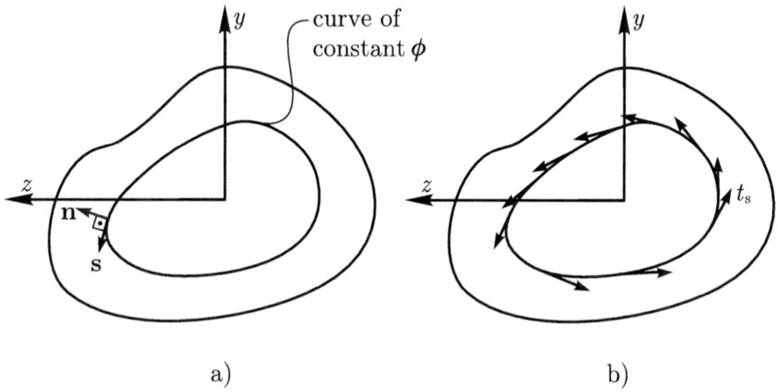

Fig. 3.68. a) A generic curve of constant ϕ; b) A typical shear stress distribution for the points on a curve of constant ϕ

We demonstrate these observations in the following example.

Example 3.9

Derive the shear stress distribution and the torsional moment of inertia for the torsion problem for a thin rectangular cross-section as shown in Figure 3.69a.

Solution

In Figure 3.69b, the deformation of the associated membrane is shown. We can see that, except in the small end regions, the membrane displacements are such that

$$\frac{\partial u}{\partial y} = 0.$$

Therefore, equation (3.150) simplifies to

$$\frac{\partial^2 u}{\partial z^2} = -\frac{p}{T}$$

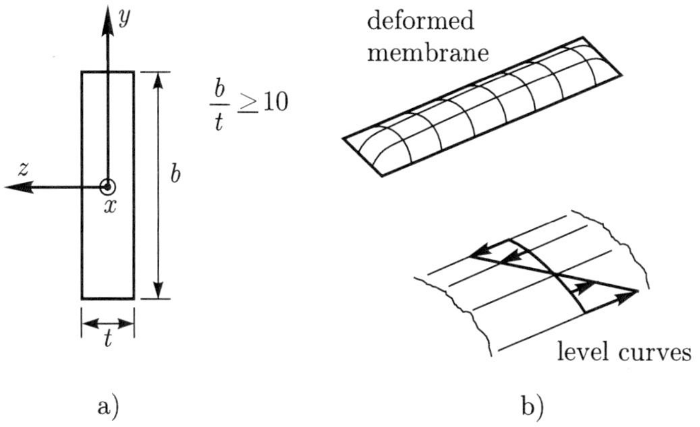

Fig. 3.69. a) Thin rectangular cross section; b) Associated deformed membrane

which upon integration leads to

$$u = \frac{p}{8T}\left(t^2 - 4z^2\right).$$

Using relation (3.151), we have

$$\phi = \frac{G\theta'}{4}\left(t^2 - 4z^2\right).$$

The shear stresses can be evaluated from (3.138) which leads to

$$\tau_{xz} = -\frac{\partial \phi}{\partial y} = 0, \quad \tau_{xy} = \frac{\partial \phi}{\partial z} = -2G\theta' z.$$

Note the shear stresses are tangent to the level curves of the membrane as shown in Figure 3.69b. The torsional moment of inertia can be calculated using (3.144) and (3.146)

$$I_t = \frac{bt^3}{3}.$$

Of course, the derived quantities are approximate since we assumed that $\frac{\partial u}{\partial y} = 0$ also in the end regions.

□

4. Mathematical models used in engineering structural analysis

In this chapter we pursue a formidable task – to present the most important mathematical models in structural mechanics. In order to best situate our present objectives, let us review some previous developments.

In Chapter 1, the hierarchical modeling process was introduced. The central idea of this process is to provide a rational framework to select appropriate mathematical models to address well defined questions of engineering interest. For structural analysis, there are a number of mathematical models and there is the need to "know" these models to perform the hierarchical modeling process (see Chapter 1).

In Chapter 2, we discussed fundamental conditions that should be met whichever structural mechanics mathematical model is established and solved.

In Chapter 3, we motivated the need for a 3-D formulation of structural problems and presented the highest hierarchical mathematical model for an isotropic material and linear analysis – the 3-D elasticity model. In the current chapter, based on our earlier discussion of the 3-D elasticity model, we present the remaining most relevant mathematical models of structural mechanics. We start with plane elasticity and then move on to bars, plates and shells.

4.1 Plane elasticity

We refer to plane elasticity as the set of mathematical models which describe the behavior of a body using only displacements in a plane. The out-of-plane behavior is assumed or inferred from the in-plane behavior. In the following, we present the plane strain, the plane stress and the axisymmetric mathematical models.

4.1.1 The plane strain model

The plane strain model could have been discussed at the end of Chapter 3 as an example of a model which leads to the exact solution of the 3-D elasticity problem when some geometric and loading restrictions apply.

In order to motivate the plane strain assumptions let us consider the dam schematically shown in Figure 4.1. The dam corresponds geometrically to a prismatic solid. The loading due to the water and gravity is the same for every cross-sectional plane. Now, suppose that at the end sections the out-of-plane displacements $w(x, y)$ are prevented but the in-plane displacements $u(x, y)$ and $v(x, y)$ are free. At the base all displacements are prevented. Under these conditions, by symmetry, the displacements are clearly the same for every section of the dam. Hence, the complete behavior of the dam can be described by the displacements $u(x, y)$ and $v(x, y)$ at a typical section.

Much aligned with the semi-inverse method, which was introduced for the torsion problem, we formulate the plane strain problem by introducing displacement assumptions for a prismatic solid.

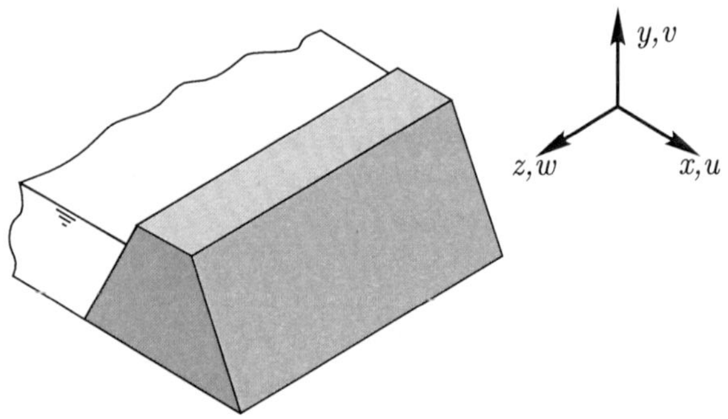

Fig. 4.1. Schematic representation of a dam

Kinematics

Formally, let us consider a prismatic solid as shown in Figure 4.2, whose cross-sections are parallel to the xy plane. Motivated by the above discussion, we introduce the following displacement assumptions

$$u = u(x, y) \tag{4.1}$$

$$v = v(x, y) \tag{4.2}$$

$$w = 0. \tag{4.3}$$

Strain compatibility

Using the compatibility relations

$$\varepsilon_{xx} = \frac{\partial u}{\partial x} \qquad \varepsilon_{yy} = \frac{\partial v}{\partial y} \qquad \gamma_{xy} = \frac{\partial u}{\partial y} + \frac{\partial v}{\partial x}$$

we obtain

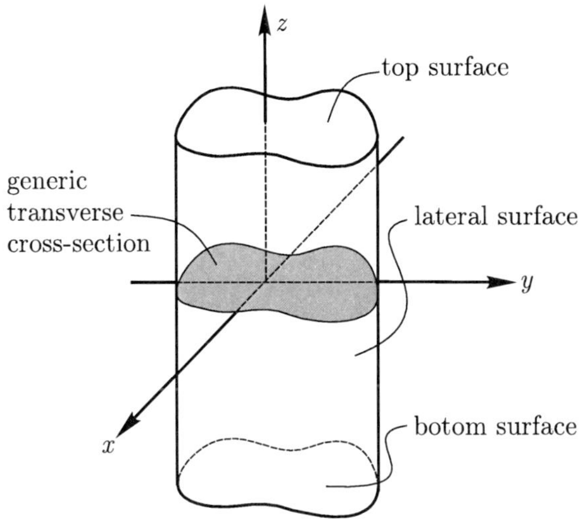

Fig. 4.2. Generic prismatic solid

$$\varepsilon_{xx} = \varepsilon_{xx}(x, y) \tag{4.4}$$

$$\varepsilon_{yy} = \varepsilon_{yy}(x, y) \tag{4.5}$$

$$\gamma_{xy} = \gamma_{xy}(x, y) \tag{4.6}$$

due to the assumptions implicitly given in equations (4.1) and (4.2). Still considering the strain compatibility relations and equations (4.1) to (4.3), we obtain

$$\gamma_{yz} = \frac{\partial v}{\partial z} + \frac{\partial w}{\partial y} = 0$$

$$\gamma_{zx} = \frac{\partial w}{\partial x} + \frac{\partial u}{\partial z} = 0$$

$$\varepsilon_{zz} = \frac{\partial w}{\partial z} = 0.$$

Constitutive relations

Let us consider next the constitutive equations, *i.e.*, the generalized Hooke's law (recall equation (3.110)). Starting with the component ε_{zz}

$$\varepsilon_{zz} = 0 = \frac{\tau_{zz}}{E} - \frac{\nu}{E}(\tau_{xx} + \tau_{yy})$$

which leads to

$$\tau_{zz} = \nu(\tau_{xx} + \tau_{yy}) \tag{4.7}$$

and considering the in-plane normal strain components, we can write

$$\varepsilon_{xx} = \frac{\tau_{xx}}{E} - \frac{\nu}{E}(\tau_{yy} + \tau_{zz}) = \frac{(1-\nu^2)}{E}\tau_{xx} - \frac{\nu(1+\nu)}{E}\tau_{yy} \tag{4.8}$$

$$\varepsilon_{yy} = \frac{\tau_{yy}}{E} - \frac{\nu}{E}(\tau_{xx} + \tau_{zz}) = \frac{(1-\nu^2)}{E}\tau_{yy} - \frac{\nu(1+\nu)}{E}\tau_{xx}. \tag{4.9}$$

Of course, we have used (4.7) to derive the final forms of equations (4.8) and (4.9). Considering the shear strains and stresses, we obtain

$$\gamma_{xy} = \frac{\tau_{xy}}{G} = \frac{2(1+\nu)}{E}\tau_{xy} \tag{4.10}$$

and

$$\tau_{xz} = 0 \tag{4.11}$$

$$\tau_{yz} = 0 \tag{4.12}$$

since $\gamma_{xz} = 0$ and $\gamma_{yz} = 0$. It is convenient to define the following column matrices

$$\tau = \begin{bmatrix} \tau_{xx} \\ \tau_{yy} \\ \tau_{xy} \end{bmatrix}, \qquad \varepsilon = \begin{bmatrix} \varepsilon_{xx} \\ \varepsilon_{yy} \\ \gamma_{xy} \end{bmatrix}.$$

Therefore, from (4.8), (4.9) and (4.10)

$$\varepsilon = \mathbf{D}\tau \tag{4.13}$$

where

$$\mathbf{D} = \frac{(1+\nu)}{E}\begin{bmatrix} (1-\nu) & -\nu & 0 \\ -\nu & (1-\nu) & 0 \\ 0 & 0 & 2 \end{bmatrix}$$

which can be inverted leading to

$$\tau = \mathbf{D}^{-1}\varepsilon = \mathbf{C}\varepsilon \tag{4.14}$$

where

$$\mathbf{C} = \frac{E(1-\nu)}{(1+\nu)(1-2\nu)}\begin{bmatrix} 1 & \frac{\nu}{1-\nu} & 0 \\ \frac{\nu}{1-\nu} & 1 & 0 \\ 0 & 0 & \frac{1-2\nu}{2(1-\nu)} \end{bmatrix}.$$

We note that the same notation (τ, ε, $\mathbf{D}$ and $\mathbf{C}$) was used for the 3-D problem for which we have, in general, all non zero stress and strain components and additional entries in the constitutive matrices. Of course, the correct meaning of these matrices is implied by the context. We also note that τ_{zz}, ε_{zz} were not considered in the above definitions, since $\varepsilon_{zz} = 0$ and τ_{zz} is obtained from τ_{xx} and τ_{yy} by equation (4.7). Therefore τ and ε, as defined, fully characterize the stress and strain states. We remark that equations (4.14), (4.4), (4.5) and (4.6) imply that the stress components τ_{xx}, τ_{yy} and τ_{xy} are functions of x, y only, *i.e.*, $\tau_{xx} = \tau_{xx}(x, y)$, $\tau_{yy} = \tau_{yy}(x, y)$ and $\tau_{xy} = \tau_{xy}(x, y)$.

Equilibrium

We need to enforce the equilibrium conditions which read

$$\frac{\partial \tau_{xx}}{\partial x} + \frac{\partial \tau_{xy}}{\partial y} + \frac{\partial \tau_{xz}}{\partial z} + f_x^B = 0 \tag{4.15}$$

$$\frac{\partial \tau_{xy}}{\partial x} + \frac{\partial \tau_{yy}}{\partial y} + \frac{\partial \tau_{yz}}{\partial z} + f_y^B = 0 \tag{4.16}$$

$$\frac{\partial \tau_{xz}}{\partial x} + \frac{\partial \tau_{yz}}{\partial y} + \frac{\partial \tau_{zz}}{\partial z} + f_z^B = 0. \tag{4.17}$$

Introducing (4.7), (4.11), (4.12) and taking into account that $\tau_{xx} = \tau_{xx}(x, y)$, $\tau_{yy} = \tau_{yy}(x, y)$, $\tau_{xy} = \tau_{xy}(x, y)$, we conclude that $f_x^B = f_x^B(x, y)$, $f_y^B = f_y^B(x, y)$ and $f_z^B = 0$. Otherwise, we would not be able to satisfy the equilibrium conditions, which are then expressed by

$$\frac{\partial \tau_{xx}}{\partial x} + \frac{\partial \tau_{xy}}{\partial y} + f_x^B = 0$$

$$\frac{\partial \tau_{xy}}{\partial x} + \frac{\partial \tau_{yy}}{\partial y} + f_y^B = 0.$$

Note that as long as $f_z^B = 0$, equation (4.17) is identically satisfied. Also, we should interpret the conditions on f_x^B, f_y^B and f_z^B as restrictions on the loading such that the displacement assumptions given in (4.1) to (4.3) are appropriate.

Boundary conditions

We need to consider next the boundary conditions. Although, when we introduced the plane strain problem, we did not distinguish between displacement and force boundary conditions, we can now consider the most general set of boundary conditions which would be compatible with the basic assumptions expressed by equations (4.1) to (4.3). Referring to Figure 4.2, let us consider first the lateral surface. On part of the lateral surface, S_u, we can prescribe displacements as long as S_u is given by the extrusion along the z

direction of a curve L_u defined at a cross-section boundary as schematically shown in Figure 4.3. The prescribed displacements are defined by

$$u(x,y,z) = \hat{u}(x,y) \tag{4.18}$$

$$v(x,y,z) = \hat{v}(x,y) \tag{4.19}$$

for any point on S_u.

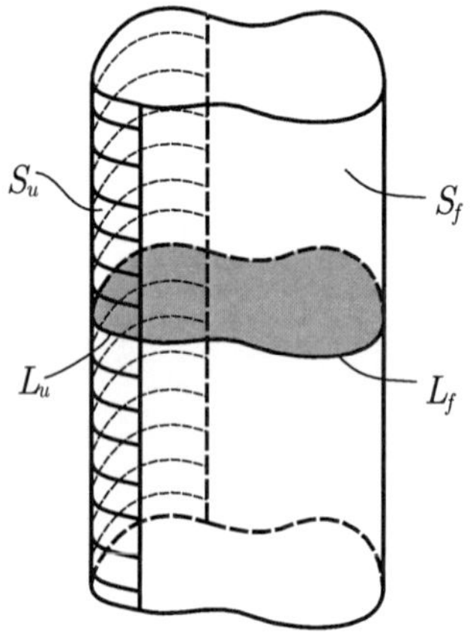

Fig. 4.3. Schematic representation of S_u, S_f and L_u, L_f

Let S_f be the complementary part of the lateral surface defined by the extrusion along z of L_f, the complementary curve to L_u, as also shown in Figure 4.3. Since the solid is prismatic, the normal unit vector at every point on the lateral surface is given by $\mathbf{n} = n_x \mathbf{e}_x + n_y \mathbf{e}_y$ and the force boundary condition reads

$$\mathbf{Tn} = \mathbf{f}^S$$

for every point on S_f, which in components is given by

$$\begin{bmatrix} \tau_{xx} & \tau_{xy} & 0 \\ \tau_{xy} & \tau_{yy} & 0 \\ 0 & 0 & \tau_{zz} \end{bmatrix} \begin{bmatrix} n_x \\ n_y \\ 0 \end{bmatrix} = \begin{bmatrix} f_x^S \\ f_y^S \\ f_z^S \end{bmatrix}.$$

Hence

$$f_x^S = \tau_{xx} n_x + \tau_{xy} n_y \tag{4.20}$$

$$f_y^S = \tau_{xy} n_x + \tau_{yy} n_y \tag{4.21}$$

$$f_z^S = 0. \tag{4.22}$$

The above equations establish additional restrictions for the external load. In fact, besides the restriction given by equation (4.22), equations (4.20) and (4.21) imply that $f_x^S = f_x^S(x,y)$ and $f_y^S = f_y^S(x,y)$ since τ_{xx} and τ_{yy} are functions of x, y only and the normal unit vector at a point on the lateral surface does not change with the coordinate z.

The top and bottom surfaces are peculiar[1] with respect to the imposition of boundary conditions. In fact, in order to be compatible with the displacement assumption $w = 0$ we should consider that $w = \hat{w} = 0$ at the top and bottom surfaces. The in-plane displacements, however, can not be restrained. For example, on the top surface $\mathbf{n} = \mathbf{e}_z$ and the surface tractions $\mathbf{f}^S = \mathbf{Tn}$ are given by

$$\begin{bmatrix} \tau_{xx} & \tau_{xy} & 0 \\ \tau_{xy} & \tau_{yy} & 0 \\ 0 & 0 & \tau_{zz} \end{bmatrix} \begin{bmatrix} 0 \\ 0 \\ 1 \end{bmatrix} = \begin{bmatrix} f_x^S \\ f_y^S \\ f_z^S \end{bmatrix}.$$

Therefore

$$\begin{aligned} f_x^S &= 0 \\ f_y^S &= 0 \\ f_z^S &= \tau_{zz} = \nu(\tau_{xx} + \tau_{yy}) \end{aligned} \tag{4.23}$$

which shows that there can not be any surface tractions applied tangentially to the plane and the surface traction in the direction $\mathbf{e}_z$ is given by (4.23). Here f_z^S can be interpreted as a reactive surface traction compatible with the restriction given by $w = 0$. Analogous derivations would lead for the bottom surface for which $\mathbf{n} = -\mathbf{e}_z$ to

$$\begin{aligned} f_x^S &= 0 \\ f_y^S &= 0 \\ f_z^S &= -\tau_{zz} = -\nu(\tau_{xx} + \tau_{yy}). \end{aligned}$$

Hence, we may say that the top and bottom surfaces behave as if supported on rollers, free to roll into the x and y directions.

Differential formulation

Now we can summarize the formulation of the plane strain problem. Let us consider a prismatic solid as shown in Figure 4.3. On the lateral surface S_u we have prescribed displacements as given by (4.18) and (4.19). On the lateral surface S_f we have prescribed forces defined by

[1] In fact, this is an example in which at a point on the surface the displacement is restricted in a direction and the surface tractions are prescribed in the remaining directions (see Section 2.1.1)

$$
\begin{aligned}
f_x^S &= \hat{f}_x^S(x,y) \\
f_y^S &= \hat{f}_y^S(x,y) \\
f_z^S &= 0
\end{aligned}
\tag{4.24}
$$

and on the top and bottom surfaces w is prescribed to be zero and $f_x^S = f_y^S = 0$.

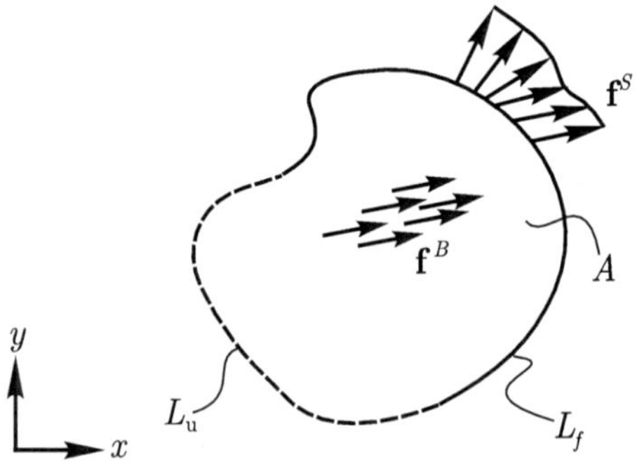

Fig. 4.4. Domain of unit thickness representing the prismatic solid, A = cross-section, V = volume $= 1 \times A$, $S_u = L_u \times 1$, $S_f = L_f \times 1$

Under these conditions the solution of the 3-D elasticity problem can be formulated in a plane as defined in Figure 4.4, for the indicated solid of unit thickness.

Differential formulation of the plane strain model

Given $f_x^B = f_x^B(x,y)$ and $f_y^B = f_y^B(x,y)$ defined in V, find $u(x,y)$, $v(x,y)$; $\tau_{xx} = \tau_{xx}(x,y)$, $\tau_{yy} = \tau_{yy}(x,y)$ and $\tau_{xy} = \tau_{xy}(x,y)$; $\varepsilon_{xx} = \varepsilon_{xx}(x,y)$, $\varepsilon_{yy} = \varepsilon_{yy}(x,y)$ and $\gamma_{xy} = \gamma_{xy}(x,y)$ such that

$$\frac{\partial \tau_{xx}}{\partial x} + \frac{\partial \tau_{xy}}{\partial y} + f_x^B = 0 \tag{4.25}$$

$$\frac{\partial \tau_{xy}}{\partial x} + \frac{\partial \tau_{yy}}{\partial y} + f_y^B = 0 \tag{4.26}$$

$$\varepsilon_{xx} = \frac{\partial u}{\partial x} \tag{4.27}$$

$$\varepsilon_{yy} = \frac{\partial v}{\partial y} \tag{4.28}$$

$$\gamma_{xy} = \frac{\partial u}{\partial y} + \frac{\partial v}{\partial x} \tag{4.29}$$

$$\tau = \mathbf{C}\varepsilon \tag{4.30}$$

for every point in V;

$$f_x^S = \hat{f}_x^S(x, y) \tag{4.31}$$

$$f_y^S = \hat{f}_y^S(x, y) \tag{4.32}$$

for every point in S_f and

$$u = \hat{u}(x, y) \tag{4.33}$$

$$v = \hat{v}(x, y) \tag{4.34}$$

for every point in S_u[2].

Once the solution to this plane problem has been found, the solution for the 3-D problem is given by appending

$$w = 0 \tag{4.35}$$

$$\varepsilon_{zz} = \gamma_{xz} = \gamma_{yz} = 0 \tag{4.36}$$

$$\tau_{xz} = \tau_{yz} = 0 \tag{4.37}$$

$$\tau_{zz} = \nu(\tau_{xx} + \tau_{yy}). \tag{4.38}$$

4.1.2 The plane stress model

The motivation for the formulation of the plane stress model is the analysis of a thin plate subjected to loading in its own plane. Consider the plate shown in Figure 4.5 with its mid-surface in the xy plane. We assume that both $\mathbf{f}^B$ and $\mathbf{f}^S$ have no component into the z direction, the top and the bottom surfaces are free from any imposed surface tractions and the thickness of the plate, denoted by h, is small when compared to a characteristic length dimension on the plane of the plate that is $\frac{h}{L} \ll 1$.

[2] Note that in each of the problem formulations given in Section 3.5 and Chapter 4, we assume that continuous displacements are sought and that sufficient boundary conditions on S_u are prescribed to make the solution possible

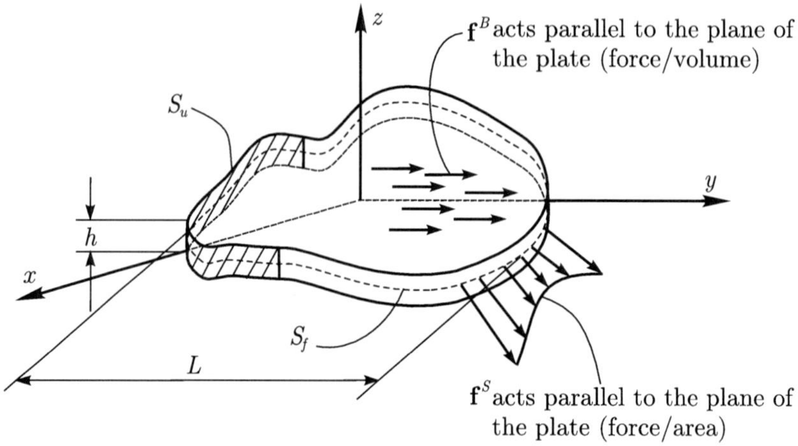

Fig. 4.5. Thin plate subjected to loading in its plane

Based on this problem description, specific assumptions can now be introduced regarding the stress field and the loading.

Stress assumptions

For the plate characterized in Figure 4.5, we assume that

$$\tau_{zz} = \tau_{xz} = \tau_{yz} = 0 \tag{4.39}$$

and taking into account that the plate is thin, we further assume that

$$\tau_{xx} = \tau_{xx}(x, y), \ \tau_{yy} = \tau_{yy}(x, y), \ \tau_{xy} = \tau_{xy}(x, y). \tag{4.40}$$

Equilibrium

The equilibrium equations given in (3.114) can be simplified and become

$$\frac{\partial \tau_{xx}}{\partial x} + \frac{\partial \tau_{xy}}{\partial y} + f_x^B = 0 \tag{4.41}$$

$$\frac{\partial \tau_{xy}}{\partial x} + \frac{\partial \tau_{yy}}{\partial y} + f_y^B = 0. \tag{4.42}$$

Therefore equations (4.41) and (4.42) are now phrased in a two-dimensional domain leading also to the restrictions $f_x^B = f_x^B(x, y)$ and $f_y^B = f_y^B(x, y)$.

Constitutive relations

In light of the stress assumptions (4.39), the constitutive equation can be simplified leading to

$$
\begin{aligned}
\varepsilon_{xx} &= \frac{\tau_{xx}}{E} - \frac{\nu}{E}\tau_{yy} \\
\varepsilon_{yy} &= \frac{\tau_{yy}}{E} - \frac{\nu}{E}\tau_{xx} \\
\varepsilon_{zz} &= -\frac{\nu}{E}(\tau_{xx} + \tau_{yy}) \\
\gamma_{xy} &= \frac{\tau_{xy}}{G} = \frac{2(1+\nu)}{E}\tau_{xy} \\
\gamma_{xz} &= \frac{\tau_{xz}}{G} = 0 \\
\gamma_{yz} &= \frac{\tau_{yz}}{G} = 0.
\end{aligned}
$$

We can organize the constitutive relations in matrix form as given by (4.13) and (4.14) where

$$
\mathbf{D} = \frac{1}{E}\begin{bmatrix} 1 & -\nu & 0 \\ -\nu & 1 & 0 \\ 0 & 0 & 2(1+\nu) \end{bmatrix} \tag{4.43}
$$

and

$$
\mathbf{C} = \frac{E}{1-\nu^2}\begin{bmatrix} 1 & \nu & 0 \\ \nu & 1 & 0 \\ 0 & 0 & \frac{1-\nu}{2} \end{bmatrix}. \tag{4.44}
$$

Note that we use the same stress and strain column matrix definitions as for the plane strain problem.

Differential formulation

Now we are ready to characterize the plane stress problem. Let us consider the 3-D problem described in Figure 4.5 where a plate is subjected to a field of body forces

$$
\begin{aligned}
f_x^B &= f_x^B(x,y) \\
f_y^B &= f_y^B(x,y) \\
f_z^B &= 0.
\end{aligned}
$$

Displacements are prescribed in S_u according to

$$
\begin{aligned}
u(x,y,z) &= \hat{u}(x,y) \\
v(x,y,z) &= \hat{v}(x,y)
\end{aligned}
$$

and the tractions are prescribed on S_f as

$$
\begin{aligned}
f_x^S &= \hat{f}_x^S(x,y) \\
f_y^S &= \hat{f}_y^S(x,y) \\
f_z^S &= 0.
\end{aligned}
$$

The top and bottom surfaces are free from any surface tractions.

The plane stress problem associated with the 3-D problem described above admits a formulation in a plane domain as generically described in Figure 4.4, but now the thickness is h.

Differential formulation of the plane stress model

Given $f_x^B = f_x^B(x,y)$ and $f_y^B = f_y^B(x,y)$ defined in $V = A \cdot h$, find $u(x,y)$, $v(x,y)$; $\tau_{xx} = \tau_{xx}(x,y)$, $\tau_{yy} = \tau_{yy}(x,y)$ and $\tau_{xy} = \tau_{xy}(x,y)$; $\varepsilon_{xx} = \varepsilon_{xx}(x,y)$, $\varepsilon_{yy} = \varepsilon_{yy}(x,y)$, $\gamma_{xy} = \gamma_{xy}(x,y)$ such that

$$\frac{\partial \tau_{xx}}{\partial x} + \frac{\partial \tau_{xy}}{\partial y} + f_x^B = 0 \tag{4.45}$$

$$\frac{\partial \tau_{xy}}{\partial x} + \frac{\partial \tau_{yy}}{\partial y} + f_y^B = 0 \tag{4.46}$$

$$\varepsilon_{xx} = \frac{\partial u}{\partial x} \tag{4.47}$$

$$\varepsilon_{yy} = \frac{\partial v}{\partial y} \tag{4.48}$$

$$\gamma_{xy} = \frac{\partial u}{\partial y} + \frac{\partial v}{\partial x} \tag{4.49}$$

$$\boldsymbol{\tau} = \mathbf{C}\boldsymbol{\varepsilon} \tag{4.50}$$

for every point in V;

$$f_x^S = \hat{f}_x^S(x,y) \tag{4.51}$$

$$f_y^S = \hat{f}_y^S(x,y) \tag{4.52}$$

for every point in S_f; and

$$u = \hat{u}(x,y) \tag{4.53}$$

$$v = \hat{v}(x,y) \tag{4.54}$$

for every point in S_u.

We note that the solution of the plane stress problem appended by

$$\tau_{zz} = \tau_{xz} = \tau_{yz} = 0 \tag{4.55}$$

$$\gamma_{xz} = \gamma_{yz} = 0 \tag{4.56}$$

$$\varepsilon_{zz} = -\frac{\nu}{E}(\tau_{xx} + \tau_{yy}) \tag{4.57}$$

and by $w(x, y, z)$ obtained by the integration of (4.57) with respect to z is not the exact solution of the 3-D problem just described. The reason is that we started from some assumptions on the stress field and did not impose all the compatibility relations (3.115). The 3-D strain field given by the ε_{xx}, ε_{yy} and γ_{xy} solution of the plane stress problem appended by (4.56) and (4.57) does not always lead to a compatible 3-D displacement field, that is, continuous displacements $u(x, y, z)$, $v(x, y, z)$ and $w(x, y, z)$ that satisfy the displacement boundary conditions (with w on S_u free).

Namely, in general, the solution of the plane stress problem leads to stresses τ_{xx} and τ_{yy} which vary with (x, y). Hence, from (4.57)

$$\varepsilon_{zz} = \varepsilon_{zz}(x, y)$$

and

$$\frac{\partial w}{\partial z} = \varepsilon_{zz}(x, y)$$

which integrated with respect to z leads to

$$w(x, y, z) = z\varepsilon_{zz}(x, y) + f(x, y)$$

where $f(x, y)$ is an arbitrary function of x and y. Assuming that $w(x, y, z)$ is zero at $z = 0$, we obtain $f(x, y) = 0$ and

$$w(x, y, z) = z\varepsilon_{zz}(x, y) .$$

Now, we can evaluate

$$\gamma_{yz} = \frac{\partial w}{\partial y} + \frac{\partial v}{\partial z} = z\frac{\partial \varepsilon_{zz}}{\partial y}$$

which is, in general, zero only for $z = 0$. Analogously

$$\gamma_{xz} = \frac{\partial w}{\partial x} + \frac{\partial v}{\partial z} = z\frac{\partial \varepsilon_{zz}}{\partial x} .$$

Hence the transverse shear strains obtained from the displacement field are not zero showing that, in general, the plane stress solution is not the exact solution of the 3-D elasticity problem. However, it is possible to show (Timoshenko and Goodier, 1970) that the solutions of the plane stress problem and of the related 3-D problem are "close" as long as the plate is thin.

The 3-D solution has additional terms which are proportional to z^2 leading to small differences for the solution variables when the thickness is small.

We can appreciate from the above description that the plane strain and plane stress models represent quite distinct behaviors and they each need to be used considering the assumptions contained in the models. In particular, the kind of restrictions imposed on the out-of-plane displacements, *i.e.*, w in the notation used above, helps to identify which model reflects best the behavior of the physical problem. However, although there are distinctly different behaviors, the mathematical formulations of these models lend themselves to a unified presentation.

Consider the differential formulations of the plane strain and plane stress problems given by equations (4.25) to (4.34) and (4.45) to (4.54), respectively. We recognize that the equations for $u(x,y)$, $v(x,y)$, $\varepsilon_{xx}(x,y)$, $\varepsilon_{yy}(x,y)$, $\gamma_{xy}(x,y)$, $\tau_{xx}(x,y)$, $\tau_{yy}(x,y)$ and $\tau_{xy}(x,y)$ are identical except that $\mathbf{C}$, which expresses the constitutive equation, is different. However, if we define, for $\nu < 0.5$,

$$E_* = \frac{E}{1-\nu^2} \tag{4.58}$$

and

$$\nu_* = \frac{\nu}{1-\nu}$$

as the effective Young's modulus and effective Poisson's ratio and introduce these in place of E and ν in the $\mathbf{C}$ matrix for the plane stress model, we obtain

$$\mathbf{C} = \frac{E_*}{1-\nu_*^2}\begin{bmatrix} 1 & \nu_* & 0 \\ \nu_* & 1 & 0 \\ 0 & 0 & \frac{1-\nu_*}{2} \end{bmatrix}. \tag{4.59}$$

Then introducing the definitions of E_* and ν_*, we obtain

$$\mathbf{C} = \frac{\frac{E}{1-\nu^2}}{\left(1-\frac{\nu}{1-\nu}\right)\left(1+\frac{\nu}{1-\nu}\right)}\begin{bmatrix} 1 & \frac{\nu}{1-\nu} & 0 \\ \frac{\nu}{1-\nu} & 1 & 0 \\ 0 & 0 & \frac{1-\frac{\nu}{1-\nu}}{2} \end{bmatrix}$$

which gives

$$\mathbf{C} = \frac{E(1-\nu)}{(1+\nu)(1-2\nu)}\begin{bmatrix} 1 & \frac{\nu}{1-\nu} & 0 \\ \frac{\nu}{1-\nu} & 1 & 0 \\ 0 & 0 & \frac{1-2\nu}{2(1-\nu)} \end{bmatrix}$$

i.e., the $\mathbf{C}$ matrix for the plane strain model.

Therefore, both the plane stress and the plane strain models can be formulated with the same set of equations, for example (4.45) to (4.54), using the definition of $\mathbf{C}$ given in (4.14) for plane stress and (4.59) for plane strain. Of course, having the solution for the in-plane variables of the plane stress model, we can readily obtain the solution for the in-plane variables of the plane strain model by replacing E by E_* and ν by ν_* in the analytical expressions for these variables. The solution for the remaining variables is given by (4.35) to (4.38) for the plane strain model and by (4.55) to (4.57) for the plane stress model.

Since E_* and ν_* are larger than E and ν the plane strain model is stiffer than the plane stress model. This fact is expected because the out-of-plane displacements in the plane strain model are constrained to be zero.

Example 4.1

Study the solution of a thin plate subjected to its own weight as shown in Figure 4.6. At the edge $y = a$ a uniform distribution of surface tractions $\mathbf{f}^S = f_y^S \mathbf{e}_y = \rho g a \mathbf{e}_y$ is applied (ρ is the density and g the acceleration due to gravity) and at the three edges $y = 0$, $x = -b/2$ and $x = b/2$ there are no externally applied surface tractions. To suppress rigid body motions the displacements at point P ($x = 0$, $y = a$) are prevented and the plate is not allowed to rotate about P.

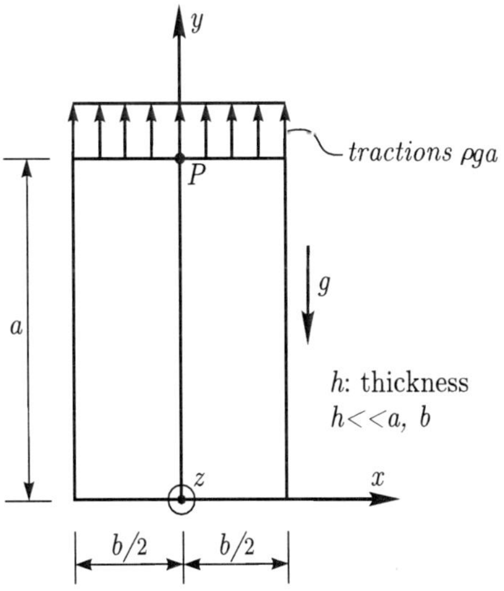

Fig. 4.6. Plate subjected to gravity

Solution

We consider a plane stress model and derive the solution. Due to gravity

$$f_x^B = 0,\ f_y^B = -\rho g,\ f_z^B = 0.$$

Since there are no externally applied surface tractions at the three edges $x = -b/2$ and $x = b/2$ the stress components τ_{xx} and τ_{xy} should be zero at these edges. Taking

$$\tau_{xx} = 0 \text{ and } \tau_{xy} = 0$$

for any point in the domain, the equilibrium equation

$$\frac{\partial \tau_{xx}}{\partial x} + \frac{\partial \tau_{xy}}{\partial y} + f_x^B = 0$$

is identically satisfied and

$$\frac{\partial \tau_{xy}}{\partial x} + \frac{\partial \tau_{yy}}{\partial y} + f_y^B = 0$$

leads to

$$\tau_{yy} = \rho g y + f(x)$$

where $f(x)$ is a function of x only. The boundary condition at $y = a$ implies

$$f_y^S = \rho g a = \tau_{yy}(y = a) = \rho g a + f(x) \qquad \Rightarrow \qquad f(x) = 0$$

and, hence

$$\tau_{yy} = \rho g y. \tag{4.60}$$

The free edge condition at the lower horizontal edge

$$f_y^S(x, 0) = 0 = \tau_{yy}(x, 0)$$

is verified by the τ_{yy} given in (4.60).

We can obtain the strains using the constitutive equations

$$\begin{aligned}
\varepsilon_{xx} &= \frac{\tau_{xx}}{E} - \frac{\nu}{E}\tau_{yy} = -\frac{\nu}{E}\tau_{yy} = -\frac{\nu}{E}\rho g y \\
\varepsilon_{yy} &= \frac{\tau_{yy}}{E} - \frac{\nu}{E}\tau_{xx} = \frac{\rho g y}{E} \\
\gamma_{xy} &= \frac{\tau_{xy}}{G} = 0
\end{aligned}$$

where E, ν and $G = \frac{E}{2(1+\nu)}$ are respectively, as usual, the Young modulus, the Poisson ratio and the shear modulus.

To obtain the displacement field, we need to integrate the compatibility relations. Consider

$$\varepsilon_{xx} = \frac{\partial u}{\partial x}$$

which leads to

$$\frac{\partial u}{\partial x} = -\frac{\nu}{E}\rho g y \qquad \Rightarrow \qquad u(x,y) = -\frac{\nu}{E}\rho g x y + f_1(y) \tag{4.61}$$

where $f_1(y)$ is a function of y only. Also

$$\varepsilon_{yy} = \frac{\partial v}{\partial y}$$

leading to

$$\frac{\partial v}{\partial y} = \frac{\rho g y}{E} \qquad \Rightarrow \qquad v(x,y) = \frac{\rho g y^2}{2E} + f_2(x) \tag{4.62}$$

where $f_2(x)$ is a function of x only. The shear strain is given by

$$\gamma_{xy} = \frac{\partial u}{\partial y} + \frac{\partial v}{\partial x} = 0 = -\nu k x + \frac{df_1}{dy} + \frac{df_2}{dx}, \quad k = \frac{\rho g}{E}. \tag{4.63}$$

Let us define

$$g_1(x) = -\nu k x + \frac{df_2}{dx}, \quad g_2(y) = \frac{df_1}{dy}. \tag{4.64}$$

Rewriting equation (4.63) using the definitions given in (4.64), we obtain

$$g_1(x) + g_2(y) = 0. \tag{4.65}$$

We note that (4.65) has to hold for any (x, y) in the domain. Therefore

$$g_1 = C_2, \; g_2 = -C_2$$

where C_2 is a constant. Using (4.64) yields

$$-\nu k x + \frac{df_2}{dx} = C_2$$

which by integration leads to

$$f_2(x) = C_2 x + \frac{\nu k x^2}{2} + C_3$$

and from (4.64)

$$\frac{df_1}{dy} = -C_2 \qquad \Rightarrow \qquad f_1(y) = -C_2 y + C_4.$$

Therefore

$$u = -\nu kxy - C_2 y + C_4 \tag{4.66}$$

$$v = \frac{k}{2}y^2 + C_2 x + \frac{\nu k x^2}{2} + C_3. \tag{4.67}$$

Let us impose the kinematic boundary condition at point P

$$u(0,a) = -C_2 a + C_4 = 0 \tag{4.68}$$

and

$$v(0,a) = \frac{ka^2}{2} + C_3 = 0. \tag{4.69}$$

In order to impose that there is no rigid body rotation about point P, we enforce that the horizontal infinitesimal fiber with origin at point P remains horizontal. This condition is given by

$$\frac{\partial v}{\partial x}(0,a) = 0.$$

From equation (4.67)

$$\frac{\partial v}{\partial x}(0,a) = C_2 \qquad \Rightarrow \qquad C_2 = 0.$$

Therefore equation (4.68) gives

$$C_4 = 0$$

and from (4.69)

$$C_3 = -\frac{ka^2}{2}.$$

Introducing all the determined constant values into (4.66) and (4.67) leads to

$$u = -\frac{\nu \rho g}{E} xy$$

$$v = \frac{\rho g}{2E}\left(y^2 + \nu x^2 - a^2\right).$$

We note, as physically expected, the thickness h of the plate does not enter the solution.

□

Example 4.2

Consider now that Figure 4.6 defines the section of a 3-D prismatic solid of length L in the z direction. At the end sections $z = L/2$ and $z = -L/2$ the displacements $w(x,y) = 0$ and the surface tractions $f_x^S = f_y^S = 0$. The rotation about the line $(x = 0, y = a, z)$ is prevented and the displacements u and v of points on this line are also prevented. At the lateral surfaces $y = 0$, $x = -b/2$ and $x = b/2$ there are no surface tractions. Find the displacement field and the tractions at the end sections $z = L/2$ and $z = -L/2$ when the solid is subjected to its own weight and to a uniform distribution of surface tractions $\mathbf{f}^S = f_y^S \mathbf{e}_y = \rho g a \mathbf{e}_y$ at the surface $y = a$ (ρ is the density and g the acceleration due to gravity).

Solution

Under the stated conditions, we have a plane strain situation with the same load and boundary conditions as for the plane stress problem of Figure 4.6. Therefore, the in-plane solution is directly obtained from the plane stress solution using the effective elastic constants $E_* = \frac{E}{1-\nu^2}$ and $\nu_* = \frac{\nu}{1-\nu}$ in place of E and ν. Hence

$$
\begin{aligned}
u &= -\frac{\nu(1+\nu)}{E}\rho g x y \\
v &= \frac{(1-\nu^2)\rho g}{2E}\left(y^2 + \frac{\nu}{1-\nu}x^2 - a^2\right)
\end{aligned}
$$

and, of course, $w = 0$. At the end section with normal $\mathbf{e}_z$, we obtain

$$
\mathbf{f}^S = \tau_{zz}\mathbf{e}_z = \nu(\tau_{xx} + \tau_{yy})\mathbf{e}_z = \nu\rho g y \mathbf{e}_z
$$

and at the opposite end section

$$
\mathbf{f}^S = -\tau_{zz}\mathbf{e}_z = -\nu\rho g y \mathbf{e}_z.
$$

□

4.1.3 The axisymmetric model

Cylindrical coordinate system

For the definition of the axisymmetric model a cylindrical coordinate system is commonly used and effective. In Figure 4.7, a Cartesian and a cylindrical coordinate system are given. As long as the z axis, the horizontal plane and the line from which the angle θ is measured are defined, the location of every point in 3-D space is uniquely given by the coordinates r, θ and z.

Associated with every point, we define orthonormal base vectors $\mathbf{e}_r$, $\mathbf{e}_\theta$ and $\mathbf{e}_z$. The unit vectors $\mathbf{e}_r$ and $\mathbf{e}_z$ have the direction of r and z, respectively, and $\mathbf{e}_\theta$ is orthogonal to the plane of $\mathbf{e}_r$ and $\mathbf{e}_z$. Therefore, from point to point, $\mathbf{e}_r$ and $\mathbf{e}_\theta$ change directions, different from using a Cartesian coordinate system for which $\mathbf{e}_x$, $\mathbf{e}_y$ and $\mathbf{e}_z$ are always the same for every point.

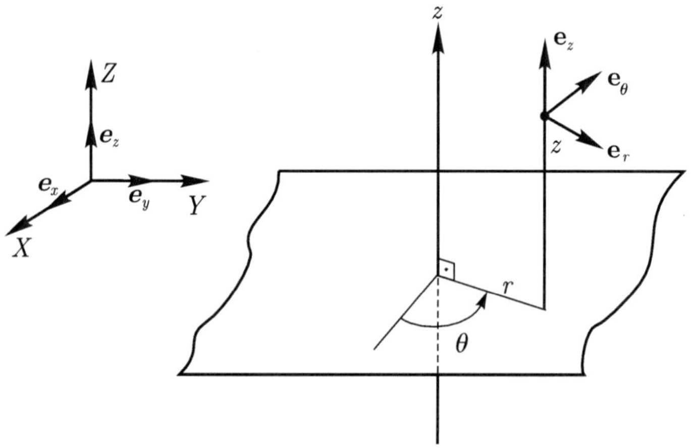

Fig. 4.7. Definition of a cylindrical coordinate system

The stress components in a cylindrical system are referred to the local system $(\mathbf{e}_r, \mathbf{e}_\theta, \mathbf{e}_z)$ which changes from point to point.

Equilibrium

A differential element of a solid at a given point corresponding to infinitesimal increments dr, $d\theta$ and dz and the stress components are shown in Figure 4.8. Further, in Figure 4.9 a top view is given. We have also in this system that $\tau_{r\theta} = \tau_{\theta r}$, $\tau_{rz} = \tau_{zr}$ and $\tau_{\theta z} = \tau_{z\theta}$. Equilibrium in the direction $\mathbf{e}_r$ requires

$$\left(\tau_{rr} + \frac{\partial \tau_{rr}}{\partial r} dr\right)(r + dr) d\theta dz - \tau_{rr} r d\theta dz$$

$$- \left(2\tau_{\theta\theta} + \frac{\partial \tau_{\theta\theta}}{\partial \theta} d\theta\right) dr dz \frac{d\theta}{2}$$

$$+ \left(\tau_{r\theta} + \frac{\partial \tau_{r\theta}}{\partial \theta} d\theta\right) dr dz - \tau_{r\theta} dr dz$$

$$+ \left(\tau_{rz} + \frac{\partial \tau_{rz}}{\partial z} dz\right)\left(r + \frac{dr}{2}\right) d\theta dr - \tau_{rz}\left(r + \frac{dr}{2}\right) d\theta dr$$

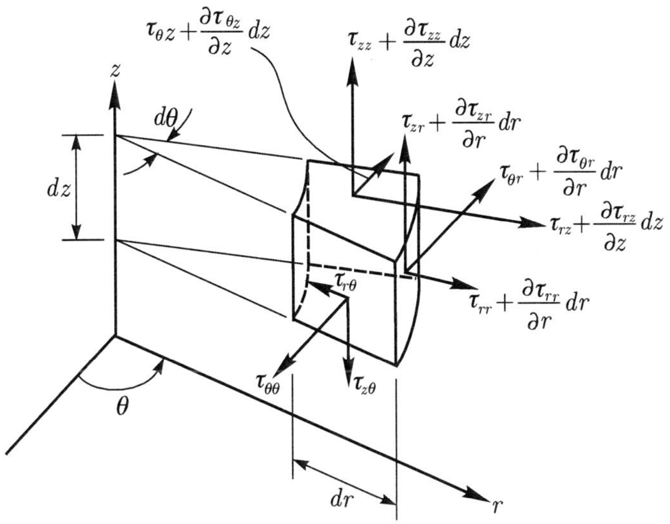

Fig. 4.8. Stresses on a differential element described by cylindrical coordinates. Stresses are shown only on visible faces

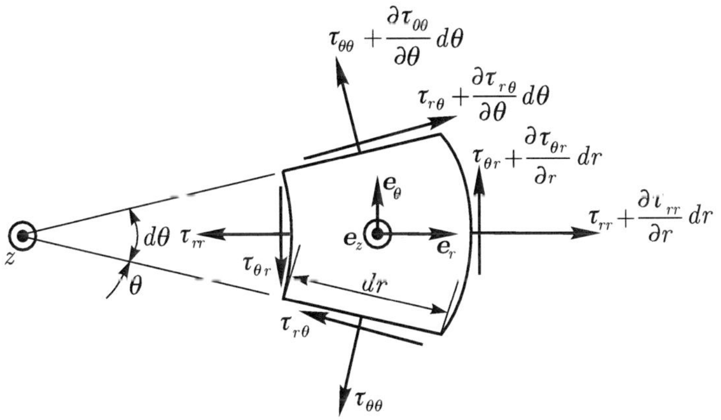

Fig. 4.9. Stresses on a differential element described by cylindrical coordinates. Stresses are shown only on visible faces

$$+f_r^B\left(r+\frac{dr}{2}\right)d\theta dr dz = 0.$$

Neglecting infinitesimals of fourth order, we obtain

$$\frac{\partial \tau_{rr}}{\partial r}+\frac{1}{r}\frac{\partial \tau_{r\theta}}{\partial \theta}+\frac{\partial \tau_{rz}}{\partial z}+\frac{\tau_{rr}-\tau_{\theta\theta}}{r}+f_r^B = 0. \tag{4.70}$$

Equilibrium in the direction $\mathbf{e}_\theta$ requires

$$\left(\tau_{\theta r} + \frac{\partial \tau_{\theta r}}{\partial r} dr\right)(r + dr) d\theta dz - \tau_{\theta r} r d\theta dz$$
$$+ \left(\tau_{\theta\theta} + \frac{\partial \tau_{\theta\theta}}{\partial \theta} d\theta\right) dr dz - \tau_{\theta\theta} dr dz$$
$$+ \left(\tau_{r\theta} + \frac{\partial \tau_{r\theta}}{\partial \theta} d\theta\right) dr dz \frac{d\theta}{2} - \tau_{r\theta} dr dz \frac{d\theta}{2}$$
$$+ \left(\tau_{\theta z} + \frac{\partial \tau_{\theta z}}{\partial z} dz\right)\left(r + \frac{dr}{2}\right) d\theta dr - \tau_{\theta z}\left(r + \frac{dr}{2}\right) d\theta dr$$
$$+ f_\theta^B \left(r + \frac{dr}{2}\right) d\theta dr dz = 0$$

leading to

$$\frac{\partial \tau_{\theta r}}{\partial r} + \frac{1}{r}\frac{\partial \tau_{\theta\theta}}{\partial \theta} + \frac{\partial \tau_{\theta z}}{\partial z} + \frac{\tau_{\theta r}}{r} + \frac{\tau_{r\theta}}{r} + f_\theta^B = 0. \tag{4.71}$$

And for the direction z

$$\left(\tau_{zr} + \frac{\partial \tau_{zr}}{\partial r} dr\right)(r + dr) d\theta dz - \tau_{zr} r d\theta dz$$
$$+ \left(\tau_{z\theta} + \frac{\partial \tau_{z\theta}}{\partial \theta} d\theta\right) dr dz - \tau_{z\theta} dr dz$$
$$+ \left(\tau_{zz} + \frac{\partial \tau_{zz}}{\partial z} dz\right)\left(r + \frac{dr}{2}\right) d\theta dr - \tau_{zz}\left(r + \frac{dr}{2}\right) d\theta dr$$
$$+ f_z^B \left(r + \frac{dr}{2}\right) d\theta dr dz = 0$$

which leads to

$$\frac{\partial \tau_{zr}}{\partial r} + \frac{1}{r}\frac{\partial \tau_{z\theta}}{\partial \theta} + \frac{\partial \tau_{zz}}{\partial z} + \frac{\tau_{zr}}{r} + f_z^B = 0. \tag{4.72}$$

Summarizing, the equilibrium conditions in the cylindrical coordinate system are

$$\begin{aligned}
\frac{\partial \tau_{rr}}{\partial r} + \frac{1}{r}\frac{\partial \tau_{r\theta}}{\partial \theta} + \frac{\partial \tau_{rz}}{\partial z} + \frac{\tau_{rr} - \tau_{\theta\theta}}{r} + f_r^B &= 0 \\
\frac{\partial \tau_{\theta r}}{\partial r} + \frac{1}{r}\frac{\partial \tau_{\theta\theta}}{\partial \theta} + \frac{\partial \tau_{\theta z}}{\partial z} + 2\frac{\tau_{\theta r}}{r} + f_\theta^B &= 0 \\
\frac{\partial \tau_{zr}}{\partial r} + \frac{1}{r}\frac{\partial \tau_{z\theta}}{\partial \theta} + \frac{\partial \tau_{zz}}{\partial z} + \frac{\tau_{zr}}{r} + f_z^B &= 0.
\end{aligned}$$

Let us now introduce axisymmetric conditions, *i.e.*, we consider a solid of revolution which is subjected to an axisymmetric load distribution. A typical situation is presented in Figure 4.10. We note that under these conditions

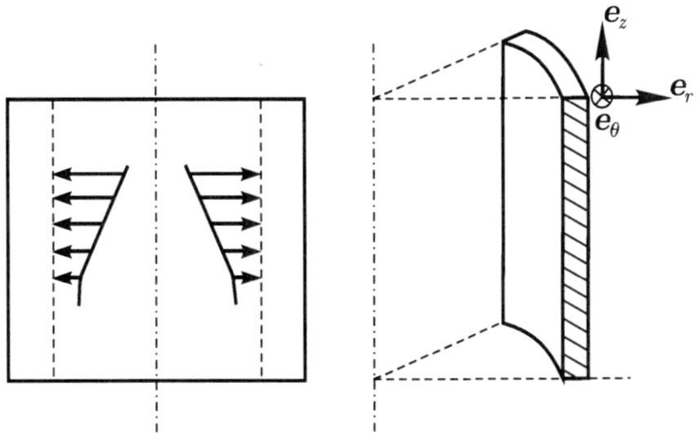

Fig. 4.10. A typical solid with an axisymmetric geometry and loading

$$f_\theta^B = 0$$

$$\frac{\partial(\cdot)}{\partial\theta} = 0 \text{ for any stress component}$$

$$\tau_{r\theta} = \tau_{z\theta} = 0$$

and the equilibrium conditions simplify to

$$\frac{\partial\tau_{rr}}{\partial r} + \frac{\partial\tau_{rz}}{\partial z} + \frac{\tau_{rr} - \tau_{\theta\theta}}{r} + f_r^B = 0 \tag{4.73}$$

$$\frac{\partial\tau_{zr}}{\partial r} + \frac{\partial\tau_{zz}}{\partial z} + \frac{\tau_{zr}}{r} + f_z^B = 0 \tag{4.74}$$

and, of course, equation (4.71) is identically satisfied.

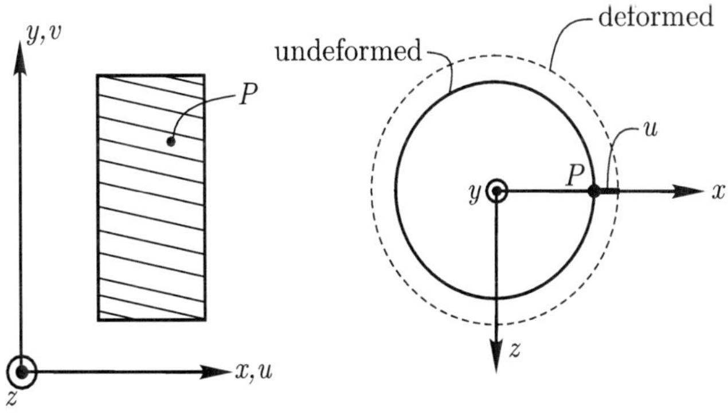

Fig. 4.11. Displacement of a generic point under axisymmetric conditions

When we have axisymmetric conditions, the displacements, stresses and strains are the same for any plane which contains the axis of symmetry.

Therefore, the problem can be formulated in a plane and it is customary to use a Cartesian coordinate system for that plane, as shown in Figure 4.11. Here y is now the axis of symmetry. In the coordinate system of Figure 4.11, the equilibrium equations (4.73) and (4.74) are

$$\frac{\partial \tau_{xx}}{\partial x} + \frac{\partial \tau_{xy}}{\partial y} + \frac{\tau_{xx} - \tau_{zz}}{x} + f_x^B = 0 \tag{4.75}$$

$$\frac{\partial \tau_{xy}}{\partial x} + \frac{\partial \tau_{yy}}{\partial y} + \frac{\tau_{xy}}{x} + f_y^B = 0. \tag{4.76}$$

Note that these equilibrium equations are the plane stress and plane strain equilibrium equations with $\frac{\tau_{xx}-\tau_{zz}}{x}$ added in the first equation and $\frac{\tau_{xy}}{x}$ in the second equation.

Strain compatibility

Due to the axisymmetric conditions, the displacements for material particles in the xy plane in Figure 4.11 are given by

$$\begin{aligned} u &= u(x, y) \\ v &= v(x, y) \\ w &= 0 \end{aligned}$$

and the strains

$$\begin{aligned} \varepsilon_{xx} &= \frac{\partial u}{\partial x} \\ \varepsilon_{yy} &= \frac{\partial v}{\partial y} \\ \gamma_{xy} &= \frac{\partial u}{\partial y} + \frac{\partial v}{\partial x} \\ \gamma_{xz} &= \gamma_{yz} = 0 \end{aligned}$$

Figure 4.11 also shows that a displacement $u(x, y)$ actually means that a circumference of radius x of material points of the solid deforms into a circumference of radius $x + u$ as shown. Therefore, we can evaluate the *circumferential* or *hoop* strain ε_{zz} by

$$\varepsilon_{zz} = \frac{2\pi(x+u) - 2\pi x}{2\pi x} = \frac{u}{x}.$$

Constitutive relations

It is convenient to define stress and strain column matrices without including the zero stress and strain components, *i.e.*,

$$\tau = \begin{bmatrix} \tau_{xx} \\ \tau_{yy} \\ \tau_{xy} \\ \tau_{zz} \end{bmatrix}, \qquad \varepsilon = \begin{bmatrix} \varepsilon_{xx} \\ \varepsilon_{yy} \\ \gamma_{xy} \\ \varepsilon_{zz} \end{bmatrix}$$

The constitutive equation is given by $\tau = \mathbf{C}\varepsilon$ with

$$\mathbf{C} = \frac{E(1-\nu)}{(1+\nu)(1-2\nu)} \begin{bmatrix} 1 & \frac{\nu}{1-\nu} & 0 & \frac{\nu}{1-\nu} \\ \frac{\nu}{1-\nu} & 1 & 0 & \frac{\nu}{1-\nu} \\ 0 & 0 & \frac{1-2\nu}{2(1-\nu)} & 0 \\ \frac{\nu}{1-\nu} & \frac{\nu}{1-\nu} & 0 & 1 \end{bmatrix}. \tag{4.77}$$

which is obtained from the $\mathbf{C}$ for 3-D conditions given by (3.113).

Differential formulation

Let us consider a solid of revolution for which a generic cross-section A is shown in Figure 4.12 subjected to axisymmetric loads and displacement boundary conditions. The volume V of the solid corresponds to revolving A about the y axis. Usually one radian is considered, see Section 5.2.4. On the lateral surface S_u, which corresponds to revolving the boundary line L_u, the displacements are prescribed. On the lateral surface S_f, which corresponds to revolving the boundary line L_f, surface tractions are applied. Body forces $\mathbf{f}^B$ are also present. Under these conditions the solution of the 3-D elasticity problem can be formulated in the plane domain described in Figure 4.12.

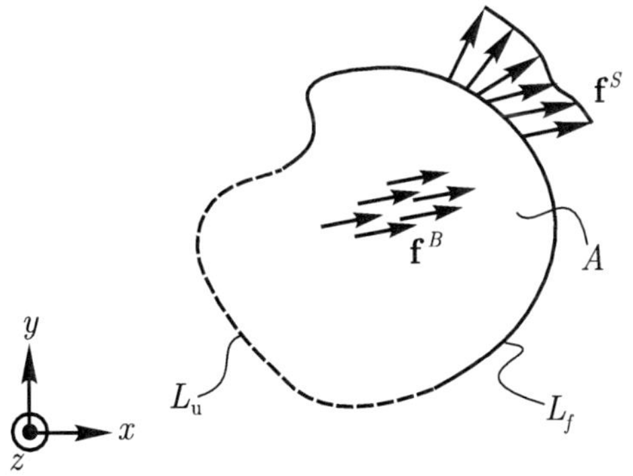

Fig. 4.12. Domain representing a generic cross-section of the solid of revolution

Differential formulation of the axisymmetric model

Given $f_x^B = f_x^B(x,y)$ and $f_y^B = f_y^B(x,y)$ defined in V, find $u(x,y)$, $v(x,y)$; $\tau_{xx} = \tau_{xx}(x,y)$, $\tau_{yy} = \tau_{yy}(x,y)$, $\tau_{xy} = \tau_{xy}(x,y)$ and $\tau_{zz} = \tau_{zz}(x,y)$; $\varepsilon_{xx} = \varepsilon_{xx}(x,y)$, $\varepsilon_{yy} = \varepsilon_{yy}(x,y)$, $\gamma_{xy} = \gamma_{xy}(x,y)$ and $\varepsilon_{zz} = \varepsilon_{zz}(x,y)$ such that

$$\frac{\partial \tau_{xx}}{\partial x} + \frac{\partial \tau_{xy}}{\partial y} + \frac{\tau_{xx} - \tau_{zz}}{x} + f_x^B = 0$$

$$\frac{\partial \tau_{xy}}{\partial x} + \frac{\partial \tau_{yy}}{\partial y} + \frac{\tau_{xy}}{x} + f_y^B = 0$$

$$\varepsilon_{xx} = \frac{\partial u}{\partial x}, \; \varepsilon_{yy} = \frac{\partial v}{\partial y}$$

$$\gamma_{xy} = \frac{\partial u}{\partial y} + \frac{\partial v}{\partial x}, \; \varepsilon_{zz} = \frac{u}{x}$$

$$\tau = \mathbf{C}\varepsilon$$

for every point in V and for the $\mathbf{C}$ given in (4.77). The boundary conditions are

$$f_x^S = \hat{f}_x^S(x,y), \; f_y^S = \hat{f}_y^S(x,y)$$

for every point in S_f and

$$u = \hat{u}(x,y), \; v = \hat{v}(x,y)$$

for every point in S_u.

We note that this solution gives the exact solution of the 3-D problem with

$$w = 0$$

and

$$\tau_{xz} = \tau_{yz} = 0$$

$$\gamma_{xz} = \gamma_{yz} = 0.$$

Example 4.3

Consider a disc with a central circular hole. The disc is subjected to internal pressure p_i and external pressure p_e. The problem description is given in Figure 4.13. Determine the stress distribution and the displacements.

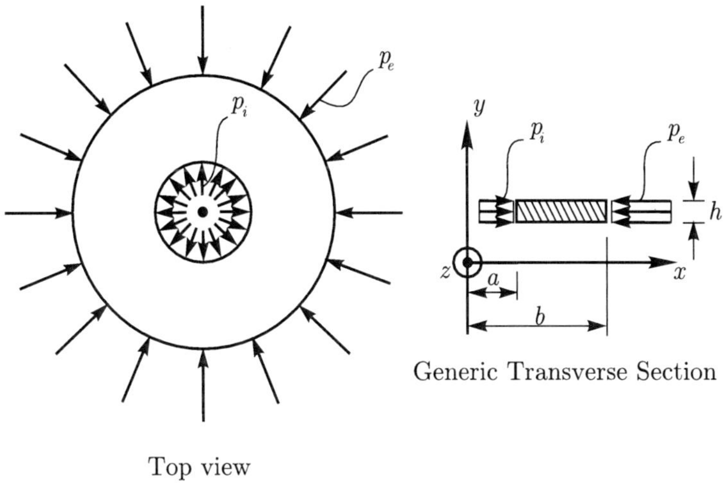

Fig. 4.13. Schematic description of disc problem

Solution

The disc is a solid of revolution subjected to axisymmetric loading. Considering the load given and that the disc is thin, plane stress conditions also apply.

Therefore

$$\tau_{xy} = 0, \; \tau_{yy} = 0$$

and the equilibrium condition given in equation (4.75) simplifies to

$$\frac{d\tau_{xx}}{dx} + \frac{\tau_{xx} - \tau_{zz}}{x} = 0. \tag{4.78}$$

Since there are no body forces, Equation (4.76) is identically satisfied. The relevant strain components are

$$\varepsilon_{xx} = \frac{du}{dx} \tag{4.79}$$

and

$$\varepsilon_{zz} = \frac{u}{x}. \tag{4.80}$$

The plane stress constitutive equation with the shear strain (stress) zero can be directly used, *i.e.*

$$\tau_{xx} = \frac{E}{1-\nu^2}\left(\varepsilon_{xx} + \nu\varepsilon_{zz}\right) \tag{4.81}$$

$$\tau_{zz} = \frac{E}{1-\nu^2}\left(\varepsilon_{zz} + \nu\varepsilon_{xx}\right). \tag{4.82}$$

Introducing (4.79) and (4.80) into (4.81) and (4.82) yields

$$\tau_{xx} = \frac{E}{1-\nu^2}\left(\frac{du}{dx} + \nu\frac{u}{x}\right) \tag{4.83}$$

$$\tau_{zz} = \frac{E}{1-\nu^2}\left(\frac{u}{x} + \nu\frac{du}{dx}\right) \tag{4.84}$$

and substituting (4.83) and (4.84) into the equilibrium equation (4.78) leads to

$$\frac{d^2u}{dx^2} + \frac{1}{x}\frac{du}{dx} - \frac{u}{x^2} = 0.$$

The above ordinary differential equation has a general solution given by

$$u = C_1 x + \frac{C_2}{x} \tag{4.85}$$

where C_1 and C_2 are constants to be determined. The general expression for the stresses can be obtained by introducing (4.85) into (4.83) and (4.84) leading to

$$\tau_{xx} = \frac{E}{1-\nu^2}\left[C_1(1+\nu) - C_2\frac{(1-\nu)}{x^2}\right] \tag{4.86}$$

$$\tau_{zz} = \frac{E}{1-\nu^2}\left[C_1(1+\nu) + C_2\frac{(1-\nu)}{x^2}\right]. \tag{4.87}$$

We are now ready to impose the force boundary conditions. At the internal surface of the disc we have

$$\tau_{xx}\big|_{x=a} = -p_i \tag{4.88}$$

and at the external surface

$$\tau_{xx}\big|_{x=b} = -p_e. \tag{4.89}$$

Equation (4.86) subjected to the conditions given by (4.88) and (4.89) leads to

$$C_1 = \frac{1-\nu}{E}\frac{a^2p_i - b^2p_e}{b^2-a^2}$$

$$C_2 = \frac{1+\nu}{E}\frac{a^2b^2(p_i-p_e)}{b^2-a^2}.$$

Hence, the stresses are given by

$$\tau_{xx} = \frac{a^2p_i - b^2p_e}{b^2-a^2} - \frac{(p_i-p_e)a^2b^2}{x^2\,(b^2-a^2)}$$

$$\tau_{zz} = \frac{a^2p_i - b^2p_e}{b^2-a^2} + \frac{(p_i-p_e)a^2b^2}{x^2\,(b^2-a^2)}.$$

Note that

$$\tau_{xx} + \tau_{zz} = 2\left(\frac{a^2p_i - b^2p_e}{b^2-a^2}\right)$$

is a constant, *i.e.*, independent of the point where it is evaluated. Therefore

$$\varepsilon_{yy} = -\nu\,(\tau_{xx} + \tau_{zz}) \tag{4.90}$$

is also a constant. Hence, the out-of-plane displacements v are independent of x and any horizontal plane remains horizontal after deformation (this also means that the distortion γ_{xy} is indeed zero and, of course, $\gamma_{yz} = \gamma_{xz} = 0$ due to the axisymmetric conditions). Hence, we conclude that we obtained the exact 3-D solution and the assumptions of the plane stress model in this case did not lead to an approximate solution.

We also note that, since horizontal planes remain horizontal after deformation the derived solution is also valid for any thickness of the disc, *i.e.*, it is valid not only for a thin disc but also for long cylinders, as long as the out-of-plane displacements are not restricted. In case we prevent the out-of-plane displacements, a plane strain condition develops and the solution can be readily obtained by using the effective material constants, *i.e.*, by replacing E by E_* and ν by ν_* in the solution of the in-plane variables, that is, $\tau_{xx}(x)$, $\tau_{zz}(x)$ and $u(x)$. Note that ε_{yy} is not an in-plane variable for the plane stress problem considered here and hence it is not obtained by replacing ν by ν_* in (4.90). Of course, $\varepsilon_{yy} = 0$ for plane strain conditions.

To conclude this example solution the radial displacements can be evaluated by substituting the value of the constants C_1 and C_2 into equation (4.85) leading to

$$u(x) = \frac{1-\nu}{E}\frac{a^2p_i - b^2p_e}{b^2-a^2}x + \frac{1+\nu}{E}\frac{a^2b^2(p_i-p_e)}{(b^2-a^2)\,x}$$

□

Before we move to the next section and embark on the discussion of the mathematical models of bars, it is timely to place such bar models and the forthcoming plate and shell models in the larger context of structural mechanics mathematical modeling.

If we look at the developments presented in this section and compare, from the hierarchical modeling perspective, the 2-D models with the 3-D elasticity model – our highest order mathematical model in linear analysis – there is an important point. As long as the geometric, loading and boundary condition restrictions are satisfied, the plane strain and axisymmetric models lead to the exact solution of the 3-D elasticity problem. Therefore, the modeling considerations should only focus on judging whether the assumptions used in the plane strain or, respectively the axisymmetric model, with respect to the 3-D model are (sufficiently) satisfied in the actual physical problem.

Such situation contrasts with the plane stress model. In such model even if the geometric, loading and boundary condition restrictions are exactly satisfied , its solution is not, in general, the solution of the associated 3-D model, *i.e.* that for which the plate is modeled as a 3-D solid. Hence, the plane stress model is what we refer to be a structural model. In what follows we will be formulating bars, plates and shell models which considering the interpretations given above are characterized also as structural models. We summarize these observations in Figure 4.14.

4.2 Bar models

There are a number of bar models which are associated with different geometric, loading and kinematic boundary conditions. Also, there are different nomenclatures in the technical literature associated with bar models, and hence we adopt a terminology which best fits the aim to use the models in the hierarchical modeling process.

Generically, a bar can be understood to be a slender 3-D solid which has one dimension much larger than the other two dimensions, which are of the same order of magnitude. We could say that if a, b and c are characteristic dimensions of a solid in three orthogonal directions, such solid can be assumed to be a bar if a is of the order of b, *i.e.*, $\frac{1}{5} \leq \frac{a}{b} \leq 5$ and $\frac{c}{\max(a,b)} \geq 10$. Then, it is usual to characterize the bar's geometry from the definition of a curve called the bar axis. At any point of the bar axis, a plane region, orthogonal to the bar axis, is defined. Such plane region is referred to as the transverse cross-section of the bar at this point. The collection of these transverse cross-sections characterize the bar's geometry, where the bar axis is taken to pass through the centers of gravity of the cross-sections. In Figure 4.15, a generic bar is shown.

We can identify a number of different situations. For example, the bar axis can be a three-dimensional curve, a planar curve or a straight line. The cross-sections may have a constant or varying shape. Moreover, of course, the bar can be loaded by various different external forces. Depending on the

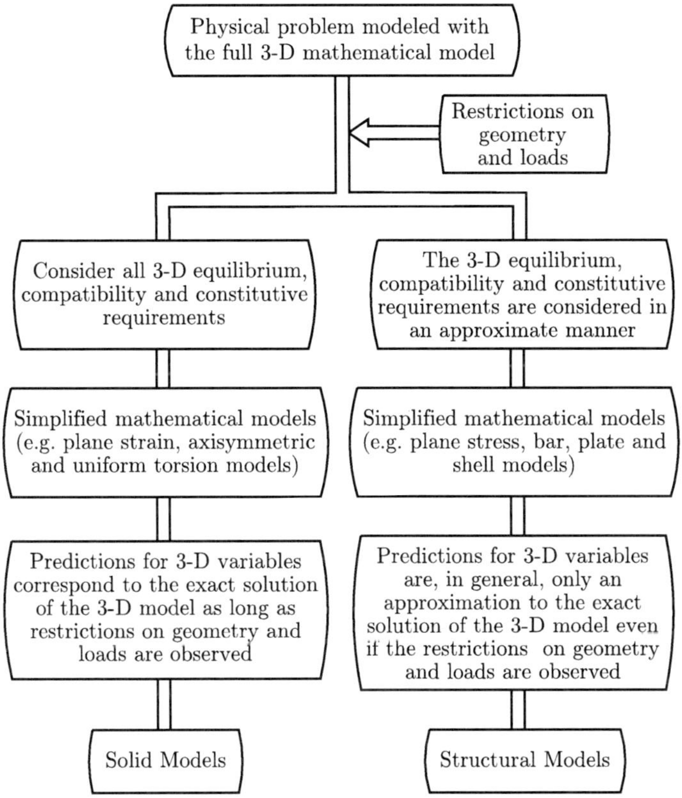

Fig. 4.14. Summary of mathematical modeling in solid and structural mechanics

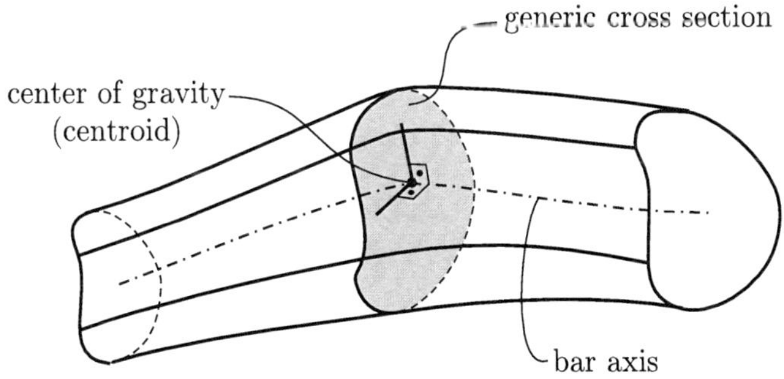

Fig. 4.15. A generic bar geometry

combination of the geometric and loading characteristics of the problem, the bar structural behavior may be different. Therefore, there are a number of mathematical models for bars which are associated with specific geometric and loading conditions.

As a first approach to formulate bar models, we consider the bar as a 3-D solid and impose geometric and loading restrictions. Also, we introduce assumptions for the stress and displacement fields. Then, the 3-D elasticity equations are used to obtain the differential formulation of the bar model.

This is the approach that we followed to obtain the plane elasticity models and which we now use to derive the bar models. The approach gives insight into how well the bar solutions satisfy the 3-D equations.

4.2.1 Prismatic bar subjected to axial loading

The assumptions used for the mathematical model are (see also Figure 4.16):

- *Geometry*: The solid is a prismatic bar (a bar of constant cross-section and straight axis).
- *Kinematics*: The cross-sections remain plane and displace only in the axial direction (they do not rotate). The section displacements are given by $u = u(x)$.
- *External loading and boundary conditions*: The body forces per unit of volume are given by $\mathbf{f}^B = f_x^B(x)\mathbf{e}_x$.
  At the section $x = 0$ either $f_x^S = f_0^S$ or $u = u_0$, and $f_y^S = f_z^S = 0$.
  At the section $x = L$ either $f_x^S = f_L^S$ or $u = u_L$, and $f_y^S = f_z^S = 0$.
  On the lateral surfaces of the bar $\mathbf{f}^S = \mathbf{0}$.
- *Stresses*: The normal stress τ_{xx} is the only nonzero stress component.

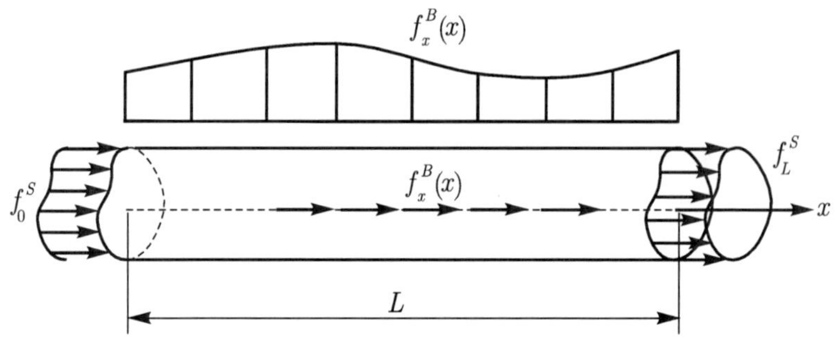

Fig. 4.16. Schematic representation of the model. Either displacement or force boundary conditions should be specified at the end sections. Bar cross-sectional area A

Now let us consider the 3-D elasticity equations.

Equilibrium

Considering the differential equilibrium equations (3.114), the first equation leads to

$$\frac{\partial \tau_{xx}}{\partial x} + f_x^B = 0 \tag{4.91}$$

and the remaining equations are identically satisfied.

Constitutive equations
Considering the generalized Hooke's law given in (3.116), we obtain

$$\varepsilon_{xx} = \frac{\tau_{xx}}{E}, \ \varepsilon_{yy} = -\frac{\nu}{E}\tau_{xx}, \ \varepsilon_{zz} = -\frac{\nu}{E}\tau_{xx} \tag{4.92}$$

$$\gamma_{xy} = 0, \ \gamma_{yz} = 0, \ \gamma_{zx} = 0. \tag{4.93}$$

Strain-displacement relations

Introducing (4.92) and (4.93) into (3.115) we obtain

$$\frac{\partial u}{\partial x} = \frac{\tau_{xx}}{E} \tag{4.94}$$

$$\frac{\partial v}{\partial y} = -\frac{\nu}{E}\tau_{xx} \tag{4.95}$$

$$\frac{\partial w}{\partial z} = -\frac{\nu}{E}\tau_{xx} \tag{4.96}$$

$$\frac{\partial u}{\partial y} + \frac{\partial v}{\partial x} = 0 \tag{4.97}$$

$$\frac{\partial v}{\partial z} + \frac{\partial w}{\partial y} = 0. \tag{4.98}$$

$$\frac{\partial u}{\partial z} + \frac{\partial w}{\partial x} = 0 \tag{4.99}$$

Boundary conditions

On the surface given by $x = 0$
either $f_0^S = -\tau_{xx}(0)$ or $u(0) = u_0$
$f_y^S = -\tau_{yx} = 0,\ f_z^S = -\tau_{zx} = 0$ (identically satisfied).

On that given by $x = L$
either $f_L^S = \tau_{xx}(L)$ or $u(L) = u_L$
$f_y^S = \tau_{yx} = 0,\ f_z^S = \tau_{zx} = 0$ (identically satisfied).

On the lateral surface

$$\mathbf{Tn} = \mathbf{0}$$

or

$$\begin{bmatrix} \tau_{xx} & 0 & 0 \\ 0 & 0 & 0 \\ 0 & 0 & 0 \end{bmatrix} \begin{bmatrix} 0 \\ n_y \\ n_z \end{bmatrix} = \begin{bmatrix} 0 \\ 0 \\ 0 \end{bmatrix}$$

which is identically satisfied.

We can obtain a solution solving (4.91) for τ_{xx} and then (4.94) for $u(x)$. Note, however, that when $f_x^B(x) \neq 0$, we obtain $\tau_{xx} = \tau_{xx}(x)$ and from (4.92)

$$\varepsilon_{yy} = \frac{\partial v}{\partial y} = -\frac{\nu}{E}\tau_{xx}(x) \tag{4.100}$$

$$\varepsilon_{zz} = \frac{\partial w}{\partial z} = -\frac{\nu}{E}\tau_{xx}(x) \tag{4.101}$$

which for varying x means a varying Poisson effect with x, unless $\nu = 0$. Since this induced extension/contraction of the fibers in the plane of the section is different for neighboring sections, the sections have to warp to keep the transverse shear strains ($\gamma_{yz} = \gamma_{xz} = 0$) equal to zero. Then, of course, $u = u(x)$ is violated, that is, the kinematic assumption "the cross-sections remain plane and displace only in the axial direction (they do not rotate)" is violated. However, in the bar model this effect is neglected and, hence, using this model we in essence assume that the physical link between two neighboring sections is given by rollers *i.e.* the contraction/extension of the fibers in the section does not affect the deformation of neighboring sections.

More formally, integrating of (4.100) with respect to y leads to

$$v = -\frac{\nu}{E}\tau_{xx}(x)y + F(x,z)$$

where $F(x, z)$ is a function of x and z only. This equation substituted into (4.97) gives

$$\frac{\partial u}{\partial y} = \frac{\nu}{E}\frac{\partial \tau_{xx}}{\partial x} y - \frac{\partial F}{\partial x}(x, z)$$

which can not be satisfied if $u = u(x)$ unless $\nu = 0$. A similar conclusion would arise if (4.101) and (4.99) were considered.

Now the 1-D bar model can be detailed.

Let $\tau = \tau_{xx}$ and $f = f_x^B A$ be the distributed axial force per unit of length. Also let $N = \tau A$ be the axial force. Hence, the equilibrium equation (4.91) can be re-written as

$$\frac{dN}{dx} + f = 0.$$

Defining $\varepsilon = \varepsilon_{xx}$, the strain compatibility relation is given by

$$\varepsilon = \frac{du}{dx}$$

and the constitutive relation by

$$\tau = E\varepsilon.$$

The boundary condition at $x = 0$ is either a displacement boundary condition

$$u(0) = u_0$$

where u_0 is the prescribed displacement, or a force boundary condition

$$N(0) = \tau(0)A = -f_0^S A = -R_0$$

where R_0 is the prescribed concentrated force with positive sense given by the x axis. Analogously, at $x = L$, we have either

$$u(L) = u_L$$

where u_L is the prescribed displacement at L, or

$$N(L) = \tau(L)A = f_L^S A = R_L$$

where R_L is the prescribed concentrated force at $x = L$ with the same sense convention as that adopted for R_0. The differential formulation is summarized below.

Differential formulation of bar subjected to axial loading

Given the axial distributed loading $f(x)$, find $N(x)$, $\varepsilon(x)$, $u(x)$ such that

$$\frac{dN}{dx} + f = 0 \tag{4.102}$$

$$\varepsilon = \frac{du}{dx} \tag{4.103}$$

$$N = EA\varepsilon \tag{4.104}$$

for all x within the bar. At $x = 0$ we have either

$$u(0) = u_0 \qquad \text{or} \qquad N(0) = -R_0$$

and at $x = L$, either

$$u(L) = u_L \qquad \text{or} \qquad N(L) = R_L.$$

It is usual to write the differential formulation in terms of displacements only, by introducing (4.103) and (4.104) into (4.102) and using that

$$N(0) = EA\varepsilon(0) = EA\frac{du}{dx}(0)$$

and

$$N(L) = EA\varepsilon(L) = EA\frac{du}{dx}(L).$$

Differential formulation of bar subjected to axial loading in terms of displacements only

---

Given $f(x)$, find $u(x)$ such that

$$EA\frac{d^2u}{dx^2} + f = 0 \tag{4.105}$$

for all x within the bar. At $x = 0$, we have either

$$u(0) = u_0 \qquad \text{or} \qquad \frac{du}{dx}(0) = -\frac{R_0}{EA} \tag{4.106}$$

and at $x = L$, either

$$u(L) = u_L \qquad \text{or} \qquad \frac{du}{dx}(L) = \frac{R_L}{EA}. \tag{4.107}$$

Of course, when the problem is solved for $u(x)$, we can obtain $\varepsilon(x)$ and $N(x)$ using equations (4.103) and (4.104).

Once the 1-D differential formulation is solved, the solution for the 3-D problem based on the 1-D solution can be obtained:

$$u = u(x)$$

$$\tau_{xx} = \tau,\ \tau_{yy} = \tau_{zz} = \tau_{xy} = \tau_{xz} = \tau_{yz} = 0$$

$$\varepsilon_{xx} = \varepsilon = \frac{du}{dx},\ \varepsilon_{yy} = -\frac{\nu}{E}\tau,\ \varepsilon_{zz} = -\frac{\nu}{E}\tau$$

$$\gamma_{xy} = \gamma_{xz} = \gamma_{yz} = 0$$

and the displacements associated with the extension/contraction of the fibers in the cross-section can be evaluated using

$$v = -\frac{\nu}{E}\tau y \tag{4.108}$$

$$w = -\frac{\nu}{E}\tau z \tag{4.109}$$

which satisfies (4.95), (4.96), (4.98) and the condition that the bar axis has no transverse displacements.

Note that when $f_x^B(x) = 0$ or $\nu = 0$ the 1-D model leads to the exact solution of the 3-D problem.

Example 4.4

Consider a steel bar subjected to its own weight as shown in Figure 4.17.

The solution of the 3-D elasticity problem is given by

$$u = \frac{\rho g}{2E}(2Lx - x^2 - \nu(y^2 + z^2))$$

$$v = -\nu\frac{\rho g}{E}(L - x)y$$

$$w = -\nu\frac{\rho g}{E}(L - x)z$$

(i) Find the solution for the bar using the 1-D model.
(ii) Compare the solution obtained in (i) with the exact solution.

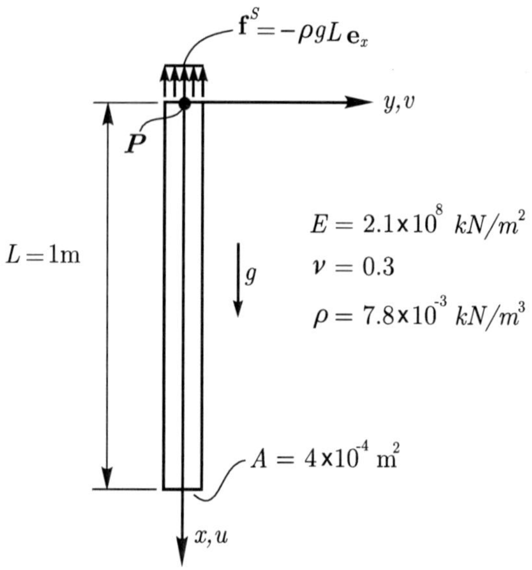

Fig. 4.17. Bar problem definition. Point P is fixed and the bar cannot rotate with respect to P

Solution

(i) The differential formulation for the 1-D model in terms of displacements is given by

$$EA\frac{d^2u}{dx^2} + \rho g A = 0$$

$$u(0) = 0$$

$$\frac{du}{dx}(L) = 0.$$

Solving, we obtain

$$u(x) = \frac{\rho g}{2E}(2Lx - x^2) \tag{4.110}$$

and the axial stress in given by

$$\tau = E\frac{du}{dx} = \rho g(L-x)$$

using equations (4.108) and (4.109), we obtain

$$v = -\nu\frac{\rho g}{E}(L-x)y \tag{4.111}$$

$$w = -\nu\frac{\rho g}{E}(L-x)z. \tag{4.112}$$

(ii) Comparing the 1-D solution with the exact solution, we observe that the predictions for v and w are exactly the same. The u displacements differ by the warping of the cross-section. In fact, we can write

$$u_{exact} = u_{one-dimensional} + u_{warp}$$

where $u_{one-dimensional}$ is the solution of the 1-D bar model and

$$u_{warp} = -\nu\frac{\rho g}{2E}(y^2+z^2)$$

is the warping displacement. Of course, when $\nu = 0$ the solutions of both models coincide. We note that for bar problems the warping displacements are small when compared to those predicted by the 1-D model. For example, considering the square cross-section for the bar in Figure 4.17, the ratio between the maximum warping displacement and that of the 1-D model at $x = L$ is 6×10^{-5}.

□

4.2.2 Prismatic bar subjected to transverse loading; the Bernoulli-Euler beam model.

This model plays a fundamental role in the formulation and understanding of mathematical models for structural analysis. The model serves as a reference for many structural models.

The external loading and geometry are selected to lead to bending in one plane only and to not induce torsion. The term beam is used to describe a bar when there is transverse loading that is transferred to the supports by bending.

The assumptions used for the mathematical model are (see also Figure 4.18):

- *Geometry*: The solid is a prismatic bar (a bar of constant cross-section and straight axis). The plane xz is a plane of symmetry.

- *Kinematics*: The cross-sections remain plane and orthogonal to the deformed bar axis. The bar axis goes through the centroid of the section, remains in the xz plane and the in-plane deformation of the cross-sections is neglected.
- *External loading and boundary conditions*: The load is transverse to the bar, that is, in the z direction and we choose to model it as body forces per unit of volume given by $\mathbf{f}^B = f_z^B(x)\mathbf{e}_z$.
  At the end sections either displacements or surface tractions are prescribed. On the lateral surfaces of the bar $\mathbf{f}^S = \mathbf{0}$.
- *Stresses*: The normal stress τ_{xx} and the transverse shear stress τ_{xz} are the only nonzero stress components.

We choose a bar of rectangular cross-section (see Figure 4.18) and consider the 3-D elasticity equations.

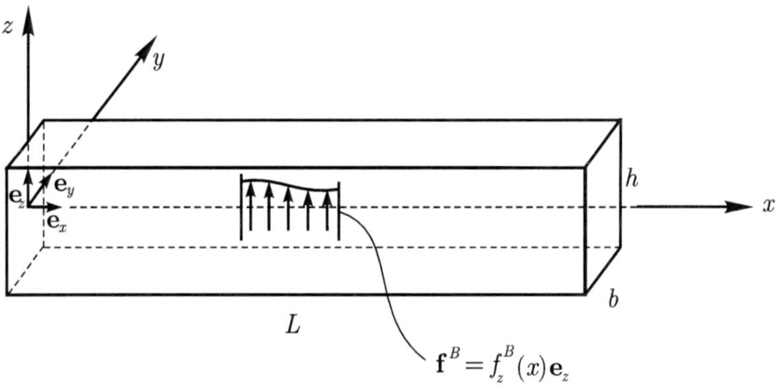

Fig. 4.18. Bar subjected to transverse loading

Equilibrium

Considering the differential equilibrium equations (3.114), the first and third equations lead to

$$\frac{\partial \tau_{xx}}{\partial x} + \frac{\partial \tau_{xz}}{\partial z} = 0 \tag{4.113}$$

$$\frac{\partial \tau_{xz}}{\partial x} + f_z^B(x) = 0 \tag{4.114}$$

and the second equation is identically satisfied.

Constitutive equations

Considering the generalized Hooke's law given in (3.116), we obtain

$$\varepsilon_{xx} = \frac{\tau_{xx}}{E}, \; \varepsilon_{yy} = -\frac{\nu}{E}\tau_{xx}, \; \varepsilon_{zz} = -\frac{\nu}{E}\tau_{xx} \tag{4.115}$$

$$\gamma_{xy} = 0, \; \gamma_{yz} = 0, \; \gamma_{zx} = \frac{\tau_{xz}}{G} \tag{4.116}$$

Strain-displacement relations

Introducing (4.115) and (4.116) into (3.115) we obtain

$$\frac{\partial u}{\partial x} = \frac{\tau_{xx}}{E} \tag{4.117}$$

$$\frac{\partial v}{\partial y} = -\frac{\nu}{E}\tau_{xx} \tag{4.118}$$

$$\frac{\partial w}{\partial z} = -\frac{\nu}{E}\tau_{xx} \tag{4.119}$$

$$\frac{\partial u}{\partial y} + \frac{\partial v}{\partial x} = 0 \tag{4.120}$$

$$\frac{\partial v}{\partial z} + \frac{\partial w}{\partial y} = 0. \tag{4.121}$$

$$\frac{\partial u}{\partial z} + \frac{\partial w}{\partial x} = \frac{\tau_{xz}}{G} \tag{4.122}$$

Let us now consider the kinematic assumptions which are pictorially described in Figure 4.19. Since we are assuming infinitesimally small displacements and we neglect the in-plane deformation of the cross-sections

$$w = w(x) \tag{4.123}$$

and from the hypothesis "the cross-sections remain plane and orthogonal to the deformed bar axis", we obtain

$$u = -z\frac{dw}{dx} \tag{4.124}$$

where we have also used that the displacements are infinitesimally small. Considering that "the bar axis remains in the xz plane" and again that "the in-plane deformation of the cross-sections is neglected"

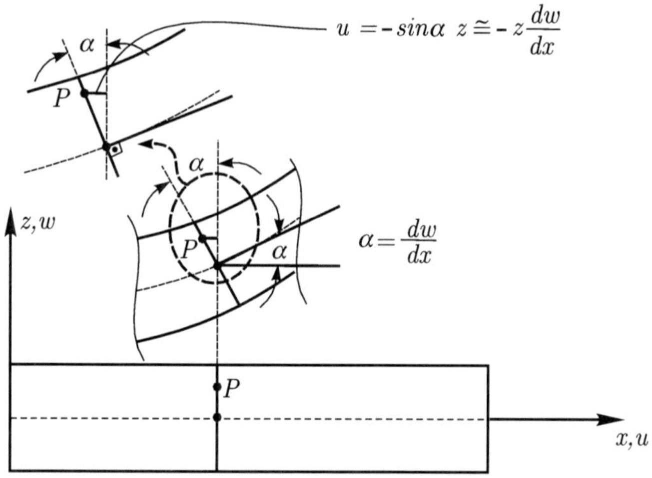

Fig. 4.19. Kinematics of beam sections

$$v = 0.$$

Note that considering the kinematic assumptions, equations (4.118), (4.119) and (4.122) can not be exactly satisfied. In fact, the violation of (4.118) and (4.119) means that we are neglecting the contraction/extension of the fibers in the plane of the cross-section due to Poisson's effect. And from (4.123) and (4.124)

$$\gamma_{xz} = \frac{\partial u}{\partial z} + \frac{\partial w}{\partial x} = -\frac{dw}{dx} + \frac{dw}{dx} = 0 \tag{4.125}$$

which corresponds to a violation of (4.122). This means that we are neglecting the induced shear strain γ_{xz} due to the shear stress τ_{xz}.

We also note that the displacements are fully determined once $w(x)$ is known.

From (4.117) and (4.124), we obtain

$$\tau_{xx} = -zE\frac{d^2w}{dx^2} \tag{4.126}$$

which substituted into (4.113) yields

$$-zE\frac{d^3w}{dx^3} + \frac{\partial \tau_{xz}}{\partial z} = 0.$$

Integrating the above equation with respect to z, we obtain

$$\tau_{xz} = \frac{z^2}{2}E\frac{d^3w}{dx^3} + f(x, y) \tag{4.127}$$

where $f(x,y)$ is a function of x and y only. Since the top and bottom surfaces of the bar are free of surface tractions, we have

$$\tau_{xz}\left(x,y,\frac{h}{2}\right)=0 \tag{4.128}$$

$$\tau_{xz}\left(x,y,-\frac{h}{2}\right)=0. \tag{4.129}$$

Considering (4.128), that is, the condition at the top surface and (4.127), we obtain

$$\frac{h^2}{8}E\frac{d^3w}{dx^3}+f(x,y)=0$$

which leads to

$$f(x,y)=g(x)=-\frac{h^2}{8}E\frac{d^3w}{dx^3}$$

and with this $f(x,y)=g(x)$, (4.127) also satisfies (4.129). Hence

$$\tau_{xz}=\frac{E}{2}\frac{d^3w}{dx^3}\left(z^2-\left(\frac{h}{2}\right)^2\right) \tag{4.130}$$

Substituting (4.130) into (4.114) yields

$$\frac{E}{2}\frac{d^4w}{dx^4}\left(z^2-\left(\frac{h}{2}\right)^2\right)+f_z^B(x)=0$$

which can be satisfied pointwise only at $z=0$, that is, at the bar axis. We can integrate the above equation over the cross-section to impose that it should be satisfied in average to obtain

$$\frac{d^4w}{dx^4}=\frac{f_z^B(x)\,A}{EI} \tag{4.131}$$

where $A=bh$ is the area of the cross-section and $I=\frac{bh^3}{12}$ is the moment of inertia of the cross-section with respect to the y axis. The boundary conditions at the lateral surfaces $\left(y=\pm\frac{b}{2}\right)$ are satisfied since there are no surface tractions applied and $\tau_{yy}=\tau_{yz}=0$.

We still have to consider the boundary conditions at the end surfaces given by $x=0$ and $x=L$.

Since the only nonzero stress components are τ_{xx} and τ_{xz}, the surface tractions at $x=0$ are $f_x^S=-\tau_{xx}$, $f_y^S=0$, $f_z^S=-\tau_{xz}$ and at $x=L$ $f_x^S=\tau_{xx}$, $f_y^S=0$, $f_z^S=\tau_{xz}$. Note that the stress components τ_{xx} and τ_{xz} are given in (4.126) and (4.130), and they are fully determined when the values of $\frac{d^2w}{dx^2}$ and $\frac{d^3w}{dx^3}$ have been obtained.

Hence, the differential formulation of the Bernoulli-Euler beam model is based on the solution of (4.131) for $w(x)$ subject to boundary conditions at $x = 0$ and $x = L$. Before we detail this differential formulation we define the section stress resultants

$$M(x) = \int_A -\tau_{xx} z \, dA \tag{4.132}$$

$$V(x) = \int_A -\tau_{xz} \, dA \tag{4.133}$$

where, of course, $M(x)$ is the bending moment and $V(x)$ is the shear force. Using (4.126) and (4.130), we obtain

$$M(x) = EI \frac{d^2 w}{dx^2} \tag{4.134}$$

$$V(x) = EI \frac{d^3 w}{dx^3} \tag{4.135}$$

and, hence,

$$\frac{dM}{dx} = V.$$

Also, we can define the transverse loading per unit of length

$$p(x) = f_z^B (x) \, A$$

which using (4.131) and (4.134) yields

$$\frac{d^2 M}{dx^2} = p.$$

We summarize below the differential formulation of the beam mathematical model.

Differential formulation of the Bernoulli-Euler beam model in terms of displacement only

---

Given the transverse distributed loading $p(x)$, find $w = w(x)$ such that

$$\frac{d^4 w}{dx^4} = \frac{p(x)}{EI} \tag{4.136}$$

for all x within the bar. At $x = 0$, we have either

$$w(0) = w_0 \qquad \text{or} \qquad EI \frac{d^3 w}{dx^3}(0) = Q_0 \tag{4.137}$$

and either

$$\frac{dw}{dx}(0) = \varphi_0 \qquad \text{or} \qquad EI\frac{d^2w}{dx^2}(0) = M_0. \tag{4.138}$$

At $x = L$, we have either

$$w(L) = w_L \qquad \text{or} \qquad EI\frac{d^3w}{dx^3}(L) = -Q_L \tag{4.139}$$

and either

$$\frac{dw}{dx}(L) = \varphi_L \qquad \text{or} \qquad EI\frac{d^2w}{dx^2}(L) = -M_L. \tag{4.140}$$

where M_0 and M_L are externally applied moments measured positive as moments about the y-axis, and Q_0 and Q_L are externally applied forces measured positive into the z-direction.

We note that

$$M_0\mathbf{e}_y = \left(\int_A f_x^S(0, y, z)\, z\, dA\right)\mathbf{e}_y$$

that is, $M_0\mathbf{e}_y$ is the moment which is equivalent to the surface tractions $f_x^S(0, y, z)$. Note the since $f_x^S(0, y, z) = -\tau_{xx}(0, y, z)$

$$M_0 = \int_A -\tau_{xx}(0, y, z)\, z\, dA = M(0) = EI\frac{d^2w}{dx^2}(0)$$

which gives the second equation of (4.138).

Also

$$Q_0\mathbf{e}_z = \left(\int_A f_z^S(0, y, z)\; dA\right)\mathbf{e}_z$$

that is, $Q_0\mathbf{e}_z$ is the force which is equivalent to the surface tractions $f_z^S(0, y, z)$. Note that since $f_z^S(0, y, z) = -\tau_{xz}(0, y, z)$

$$Q_0 = \int_A -\tau_{xz}(0, y, z)\; dA = V(0) = EI\frac{d^3w}{dx^3}(0)$$

which is the second equation of (4.137). Analogous interpretations hold for $x = L$. In summary, at the bar end sections we should prescribe either a displacement or a force and either a rotation or a moment.

Note that once the differential formulation is solved and $w(x)$ determined, we can obtain the 3-D predictions of the bar mathematical model using (4.124), (4.126), (4.130) and the assumptions $w(x, y, z) = w(x)$, $v(x, y, z) = 0$, $\tau_{yy} = \tau_{zz} = \tau_{xy} = \tau_{yz} = 0$. If the in-plane extension/contraction of the fibers are of interest, we can improve the predictions for $w(x, y, z)$ and $v(x, y, z)$ substituting τ_{xx} given by (4.126) into (4.118), (4.119) and considering (4.121).

The approach we used to derive the 1-D differential formulation for the Bernoulli-Euler beam model, and also for the bar subjected to axial loading only, in which we considered the 3-D elasticity equations, permits to clearly identify what the model assumptions are and how they affect the satisfaction of the 3-D equations. However, as we consider more complex structural mathematical models such as curved beam, plate and shell models this approach becomes difficult to follow and we consider for the formulation of the remaining models a classical approach.

In this classical approach we also start with a 3-D solid and consider geometric, kinematic and mechanical assumptions. Then, we enforce equilibrium, constitutive and compatibility conditions. These conditions are related to the analogous 3-D elasticity conditions but are not exactly those. For example, equilibrium is imposed in terms of stress resultants and constitutive and compatibility conditions are selectively enforced. We exemplify the use of this classical approach with the Bernoulli-Euler model considered already.

We start from the same geometric and kinematic assumptions given above and consider a load $p(x)$ per unit longitudinal length acting into the z direction. Then, we use (4.124) and (4.117) which implicitly consider the compatibility relation

$$\varepsilon_{xx} = \frac{\partial u}{\partial x} \tag{4.141}$$

and the constitutive relation

$$\varepsilon_{xx} = \frac{\tau_{xx}}{E}$$

to obtain (4.126). Note that in using the above constitutive relation, we implicitly assumed that $\tau_{yy} = \tau_{zz} = 0$.

Then we consider equilibrium in terms of the stress resultants of a differential element as summarized in Figure 4.20[3].

We note that to represent the actions of the rest of the beam on the end cross-sections of this differential element we have introduced the bending moment $M(x)$ and the shear force $V(x)$ which are necessary to equilibrate the transverse load as detailed below.

Equilibrium of the differential element in the z direction yields

$$V - (V + dV) + p(x)dx = 0$$

$$\frac{dV}{dx} = p(x) \tag{4.142}$$

[3] Note that this sign convention for the transverse shear V for the beam models (see also Section 4.2.8) is quite common but is opposite to the convention used for plates and shells

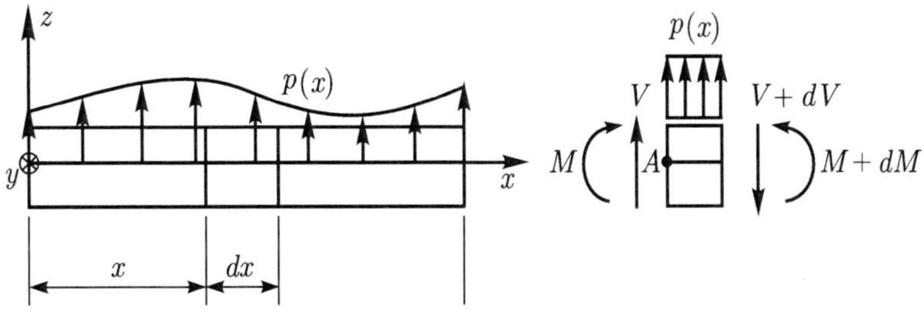

Fig. 4.20. Equilibrium of a generic differential element

and moment equilibrium, which is enforced about point A for instance, gives

$$-(V + dV)dx + p(x)dx\frac{dx}{2} - M + (M + dM) = 0.$$

Neglecting infinitesimals of higher order, we obtain

$$\frac{dM}{dx} = V. \tag{4.143}$$

Equations (4.142) and (4.143) are the bar equilibrium conditions. Of course, equilibrium in the axial direction is trivially satisfied since we are considering transverse loading only and the axial forces are zero.

Taking derivatives of (4.143) with respect to x and substituting the result in (4.142) yields

$$\frac{d^2M}{dx^2} = p(x). \tag{4.144}$$

Using (4.126) and (4.132), we obtain (4.134) which substituted in (4.144) yields (4.136). Now we could write the same differential formulation obtained above and given in (4.136) to (4.140). Hence, both approaches lead to the same differential formulation.

We state below some classical equations and definitions for the beam model. From (4.126) and (4.134), we obtain

$$\tau_{xx} = -\frac{M}{I}z \tag{4.145}$$

which gives the linear normal stress distribution at the beam cross-section. We note that, since z is measured from the bar axis going through the centroid of the section,

$$\int_A \tau_{xx}\, dA = 0.$$

Indeed, we want this property and have chosen the location of the bar axis correspondingly. We also define

$$\kappa = \frac{d^2 w}{dx^2} \tag{4.146}$$

where κ gives the curvature of the bar axis since we are considering infinitesimally small displacements. Then, (4.134) gives

$$\kappa = \frac{M}{EI} \tag{4.147}$$

which leads to the important physical interpretation that the bending moment $M(x)$ induces curvature of the beam axis. The product EI is called the bending rigidity of the section (or of the beam, when the beam has a constant cross-section). It gives the stiffness of the section for bending induced curvature.

We can rewrite the differential formulation given in (4.136) to (4.140), now considering (4.144), (4.146) and (4.147) to obtain:
Differential formulation of the Bernoulli-Euler beam model

Given the transverse distributed loading $p(x)$, find $M(x)$, $\kappa(x)$, $w(x)$ such that

$$\frac{d^2 M}{dx^2} = p(x) \tag{4.148}$$

$$\kappa(x) = \frac{d^2 w}{dx^2} \tag{4.149}$$

$$\kappa(x) = \frac{M(x)}{EI} \tag{4.150}$$

for all x within the bar. At $x = 0$ we have

$$w(0) = w_0 \qquad \text{or} \qquad \frac{dM}{dx}(0) = Q_0 \tag{4.151}$$

and

$$\frac{dw}{dx}(0) = \varphi_0 \qquad \text{or} \qquad M(0) = M_0. \tag{4.152}$$

At $x = L$

$$w(L) = w_L \qquad \text{or} \qquad \frac{dM}{dx}(L) = -Q_L \tag{4.153}$$

and

$$\frac{dw}{dx}(L) = \varphi_L \qquad \text{or} \qquad M(L) = -M_L. \tag{4.154}$$

Of course, the solution of both differential formulations leads to the same transverse displacements and, hence, the same model predictions. We note that the differential formulation given by (4.148)—(4.154) explicitly shows the conditions of equilibrium, compatibility and constitutive behavior. In fact, (4.148) is the equilibrium equation, (4.149) the compatibility relation and (4.150) the constitutive equation.

Considering the variables in the formulation of this bar model and the variables of the bar under axial loading model, we recognize that some of these variables are quite different from those of the 2-D and 3-D elasticity models. For example, in the bending bar model we have stress resultants and curvature variables. This is usually the case for structural mathematical models.

In order to give a unified framework for the presentation of the structural mathematical models, let us make some definitions. There are always *kinematic variables* which fully characterize the kinematics of the model. The kinematic variable for this model is $w = w(x)$. Note that all displacement components for any point in the bar can be obtained from $w(x)$. In fact,

$$\begin{aligned} u(x,y,z) &= -z\frac{dw}{dx} \\ v(x,y,z) &= 0 \\ w(x,y,z) &= w(x). \end{aligned}$$

There are strain type variables which are linked to the straining of the beam fibers. They are referred to as *generalized strain variables*. The curvature $\kappa(x)$ is the generalized strain variable for this model.

There are variables which are linked to the internal actions and are referred to as *generalized internal force variables* or *generalized stress variables*. The moment $M(x)$ is such variable for the beam model.

We note that there is a correspondence between the above defined generalized strain and the usual strains of the 2-D and 3-D elasticity models. Both are obtained from the kinematic variables and are related to the straining of the material fibers.

The correspondence also holds for the generalized internal force or generalized stress in the beam model and the stresses for the 2-D and 3-D elasticity models since both reflect the internal transfer of forces. Finally, the generalized strain and internal force are related by a generalized constitutive equation (see equation (4.150)).

The kinematic, the generalized strain and the generalized stress variables are called *primary variables* of the model since they completely characterize the model and its mathematical formulation can be fully stated considering these variables. Also, the primary variables are used to express the conditions of equilibrium, compatibility and constitutive behavior.

Note that the shear force does not enter directly in the differential formulation of the problem and can be obtained by equation (4.143) after $M(x)$ has been found. It is termed an *auxiliary variable.*

Example 4.5

Obtain the transverse shear stress distribution at a generic cross-section of a beam.

Solution

Due to the kinematic assumptions of the Bernoulli-Euler beam model the transverse shear strain γ_{xz} is zero, see (4.125). Therefore, the constitutive relation would lead to $\tau_{xz} = \tau_{zx} = 0$. However, considering the equilibrium condition (4.143) a nonzero shear force V is in general required. Of course, the shear force is the stress resultant associated with the distribution of τ_{xz}. At a given cross-section

$$V = \int_A (-\tau_{zx})\ dA \tag{4.155}$$

where the minus sign is a result of the sign convention. Equation (4.155) shows that τ_{zx} can not be zero throughout the cross-section when V is different from zero. This apparent inconsistency is resolved when we recognize that in the Bernoulli-Euler model the shear deformations which would be induced by τ_{zx} are neglected. This assumption is increasingly appropriate as h/L becomes smaller, because for slender beams this shear deformation contributes very little to the overall beam deformation. However, the contribution of τ_{zx} for equilibrium can not be neglected.

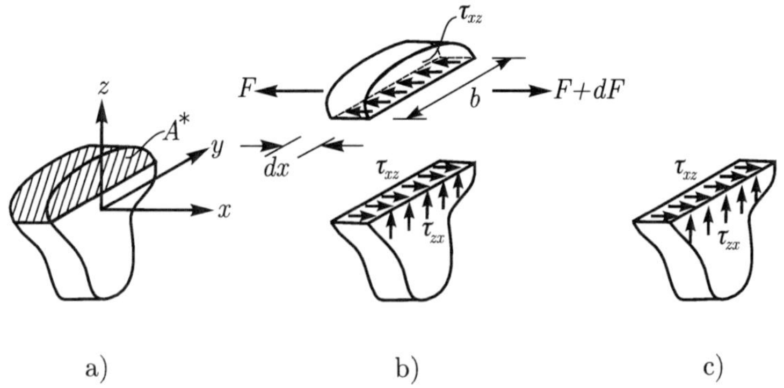

Fig. 4.21. Transverse shear stresses in beams

In order to evaluate an approximate distribution of τ_{zx} (approximate when compared to the 3-D elasticity solution), consider the part of the beam shown in Figure 4.21a which is obtained by cutting the beam at two cross-sections dx apart. In Figure 4.21b, the equilibrium in the x direction of the highlighted part of Figure 4.21a is shown. Of course

$$F = \int_{A^*} \tau_{xx}\, dA = \int_{A^*} -\frac{M}{I} z\, dA = -\frac{M}{I}\int_{A^*} z\, dA. \tag{4.156}$$

Note that a constant distribution of τ_{zx} is assumed at the horizontal plane defined by the cut. Recalling a result for shear stresses acting on orthogonal planes, we note that the shear stress acting on this horizontal plane defines the shear stress acting on the beam cross-section as schematically shown in Figure 4.21c. Considering Figure 4.21b the equilibrium in the x direction leads to

$$dF = \tau_{xz} b dx$$

or

$$\tau_{xz} = \frac{1}{b}\frac{dF}{dx}. \tag{4.157}$$

Let us define

$$\mathfrak{M}_y = \int_{A^*} z\, dA \tag{4.158}$$

which is the static moment of the area A^* with respect to the y axis. From equations (4.143), (4.156), (4.157) and (4.158) we obtain

$$\tau_{xz} = \tau_{zx} = -\frac{\mathfrak{M}_y}{bI}\frac{dM}{dx} = -\frac{V\mathfrak{M}_y}{bI}. \tag{4.159}$$

In order to obtain insight into the shear stress distribution, consider a beam of rectangular cross-section. Referring to Figure 4.22a, we can write

$$\begin{aligned}\mathfrak{M}_y &= \int_{A^*} z\, dA = \int_{z^*}^{h/2}\int_{-b/2}^{b/2} z\, dydz = b\int_{z^*}^{h/2} z\, dz = b\frac{z^2}{2}\bigg|_{z^*}^{h/2}\\ &= \frac{b}{2}\left(\frac{h^2}{4} - (z^*)^2\right)\end{aligned}$$

where z^* is the z coordinate associated with the cutting horizontal plane as shown in Figure 4.22a. Then, from (4.159) we obtain

$$\tau_{xz} = -\frac{V}{2I}\left(\frac{h^2}{4} - (z^*)^2\right) = -\frac{3}{2}\frac{V}{bh}\left(1 - \left(\frac{z^*}{h/2}\right)^2\right) \tag{4.160}$$

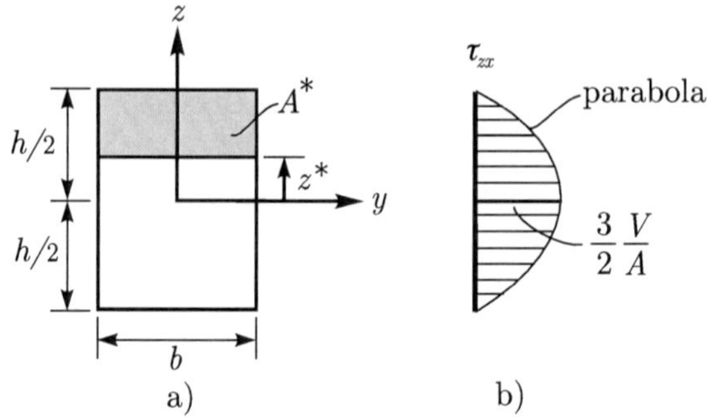

Fig. 4.22. a) Geometric definitions; b) Transverse shear stress distribution for a rectangular cross-section

where we used that $I = bh^3/12$. Of course, we note the agreement between (4.130) and (4.160).

As summarized in Figure 4.22b, the shear stress distribution for a rectangular cross-section is parabolic with the maximum value at the centroid given by 3/2 times the average shear stress V/A. Note that the condition of zero shear stresses at the top and bottom surfaces of the beam is satisfied.

□

Demonstrative solutions

We give below some example solutions. These examples, besides giving some insight into the use and behavior of the beam model, present solutions in a convenient form to introduce the matrix method of analysis for frames.

Example 4.6

Find the solution of the beam problem described in Figure 4.23.

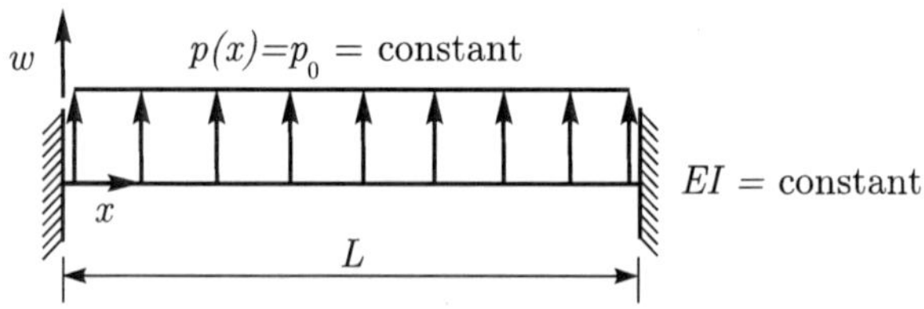

Fig. 4.23. Built-in beam subjected to constant transverse load

Solution

We use the differential formulation

$$\frac{d^4w}{dx^4} = \frac{p_0}{EI} \tag{4.161}$$

with the following boundary conditions

$$w(0) = \frac{dw}{dx}(0) = w(L) = \frac{dw}{dx}(L) = 0$$

to enforce the kinematic restraints to displacements and rotations at both ends. Integrating equation (4.161) with respect to x yields

$$w(x) = \frac{p_0 x^4}{24EI} + C_1x^3 + C_2x^2 + C_3x + C_4 \tag{4.162}$$

where C_1, C_2, C_3 and C_4 are constants to be determined. Imposing the boundary conditions gives

$$w(x) = -\frac{p_0 x^4}{24EI} + \frac{p_0 L}{12EI}x^3 - \frac{p_0 L^2}{24EI}x^2. \tag{4.163}$$

The end-forces and moments are obtained by using (4.134) and (4.135)

$$M(0) = M(L) = -\frac{p_0 L^2}{12},\ V(0) = -V(L) = \frac{p_0 L}{2}.$$

□

Example 4.7

Find the solution to the beam problem subjected to an imposed end-displacement as summarized in Figure 4.24.

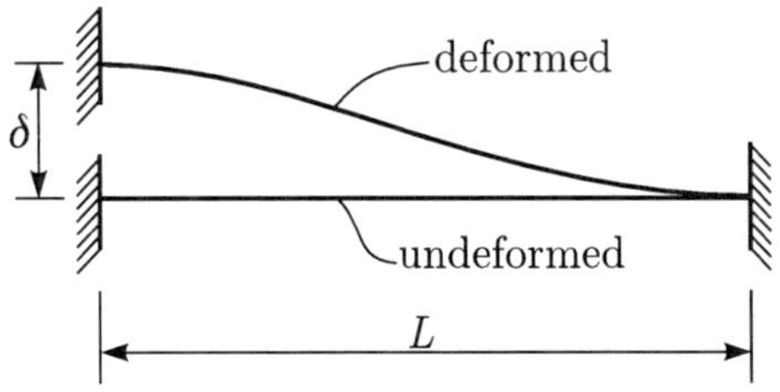

Fig. 4.24. Built-in beam subjected to imposed end transverse displacement

Solution

We use

$$\frac{d^4w}{dx^4} = 0 \tag{4.164}$$

with the following boundary conditions

$$w(0) = \delta, \ \frac{dw}{dx}(0) = w(L) = \frac{dw}{dx}(L) = 0. \tag{4.165}$$

Integrating (4.164) leads to

$$w(x) = C_1x^3 + C_2x^2 + C_3x + C_4. \tag{4.166}$$

The constants can be determined by imposing the boundary conditions given in (4.165) leading to

$$\begin{aligned} w(x) &= \left(\frac{2x^3}{L^3} - \frac{3x^2}{L^2} + 1\right)\delta \\ &= h_2(x)\delta \end{aligned} \tag{4.167}$$

where

$$h_2(x) = \frac{2x^3}{L^3} - \frac{3x^2}{L^2} + 1.$$

The function $h_2(x)$ gives the transverse displacement of the beam axis when a unit transverse displacement is imposed at the left end and the remaining end-displacement and rotations are fixed. The corresponding end-forces and moments are

$$M(0) = -M(L) = -\frac{6EI}{L^2}, \ V(0) = V(L) = \frac{12EI}{L^3}.$$

□

Proceeding as in Example 4.7, we can derive the solutions of the beam model for other unit end displacement/rotation conditions as summarized in Table 4.1. The h_i functions are referred to as the beam Hermitian functions. These solutions are very useful since they can be used to find solutions to arbitrarily chosen end displacement/rotation conditions. In fact, let us consider the beam without transverse loading but with imposed end displacement/rotation conditions given by

$$w(0) = w_0, \ w(L) = w_L, \ \frac{dw}{dx}(0) = w_0' \text{ and } \frac{dw}{dx}(L) = w_L'. \tag{4.168}$$

The solution to this problem is given by

Table 4.1. Exact solution for unit end displacement/rotation conditions

Problem description	Bar end-forces and solution for $w(x)$
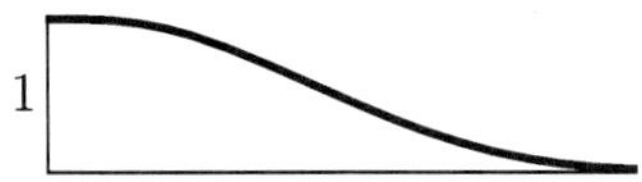 $w(0) = 1;\ w(L) = 0;\ \frac{dw(0)}{dx} = 0;\ \frac{dw(L)}{dx} = 0$	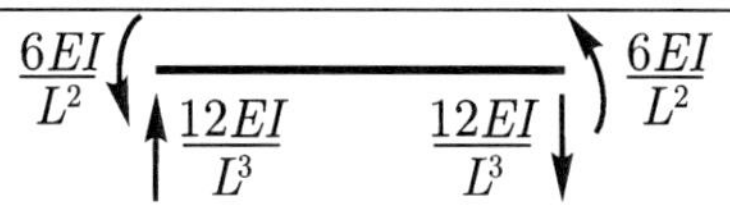 $h_2(x) = \frac{2x^3}{L^3} - \frac{3x^2}{L^2} + 1$
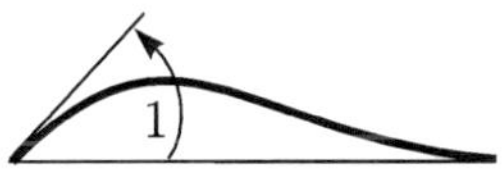 $w(0) = 0;\ w(L) = 0;\ \frac{dw(0)}{dx} = 1;\ \frac{dw(L)}{dx} = 0$	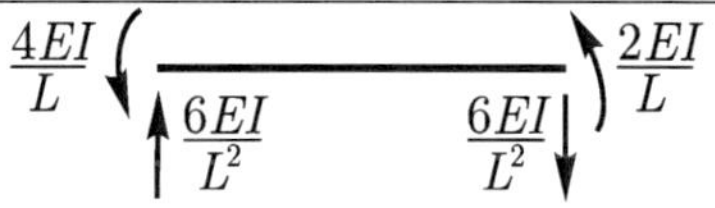 $h_3(x) = \frac{x^3}{L^2} - \frac{2x^2}{L} + x$
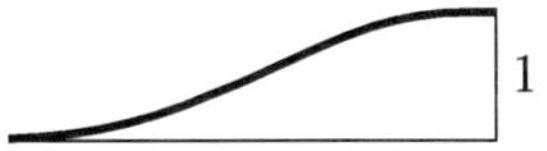 $w(0) = 0;\ w(L) = 1;\ \frac{dw(0)}{dx} = 0;\ \frac{dw(L)}{dx} = 0$	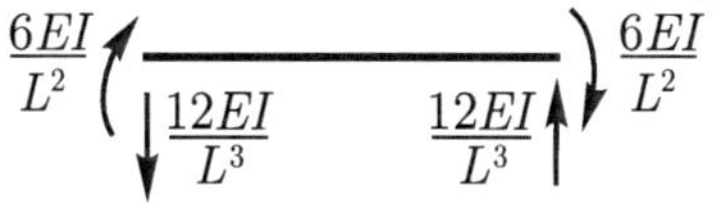 $h_5(x) = -\frac{2x^3}{L^3} + \frac{3x^2}{L^2}$
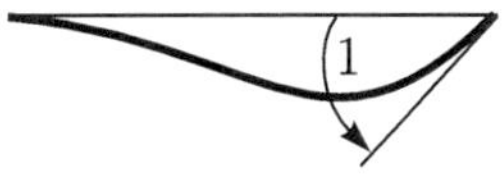 $w(0) = 0;\ w(L) = 0;\ \frac{dw(0)}{dx} = 0;\ \frac{dw(L)}{dx} = 1$	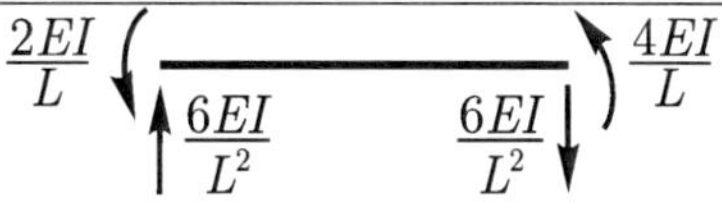 $h_6(x) = \frac{x^3}{L^2} - \frac{x^2}{L}$

$$w(x) = h_2(x)w_0 + h_3(x)w_0' + h_5(x)w_L + h_6(x)w_L'. \tag{4.169}$$

The $w(x)$ defined in (4.169) satisfies (4.136) for $p(x) = 0$ since $h_2(x)$, $h_3(x)$, $h_5(x)$ and $h_6(x)$ also verify (4.136) for $p(x) = 0$, and the $w(x)$ satisfies the boundary conditions given in (4.168) since the $h_i(x)$, $\frac{dh_i(x)}{dx}$ assume the value of 1 for the corresponding end displacement/rotation conditions (see Table 4.1) and the value of zero for the other end conditions. The solution in (4.169) represents of course an application of the principle of superposition (See Section 2.3.6).

Example 4.8

Assume that the end conditions for the problem in Figure 4.23 are those stated in (4.168). Find the solution of the problem.

Solution

Denoting by $w_1(x)$ the solution for the transverse load with zero end displacements/rotations derived in Example 4.6 given by (4.163) and by $w_2(x)$ the solution for the specified displacement/rotation end conditions with no transverse load given by (4.169), the function

$$w(x) = w_1(x) + w_2(x)$$

is the solution sought. In fact

$$\frac{d^4 w}{dx^4} = \frac{d^4}{dx^4}(w_1(x) + w_2(x)) = \frac{d^4 w_1}{dx^4} + \frac{d^4 w_2}{dx^4} = \frac{d^4 w_1}{dx^4} = \frac{p(x)}{EI}$$

and since $w_1(x)$ corresponds to the zero end conditions (it is the solution to the problem of Example 4.6) $w(x)$ satisfies the given boundary conditions.

□

Note that we can obtain the solution to the problem of a beam subjected to a transverse load $p(x)$ and imposed end displacements/rotations by solving a beam clamped at both ends subjected to the $p(x)$ (see Example 4.6) and adding to this solution the response due to the imposed end displacements/rotations (see (4.169)).

In the next example, we examine a situation in which at one end we have a prescribed displacement and no restriction on the section rotation.

Example 4.9

Solve the problem described in Figure 4.25. We note that there is no restric-

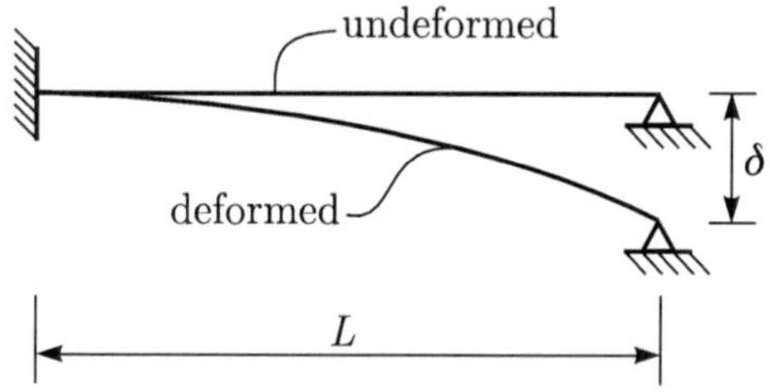

Fig. 4.25. Beam subjected to imposed end transverse displacement. Section rotation at right end is left free

tion on the rotation at the end $x = L$.

Solution

Although we do not have a prescribed rotation at the end $x = L$, we can still take advantage of the general solution for $p(x) = 0$ which is given by (4.169). Of course $w_0 = w_0' = 0$, $w_L = -\delta$ and w_L' is to be determined. The condition of leaving the section rotation at $x = L$ free is represented by imposing that the bending moment at that section is zero. Hence the solution can be written as

$$w(x) = h_5(x)(-\delta) + h_6(x)w_L'$$

with $M(L) = 0$. This condition corresponds to

$$M(L) = EI\frac{d^2w}{dx^2}(L) = 0$$

and hence

$$\frac{d^2w}{dx^2}(L) = \left.\frac{d^2h_5(x)}{dx^2}\right|_L(-\delta) + \left.\frac{d^2h_6(x)}{dx^2}\right|_L w_L' = 0$$

to give

$$w_L' = -\frac{3\delta}{2L}.$$

Therefore the final solution is

$$w(x) = -\left(h_5(x) + \frac{3}{2L}h_6(x)\right)\delta.$$

□

The *principle of superposition* used in the above solutions is employed abundantly, and we choose to detail it further in the next example.

Example 4.10

Consider the problem described in Figure 4.26a. Show how the principle of superposition can be used to solve this problem.

Solution

In Figure 4.26b we show four simpler problems whose solutions *superimposed* give the solution to the original problem.

We note that for each problem the restraints are those of the original problem and an external action – either a part of the total loading or a support settlement/rotation – is introduced. Of course, when considered together, the external actions in the simpler problems should represent all external actions on the original structure.

The validity of the principle of superposition rests on the linearity of the differential formulation of the problem. Referring to the differential formulation of the Bernoulli-Euler beam model, we summarize in Table 4.2 the differential formulation of each simple problem. Due to the linearity of the differential formulation

$$w(x) = w_1(x) + w_2(x) + w_3(x) + w_4(x)$$

is the solution of the original problem.

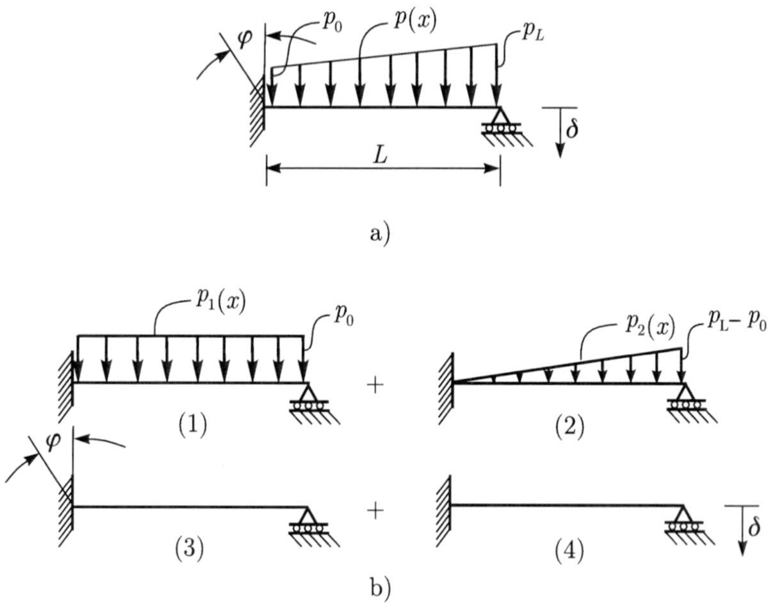

Fig. 4.26. a) Original problem; b) Simpler problems used for the superposition

In fact,

$$\begin{aligned} \frac{d^4 w}{dx^4} &= \frac{d^4 w_1}{dx^4} + \frac{d^4 w_2}{dx^4} + \frac{d^4 w_3}{dx^4} + \frac{d^4 w_4}{dx^4} = \frac{p_1(x)}{EI} + \frac{p_2(x)}{EI} = \frac{p(x)}{EI} \\ w(0) &= w_1(0) + w_2(0) + w_3(0) + w_4(0) = 0 \end{aligned}$$

Table 4.2. Differential formulations of simpler problems

(1) $\frac{d^4 w_1}{dx^4} = \frac{p_1(x)}{EI}$ $w_1(0) = 0, \quad w_1(L) = 0$ $\frac{dw_1}{dx}(0) = 0, \quad EI\frac{d^2 w_1}{dx^2}(L) = 0$	(2) $\frac{d^4 w_2}{dx^4} = \frac{p_2(x)}{EI}$ $w_2(0) = 0, \quad w_2(L) = 0$ $\frac{dw_2}{dx}(0) = 0, \quad EI\frac{d^2 w_2}{dx^2}(L) = 0$
(3) $\frac{d^4 w_3}{dx^4} = 0$ $w_3(0) = 0, \quad w_3(L) = 0$ $\frac{dw_3}{dx}(0) = \varphi, \quad EI\frac{d^2 w_3}{dx^2}(L) = 0$	(4) $\frac{d^4 w_4}{dx^4} = 0$ $w_4(0) = 0, \quad w_4(L) = -\delta$ $\frac{dw_4}{dx}(0) = 0, \quad EI\frac{d^2 w_4}{dx^2}(L) = 0$

$$
\begin{aligned}
\frac{dw}{dx}(0) &= \frac{dw_1}{dx}(0) + \frac{dw_2}{dx}(0) + \frac{dw_3}{dx}(0) + \frac{dw_4}{dx}(0) = \varphi \\
w(L) &= w_1(L) + w_2(L) + w_3(L) + w_4(L) = -\delta \\
EI\frac{d^2 w}{dx^2}(L) &= EI\frac{d^2 w_1}{dx^2}(L) + EI\frac{d^2 w_2}{dx^2}(L) + EI\frac{d^2 w_3}{dx^2}(L) + EI\frac{d^2 w_4}{dx^2}(L) = 0
\end{aligned}
$$

which show that $w(x)$ satisfies all field and boundary conditions of the original problem. The solution of the seemingly complex problem has therefore been reduced to the solution of the four simpler problems.

□

Bars subjected to axial and transverse loading

In Section 4.1 we examined the solution of bars subjected to axial loading only and in this section we obtained the solution of bars subjected to transverse loading only. In the linear analysis considered, the axial loading does not induce any rotations of the bar sections and the transverse loading does not induce any axial displacement at the bar axis. Therefore, in the context of infinitesimally small displacements, the solution of a bar subjected to the simultaneous actions of axial and transverse loadings can be obtained by the superposition of the solutions to the axial and bending problems.

Example 4.11

Find the solution of a bar of length L and cross-sectional area A subjected to imposed axial end displacements u_0 at $x = 0$ and u_L at $x = L$.

Solution

Considering the differential formulation of the bar problem subjected to axial loading given by (4.105) to (4.107), we have

$$EA\frac{d^2u}{dx^2} = 0$$

where E is Young's modulus. Hence,

$$u(x) = C_2 x + C_1$$

and imposing $u(0) = u_0$ and $u(L) = u_L$, we obtain

$$u(x) = u_0\left(1 - \frac{x}{L}\right) + u_L\frac{x}{L}$$

or

$$u(x) = h_1(x)u_0 + h_4(x)u_L.$$

We note that $h_1(x)$ gives the solution for an imposed unit axial displacement at $x = 0$ with the other end fixed. Analogously, the function $h_4(x)$ gives the solution for an imposed axial unit displacement at $x = L$ with the end $x = 0$ fixed.

□

Let us consider a generic bar and number the end degrees of freedom as schematically shown in Figure 4.27.

Fig. 4.27. Numbering of end section degrees of freedom. Young's modulus E, the moment of inertia I and the cross-sectional area A are all constant

Consider that these end displacements are imposed and that there is no distributed axial and transverse loading. Using Examples 4.7 and 4.11, the solution can be written as[4]

$$u(x) = h_1(x)u_1 + h_4(x)u_4 \tag{4.170}$$

[4] Note that $u(x)$ in (4.170) represents a uniform section displacement (independent of z used in (4.124)) and $w(x)$ in (4.171) results in an additional u-displacement as given in (4.124)

and

$$w(x) = h_2(x)u_2 + h_3(x)u_3 + h_5(x)u_5 + h_6(x)u_6 \tag{4.171}$$

and the numbering used for the functions h_i is now obvious. If in addition a transverse or axial loading is applied, the solutions can be obtained by superposition (see Example 4.10).

4.2.3 Bar models obtained by an assemblage of bars

We have studied so far bar models of only one bar. When bar models are used in structural engineering to model real structures, the resulting models almost always involve an assemblage of bars.

In Chapter 2, we studied truss structures made of several bars. The truss models are simpler than those we consider next since in truss structures each bar only carries an axial force due to pin end conditions. However, the same concepts can be used to analyze structures made of an assemblage of bars carrying bending and axial forces.

Consider the structure in Figure 4.28a. The structure is clearly properly supported since, due to the clamped condition at section A, the structure can not undergo any motion when considered rigid.

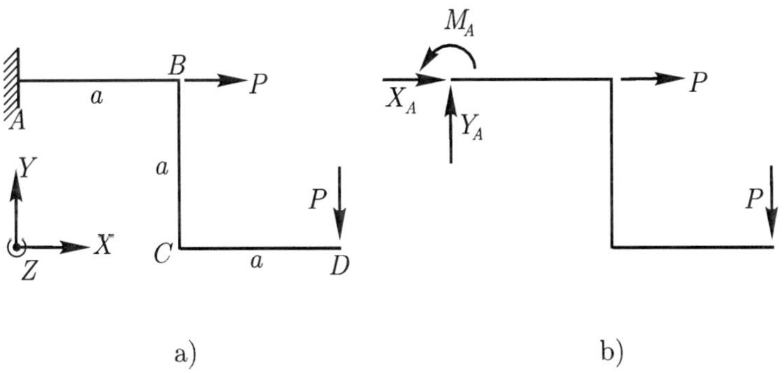

Fig. 4.28. a) Three-bar structure model. Bars AB, BC and CD have constant EI and EA; b) External actions on the structure with reactions included

This structure is statically determinate since the suppression of any of its restraints would yield a structure that is no longer properly supported.

Associated with each restraint, a reaction force is introduced as schematically shown in Figure 4.28b. Since the structure is statically determinate the reactions can be determined using the global equilibrium conditions below

$$\sum F_X = 0 \quad \Rightarrow \quad X_A = -P$$

$$\sum F_Y = 0 \quad \Rightarrow \quad Y_A = P$$

$$M_A - P \cdot 2a = 0 \qquad \Rightarrow \qquad M_A = 2Pa.$$

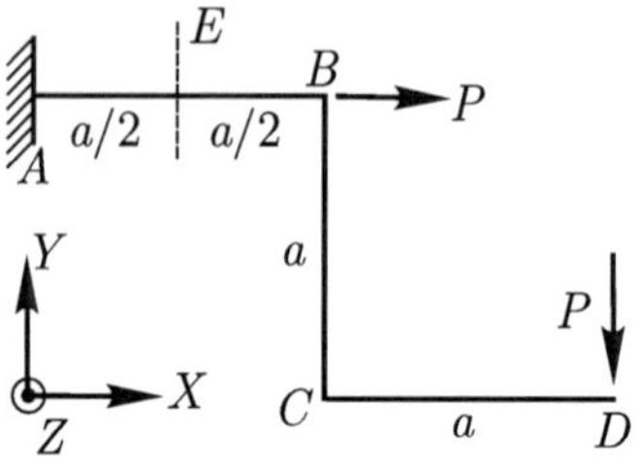

Fig. 4.29. Definition of section E

Let us discuss how to determine the internal forces and moment at any section by examining the section E shown in Figure 4.29. The structure is conceptually cut at this section and we introduce the section forces and moments acting onto the two resulting parts as shown in Figure 4.30.

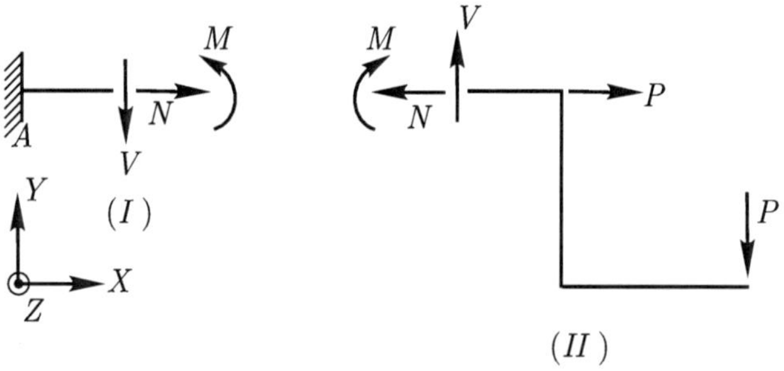

Fig. 4.30. Internal forces at section E

Of course, the internal forces and moments at section E can be evaluated by enforcing equilibrium of either part (I) or (II). Since in part (II) we have only known external actions, its equilibrium directly gives these internal forces and moments. We obtain

$$\begin{aligned} N &= P \\ V &= P \\ M &= -\frac{3Pa}{2}. \end{aligned}$$

In this way, we can determine N, V and M for any section and obtain the axial, shear and moment diagrams which are shown in Figure 4.31. For a generic point on the bar axis, the magnitude shown orthogonal to the axis indicates the value of the particular internal force/moment at that bar section. For the axial and shear force diagrams the sign convention is indicated next to the diagrams. For the bending moment, the convention is to draw the diagram on the side where the fibers are tensioned due to the action of the bending moment.

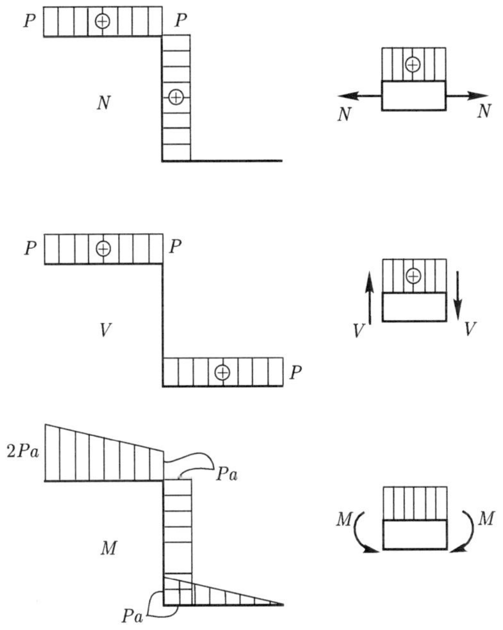

Fig. 4.31. Axial, shear and bending moment diagrams

To illustrate how displacements can be evaluated for an assemblage of bars, we consider the example below.

Example 4.12

Calculate the section displacement indicated in Figure 4.32.

Solution

We need to consider the contributions due to the axial force and the bending moment. Note that the displacement δ_C depends not only on the deforma-

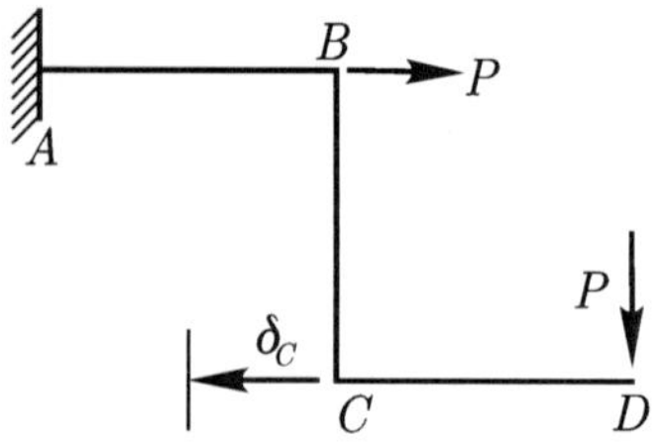

Fig. 4.32. Section displacement to be evaluated

tion of bar BC but also on that of bar AB. Since we know the displacements/rotation at A and the internal forces of bar AB, we can evaluate the displacements/rotation at point B by solving the applicable differential equations for bar AB.

For every bar we adopt a local coordinate system such that the x axis is always axial as described in Figure 4.33a.

The differential formulation for the bending problem of bar AB reads

$$\frac{d^2w}{dx^2} = \frac{M(x)}{EI} = \frac{1}{EI}(Px - 2Pa) \tag{4.172}$$

$$w(0) = 0; \qquad w'(0) = 0$$

and for the axial problem

$$\frac{du}{dx} = \frac{N(x)}{EA} = \frac{P}{EA} \tag{4.173}$$

$$u(0) = 0.$$

Solving equations (4.172) and (4.173), we obtain for section B

$$u_B = \frac{Pa}{EA}$$

$$w_B = -\frac{5Pa^3}{6EI}$$

$$w'_B = -\frac{3Pa^2}{2EI}.$$

For bar BC, we have

$$\frac{d^2w}{dx^2} = \frac{M(x)}{EI} = -\frac{Pa}{EI}$$

$$w(0) = u_B = \frac{Pa}{EA}; \qquad w'(0) = w'_B = -\frac{3Pa^2}{2EI}$$

and

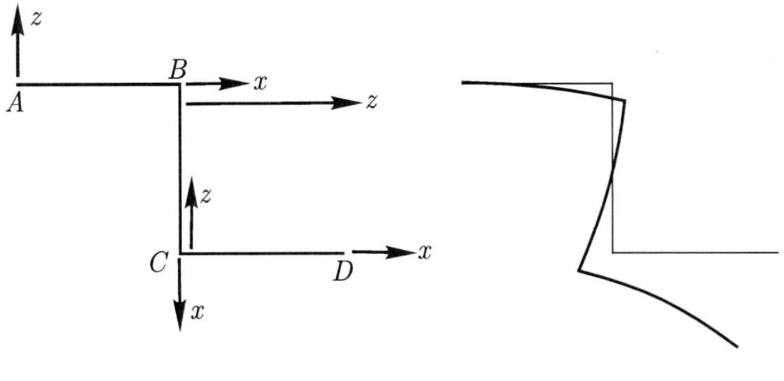

Fig. 4.33. a) Choices of local axes; b) Magnified deformed shape

$$\frac{du}{dx} = \frac{N(x)}{EA} = \frac{P}{EA}$$

$$u(0) = -w_B = \frac{5Pa^3}{6EI}.$$

Of course, the end conditions at $x = 0$ for bar BC have been obtained from the displacements and the rotation at section B considered as the end section of bar AB.

Solving the above equations, we obtain

$$u_C = \frac{Pa}{EA} + \frac{5Pa^3}{6EI} = \frac{5Pa^3}{6EI}\left(1 + \frac{6I}{5Aa^2}\right)$$

and

$$w_C = -\frac{2Pa^3}{EI} + \frac{Pa}{EA} = -\frac{2Pa^3}{EI}\left(1 - \frac{I}{2Aa^2}\right).$$

Therefore

$$\delta_C = -w_C = \frac{2Pa^3}{EI}\left(1 - \frac{I}{2Aa^2}\right). \tag{4.174}$$

A magnified deformed shape of the structure is shown in Figure 4.33b.

We compare the relative contribution of the axial and bending deformations to the calculated displacement. Assuming that the bars are of a rectangular cross-section of height h, expression (4.174) becomes

$$\delta_C = \frac{2Pa^3}{EI}\left(1 - \frac{1}{24}\left(\frac{h}{a}\right)^2\right)$$

and we see that for a usual range of values for h and a the contribution of the axial deformation is negligible when compared to that of the bending deformation.

□

Although the methodology used in the above example to find displacements/rotations for an assemblage of bars provides insight into the kinematics of the deformation, it is not efficient for a structure of many bars. Moreover, the solution would become even more cumbersome for statically indeterminate structures. Consider, for example, the structure described in Figure 4.34 which was obtained by adding a support at point C to the structure of Figure 4.28.

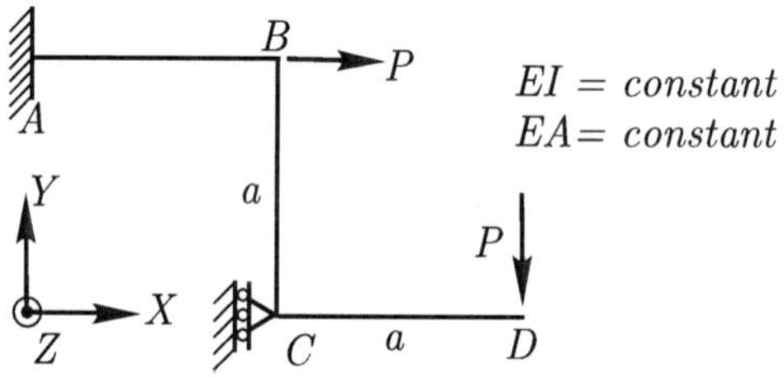

Fig. 4.34. Modified three-bar structure

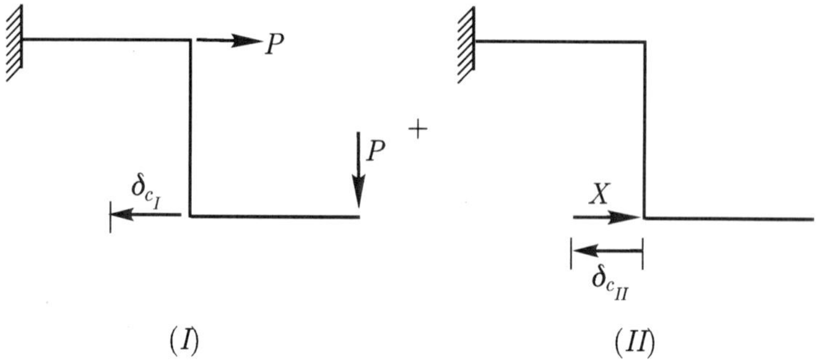

Fig. 4.35. Superposition for three-bar structure

Before we can solve for the displacements we need to find the reaction at C. For that purpose we can use the superposition of effects as shown in Figure 4.35 and impose the compatibility relation

$$\delta_{C_I} + \delta_{C_{II}} = 0.$$

If we define $\delta^1_{C_{II}}$ as $\delta_{C_{II}}$ for $X = 1$ then

$$X = -\frac{\delta_{C_I}}{\delta^1_{C_{II}}}.$$

Of course, δ_{C_I} is the δ_C evaluated above and $\delta^1_{C_{II}}$ can be calculated in a similar way.

The discussion above reinforces the need for a systematic approach to solve engineering structures made of many bars and provides the motivation for the next section.

4.2.4 Matrix displacement method for frames

In this section we introduce a systematic approach to solve structures made of several bars. The methods described in the previous section, which are very useful to obtain insight into the behavior of simple bar structures, are not adequate for solving structures made of several bars.

Analogous to the developments for truss structures, presented in Chapter 2, the matrix method provides a very efficient approach to analyze frame structures of arbitrary complexity.

The main ingredients of the matrix method have already been presented in Section 2.3 and using Table 4.1 we can now directly assemble the bar element stiffness matrix including bending effects.

Stiffness matrix for a bar in a local system

The stiffness matrix of a bar in the local system of axes and corresponding to the nodal degrees of freedom shown in Figure 4.36 is given by the equation

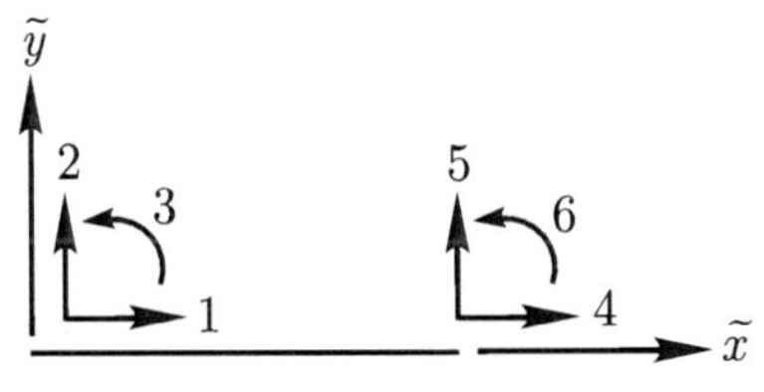

Fig. 4.36. Bar local axes and degrees of freedom

$$\tilde{\mathbf{k}}\tilde{\mathbf{u}} = \tilde{\mathbf{f}} \tag{4.175}$$

where

$$\tilde{\mathbf{u}} = \begin{bmatrix} \tilde{u}_1 \\ \tilde{u}_2 \\ \tilde{u}_3 \\ \tilde{u}_4 \\ \tilde{u}_5 \\ \tilde{u}_6 \end{bmatrix}, \qquad \tilde{\mathbf{f}} = \begin{bmatrix} \tilde{f}_1 \\ \tilde{f}_2 \\ \tilde{f}_3 \\ \tilde{f}_4 \\ \tilde{f}_5 \\ \tilde{f}_6 \end{bmatrix}$$

$$\tilde{\mathbf{k}} = \begin{bmatrix} \frac{EA}{L} & 0 & 0 & -\frac{EA}{L} & 0 & 0 \\ 0 & \frac{12EI}{L^3} & \frac{6EI}{L^2} & 0 & -\frac{12EI}{L^3} & \frac{6EI}{L^2} \\ 0 & \frac{6EI}{L^2} & \frac{4EI}{L} & 0 & -\frac{6EI}{L^2} & \frac{2EI}{L} \\ -\frac{EA}{L} & 0 & 0 & \frac{EA}{L} & 0 & 0 \\ 0 & -\frac{12EI}{L^3} & -\frac{6EI}{L^2} & 0 & \frac{12EI}{L^3} & -\frac{6EI}{L^2} \\ 0 & \frac{6EI}{L^2} & \frac{2EI}{L} & 0 & -\frac{6EI}{L^2} & \frac{4EI}{L} \end{bmatrix}. \tag{4.176}$$

The elements in the matrix $\tilde{\mathbf{k}}$ corresponding to bending are, of course, the forces and moments listed in Table 4.1. To illustrate this observation, consider that we impose $\tilde{u}_2 = 1$ and $\tilde{u}_1 = \tilde{u}_3 = \tilde{u}_4 = \tilde{u}_5 = \tilde{u}_6 = 0$. The solution to this problem is given in Table 4.1. and is summarized in Figure 4.37.

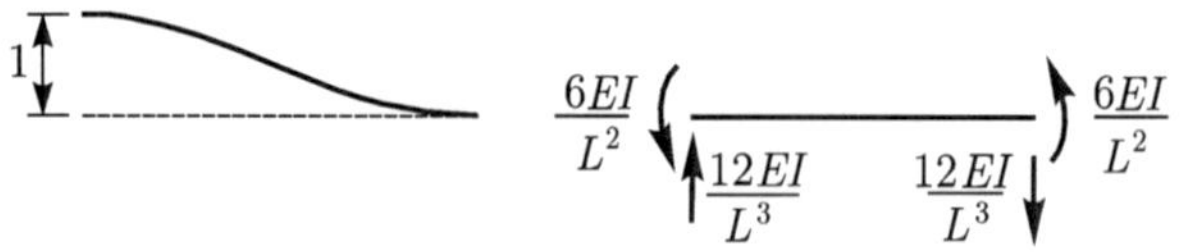

Fig. 4.37. Solution that leads to the second column of $\tilde{\mathbf{k}}$

Of course, from (4.175), we obtain

$$\tilde{\mathbf{f}} = \left[\tilde{k}_{i2}\right] = \begin{bmatrix} 0 \\ \frac{12EI}{L^3} \\ \frac{6EI}{L^2} \\ 0 \\ -\frac{12EI}{L^3} \\ \frac{6EI}{L^2} \end{bmatrix}$$

exemplifying how the second column of $\tilde{\mathbf{k}}$ is obtained.

Stiffness matrix of bar in the global system

The stiffness matrix of an arbitrarily inclined bar is now obtained in the global system as for a truss element in Chapter 2. Using the nodal displacement/rotation and force/moment conventions in Figure 4.38, we obtain

$$u_3 = \tilde{u}_3,\ u_6 = \tilde{u}_6,\ f_3 = \tilde{f}_3,\ f_6 = \tilde{f}_6.$$

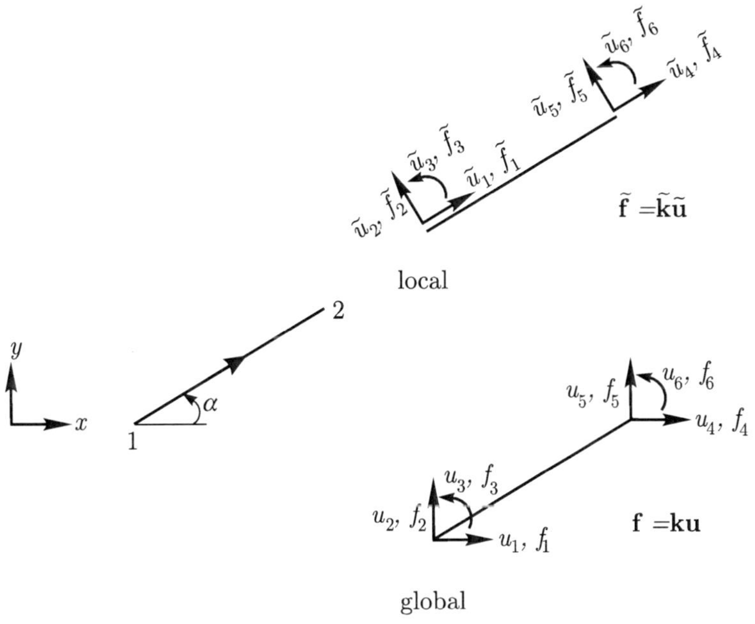

Fig. 4.38. Local and global degrees of freedom of an arbitrarily oriented bar element

We have

$$\mathbf{f} = \mathbf{k}\mathbf{u}$$

with

$$\mathbf{k} = \mathbf{T}^T\tilde{\mathbf{k}}\mathbf{T}$$

$$\mathbf{f} = \begin{bmatrix} f_1 \\ f_2 \\ f_3 \\ f_4 \\ f_5 \\ f_6 \end{bmatrix}, \qquad \mathbf{u} = \begin{bmatrix} u_1 \\ u_2 \\ u_3 \\ u_4 \\ u_5 \\ u_6 \end{bmatrix}$$

and

$$\tilde{\mathbf{u}} = \mathbf{T}\,\mathbf{u}, \qquad \tilde{\mathbf{f}} = \mathbf{T}\,\mathbf{f}$$

where

$$\mathbf{T} = \begin{bmatrix} \cos\alpha & \sin\alpha & 0 & 0 & 0 & 0 \\ -\sin\alpha & \cos\alpha & 0 & 0 & 0 & 0 \\ 0 & 0 & 1 & 0 & 0 & 0 \\ 0 & 0 & 0 & \cos\alpha & \sin\alpha & 0 \\ 0 & 0 & 0 & -\sin\alpha & \cos\alpha & 0 \\ 0 & 0 & 0 & 0 & 0 & 1 \end{bmatrix}. \tag{4.177}$$

Formulation of the matrix method for frames and a demonstrative example

From this point onwards, the formulation of the matrix method is as in the analysis of truss structures. In other words, we can define $\mathbf{F}^{(m)}$, $\mathbf{U}^{(m)}$ and $\mathbf{K}^{(m)}$ in an analogous manner considering that for the frame bar we have six degrees of freedom per element. Then equilibrium at all degrees of freedom is enforced by

$$\mathbf{R} = \sum_{m=1}^{n_e} \mathbf{F}^{(m)}$$

where n_e = number of elements in the structure. Element equilibrium, compatibility, the stress strain behavior, and the nodal compatibility are enforced by using

$$\mathbf{F}^{(m)} = \mathbf{K}^{(m)}\mathbf{U}.$$

Hence, we arrive at

$$\mathbf{KU} = \mathbf{R}$$

with

$$\mathbf{K} = \sum_{m=1}^{n_e} \mathbf{K}^{(m)}.$$

The assemblage process is implemented using the $\mathbf{LM}^{(m)}$ array which, for this case, has six entries instead of four. In the next example we explore the definitions above.

Example 4.13

Consider the bar structure defined in Figure 4.39. Find the nodal displacements, the reactions and draw the internal force diagrams for the structure.

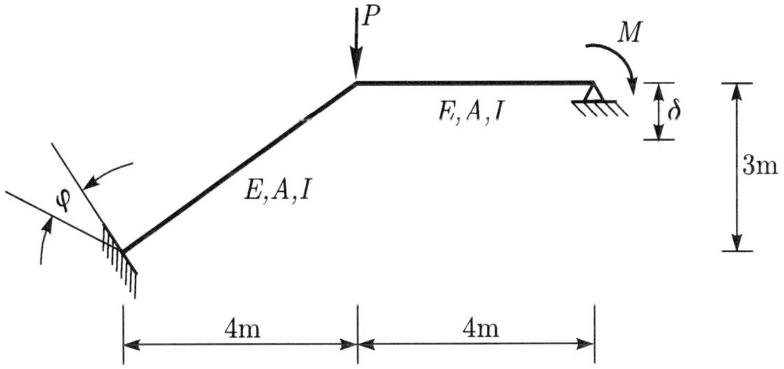

$E = 20 \times 10^6 \text{kN/m}^2$, $A = 0.12\text{m}^2$, $I = 36 \times 10^{-4} \text{m}^4$
$P = 50\text{kN}$, $M = 100\text{kN.m}$, $\delta = 5 \times 10^{-3}\text{m}$, $\varphi = 10^{-3}$ rad

Fig. 4.39. Problem description. $E \equiv$ Young's modulus, $A \equiv$ cross-sectional area and $I \equiv$ moment of inertia. The δ is a vertical settlement and φ is an imposed support rotation

Solution

The matrix method will be used and, since the objective of this example is to illustrate this method, the solution is presented in detail.

Step 1 – Number nodes and bars. Also number degrees of freedom, numbering first the free ones. Establish bar orientations.

The result of this step is summarized in Figure 4.40.

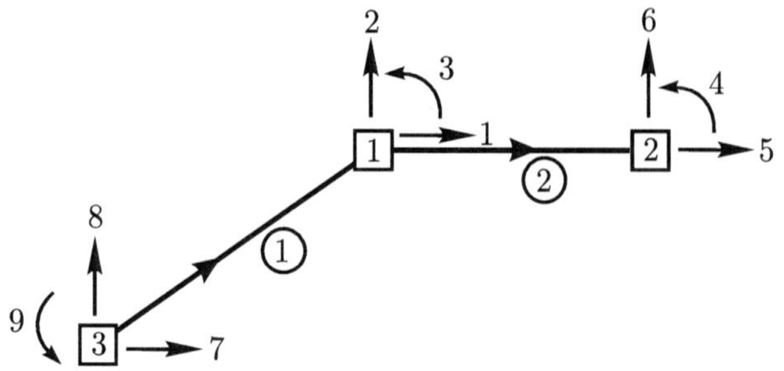

Fig. 4.40. Numbering definitions for the structure and bar orientations

Step 2 – Evaluate $\tilde{\mathbf{k}}$ and $\mathbf{T}$ for every bar.

Note that the $\tilde{\mathbf{k}}$ matrix has always the form given in (4.176). The only changes from bar to bar are due to different geometric and material properties of the bars, *i.e.*, E, A and L.

The matrix $\mathbf{T}$ is defined by (4.177) and for each bar the angle α has to be introduced. The bar orientation is used to arrive at the correct α where the first bar node defines the origin of the $\tilde{x}$ axis. We obtain

$$\tilde{\mathbf{k}}^{(1)} = \begin{bmatrix} 480000 & 0 & 0 & -480000 & 0 & 0 \\ 0 & 6912 & 17280 & 0 & -6912 & 17280 \\ 0 & 17280 & 57600 & 0 & -17280 & 28800 \\ -480000 & 0 & 0 & 480000 & 0 & 0 \\ 0 & -6912 & -17280 & 0 & 6912 & -17280 \\ 0 & 17280 & 28800 & 0 & -17280 & 57600 \end{bmatrix}$$

and

$$\mathbf{T}^{(1)} = \begin{bmatrix} 0.8 & 0.6 & 0 & 0 & 0 & 0 \\ -0.6 & 0.8 & 0 & 0 & 0 & 0 \\ 0 & 0 & 1 & 0 & 0 & 0 \\ 0 & 0 & 0 & 0.8 & 0.6 & 0 \\ 0 & 0 & 0 & -0.6 & 0.8 & 0 \\ 0 & 0 & 0 & 0 & 0 & 1 \end{bmatrix}$$

where we have used $\alpha_1 = arctg(3/4)$. For bar 2

$$\tilde{\mathbf{k}}^{(2)} = \begin{bmatrix} 600000 & 0 & 0 & -600000 & 0 & 0 \\ 0 & 13500 & 27000 & 0 & -13500 & 27000 \\ 0 & 27000 & 72000 & 0 & -27000 & 36000 \\ -600000 & 0 & 0 & 600000 & 0 & 0 \\ 0 & -13500 & -27000 & 0 & 13500 & -27000 \\ 0 & 27000 & 36000 & 0 & -27000 & 72000 \end{bmatrix}$$

and

$$\mathbf{T}^{(2)} = \begin{bmatrix} 1 & 0 & 0 & 0 & 0 & 0 \\ 0 & 1 & 0 & 0 & 0 & 0 \\ 0 & 0 & 1 & 0 & 0 & 0 \\ 0 & 0 & 0 & 1 & 0 & 0 \\ 0 & 0 & 0 & 0 & 1 & 0 \\ 0 & 0 & 0 & 0 & 0 & 1 \end{bmatrix}$$

where we have used $\alpha_1 = 0$. $\mathbf{T}^{(2)}$ is the identity matrix, since the global and local axes are coincident for this bar.

Step 3 – Evaluate $\mathbf{k}$ for each bar using $\mathbf{k} = \mathbf{T}^T\tilde{\mathbf{k}}\mathbf{T}$.
We obtain

$$\mathbf{k}^{(1)} = \begin{bmatrix} 309688 & 227082 & -10368 & -309688 & -227082 & -10368 \\ 227082 & 177224 & 13824 & -227082 & -177224 & 13824 \\ -10368 & 13824 & 57600 & 10368 & -13824 & 28800 \\ -309688 & -227082 & 10368 & 309688 & 227082 & 10368 \\ -227082 & -177224 & -13824 & 227082 & 177224 & -13824 \\ -10368 & 13824 & 28800 & 10368 & -13824 & 57600 \end{bmatrix}$$

$$\mathbf{k}^{(2)} = \begin{bmatrix} 600000 & 0 & 0 & -600000 & 0 & 0 \\ 0 & 13500 & 27000 & 0 & -13500 & 27000 \\ 0 & 27000 & 72000 & 0 & -27000 & 36000 \\ -600000 & 0 & 0 & 600000 & 0 & 0 \\ 0 & -13500 & -27000 & 0 & 13500 & -27000 \\ 0 & 27000 & 36000 & 0 & -27000 & 72000 \end{bmatrix}$$

Step 4 – Construct the **LM** array for each bar.
Referring to Figure 4.40

$$\mathbf{LM}^{(1)} = \begin{bmatrix} 7 & 8 & 9 & 1 & 2 & 3 \end{bmatrix}$$

$$\mathbf{LM}^{(2)} = \begin{bmatrix} 1 & 2 & 3 & 5 & 6 & 4 \end{bmatrix}$$

Step 5 – Assemble the structure stiffness matrix **K**, given below.
From the element stiffness matrices, $\mathbf{k}^{(m)}$, we obtain **K** using the $\mathbf{LM}^{(m)}$ arrays as discussed in Chapter 2. We construct explicitly only the upper part of the matrix and take advantage of symmetry to obtain the lower part.

We note that since there is no direct physical connection between nodes 2 and 3, *i.e.*, no bar linking nodes 2 and 3, the stiffness terms coupling the degrees of freedom of node 2 (4, 5, 6) and node 3 (7, 8, 9) should be zero, as obtained. The solid lines shown in the matrix below identify the partitions associated with the free and restrained degrees of freedom. Since we have first numbered all free degrees of freedom, there is no need to exchange rows to arrive at $\mathbf{K}_{aa}$, $\mathbf{K}_{ab}$, $\mathbf{K}_{ba}$ and $\mathbf{K}_{bb}$.

Step 6 – Construct the load column matrix for the free degrees of freedom $\mathbf{R}_a$.
Considering the load given, we obtain

$$\mathbf{R}_a = \begin{bmatrix} R_1 \\ R_2 \\ R_3 \\ R_4 \end{bmatrix} = \begin{bmatrix} 0 \\ -50 \\ 0 \\ -100 \end{bmatrix}$$

Step 7 – Construct the displacement column matrix for the restrained degrees of freedom $\mathbf{U}_b$.

$$\mathbf{U}_b = \begin{bmatrix} U_5 \\ U_6 \\ U_7 \\ U_8 \\ U_9 \end{bmatrix} = \begin{bmatrix} 0 \\ -0.005 \\ 0 \\ 0 \\ 0.001 \end{bmatrix}$$

$$
\mathbf{K} = \left[\begin{array}{rrrr|rrrrr}
909688 & 227082 & 10368 & 0 & -600000 & 0 & -309688 & -227082 & 10368 \\
227082 & 190724 & 13176 & 27000 & 0 & -13500 & -227082 & -177224 & -13824 \\
10368 & 13176 & 129600 & 36000 & 0 & -27000 & -10368 & 13824 & 28800 \\
0 & 27000 & 36000 & 72000 & 0 & -27000 & 0 & 0 & 0 \\
\hline
-600000 & 0 & 0 & 0 & 600000 & 0 & 0 & 0 & 0 \\
0 & -13500 & -27000 & -27000 & 0 & 13500 & 0 & 0 & 0 \\
-309688 & -227082 & -10368 & 0 & 0 & 0 & 309688 & 227082 & -10368 \\
-227082 & -177224 & 13824 & 0 & 0 & 0 & 227082 & 177224 & 13824 \\
10368 & -13824 & 28800 & 0 & 0 & 0 & -10368 & 13824 & 57600
\end{array}\right].
$$

Step 8 – Construct and solve the linear system of algebraic equations given by

$$\mathbf{K}_{aa}\mathbf{U}_a = \mathbf{R}_a - \mathbf{K}_{ab}\mathbf{U}_b$$

$$\begin{bmatrix} 909688 & 227082 & 10368 & 0 \\ 227082 & 190724 & 13176 & 27000 \\ 10368 & 13176 & 129600 & 36000 \\ 0 & 27000 & 36000 & 72000 \end{bmatrix} \begin{bmatrix} U_1 \\ U_2 \\ U_3 \\ U_4 \end{bmatrix} = \begin{bmatrix} 0 \\ -50 \\ 0 \\ -100 \end{bmatrix} -$$

$$\begin{bmatrix} -600000 & 0 & -309688 & -227082 & 10368 \\ 0 & -13500 & -227082 & -177224 & -13824 \\ 0 & -27000 & -10368 & 13824 & 28800 \\ 0 & -27000 & 0 & 0 & 0 \end{bmatrix} \begin{bmatrix} 0 \\ -0.005 \\ 0 \\ 0 \\ 0.001 \end{bmatrix}$$

$$\mathbf{U}_a = \begin{bmatrix} U_1 \\ U_2 \\ U_3 \\ U_4 \end{bmatrix} = \begin{bmatrix} 2.192 \times 10^{-5} \\ -1.144 \times 10^{-4} \\ -4.172 \times 10^{-4} \\ -3.012 \times 10^{-3} \end{bmatrix}$$

Step 9 – Evaluate the reactions given by $\mathbf{R}_b$ using

$$\mathbf{R}_b = \mathbf{K}_{ba}\mathbf{U}_a + \mathbf{K}_{bb}\mathbf{U}_b$$

$$\begin{bmatrix} R_5 \\ R_6 \\ R_7 \\ R_8 \\ R_9 \end{bmatrix} = \begin{bmatrix} -600000 & 0 & 0 & 0 \\ 0 & -13500 & -27000 & -27000 \\ -309688 & -227082 & -10368 & 0 \\ -227082 & -177224 & 13824 & 0 \\ 10368 & -13824 & 28800 & 0 \end{bmatrix} \begin{bmatrix} U_1 \\ U_2 \\ U_3 \\ U_4 \end{bmatrix} +$$

$$\begin{bmatrix} 600000 & 0 & 0 & 0 & 0 \\ 0 & 13500 & 0 & 0 & 0 \\ 0 & 0 & 309688 & 227082 & -10368 \\ 0 & 0 & 227082 & 177224 & 13824 \\ 0 & 0 & -10368 & 13824 & 57600 \end{bmatrix} \begin{bmatrix} 0 \\ -0.005 \\ 0 \\ 0 \\ 0.001 \end{bmatrix}$$

$$\mathbf{R}_b = \begin{bmatrix} -13.15 \\ 26.64 \\ 13.15 \\ 23.36 \\ 47.39 \end{bmatrix}$$

Step 10 – Evaluate the nodal forces in each bar.

From the nodal displacements of the structure, we extract, for each bar, the bar nodal displacements. This can be systematically accomplished using the **LM** arrays. In fact, for bar 1

$$u_1 = U_7, \quad u_2 = U_8, \quad u_3 = U_9$$

$$u_4 = U_1, \quad u_5 = U_2, \quad u_6 = U_3.$$

Therefore, with $\mathbf{u}^{(1)}$ determined, we can evaluate $\mathbf{f}^{(1)}$ using $\mathbf{f}^{(1)} = \mathbf{k}^{(1)}\mathbf{u}^{(1)}$ which yields

$$\mathbf{f}^{(1)^T} = \begin{bmatrix} 13.15 & 23.36 & 47.39 & -13.15 & -23.36 & 6.58 \end{bmatrix}.$$

For bar 2, using $\mathbf{LM}^{(2)}$ we directly write

$$\mathbf{u}^{(2)^T} = \begin{bmatrix} 2.192 \times 10^{-5} & -1.144 \times 10^{-4} & -4.172 \times 10^{-4} & 0 & -5 \times 10^{-3} & -3.012 \times 10^{-3} \end{bmatrix}$$

and considering $\mathbf{f}^{(2)} = \mathbf{k}^{(2)}\mathbf{u}^{(2)}$ we arrive at

$$\mathbf{f}^{(2)^T} = \begin{bmatrix} 13.15 & -26.64 & -6.58 & 13.15 & 26.64 & 100 \end{bmatrix}.$$

Step 11 – Evaluate bar nodal forces in the bar's local system.

Although we could find the internal force diagrams from the bar nodal forces in the global system, they are more easily determined from the bar nodal forces in the local system.

We have

$$\tilde{\mathbf{f}}^{(1)} = \mathbf{T}^{(1)}\mathbf{f}^{(1)} = \begin{bmatrix} 24.54 \\ 10.79 \\ 47.39 \\ -24.54 \\ -10.79 \\ 6.58 \end{bmatrix}.$$

Analogously

$$\tilde{\mathbf{f}}^{(2)} = \mathbf{T}^{(2)}\mathbf{f}^{(2)} = \begin{bmatrix} 13.15 \\ -26.64 \\ -6.58 \\ -13.15 \\ 26.64 \\ -100 \end{bmatrix}.$$

Step 12 – Construct the internal force diagrams.

These diagrams can be constructed bar by bar using $\tilde{\mathbf{f}}^{(m)}$ and they are shown in Figure 4.41. Only as a verification, we can check that the equilibrium

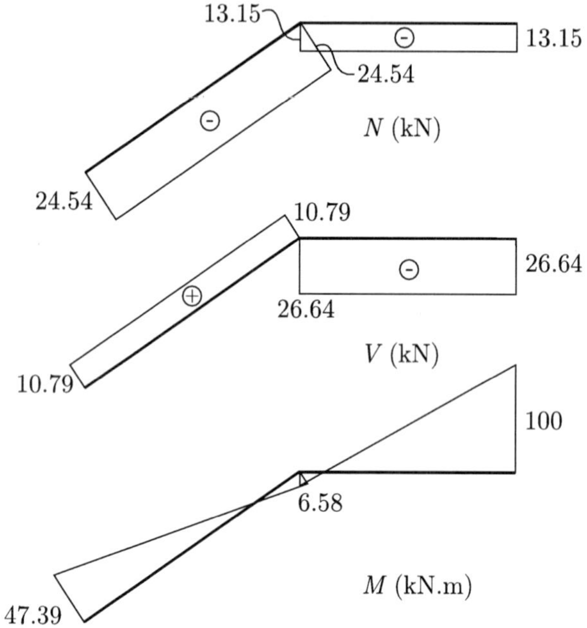

Fig. 4.41. Internal force diagrams

of node 1 holds. Figure 4.42 summarizes this situation.

Regarding the units, although we did not show these explicitly, all obtained displacements are in meters, rotations in radians, forces in kN and moments in $kN \cdot m$. These units are those used in the physical problem, see Figure 4.39, and were used throughout the solution.

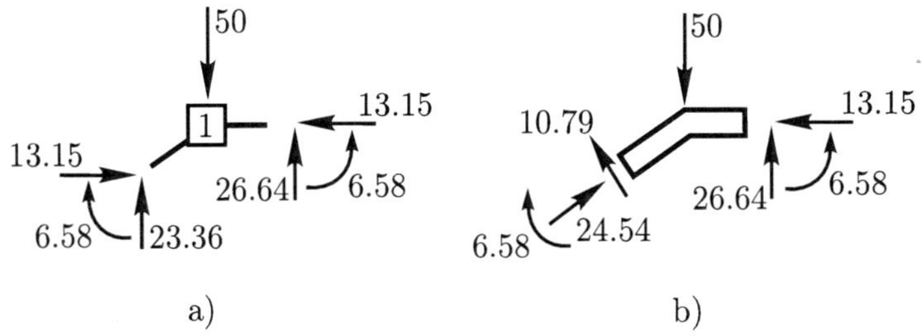

Fig. 4.42. Equilibrium of node 1. a) bar end forces in the global system acting onto node; b) Internal forces at "node" sections

□

External loads applied to the bars

So far, the matrix method was formulated considering external loads applied only to the nodes. However, in practice we frequently find structures in which loads are not applied to the nodes. Therefore, we need to extend our formulation to consider loads applied to the bars.

The central idea behind the procedure to be presented is to take advantage of the superposition of effects and construct a solution strategy based on the formulation of the matrix method which considers loads applied to the nodes only.

For that consider that the solution will be obtained by superimposing the solutions for two problems: (I) and (II). Problem (I) is defined considering the external loads which are applied directly onto the bars (*i.e.* not to the nodes) and restraining all degrees of freedom to have zero displacements/rotations. The external reactions which should be introduced to guarantee that all displacements/rotations are zero when we consider the loads applied directly onto the bars are given by $\mathbf{R}_0$ with $\mathbf{R}_{0,a}$ being the partition associated with the degrees of freedom that are free in the original structure and $\mathbf{R}_{0,b}$ with those that are restrained.

Since the loads given by $\mathbf{R}_0$ do not exist in the original structure, they are introduced in problem (II) with the reversed sense, *i.e.*, $-\mathbf{R}_0$. In problem (II), we consider in addition the external loads of the original structure which are applied directly to the nodes.

We note that problem (II) can be solved in the usual way since the external loads are applied to the nodes only. The nodal displacement/rotation solution for problem (II) is already the solution for the original structure since for structure (I) all degrees of freedom were fixed.

Now the solution strategy is apparent. The solution of problem (II) will lead to the displacements/rotations and reactions of the original problem. The matrix equation for problem (II) can be written as

$$\begin{bmatrix} \mathbf{K}_{aa} & \mathbf{K}_{ab} \\ \mathbf{K}_{ba} & \mathbf{K}_{bb} \end{bmatrix} \begin{bmatrix} \mathbf{U}_a \\ \mathbf{U}_b \end{bmatrix} = \begin{bmatrix} \mathbf{R}_a - \mathbf{R}_{0,a} \\ \mathbf{R}_b - \mathbf{R}_{0,b} \end{bmatrix}.$$

The first set of equations

$$\mathbf{K}_{aa}\mathbf{U}_a = \mathbf{R}_a - \mathbf{R}_{0,a} - \mathbf{K}_{ab}\mathbf{U}_b \tag{4.178}$$

can be solved for $\mathbf{U}_a$ and the second set leads to the reactions, *i.e.*,

$$\mathbf{R}_b = \mathbf{R}_{0,b} + \mathbf{K}_{ba}\mathbf{U}_a + \mathbf{K}_{bb}\mathbf{U}_b. \tag{4.179}$$

We emphasize that

$$\mathbf{U}^T = \begin{bmatrix} \mathbf{U}_a & \mathbf{U}_b \end{bmatrix}$$

is the solution for the original problem. Note that, if there are imposed nodal displacements/rotations they should be introduced in problem (II) only and the above equations are already taking into account this possibility. We also assumed that no external concentrated forces and moments are applied at the restraint degrees of freedom. These concentrated forces/moments would have to be added (with appropriate senses) to the reactions calculated in (4.179).

Regarding the calculation of the bar nodal forces, we need to be particularly careful. We note that, although there are no nodal displacements for problem (I), in general, the bar nodal forces are not zero since the external loads applied to the bars should be equilibrated by these bar nodal forces. Let us denote by $\tilde{\mathbf{f}}_0^{(m)}$ the bar (m) nodal forces in the local system for problem (I). Therefore, the bar nodal forces for bar (m) for the original problem are given by

$$\tilde{\mathbf{f}}^{(m)} = \tilde{\mathbf{f}}_0^{(m)} + \tilde{\mathbf{k}}^{(m)}\tilde{\mathbf{u}}^{(m)}.$$

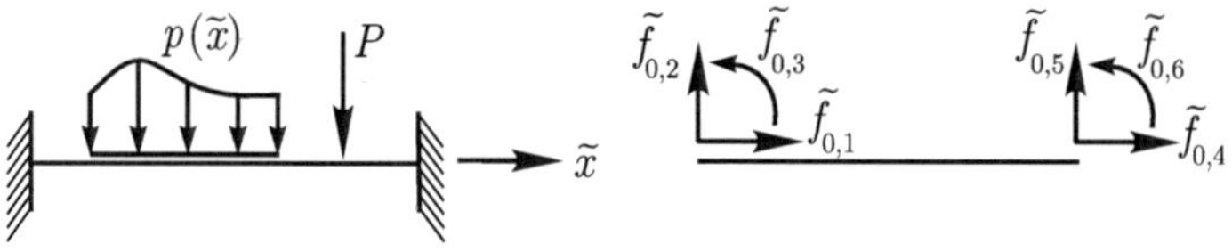

Fig. 4.43. Generic situation to evaluate $\tilde{\mathbf{f}}_0$

Let us show how to obtain $\mathbf{R}_0$ from the bar nodal forces $\tilde{\mathbf{f}}_0^{(m)}$ of the structure. We first note that $\tilde{\mathbf{f}}_0^{(m)}$ can be obtained by considering bar (m) clamped at both ends subjected to the external load applied to bar (m). A generic situation is shown in Figure 4.43. In the bar global system these forces are given by

$$\mathbf{f}_0^{(m)} = \mathbf{T}^{(m)^T} \, \tilde{\mathbf{f}}_0^{(m)}. \tag{4.180}$$

Now we define $\mathbf{F}_0^{(m)}$ as the $N \times 1$ column matrix, N being the total number of degrees of freedom of the structure, which is obtained from $\mathbf{f}_0^{(m)}$ in the same way as $\mathbf{F}^{(m)}$ is obtained from $\mathbf{f}^{(m)}$, *i.e.*, using the correspondence between the local and global numbering. Then we can enforce equilibrium of every node of problem (I) by

$$\mathbf{R}_0 = \sum_{m=1}^{n_e} \mathbf{F}_0^{(m)} \tag{4.181}$$

which shows how to obtain $\mathbf{R}_0$ from $\mathbf{F}_0^{(m)}$. Of course, since $\mathbf{F}_0^{(m)}$ has many zero entries, the summation given in (4.181) is efficiently performed directly from the nodal force column matrices $\tilde{\mathbf{f}}_0^{(m)}$ using the $\mathbf{LM}^{(m)}$ arrays.

Typical examples of $\tilde{\mathbf{f}}_0^{(m)}$ are given in Figure 4.44 and Figure 4.45 for which

$$\tilde{\mathbf{f}}_0^{(i)^T} = \begin{bmatrix} 0 & \frac{P}{2} & \frac{Pa}{8} & 0 & \frac{P}{2} & -\frac{Pa}{8} \end{bmatrix}$$

and

$$\tilde{\mathbf{f}}_0^{(j)^T} = \begin{bmatrix} 0 & \frac{pa}{2} & \frac{pa^2}{12} & 0 & \frac{pa}{2} & -\frac{pa^2}{12} \end{bmatrix}.$$

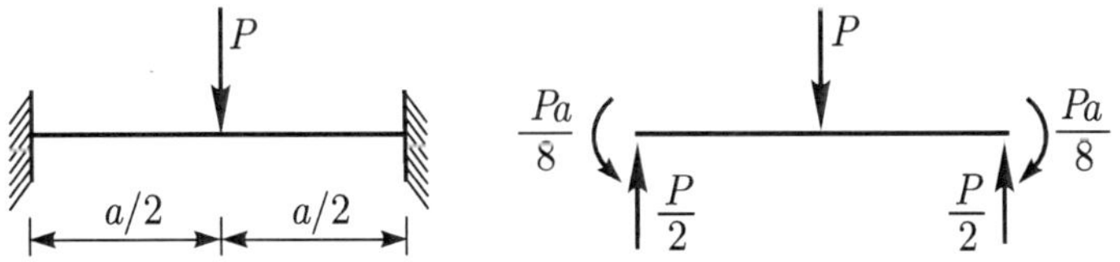

Fig. 4.44. Evaluation of $\tilde{\mathbf{f}}_0^{(i)}$

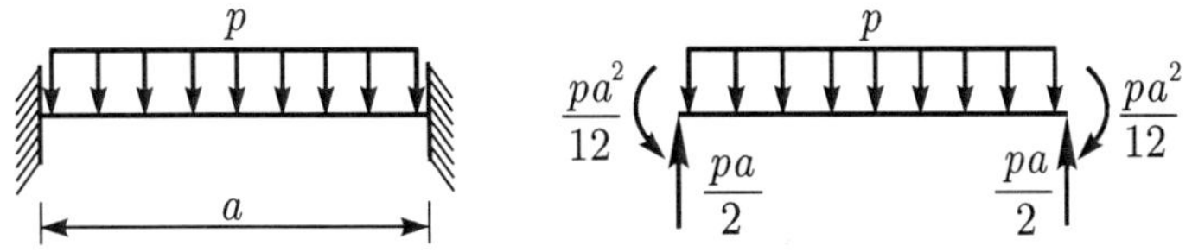

Fig. 4.45. Evaluation of $\tilde{\mathbf{f}}_0^{(j)}$

Example 4.14

Consider the structure defined in Figure 4.39 with the addition of the loads applied to the bars as summarized in Figure 4.46. Find the nodal displacements, the reactions and draw the internal force diagrams for the structure.

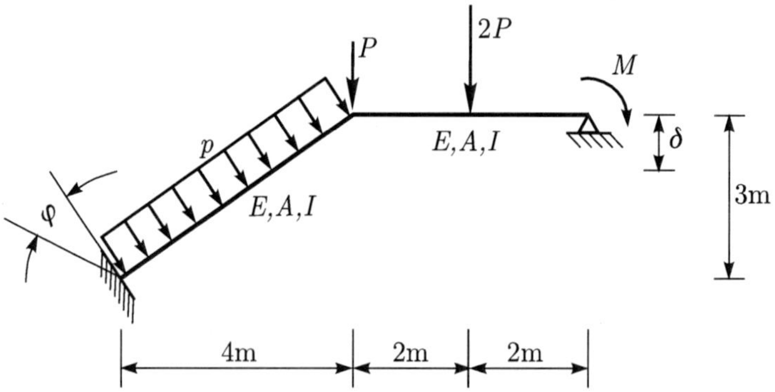

Fig. 4.46. Problem description with the added load. The data given in Figure 4.39 remains unchanged and $p = 60$ kN/m

Solution

We need to evaluate $\mathbf{R}_0$ and use the superposition detailed in this section. We keep the same definitions of Example 4.13.

Using the solutions summarized in Figures 4.44 and 4.45, we can write

$$\tilde{\mathbf{f}}_0^{(1)^T} = \begin{bmatrix} 0 & 150 & 125 & 0 & 150 & -125 \end{bmatrix}$$

$$\tilde{\mathbf{f}}_0^{(2)^T} = \begin{bmatrix} 0 & 50 & 50 & 0 & 50 & -50 \end{bmatrix}$$

and using (4.180)

$$\mathbf{f}_0^{(1)^T} = \begin{bmatrix} -90 & 120 & 125 & -90 & 120 & -125 \end{bmatrix}$$

$$\mathbf{f}_0^{(2)^T} = \begin{bmatrix} 0 & 50 & 50 & 0 & 50 & -50 \end{bmatrix}.$$

Taking advantage of the **LM** arrays already defined in Example 4.13, we write

$$
\begin{aligned}
\mathbf{R}_{0,a}^T &= \begin{bmatrix} -90 & 170 & -75 & -50 \end{bmatrix} \\
\mathbf{R}_{0,b}^T &= \begin{bmatrix} 0 & 50 & -90 & 120 & 125 \end{bmatrix}.
\end{aligned}
$$

Now $\mathbf{U}_a$ can be evaluated using (4.178) which is written as

$$
\begin{bmatrix} 909688 & 227082 & 10368 & 0 \\ 227082 & 190724 & 13176 & 27000 \\ 10368 & 13176 & 129600 & 36000 \\ 0 & 27000 & 36000 & 72000 \end{bmatrix} \begin{bmatrix} U_1 \\ U_2 \\ U_3 \\ U_4 \end{bmatrix} = \begin{bmatrix} 0 \\ -50 \\ 0 \\ -100 \end{bmatrix} - \begin{bmatrix} -90 \\ 170 \\ -75 \\ -50 \end{bmatrix} -
$$

$$
\begin{bmatrix} -600000 & 0 & -309688 & -227082 & 10368 \\ 0 & -13500 & -227082 & -177224 & -13824 \\ 0 & -27000 & -10368 & 13824 & 28800 \\ 0 & -27000 & 0 & 0 & 0 \end{bmatrix} \begin{bmatrix} 0 \\ -0.005 \\ 0 \\ 0 \\ 0.001 \end{bmatrix}. \tag{4.182}
$$

Of course, the stiffness matrices are the same as for Example 4.13. Solving (4.182), we obtain

$$
\mathbf{U}_a^T = \begin{bmatrix} 5.396 \times 10^{-4} & -1.810 \times 10^{-3} & -2.222 \times 10^{-5} & -1.880 \times 10^{-3} \end{bmatrix}.
$$

The reactions $\mathbf{R}_b$ can be obtained using (4.179)

$$\mathbf{R}_b = \begin{bmatrix} 0 \\ 50 \\ -90 \\ 120 \\ 125 \end{bmatrix} + \begin{bmatrix} -600000 & 0 & 0 & 0 \\ 0 & -13500 & -27000 & -27000 \\ -309688 & -227082 & -10368 & 0 \\ -227082 & -177224 & 13824 & 0 \\ 10368 & -13824 & 28800 & 0 \end{bmatrix} \begin{bmatrix} U_1 \\ U_2 \\ U_3 \\ U_4 \end{bmatrix} +$$

$$\begin{bmatrix} 600000 & 0 & 0 & 0 & 0 \\ 0 & 13500 & 0 & 0 & 0 \\ 0 & 0 & 309688 & 227082 & -10368 \\ 0 & 0 & 227082 & 177224 & 13824 \\ 0 & 0 & -10368 & 13824 & 57600 \end{bmatrix} \begin{bmatrix} 0 \\ -0.005 \\ 0 \\ 0 \\ 0.001 \end{bmatrix}$$

leading to

$$\mathbf{R}_b^T = \begin{bmatrix} -323.72 & 58.28 & 143.72 & 331.72 & 212.57 \end{bmatrix}.$$

From $\mathbf{U}$ we obtain the bar nodal displacements $\mathbf{u}^{(1)}$ and $\mathbf{u}^{(2)}$, and in the local system they are given by

$$\tilde{\mathbf{u}}^{(1)} = \mathbf{T}^{(1)}\mathbf{u}^{(1)} = \mathbf{T}^{(1)} \begin{bmatrix} 0 \\ 0 \\ 0.001 \\ 5.396 \times 10^{-4} \\ -1.810 \times 10^{-3} \\ -2.222 \times 10^{-5} \end{bmatrix} = \begin{bmatrix} 0 \\ 0 \\ 0.001 \\ -6.542 \times 10^{-4} \\ -1.771 \times 10^{-3} \\ -2.222 \times 10^{-5} \end{bmatrix}$$

and

$$\tilde{\mathbf{u}}^{(2)} = \mathbf{u}^{(2)} = \begin{bmatrix} 5.396 \times 10^{-4} \\ -1.810 \times 10^{-3} \\ -2.222 \times 10^{-5} \\ 0 \\ -0.005 \\ -1.880 \times 10^{-3} \end{bmatrix}$$

since $\mathbf{T}^{(2)} = \mathbf{I}$.

The bar nodal forces in the local system can be evaluated by

$$\tilde{\mathbf{f}}^{(1)} = \tilde{\mathbf{f}}_0^{(1)} + \tilde{\mathbf{k}}^{(1)}\tilde{\mathbf{u}}^{(1)} = \begin{bmatrix} 314 \\ 179.14 \\ 212.57 \\ -314 \\ 120.86 \\ -66.87 \end{bmatrix}$$

$$\tilde{\mathbf{f}}^{(2)} = \tilde{\mathbf{f}}_0^{(2)} + \tilde{\mathbf{k}}^{(2)}\tilde{\mathbf{u}}^{(2)} = \begin{bmatrix} 323.72 \\ 41.72 \\ 66.87 \\ -323.72 \\ 58.28 \\ -100 \end{bmatrix}.$$

The internal force diagrams are shown in Figure 4.47.

□

Additional bar end conditions

In the frame structures considered so far, continuity of rotations was implied for the bar sections and, in particular, at the nodes. Referring to Figures 4.39 and 4.40, as an example, we can see that the rotation of node 1 (taken as a typical node) is the same as the end section rotation of bar 1 and that of the initial section of bar 2. This continuity is taken into account in the matrix formulation through the compatibility condition

$$U_3 = u_6^{(1)} = u_3^{(2)}.$$

The stiffness matrix given in (4.176) has been derived to model such situations since it provides stiffness with respect to the bar end section rotations which, by compatibility, correspond to the nodal rotations.

In engineering structural analysis we frequently also have an internal rotational hinge (pin type connection) as in Figure 4.48. In this case the bar element sections connected to the hinge no longer have to undergo the same rotation (while the section displacements have to be the same). Hence, the

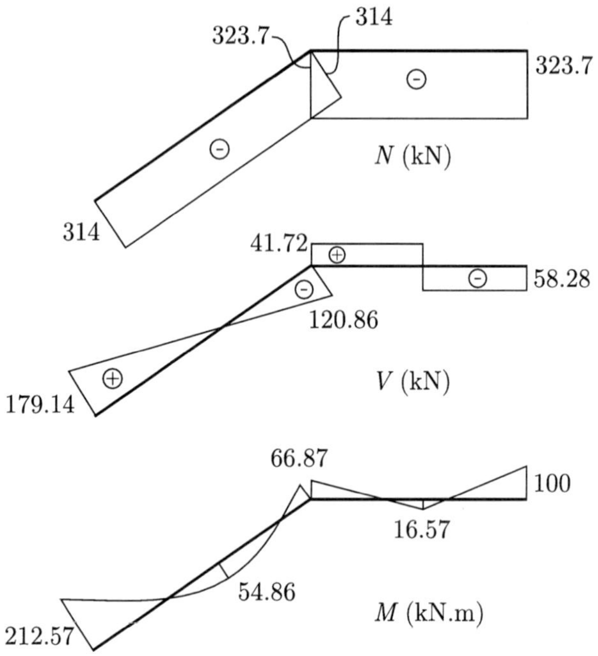

Fig. 4.47. Internal force diagrams

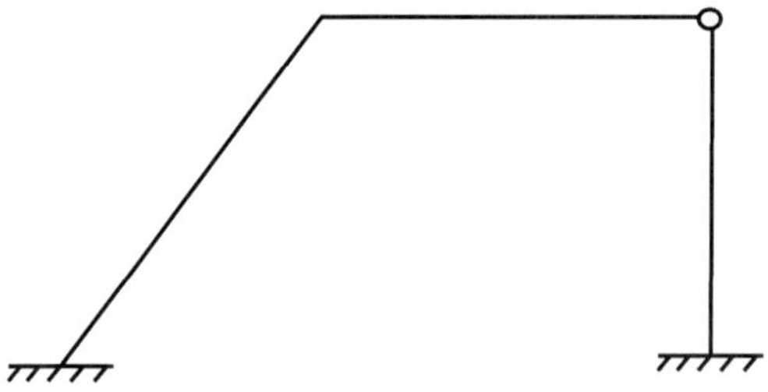

Fig. 4.48. Frame structure with a pin connection

compatibility condition for the rotation no longer exists and there is no moment transferred between the sections connected to the hinge.

There is no unique way to model the rotational hinge. For example, in Figure 4.49a we choose a node to represent each section connected to the hinge and enforce that the section translations are the same, that is, $U_4 = U_7$ and $U_5 = U_8$. This could be efficiently accomplished by choosing as degrees of freedom the translations of the hinge plus the rotations of the sections connected to the hinge, as shown in Figure 4.49b.

An efficient modeling alternative that keeps the number of degrees of freedom per node at three is shown in Figure 4.50. In Figure 4.50a, node 2

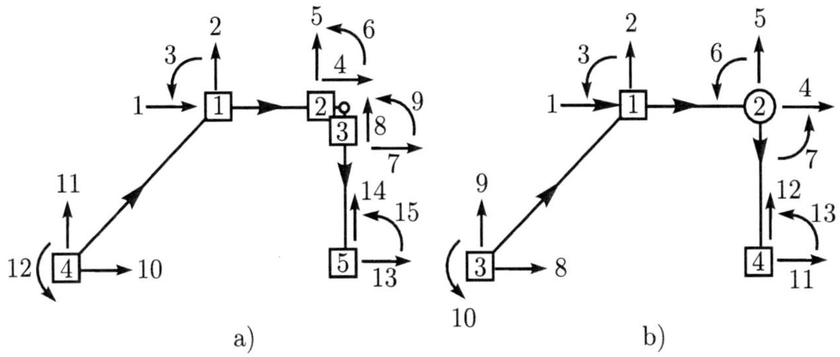

Fig. 4.49. Modeling the rotational hinge

is considered attached to bar 2, *i.e.*, the rotation of node 2 is the same as the rotation of the end section of bar 2. Therefore, the degree of freedom 6 corresponds to the rotation of the end section of bar 2. The translations of node 2, *i.e.*, degrees of freedom 4 and 5, still correspond to the translations of the sections of bars 2 and 3 which connect at node 2. The stiffness matrix of bar 3 has to be modified, as detailed shortly, and the rotation of the section of bar 3 at node 2 will no longer be a direct outcome of the solution of the matrix equations.

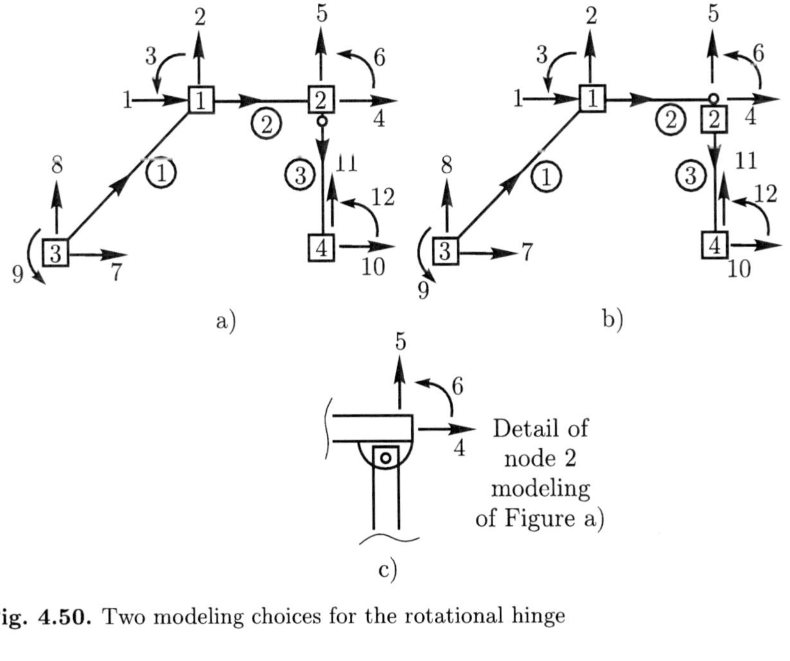

Fig. 4.50. Two modeling choices for the rotational hinge

In Figure 4.50b, node 2 is chosen to be attached to bar 3, and the degree of freedom 6 is now equal to the section rotation of bar 3 and the stiffness matrix of bar 2 should be modified.

Consider the bar described in Figure 4.51. The rotation of node 2 of the

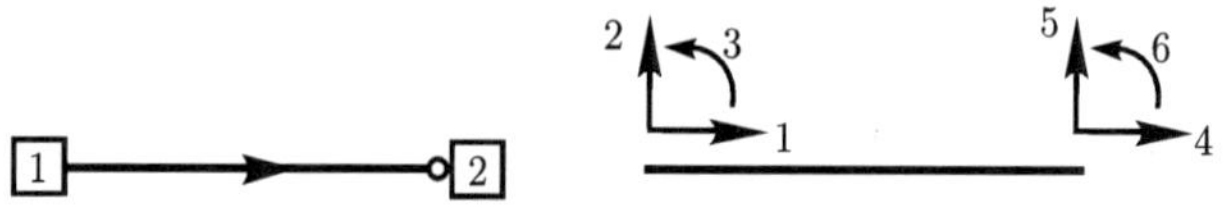

Fig. 4.51. Bar with built-in condition at local node one and pin condition at local node 2

bar is free; that is, there is no bending moment at the end section of the bar. We can construct this stiffness matrix column by column imposing unit end displacements/rotations. Columns 1 and 4 are obtained as for the stiffness matrix given in (4.176).

We detail the derivation of the fifth column which is obtained by imposing $u_5 = 1$, $u_1 = u_2 = u_3 = u_4 = 0$. We should take into account that $M(L) = 0$. The axial displacements are trivially zero as given by equation (4.170), *i.e.*,

$$u(\tilde{x}) = 0 \quad \rightarrow \quad N(\tilde{x}) = 0.$$

For the transverse displacements, we refer to the solution derived in Example 4.9 which gives the solution sought when we take $\delta = -1$ leading to

$$w(\tilde{x}) = h_5(\tilde{x}) + \frac{3}{2L} h_6(\tilde{x}) = -\frac{\tilde{x}^3}{2L^3} + \frac{3\tilde{x}^2}{2L^2}.$$

Therefore

$$\begin{aligned} M(\tilde{x}) &= EIw''(\tilde{x}) = \frac{EI}{L^3}(3L - 3\tilde{x}) \\ M(0) &= \frac{3EI}{L^2}, \qquad\qquad M(L) = 0 \\ V(\tilde{x}) &= EIw'''(\tilde{x}) = -\frac{3EI}{L^3}. \end{aligned}$$

The results are summarized in Figure 4.52. In analogous manner the remaining columns are obtained and we summarize the results in Figure 4.53. Therefore, the stiffness matrix is given by

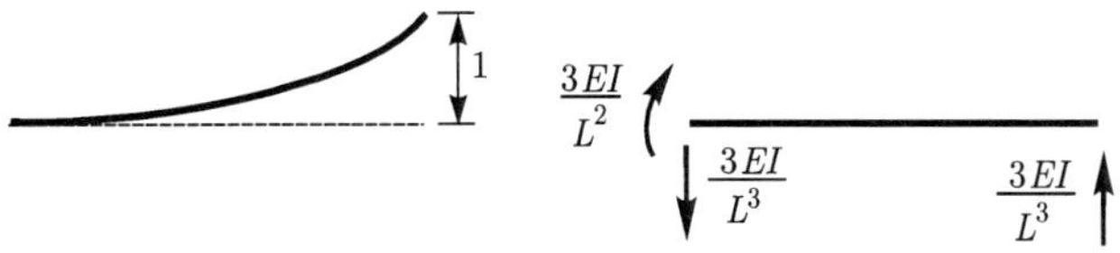

Fig. 4.52. Solution that leads to the fifth column of $\mathbf{k}$

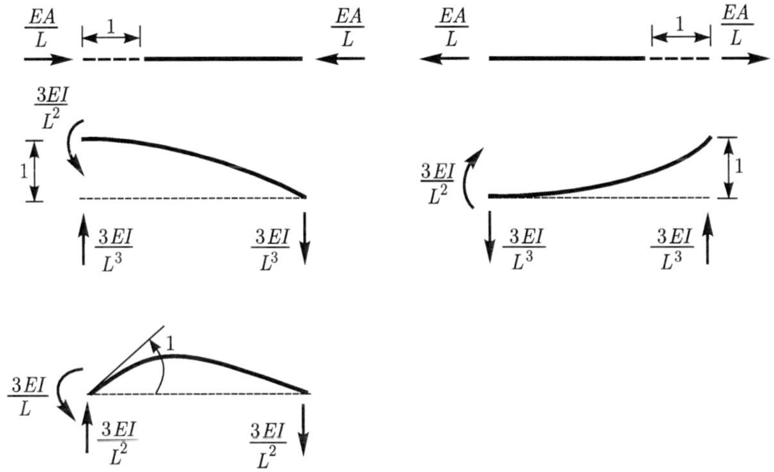

Fig. 4.53. Bar end forces for unit end displacements

$$\tilde{\mathbf{k}} = \begin{bmatrix} \frac{EA}{L} & 0 & 0 & -\frac{EA}{L} & 0 & 0 \\ 0 & \frac{3EI}{L^3} & \frac{3EI}{L^2} & 0 & -\frac{3EI}{L^3} & 0 \\ 0 & \frac{3EI}{L^2} & \frac{3EI}{L} & 0 & \frac{3EI}{L^2} & 0 \\ -\frac{EA}{L} & 0 & 0 & \frac{EA}{L} & 0 & 0 \\ 0 & -\frac{3EI}{L^3} & -\frac{3EI}{L^2} & 0 & \frac{3EI}{L^3} & 0 \\ 0 & 0 & 0 & 0 & 0 & 0 \end{bmatrix} . \tag{4.183}$$

Note that the degree of freedom 6 no longer exists for the beam element; hence column six has all zero entries. Of course, there is an explicit understanding that when such a bar is used in an assemblage, the rotational stiffness of node 2 must come from other bars.

When the local node 1 is pinned as shown in Figure 4.54 the stiffness matrix can be obtained analogously and is given by

$$
\tilde{\mathbf{k}} = \begin{bmatrix}
\frac{EA}{L} & 0 & 0 & -\frac{EA}{L} & 0 & 0 \\
0 & \frac{3EI}{L^3} & 0 & 0 & -\frac{3EI}{L^3} & \frac{3EI}{L^2} \\
0 & 0 & 0 & 0 & 0 & 0 \\
-\frac{EA}{L} & 0 & 0 & \frac{EA}{L} & 0 & 0 \\
0 & -\frac{3EI}{L^3} & 0 & 0 & \frac{3EI}{L^3} & -\frac{3EI}{L^2} \\
0 & \frac{3EI}{L^2} & 0 & 0 & -\frac{3EI}{L^2} & \frac{3EI}{L}
\end{bmatrix} . \tag{4.184}
$$

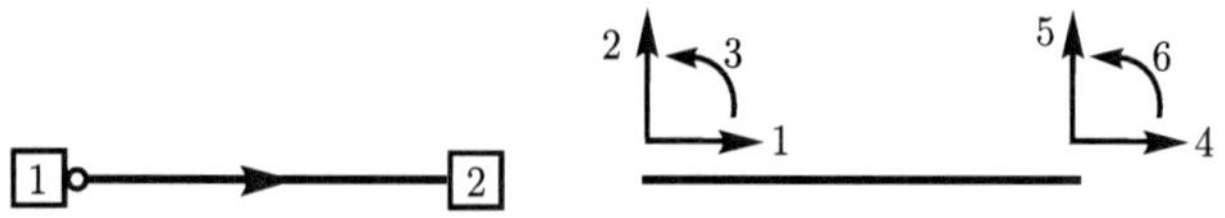

Fig. 4.54. Bar with pin condition at local node 1 and built-in condition at local node 2

Example 4.15

Consider again the structure defined in Figure 4.39, with the additional loads introduced in Example 4.14, but now with a pin connection as shown in Figure 4.55. Obtain the displacements of the free degrees of freedom.

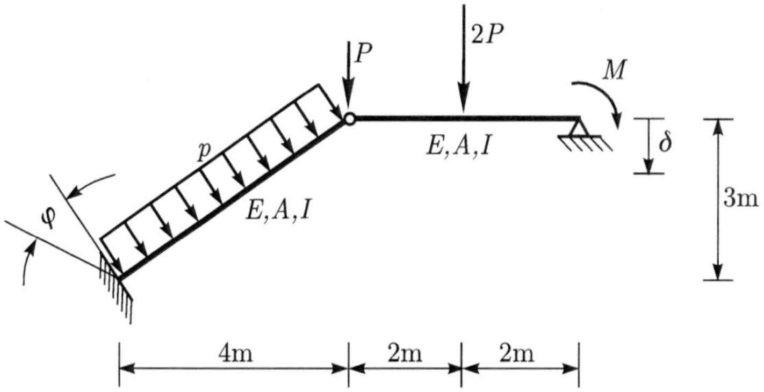

Fig. 4.55. Problem description having a pin connection.

Considering the numbering choices given in Figure 4.40 evaluate the nodal displacements. The rotation of the section of the inclined bar adjacent to the pin is to be predicted.

Solution

The modeling choices are shown in Figure 4.56. Note that the rotation of

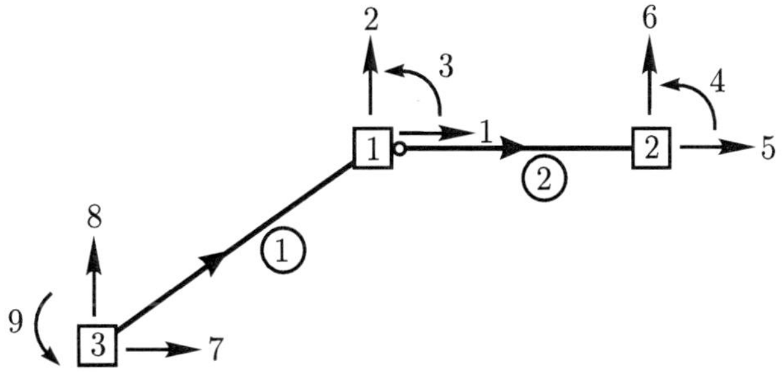

Fig. 4.56. Model definitions

node 1 given by U_3 is the rotation of the section of the inclined bar adjacent to the pin.

There are two modifications that should be introduced. The end conditions of bar 2 should be pinned – built-in (see Figure 4.54) and can be obtained using (4.184)

$$\tilde{\mathbf{k}}^{(2)} = \begin{bmatrix} 600000 & 0 & 0 & -600000 & 0 & 0 \\ 0 & 3375 & 0 & 0 & -3375 & 13500 \\ 0 & 0 & 0 & 0 & 0 & 0 \\ -600000 & 0 & 0 & 600000 & 0 & 0 \\ 0 & -3375 & 0 & 0 & 3375 & -13500 \\ 0 & 13500 & 0 & 0 & -13500 & 54000 \end{bmatrix}.$$

The other modification refers to the evaluation of $\tilde{\mathbf{f}}_0^{(2)}$. Since, in the superposition, the pin has to be taken into account. The condition that leads to $\tilde{\mathbf{f}}_0^{(2)}$ is shown in Figure 4.57 and therefore

$$\tilde{\mathbf{f}}_0^{(2)^T} = \left[\begin{array}{cccccc} 0 & \frac{10P}{16} & 0 & 0 & \frac{22P}{16} & -\frac{6Pa}{16} \end{array}\right].$$

Introducing the modifications above and following the same solution steps which have been used, we obtain

$$\mathbf{U}_a^T = \left[\begin{array}{cccc} 4.481 \times 10^{-4} & -1.501 \times 10^{-3} & 1.229 \times 10^{-3} & -1.338 \times 10^{-3} \end{array}\right].$$

□

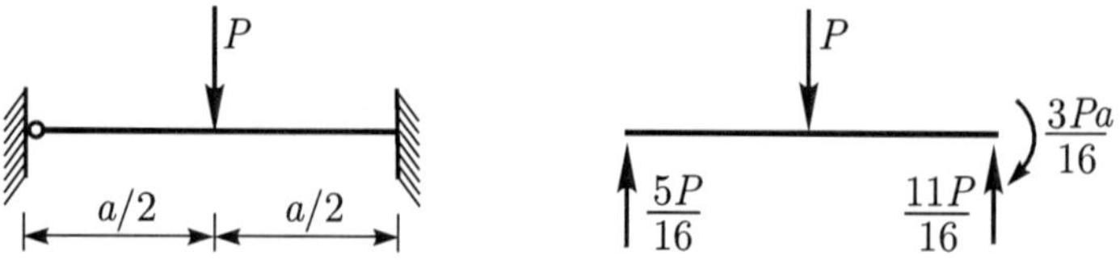

Fig. 4.57. Solution required for evaluation of $\tilde{\mathbf{f}}_0^{(2)}$

4.2.5 Bars subjected to 3-D actions

In this section we study the behavior of a straight prismatic bar subjected to arbitrary transverse and axial loads. The objective is to derive the stiffness matrix of the bar for these conditions.

Bending and axial actions

In Section 4.2.2 we introduced the Bernoulli-Euler beam model for planar conditions, *i.e.*, we assumed that the bar has a vertical plane of symmetry, that the transversely distributed load was acting in the plane of symmetry and that the bar axis remains in the plane of symmetry.

Now consider the situation summarized in Figure 4.58. The x axis is taken along the section centroid[5] and since there are no section symmetries, the y and z axes are arbitrarily chosen. The transverse load is decomposed into p_y and p_z acting along directions y and z respectively. The axial loading is denoted by f_x. The stress resultants are shown in Figure 4.58 for a generic section with outward normal given by $\mathbf{e}_x$. Actually, there are many ways in which these stress resultants may be defined. In order to obtain analogous differential equilibrium equations to those of the planar beam problem, we choose the conventions of Figure 4.58.

The differential equilibrium equations are

$$\frac{dN}{dx} + f_x = 0 \tag{4.185}$$

$$\frac{dV_y}{dx} = p_y, \qquad \frac{dV_z}{dx} = p_z \tag{4.186}$$

$$\frac{dM_y}{dx} = V_z, \qquad \frac{dM_z}{dx} = V_y. \tag{4.187}$$

The strain compatibility relations can be obtained considering the rotations of fibers on the cross-section which are aligned with the y and z axes as shown in Figure 4.59.

Note that

[5] We assume that either the shear center coincides with the centroid or it is close enough to the centroid that the induced warping displacements can be neglected (see Section 4.2.6 for the definition of the shear center)

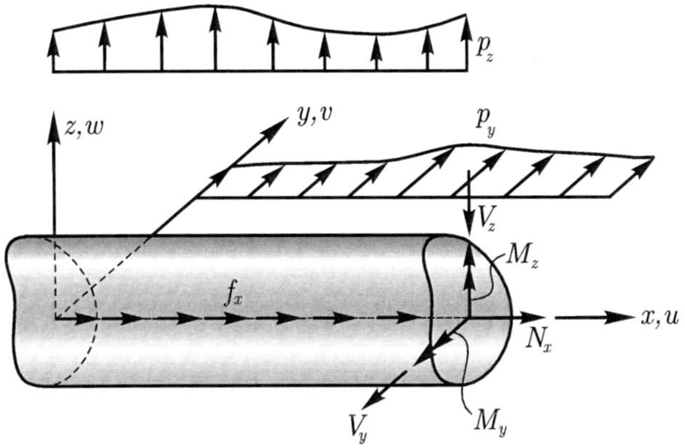

Fig. 4.58. Loading and stress resultant definitions

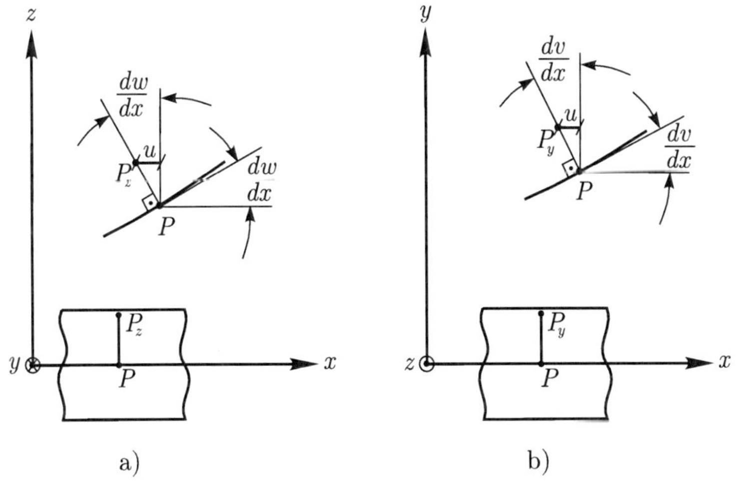

Fig. 4.59. Rotation of fibers aligned with axes y and z

$$v(x, y, z) = v(x) \tag{4.188}$$

$$w(x, y, z) = w(x). \tag{4.189}$$

The kinematics summarized in Figure 4.59 leads to

$$u(x, y, z) = u_0(x) - y\frac{dv}{dx} - z\frac{dw}{dx} \tag{4.190}$$

where $u_0(x)$ is the displacement in the x direction of points on the bar axis. Note that for a given cross-section u_0, $\frac{dv}{dx}$ and $\frac{dw}{dx}$ are constant values and equations (4.188), (4.189) and (4.190) show, as assumed, that the cross-section remains plane. Orthogonality to the deformed axis, corresponding to the Bernoulli-Euler hypothesis, can be directly inferred from Figure 4.59.

The longitudinal strain is given by

$$\varepsilon_{xx} = \frac{\partial u}{\partial x} = \frac{du_0}{dx} - y\frac{d^2 v}{dx^2} - z\frac{d^2 w}{dx^2}$$

and Hooke's law leads to

$$\tau_{xx} = E\left(\frac{du_0}{dx} - y\frac{d^2 v}{dx^2} - z\frac{d^2 w}{dx^2}\right).$$

The stress resultants can be evaluated by integration of the stresses over the cross-sections. The axial force is given by

$$N(x) = \int_A \tau_{xx}\, dA = EA\frac{du_0}{dx} \tag{4.191}$$

since the x axis passes through the centroid.

The moment M_y is given by

$$M_y = \int_A \tau_{xx}\,(-z)\; dA = -E\frac{du_0}{dx}\int_A z\, dA + E\frac{d^2 v}{dx^2}\int_A yz\, dA + E\frac{d^2 w}{dx^2}\int_A z^2\, dA. \tag{4.192}$$

We can define

$$I_y = \int_A z^2\, dA, \qquad I_{yz} = \int_A yz\, dA \tag{4.193}$$

where I_y is the moment of inertia of the section with respect to y and I_{yz} is the product of inertia of the section with respect to y and z.

Introducing (4.193) into (4.192) we obtain

$$M_y = EI_y\frac{d^2 w}{dx^2} + EI_{yz}\frac{d^2 v}{dx^2}. \tag{4.194}$$

The moment M_z is given by

$$M_z = \int_A \tau_{xx}\,(-y)\; dA = E\frac{du_0}{dx}\int_A y\, dA + E\frac{d^2 v}{dx^2}\int_A y^2\, dA + E\frac{d^2 w}{dx^2}\int_A yz\, dA$$

which upon the definition of the moment of inertia with respect to z as

$$I_z = \int_A y^2\, dA$$

leads to

$$M_z = EI_z\frac{d^2 v}{dx^2} + EI_{yz}\frac{d^2 w}{dx^2}. \tag{4.195}$$

We note that the equilibrium equations (4.185) to (4.187) and equations (4.191), (4.194) and (4.195) which represent the compatibility and constitutive relations give all the required conditions for the model and can be used to find solutions when appropriate boundary conditions are specified.

Equations (4.194) and (4.195) can be placed in a simpler form if a particular set of axes is chosen. It is possible to show that we can always find a position for the y and z axes by rotating them about the centroid of the section such that the product of inertia vanishes, that is

$$I_{yz} = 0. \tag{4.196}$$

The axes y and z for which (4.196) holds are the *principal axes of inertia* of the cross-section. For these axes, equations (4.194) and (4.195) become

$$M_y = EI_y \frac{d^2 w}{dx^2}, \qquad M_z = EI_z \frac{d^2 v}{dx^2}.$$

Comparing the above equations with equation (4.147) which is applicable for planar situations, we can see that the solution for 3-D bars in bending can be obtained as the superposition of two planar bending problems, as detailed in Section 4.4.2, which should be defined for the planes xy and xz with y and z being the principal axes of the cross-section.

Torsional actions

In Section 3.6 we studied the torsion problem of a prismatic bar. The formulation led to the exact 3-D elasticity solution as long as the torsional actions are introduced at the end sections as a specific field of shear surface forces which is mechanically equivalent to a torsional moment only. Additionally, the cross-sections should be free to warp.

These conditions are frequently violated in practical problems. For example, the bar is not free to warp at the bar ends where it is either connected to other bars or its section is restrained. However, as long as the cross-sections are not thin walled, these end perturbations affect the solution only in a small region close to the bar ends (see Timoshenko and Goodier, 1970) and we can assume that the bar section rotations are governed by

$$\frac{d\theta_x}{dx} = \frac{M_x}{EI_t} \tag{4.197}$$

where, referring to Figure 4.58, the end section torsional moments are given by $\mathbf{M} = M_x \mathbf{e}_x$ and $-\mathbf{M}$, θ_x is the section rotation about the x axis and I_t is the torsional moment of inertia of the cross-section (see Section 3.6).

Matrix formulation for a 3-D bar

We can obtain the stiffness matrix of a bar subjected to 3-D actions in an analogous manner to the 2-D bar solving the applicable differential equations for unit end displacements.

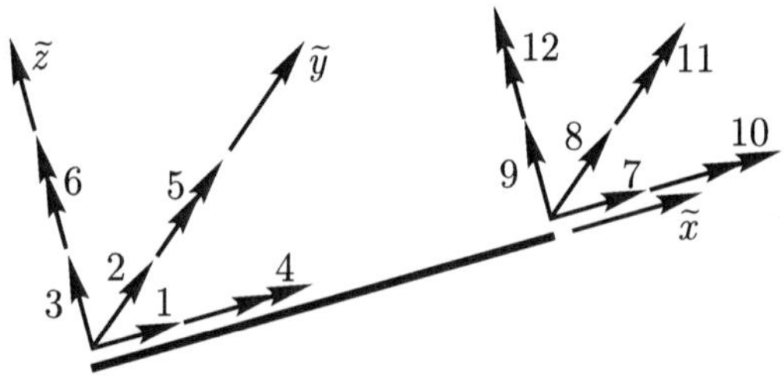

Fig. 4.60. Degrees of freedom for a 3-D bar

In Figure 4.60 the end section degrees of freedom are summarized. The local reference system is chosen such that $\tilde{x}$ is taken along the section centroid and $\tilde{y}$ and $\tilde{z}$ are the principal axes of the bar cross-section.

We do not detail the evaluation of the stiffness coefficients since the bending solutions in planes $\tilde{x}\tilde{z}$ and $\tilde{y}\tilde{z}$ are identical to that of the planar beam. The axial solution is the same and for the torsion, we use (4.197) with unit end rotations. Therefore, the stiffness matrix is given on the next page.

We note that we can solve 3-D bar structures of arbitrary complexity with the matrix formulation. Of course, all the matrix procedures that were detailed for truss structures and for planar frames are directly applicable for 3-D bar structures.

4.2.6 Thin walled bars

Thin walled bars are used widely in engineering practice due to their efficient load carrying capabilities. In this section, we discuss some basic behaviors which are important for the modeling of thin walled bar structures. These behaviors, besides being essential for the modeling of thin walled bars, can be used to establish low-order models of complex structures. For example, the gross structural behavior of some buildings can be represented by an equivalent thin walled bar.

Bending behavior

A thin walled bar may be naturally obtained as a result of modifying the distribution of material over the cross-section to improve the bar stiffness with respect to bending. In Figure 4.61a, we show a highlighted region of a rectangular cross-section for which the normal stresses due to bending are significantly smaller than in the outer regions (refer to equation (4.145) for the linear normal stress distribution in bending) and, hence, in the highlighted region the material is not being used as effectively as in the outer regions. In Figure 4.61b, we show an I section whose area is the same as that of the rectangular section of Figure 4.61a. Let I_a and I_b be the moments of inertia of the rectangular and of the I section, respectively, and let us compare the

$$
\mathbf{k} = \begin{bmatrix}
\frac{EA}{L} & 0 & 0 & 0 & 0 & 0 & -\frac{EA}{L} & 0 & 0 & 0 & 0 & 0 \\
0 & \frac{12EI_z}{L^3} & 0 & 0 & 0 & \frac{6EI_z}{L^2} & 0 & -\frac{12EI_z}{L^3} & 0 & 0 & 0 & \frac{6EI_z}{L^2} \\
0 & 0 & \frac{12EI_y}{L^3} & 0 & -\frac{6EI_y}{L^2} & 0 & 0 & 0 & -\frac{12EI_y}{L^3} & 0 & -\frac{6EI_y}{L^2} & 0 \\
0 & 0 & 0 & \frac{GI_t}{L} & 0 & 0 & 0 & 0 & 0 & -\frac{GI_t}{L} & 0 & 0 \\
0 & 0 & -\frac{6EI_y}{L^2} & 0 & \frac{4EI_y}{L} & 0 & 0 & 0 & \frac{6EI_y}{L^2} & 0 & \frac{2EI_y}{L} & 0 \\
0 & \frac{6EI_z}{L^2} & 0 & 0 & 0 & \frac{4EI_z}{L} & 0 & -\frac{6EI_z}{L^2} & 0 & 0 & 0 & \frac{2EI_z}{L} \\
-\frac{EA}{L} & 0 & 0 & 0 & 0 & 0 & \frac{EA}{L} & 0 & 0 & 0 & 0 & 0 \\
0 & -\frac{12EI_z}{L^3} & 0 & 0 & 0 & -\frac{6EI_z}{L^2} & 0 & \frac{12EI_z}{L^3} & 0 & 0 & 0 & -\frac{6EI_z}{L^2} \\
0 & 0 & -\frac{12EI_y}{L^3} & 0 & \frac{6EI_y}{L^2} & 0 & 0 & 0 & \frac{12EI_y}{L^3} & 0 & \frac{6EI_y}{L^2} & 0 \\
0 & 0 & 0 & -\frac{GI_t}{L} & 0 & 0 & 0 & 0 & 0 & \frac{GI_t}{L} & 0 & 0 \\
0 & 0 & -\frac{6EI_y}{L^2} & 0 & \frac{2EI_y}{L} & 0 & 0 & 0 & \frac{6EI_y}{L^2} & 0 & \frac{4EI_y}{L} & 0 \\
0 & \frac{6EI_z}{L^2} & 0 & 0 & 0 & \frac{2EI_z}{L} & 0 & -\frac{6EI_z}{L^2} & 0 & 0 & 0 & \frac{4EI_z}{L}
\end{bmatrix}
$$

displacements and stresses obtained for the choices of cross-sections of Figure 4.61 due to bending in the plane of symmetry. Referring to equation (4.136), the ratio between the induced maximum displacements is

$$\frac{(w_{\max})_b}{(w_{\max})_a} = \frac{I_a}{I_b} = 0.0634 \tag{4.198}$$

and due to equation (4.145) the ratio between the maximum normal stresses is

$$\frac{(\tau_{n,\max})_b}{(\tau_{n,\max})_a} = \frac{I_a}{I_b}\frac{h_b}{h_a} = 0.176 \tag{4.199}$$

where $h_a = w$ and $h_b = h+t_f$. Although these evaluations are very basic, the values given in (4.198) and (4.199) quantify the effectiveness of the I section to resist bending when compared to the rectangular section.

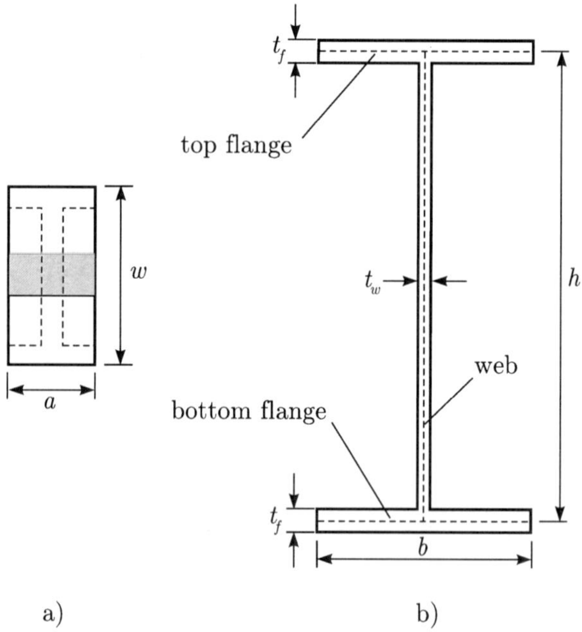

Fig. 4.61. Rectangular and I section of same area whose geometric data is given by $a = 6.615$ in, $w = 13.229$ in, $b = 16.655$ in, $h = 35.06$ in, $t_w = 0.945$ in, $t_f = 1.680$ in

The shear stress distribution due to bending for the I beam and, in fact, for any open thin walled section can be evaluated as detailed in Example 4.5. However, the "plane sectioning" (refer to Figure 4.21) should be taken orthogonal to the midlines of the flange/web and the shear stresses are assumed to be constant at these sections. We exemplify the sectioning in Figure 4.62a, and in Figure 4.62b we show the distribution of shear stresses where

$$\tau_f = \frac{bhV}{4I}, \qquad \tau_{w_1} = \frac{bht_f V}{2It_w}, \qquad \tau_{w_2} = \left(\frac{bt_f}{t_w} + \frac{h}{4}\right)\frac{Vh}{2I}.$$

Of course, I is the moment of inertia and V the shear force. For every point

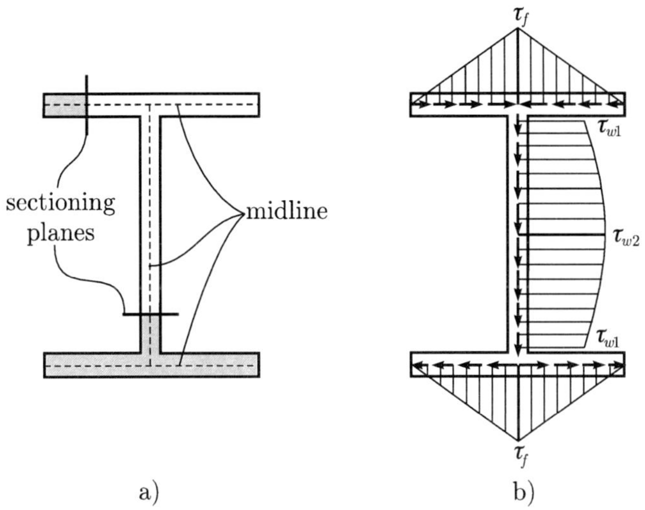

Fig. 4.62. a) Typical sectioning planes; b) Distribution of shear stresses

of the midline of the flange/web we define the shear flow as

$$f_s = \tau_s t$$

where τ_s is the magnitude of the shear stress and t is the thickness of the flange/web.

Shear center

An important concept associated with the distribution of shear stresses due to bending in a thin walled bar is the *shear center*. Consider a bar with the cross-section described in Figure 4.63a subjected to bending induced by a transverse vertical force distribution. The distribution of shear stresses is shown in Figure 4.63b where

$$\tau_f = \frac{bhV}{4I}, \qquad \tau_{w_1} = \frac{bht_f V}{2It_w}, \qquad \tau_{w_2} = \left(\frac{bt_f}{t_w} + \frac{h}{4}\right)\frac{Vh}{2I}.$$

The resultants at the web and at the flanges can be evaluated by integration of the shear stresses and are shown in Figure 4.64a.

If we reduce these forces to a generic point D on the horizontal axis of symmetry, we obtain the vertical force R_w which is equal to the shear force V and a moment given by

$$M_D = R_w d - R_f h.$$

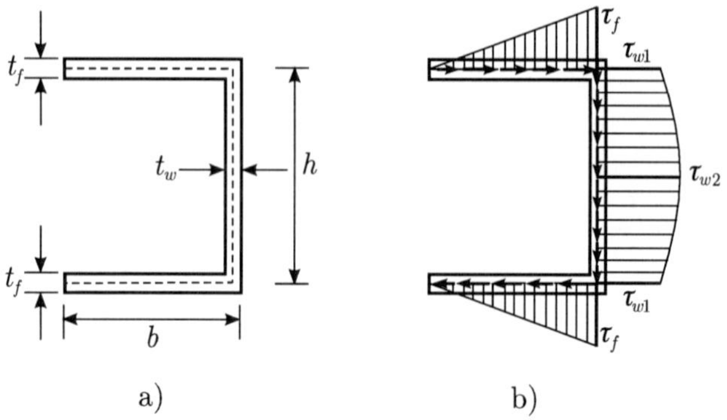

Fig. 4.63. a) Cross-section definition; b) Shear stress distribution

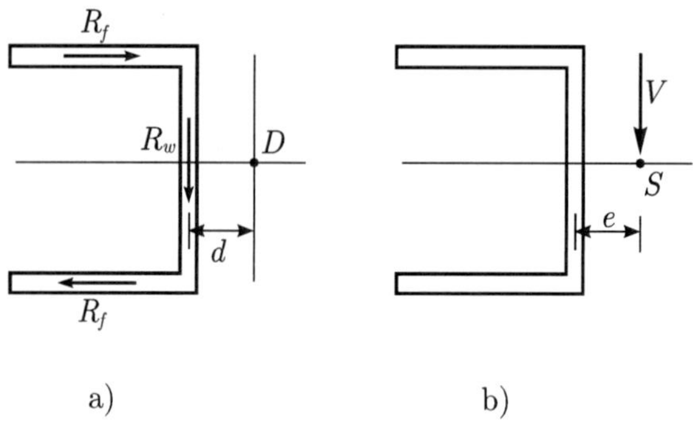

Fig. 4.64. a) Shear resultants; b) Location of the shear center

There is a position of point D for which $M_D = 0$, *i.e.*,

$$d = e = \frac{R_f h}{R_w}.$$

The value e determines the position of point D referred to as S – the shear center of the section.

Note that the shear center gives the position through which the resultant force – associated with all external loading to the section – should pass in order for there to be no twisting; that is, the resulting shear stress distribution is induced by bending alone. In Figure 4.63b, we summarize the situation for which only shear stresses induced by bending are developed.

Hence, if the resultant associated with all external loading does not pass through the shear center, shear stresses associated with torsion are induced.

We also note that, in general, the shear center and the center of gravity do not coincide. They coincide when there is a vertical plane of symmetry for the section.

Torsional behavior

The modeling of the torsional behavior of thin walled bars can be quite delicate. This behavior is drastically different depending on whether the cross-section is open or closed. Also, for open sections the restrained warping may affect the stress distribution in the whole bar. We first discuss the Saint Venant approach for open and closed sections and then introduce the modeling of restrained warping.

Open sections

The membrane analogy discussed in Section 3.6 is extremely useful to determine the shear stress distribution at an open[6] thin walled section which is free to warp.

Consider, for example, the C section ("C" for channel) discussed above. In Figure 4.65, we show schematically the deformed shape of the membrane for this section. Except for the end regions and the corners, the membrane deforms as for the thin rectangular section studied in Example 3.9 and we obtain

$$I_t = 2I_t^f + I_t^w = 2\frac{bt_f^3}{3} + \frac{ht_w^3}{3}$$

and the maximum shear stress at the flange and web are given respectively by

$$\tau_{\max}^f = \frac{3M_t}{bt_f^2}\frac{I_t^f}{I_t}, \qquad \tau_{\max}^w = \frac{3M_t}{bt_w^2}\frac{I_t^w}{I_t}.$$

In Figure 4.65b, we show schematically the shear stress distribution. We note that the methodology above can be applied for a thin walled section of n segments of length b_i and width t_i. In such a case

$$I_t = \sum_{i=1}^{n} \frac{b_i t_i^3}{3} \tag{4.200}$$

and the maximum shear stress for segment i

$$\tau_{\max}^i = \frac{M_t}{I_t^i} t_i \frac{I_t^i}{I_t} = \frac{M_t}{I_t} t_i. \tag{4.201}$$

[6] When the midline of a thin walled section is a closed curve we say that the section is closed, otherwise the section is open

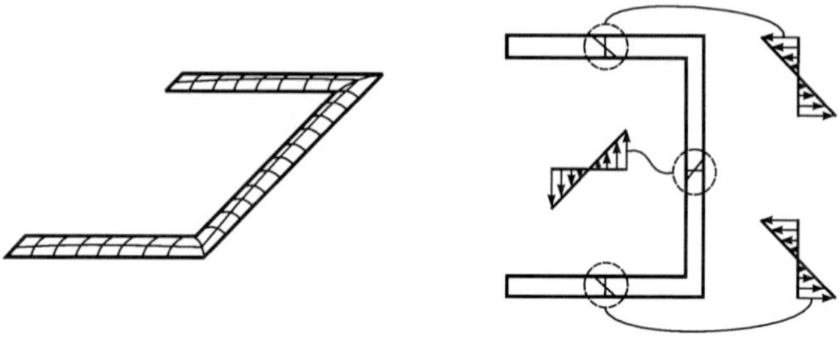

a) b)

Fig. 4.65. a) Deformed shape of membrane; b) Distribution of shear stresses

Closed sections

We recall that in Section 3.6 we studied the uniform torsion of bars of arbitrary cross-sections. However, we made the implicit assumption that there were no holes in the cross-sections, *i.e.*, the cross-sectional region is simply connected.

The uniform torsion theory can be extended to be applicable to the more general cross-sections having holes. An interesting approach is to use the membrane analogy placing a rigid plate to cover each hole[7]. For example, consider the elliptical section with an elliptical hole as shown in Figure 4.66a where the elliptical hole is highlighted to emphasize that a rigid plate is placed there and in Figure 4.66b we show a side view. For thin walled sections we can assume that a section of the deformed shape of the membrane between the outer boundary and the rigid plate is a straight line.

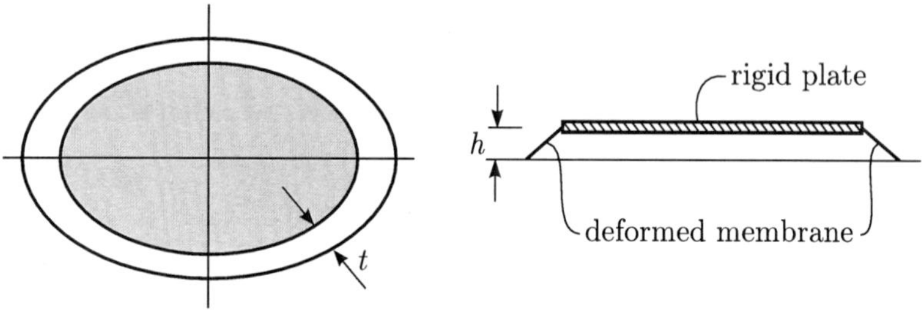

a) b)

Fig. 4.66. a) Elliptical cross section with hole; b) Deformed membrane

[7] In the general case, additional conditions should also be enforced to use the membrane analogy. However, it suffices to introduce the rigid plate for the thin walled closed sections (see Timoshenko and Gere, 1961)

The distribution of shear stresses can be directly inferred from the properties discussed in Section 3.6. In Figure 4.67a, a generic level curve of the membrane is shown and therefore the shear stresses are tangent to this curve and of constant magnitude (see equation (3.152)) and it is easily shown (see Timoshenko and Goodier, 1970) that

$$\tau_s = \frac{M_t}{2A_m t} \tag{4.202}$$

where A_m is the area enclosed by the midline of the wall as shown in Figure 4.67b. It can be shown that

$$I_t = \frac{M_t}{G\theta'} = \frac{4\left(A_m\right)^2 t}{L_m}. \tag{4.203}$$

where L_m is the length of the midline.

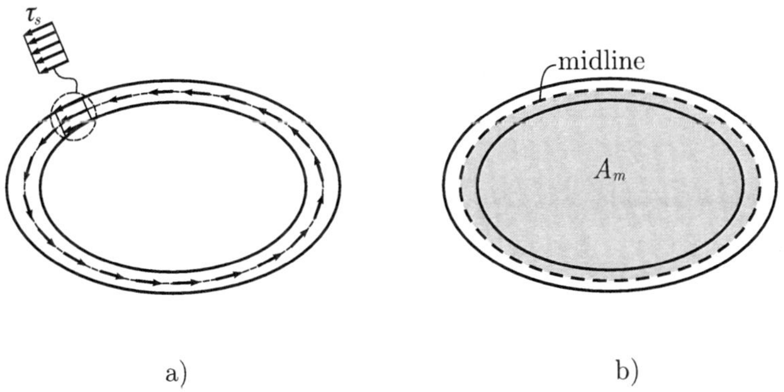

Fig. 4.67. a) Distribution of shear stresses; b) Geometric definitions

It is very interesting to compare the behavior of closed and open thin walled sections with respect to torsion. For that, we choose the simplest closed section – a thin walled circular tube. The open section is obtained by cutting the wall as shown in Figure 4.68b. For a given moment M_t, we can evaluate the I_t and the $\tau_{\max}$ for both situations. Using (4.200) we obtain for the open section

$$I_t^o = \frac{2\pi r t^3}{3}$$

and (4.203) gives for the closed section

$$I_t^c = \frac{4\left(\pi r^2\right)^2 t}{2\pi r} = \frac{4\pi^2 r^4 t}{2\pi r} = 2\pi r^3 t.$$

Therefore, the ratio between the torsional moments of inertia is

$$\frac{I_t^c}{I_t^o} = 3\left(\frac{r}{t}\right)^2.$$

This ratio is very large for usual sections. For example, considering a tubular

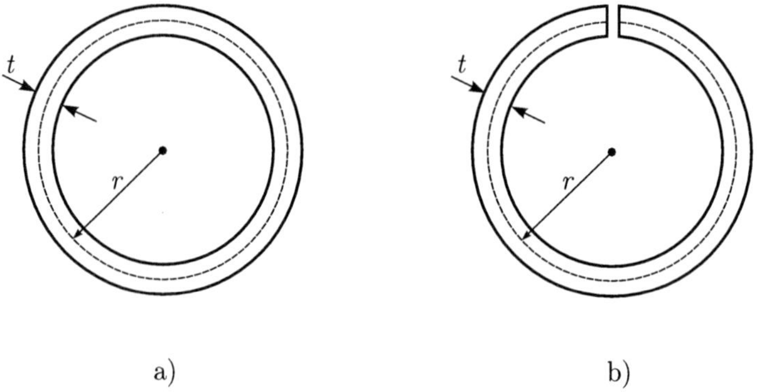

Fig. 4.68. Thin walled circular tubes: a) "closed" and b) "open"

section of $r = 12.375$ in and $t = 0.375$ in we obtain a ratio of 3267. Therefore the torsional stiffness of the closed section is much larger than that of the open section. This result could have been anticipated using the membrane analogy since I_t is proportional to the volume under the deformed membrane.

The maximum shear stress for the open section is given by (4.201)

$$\tau_{\max}^o = \frac{M_t}{I_t} t$$

and for the closed section by (4.202)

$$\tau_{\max}^c = \frac{M_t}{2\pi r^2 t}$$

leading to

$$\frac{\tau_{\max}^c}{\tau_{\max}^o} = \frac{1}{3}\frac{t}{r}$$

which is a small ratio. For the tubular section described above we obtain for this ratio 0.01.

In Figure 4.69 we summarize the distribution of shear stresses for both cases. Note that the closed section resists torsion very efficiently, because the shear flow has a lever arm of r for the resisting moment. On the other hand, for the open section, the shear flow is interrupted as the cutting plane

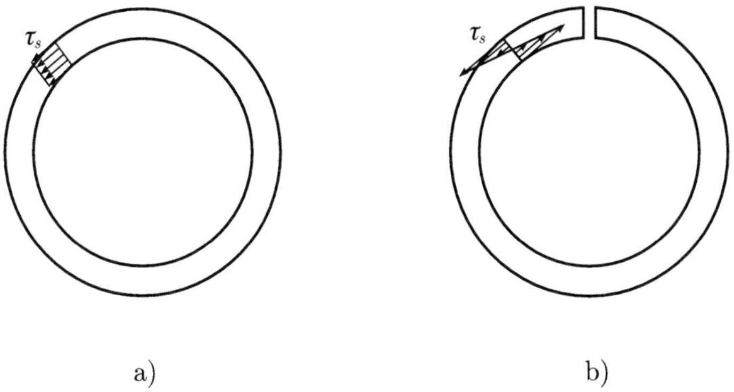

Fig. 4.69. a) Shear stresses for closed section; b) Shear stresses for open section. The magnitude of τ_s is not shown to scale

partially separates the section, and the shear flow has only a lever arm of the order of the wall thickness.

The differences in behavior with respect to torsion of closed and open sections were examined above for a particular case. However, they are representative of the qualitative behavior of general closed and open sections. Hence, for structural problems where torsion is an issue, closed sections are very efficient. For example, in bridges for which both bending and torsion are relevant, box sections are effective.

Warping effects

The modeling of torsion discussed so far is based on Saint Venant's uniform torsion model introduced in Section 3.6. In this theory it is assumed that the cross-sections are free to warp, *i.e.*, there are no kinematic restrictions preventing the out-of-plane displacements.

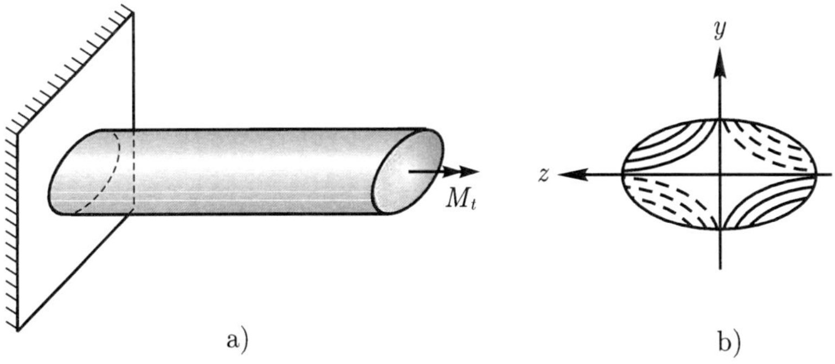

Fig. 4.70. a) Built-in bar of elliptical cross section subjected to torsion; b) Isolines of warping displacements predicted by Saint Venant's theory

Consider, for example, the case of a bar of elliptical cross-section subjected to a torsional moment at one end and built-in at the other, as summarized in Figure 4.70a. In Figure 4.70b, we show the isolines of warping displacements derived in Section 3.6 based on Saint Venant's theory. Of course, at the built-in end the warping displacements are kinematically prevented giving rise to a field of normal stresses with zero resultant. It can be shown (Timoshenko and Goodier, 1970) that for the elliptical section these normal stresses result in a perturbation of the Saint Venant solution which rapidly diminishes in magnitude as we move away from the built-in end. This is the case for solid and thin walled closed cross-sections, but not for thin walled open sections.

In order to obtain insight into the behavior of open sections consider the situation summarized in Figure 4.71. A top view of the deformation of the bar predicted by Saint Venant's theory is shown in Figure 4.71b. Of course, such deformation is incompatible with the built-in end. In Figure 4.72a we show qualitatively the distribution of normal stresses at the built-in section.

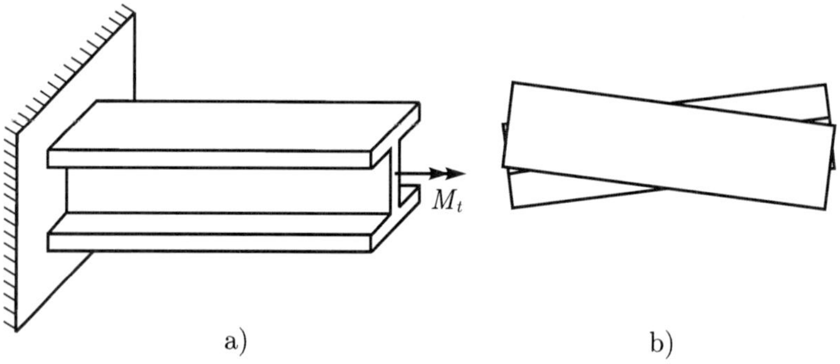

Fig. 4.71. a) Built-in bar of I section subjected to torsion; b) Top view of the deformation predicted by Saint Venant's theory

Although the overall resultant of this stress distribution at the cross-section is null, the stresses give rise to bending moments of intensity M acting on the upper and lower flanges as summarized in Figure 4.72b. In fact, these bending moments acting at the rectangular cross-sections of the flanges induce the required deformation to counter, at the built-in end, the warping of the Saint Venant solution shown in Figure 4.71b. The quantity

$$B = Mh$$

is called the bimoment. To illustrate the interaction between the bimoment and the part of the torsional moment acting at a section we show in Figure 4.73 the equilibrium of an element of infinitesimally small length extracted from the I beam. We note that shear forces V are induced in the flanges associated with the variation of M. Of course,

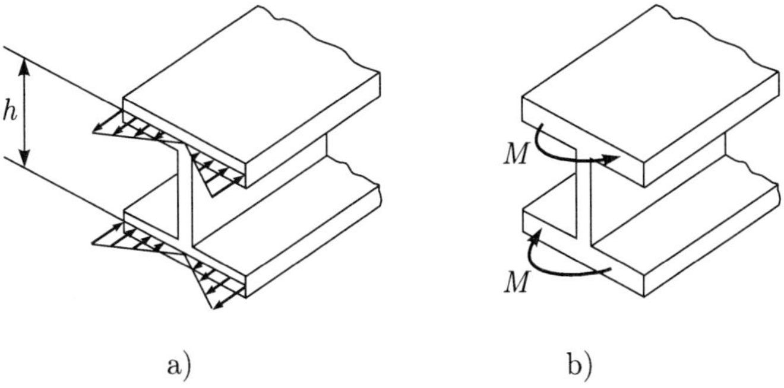

Fig. 4.72. a) Distribution of normal stresses at flange at built-in section; b) Moment resultant at flanges

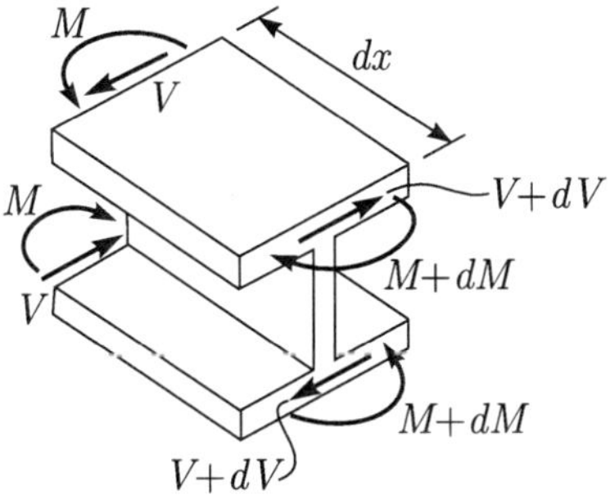

Fig. 4.73. Stress resultants at flanges

$$V = \frac{dM}{dx}.$$

Note that the shear forces acting at a section are equivalent to a torsional moment

$$(M_t)_w = Vh = h\frac{dM}{dx} = \frac{d\,(Mh)}{dx} = \frac{dB}{dx}$$

i.e., the variation of the bimoment induces a twisting moment distribution $(M_t)_w$ referred to as warping torsional moment since it is associated with restrained warping. The total torsional moment at a section can be written as

$$M_t = (M_t)_w + (M_t)_s$$

where $(M_t)_s$ is the moment of the Saint Venant part of the solution.

The complete formulation of the torsion problem of open sections with restrained warping is out of the scope of this book. For that, we refer the reader to Murray, 1985.

Our objective was to give some insight into the effect of restraining the warping for bars of open sections and how it influences the overall solution. This insight can be important when solving models and when establishing higher-order 3-D models.

4.2.7 Curved bar model

In this section we present the curved bar model and discuss basic behaviors of curved structures. The fundamental concepts introduced serve as a reference for many curved structural forms such as shell structures.

Basic facts

Referring to the straight bar model, transverse loads, such as those shown in Figure 4.74, are transferred to the supports by the beam through bending.

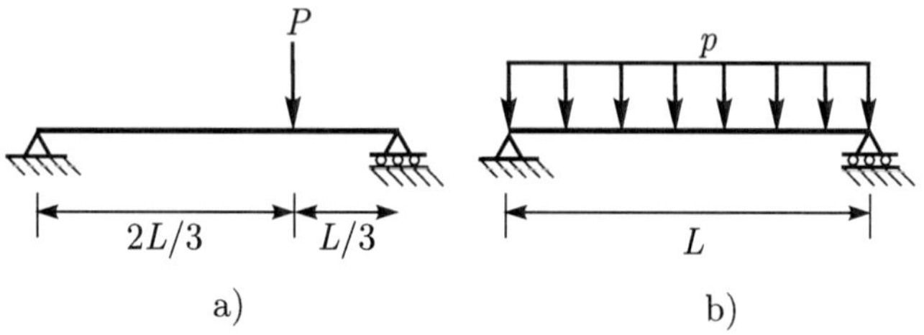

Fig. 4.74. Sample transverse loads applied to straight bars

Bending is not an efficient way of transferring load. We recall that the normal stress distribution due to bending at a bar cross-section is linear with the extreme values occurring at the outer fibers. Since the stress magnitude that can be supported by a given material is limited, in the case of bending this limit value is attained first at the outer fibers and there is a significant portion of the cross-section, around the centroid, for which the stresses are much lower than the limit values.

In contrast, when we have an axial load on the bar, the stress distribution is constant over the cross-section and all fibers simultaneously attain the stress limit value.

Suppose we consider the supports and the load of Figure 4.74a and that we would like to transfer such load to the supports without bending. We could try to do so with a cable, since cables can not sustain any bending. Although everyone has an intuitive understanding of what a cable is, in our context a cable can be understood as a very slender bar which is so flexible that it can not sustain any bending. If we place a cable linking the supports of Figure 4.74a we can not find an equilibrated position since as the section of the cable under the load starts moving downwards the right support moves left horizontally. However, if we modify the right support to be fixed, the cable will find an equilibrated position as shown in Figure 4.75a and the axial forces in the cable can be evaluated by equilibrium. In fact equilibrium in the horizontal direction leads to

$$T_1 \cos\alpha = T_2 \cos\beta = H$$

and in the vertical to

$$P = T_1 \sin\alpha + T_2 \sin\beta$$

and we obtain

$$T_1 = \frac{P}{(\sin\alpha + \tan\beta\cos\alpha)}, \quad T_2 = \frac{P}{(\sin\beta + \tan\alpha\cos\beta)}.$$

Of course, to find the value of h (or α or β) we need to consider the section

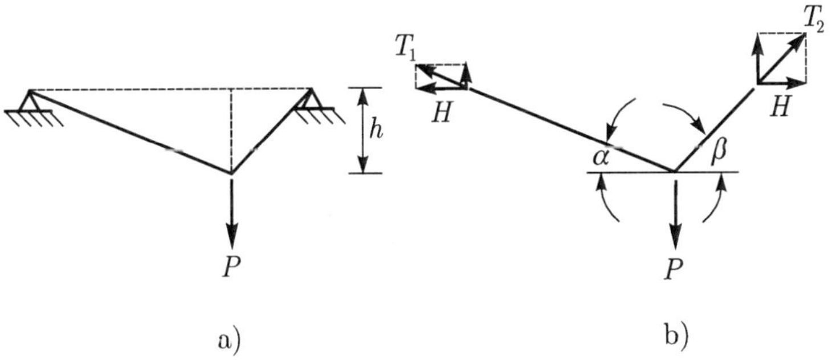

Fig. 4.75. Cable equilibrated configuration for a transverse concentrated load

properties and the material of the cable[8].

An important observation which is always valid is that, to be able to equilibrate a vertical load with axial forces only, a horizontal reactive component at the supports is required which is called the thrust, in Figure 4.75b given by H. The equilibrated shape of the cable is called a funicular shape and h, the largest vertical displacement, is referred to as the sag. Let us find the funicular shape for the cable subjected to a uniformly distributed transverse load, see Figure 4.74b. Taking advantage of symmetry, an equilibrated configuration is shown in Figure 4.76. Of course, if this shape is funicular the bending moment at a generic section has to be zero. Hence,

$$M(x) = -(px)\frac{x}{2} + \frac{pL}{2}x - Hy(x) = 0$$

[8] The value of h may be relatively large compared with the span for a rubber like cable and also for a steel cable when it is loose before the application of the load, that is, its length is greater than the span

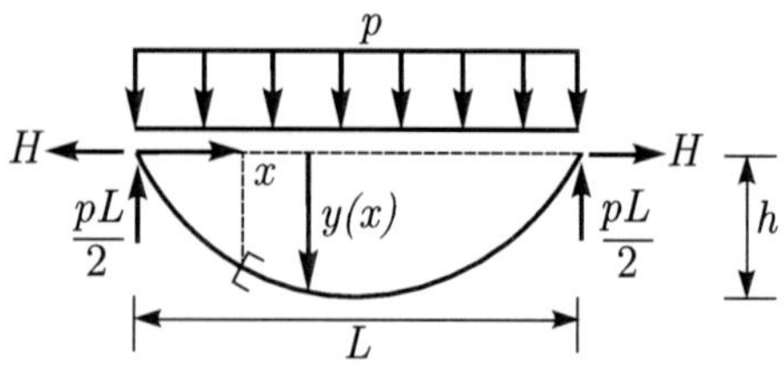

Fig. 4.76. Cable equilibrated configuration for a uniformly distributed load

which leads to

$$y(x) = \frac{p}{2H}\left(xL - x^2\right). \tag{4.204}$$

Therefore the funicular shape of a cable subjected to a uniformly distributed pressure is parabolic. We note that the sag is related to the value of the thrust. In fact, from equation (4.204)

$$h = y\left(\frac{L}{2}\right) = \frac{p}{2H}\left(\frac{L^2}{2} - \frac{L^2}{4}\right) = \frac{pL^2}{8H}$$

or

$$H = \frac{pL^2}{8h}.$$

Consider the funicular shape developed by the cable either in Figure 4.75b or 4.76. Instead of the very small cross-section of the cable, let us consider bars with a much larger cross-section whose axes coincide in its initial configuration with the funicular shape considered. Of course, in this case the cross-section is assumed to be large enough such that the displacements are small and, therefore, equilibrium is imposed in the undeformed configuration.

Now suppose we turn these bar structures over with respect to a horizontal line at the support level. In Figure 4.77 we summarize the resulting structures which are now referred to as arch structures. The structure of Figure 4.77a is called a polygonal arch since it consists of straight bars. The structure of Figure 4.77b is simply referred to as an arch; it is actually a curved bar structure.

The fundamental change that occurs when we turn the structures of Figure 4.75 over is that the axial forces developed in the bars go from tension to compression. This is a crucial distinction since structural materials behave differently in tension than in compression, especially with respect to their ultimate strength. In fact, historically, arches have had a tremendous importance in constructions since, for centuries, many construction materials have had a very low strength in tension.

The shape that leads to compression only in arch structures is referred to as "the line of pressure". We note that the line of pressure is load dependent.

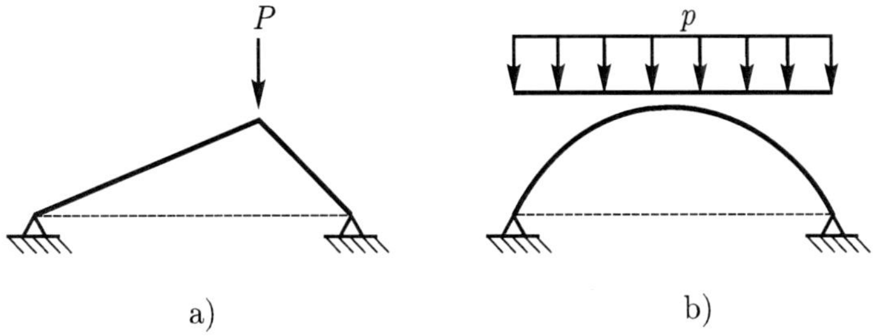

Fig. 4.77. Arch structures

For example, if we change the position of the concentrated load acting onto the arch of Figure 4.77a, bending is induced. Similarly, if a concentrated load acts onto the arch of Figure 4.77b bending is also induced. Here, cable structures behave differently, since when the load changes the cable changes its shape in such a way that the funicular shape for this new loading is always reached.

Consider the arch of Figure 4.77b whose bar axis is the line of pressure for the uniformly distributed load and is given by equation (4.204). A concentrated load is now applied as shown in Figure 4.78a. Of course, the axis of

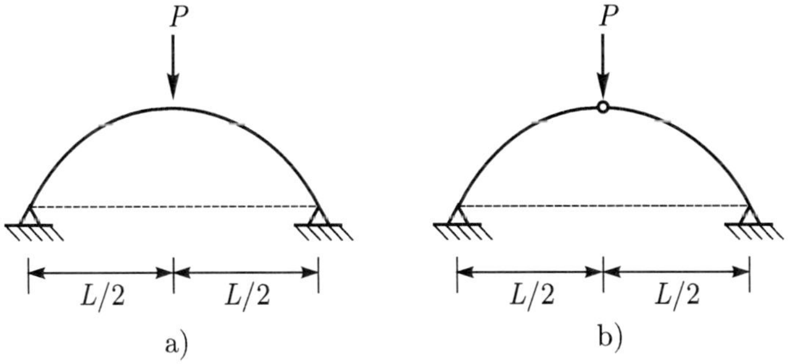

Fig. 4.78. Arches subjected to a concentrated load

this arch does not correspond to the pressure line for the concentrated load, since for this load the line of pressure is given by a polygonal arch defined by two straight bars from the point of load application to the supports.

We realize that we cannot solve this arch problem directly and find the internal forces since the structure is statically indeterminate. Therefore, to solve this problem we need to first discuss the formulation of the curved bar model which is addressed later in this section.

Let us for now transform the arch of Figure 4.78a into a statically determinate arch by introducing a pin type joint or hinge as shown in Figure 4.78b.

This kind of arch is called a three hinged arch and represents an important structural arrangement with wide engineering applications. We discuss the three hinged arch below, since insight into the behavior of arch structures is gained by means of this simple statically determinate structure.

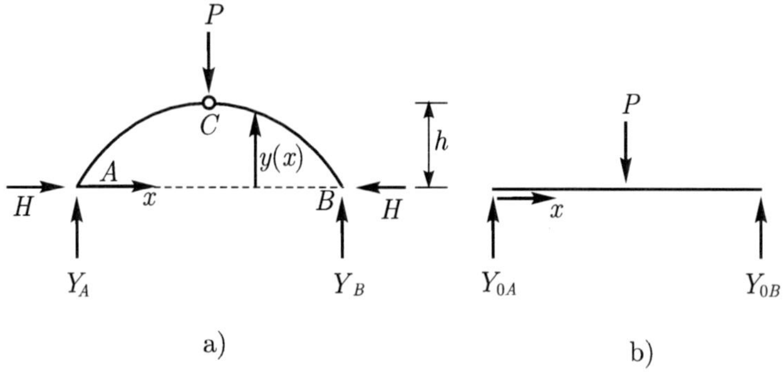

Fig. 4.79. a) Three hinged arch; b) Straight beam

Although we analyse the particular structure shown in Figure 4.78b, the solution procedure used applies for three hinged arches in general. In Figure 4.79a, we introduce the reactions taking into consideration that, as the external load is vertical, the horizontal reactions at the supports need to be self-equilibrated. To obtain insight into how the loads are transferred by an arch when compared to a straight beam, we consider the straight beam of Figure 4.79b.

To evaluate the vertical reactions for the arch we use equilibrium in the vertical direction

$$\sum F_y = 0 \qquad \Rightarrow \qquad Y_A + Y_B - P = 0$$

and moment equilibrium about A

$$\sum M_A = 0 \qquad \Rightarrow \qquad Y_B \cdot L - P\frac{L}{2} = 0.$$

These equations are the same that lead to reactions of the straight bar of Figure 4.79b. Therefore, in general,

$$\begin{aligned} Y_A &= Y_{0A} \\ Y_B &= Y_{0B}. \end{aligned}$$

The additional condition to be considered is

$$M_C = 0$$

which is given by

$$M_C = M_{0C} - Hh = 0$$

where M_{0C} is the moment at the section of the straight beam corresponding to the hinge, that is, $M_{0C} = Y_B.\frac{L}{2}$. Therefore,

$$H = \frac{M_{0C}}{h}.$$

The moment at a generic section of the arch is written as

$$M(x) = M_0(x) - Hy(x) \tag{4.205}$$

where $M_0(x)$ gives the moment distribution for the straight beam. Therefore, the moment for the arch is diminished with respect to that of the straight beam by the amount $Hy(x)$. In Figure 4.80a we graphically show the superposition given by equation (4.205) and in Figure 4.80b the moment diagram of the straight beam. The decrease in moment magnitudes in the arch with respect to the straight beam is clearly seen.

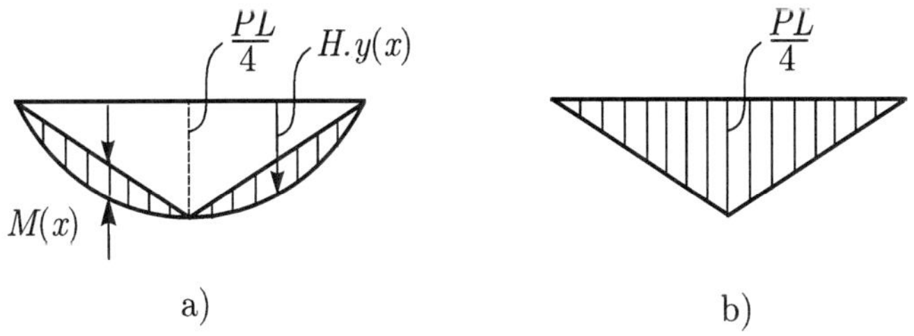

Fig. 4.80. Moment diagrams for arch and straight beam

Before we close this section, we would like to highlight some important facts. Due to the curved bar axis, it is possible to transfer certain transverse loads to the supports by developing axial forces only.

It is essential that the supports provide horizontal reactions which are called thrusts. Otherwise bending is developed.

We also note that even when the geometry of the arch does not correspond to the pressure line for a given load, the bending moment developed in the arch can be significantly lower than that of a straight bar of same span.

Differential formulation of a curved planar bar

Consider the curved bar problem described in Figure 4.81. The bar has a plane of symmetry and the bar axis is a curve in this plane. The bar axis is going through the centroid of the cross-section of the bar. The externally applied distributed loads are applied at the bar axis in the plane of symmetry and can be decomposed into a normal component p_z and a tangential component p_x. For every point along the axis, we define a local coordinate system xz, x being tangential to the axis and z orthogonal. At the bar ends either prescribed displacements/rotations or force boundary conditions are applied.

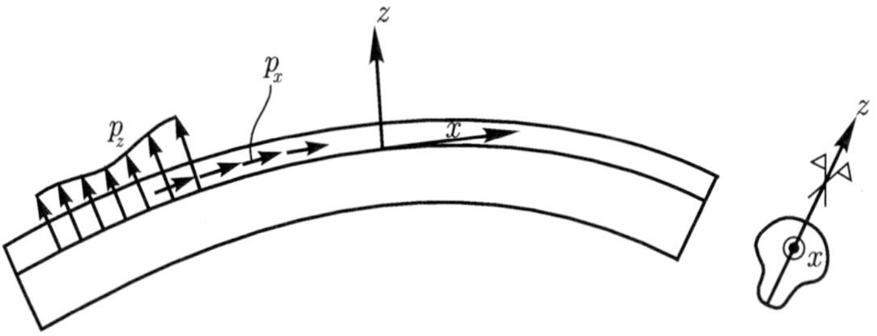

Fig. 4.81. Some definitions for curved bar problem; the axis goes through the centroid of the cross-section

Except for considering now a curved bar, the remaining definitions are analogous to those used for a straight bar. The basic kinematic hypothesis of the Bernoulli-Euler beam theory that sections initially orthogonal to the bar axis remain undeformed and orthogonal to this axis during deformations is also adopted.

Before we present the formulation, let us recall some basic facts for a planar curve. Consider a generic planar curve as shown in Figure 4.82. Let s be the arc-length coordinate along the curve. For any point on the curve, say point P, we can define a circle centered at point C located on a straight line defined by the normal to the curve at P. The radius of the circle and, hence, point C are defined to make the circle coincide with the curve at and in the vicinity of point P. A typical situation is summarized in Figure 4.82.

The radius r of the circle is called the radius of curvature of the curve at point P and the point C the center of curvature, and we have

$$ds = r d\theta. \tag{4.206}$$

The curvature κ at point P is given by

$$\kappa = \frac{d\theta}{ds} = \frac{1}{r}. \tag{4.207}$$

Kinematics

Consider a differential element of the bar, see Figure 4.83a. Since (4.206) holds we also have

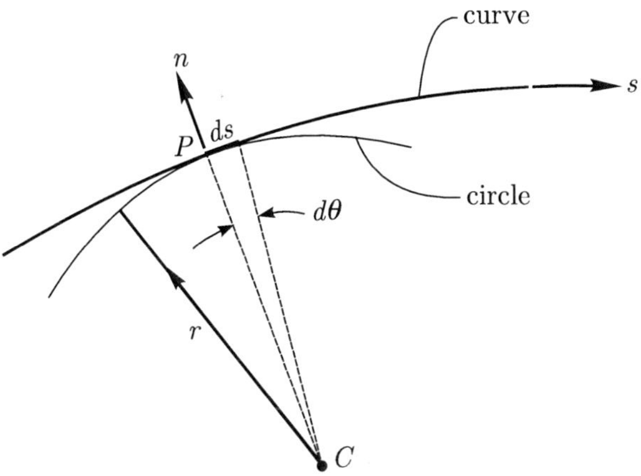

Fig. 4.82. Radius of curvature definitions for point P

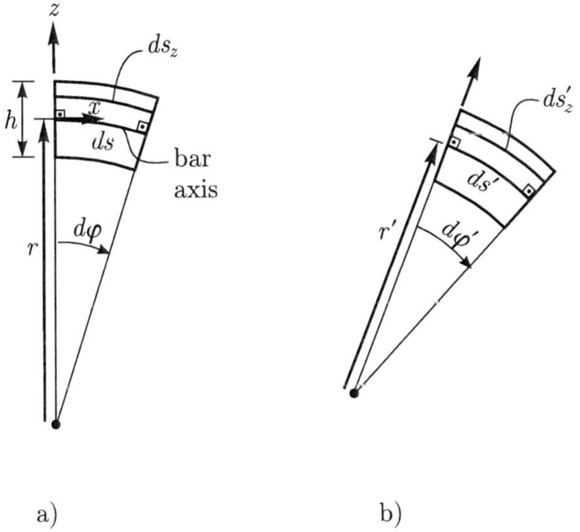

Fig. 4.83. Deformed and undeformed configurations for a differential element

$$ds_z = (r + z)d\varphi \tag{4.208}$$

where ds_z is the differential arc length at a distance z from the axis.

The deformed configuration of the differential element of Figure 4.83a is shown in Figure 4.83b. We note that the Bernoulli-Euler hypothesis is used to characterize the deformed configuration and a prime indicates that the quantities are associated with the deformed configuration.

The strain in the tangential direction at the bar axis is given by

$$\varepsilon_{xx_0} = \frac{ds' - ds}{ds} \tag{4.209}$$

and as a function of z by

$$\varepsilon_{xx} = \frac{ds'_z - ds_z}{ds_z}. \tag{4.210}$$

Using equations (4.206) and (4.208), we obtain

$$ds_z = \left(1 + \frac{z}{r}\right) ds. \tag{4.211}$$

In an analogous manner, we can write for the deformed configuration

$$ds'_z = \left(1 + \frac{z}{r'}\right) ds'. \tag{4.212}$$

Taking into account that relation (4.209) can be re-written as

$$ds' = (1 + \varepsilon_{xx_0})ds$$

and substituting (4.211) and (4.212) into (4.210) yields

$$\varepsilon_{xx} = \frac{\left(1 + \frac{z}{r'}\right)(1 + \varepsilon_{xx_0}) - \left(1 + \frac{z}{r}\right)}{\left(1 + \frac{z}{r}\right)}$$

which can be re-written as

$$\varepsilon_{xx} = \frac{\varepsilon_{xx_0}}{\left(1 + \frac{z}{r}\right)} + \frac{z}{\left(1 + \frac{z}{r}\right)} \left((1 + \varepsilon_{xx_0}) \frac{1}{r'} - \frac{1}{r}\right). \tag{4.213}$$

We now want to focus on situations where the thickness of the bar is small compared with the ratio of curvature of the bar axis, that is, $h/r << 1.0$ and also where we can neglect the stretching of the axis on the change of curvature [9] (see Timoshenko and Woinowsky-Krieger, 1959). Then expression (4.213) simplifies to

$$\varepsilon_{xx} = \varepsilon_{xx_0} + z\left(\frac{1}{r'} - \frac{1}{r}\right). \tag{4.214}$$

Let χ be the change in curvature due to deformations, *i.e.*,

$$\chi = \frac{1}{r'} - \frac{1}{r}. \tag{4.215}$$

[9] The exact linearized expression for the change in curvature actually includes the stretching of the axis and is given in Chapelle and Bathe, 2010*a*

Introducing (4.215) into (4.214) yields

$$\varepsilon_{xx} = \varepsilon_{xx_0} + z\chi. \tag{4.216}$$

Therefore, we see that the tangential strain can be interpreted as the axial strain plus a term which is proportional to the distance of the longitudinal fiber to the axis times the change in curvature.

Let us introduce the kinematic variables u and w which give the displacements of points of the axis along x and z, *i.e.*, along the tangential and normal directions, and use small strain conditions. We would like to obtain the strain ε_{xx} as a function of these kinematic variables. Refering to Figure 4.84, the contributions for ε_{xx_0} are given by

$$\varepsilon_{xx_0} = \frac{du + (w + r)d\varphi - rd\varphi}{rd\varphi} = \frac{du}{ds} + \frac{w}{r}. \tag{4.217}$$

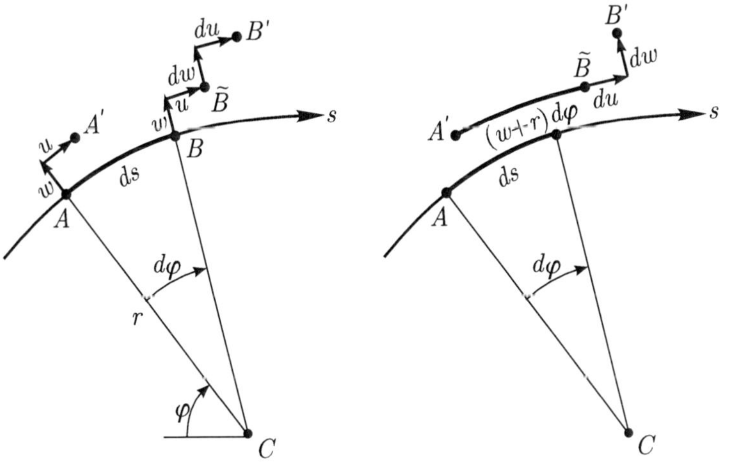

Fig. 4.84. Displacements and displacement increments for a generic differential arc length increment on the axis

To evaluate the change in curvature, we introduce an additional variable which we will eliminate later, namely the section rotation $\beta(s)$, see Figure 4.85. We also use the assumptions that lead to (4.214) and of small deformations and strains.

Then we have at a section the displacements

$$u_\varphi(\varphi, r, z) = u(\varphi) - z\beta(\varphi) \tag{4.218}$$

and of course

$$u_r(\varphi, r, z) = w(\varphi).$$

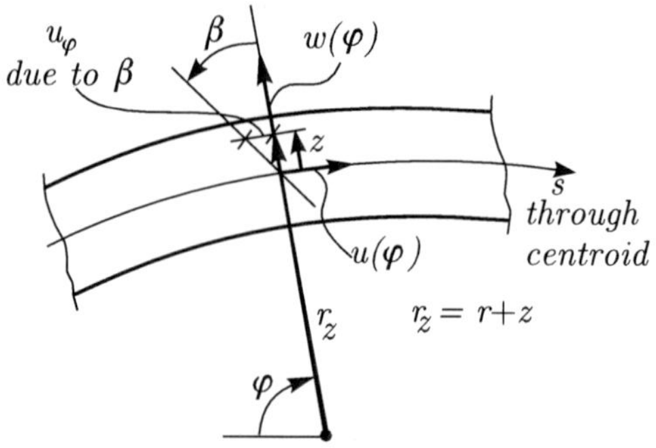

Fig. 4.85. Deformations of the beam

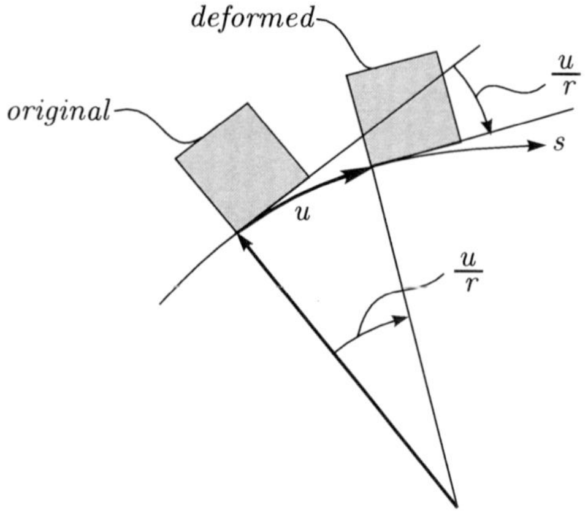

Fig. 4.86. Shear component due to displacement u

The normal strain is then given by (see Section 4.1.3)

$$\varepsilon_{xx}(\varphi, r, z) = \frac{w}{r_z} + \frac{1}{r_z}\frac{\partial u_\varphi}{\partial \varphi}$$

where we recognize the first term to be straining due to the radial displacement, and the second term to be the usual normal strain.

Since $r_z = r + z$, and using (4.218), we obtain

$$\varepsilon_{xx} = \frac{1}{1 + z/r}\left(\frac{w(\varphi)}{r} + \frac{1}{r}\frac{\partial u(\varphi)}{\partial \varphi} - \frac{z}{r}\frac{\partial \beta(\varphi)}{\partial \varphi}\right)$$

Assuming that $h/r << 1.0$, we thus have

$$\varepsilon_{xx} = \frac{w}{r} + \frac{1}{r}\frac{\partial u}{\partial \varphi} - \frac{z}{r}\frac{\partial \beta}{\partial \varphi}.$$

However, our assumption is that of Bernoulli-Euler theory, that the shear strain is zero (that is, originally straight fibers normal to the axis remain normal to the axis during deformations). This requires that the shear given by

$$\gamma = \frac{1}{r+z}\frac{\partial w}{\partial \varphi} + \frac{\partial u_\varphi}{\partial z} - \frac{u_\varphi}{r+z}$$

be zero. Here the first two terms correspond to the usual shear strain components in a Cartesian coordinate system (see Section 3.2.5). The last term corresponds to the fact that the s-axis changes direction.

Assuming again $h/r << 1.0$, we obtain

$$\gamma = \frac{1}{r}\frac{\partial w}{\partial \varphi} + \frac{\partial u_\varphi}{\partial z} - \frac{u}{r}$$

where we use u instead of u_φ in the last term, resulting into a constant shear strain at a section. We give an interpretation of the u/r term in Figure 4.86. Substituting from (4.218) and using the condition that γ be zero, we obtain

$$\beta = \frac{1}{r}\frac{\partial w}{\partial \varphi} - \frac{u}{r} \tag{4.219}$$

and hence

$$\varepsilon_{xx} = \frac{w}{r} + \frac{1}{r}\frac{du}{d\varphi} - \frac{z}{r}\frac{d}{d\varphi}\left(\frac{1}{r}\frac{dw}{d\varphi} - \frac{u}{r}\right) \tag{4.220}$$

where u and w, measured at the axis of the beam, are only a function of φ. This strain relationship corresponds to (4.216) with[10]

$$\begin{aligned} \varepsilon_{xx_0} &= \frac{w}{r} + \frac{du}{ds} \\ \chi &= -\frac{1}{r}\frac{d}{d\varphi}\left(\frac{1}{r}\frac{dw}{d\varphi} - \frac{u}{r}\right) \end{aligned} \tag{4.221}$$

and ε_{xx_0} was already given in (4.217).

We note that, both, the tangential and normal displacements contribute to the strain along the axis ($z = 0$) and to the change in curvature. This fact contrasts with the situation for the straight bar where the strain along the axis depends only on the tangential displacement and the curvature change depends only on the transverse displacement (analogous to the normal displacement in the curved bar). As a consequence curved structures – and in particular shell structures – are much more difficult to analyze than straight or flat structures (straight beams and plates).

[10] Note that, regarding the sign convention, the w-dependent term corresponds to a *decrease* in the original curvature (corresponding to a positive moment M). For a straight bar, a positive moment M causes an *increase* in curvature, but from zero curvature (see (4.134)); hence the sign convention is consistent, and we use the same convention for plates and shells

Constitutive relation and stress resultants

Hooke's law applies for the tangential direction, *i.e.*,

$$\tau_{xx} = E\varepsilon_{xx} \tag{4.222}$$

and using (4.216) we obtain

$$\tau_{xx} = E\left(\varepsilon_{xx_0} + z\chi\right) = E\varepsilon_{xx_0} + zE\chi. \tag{4.223}$$

The axial force N is given by

$$N = \int_A \tau_{xx}\, dA = EA\varepsilon_{xx_0} \tag{4.224}$$

and the bending moment by

$$M = \int_A -\tau_{xx} z\, dA = -EI\chi. \tag{4.225}$$

We note that the linear distribution of normal stress given in equation (4.223) is only valid when the section height of the beam is small compared with the radius of curvature of the axis (see the derivation from (4.218) to (4.221)). If the height of the beam is not small with respect to the radius of curvature the distribution of normal stress will no longer be linear. Actually we would obtain a hyperbolic distribution of normal stress along the beam height. However, the difference in stress predictions is only significant for situations where the ratio h/r is large. For example, when we consider the bending of a rectangular cross-section with $h/r = 1/4$, the difference in stress predictions at the extreme fibers is only about 8%.

Equilibrium

We consider the equilibrium of a differential element as shown in Figure 4.87. Equilibrium in the tangential and normal directions can be written as

$$\begin{aligned} -N + (N + dN) - V d\theta + p_x ds &= 0 \\ V - (V + dV) - N d\theta + p_z ds &= 0 \end{aligned}$$

which leads to

$$\frac{dN}{ds} - \frac{V}{r} = -p_x \tag{4.226}$$

$$\frac{dV}{ds} + \frac{N}{r} = p_z \tag{4.227}$$

and moment equilibrium about point Q can be written as

$$-M + (M + dM) - V ds = 0$$

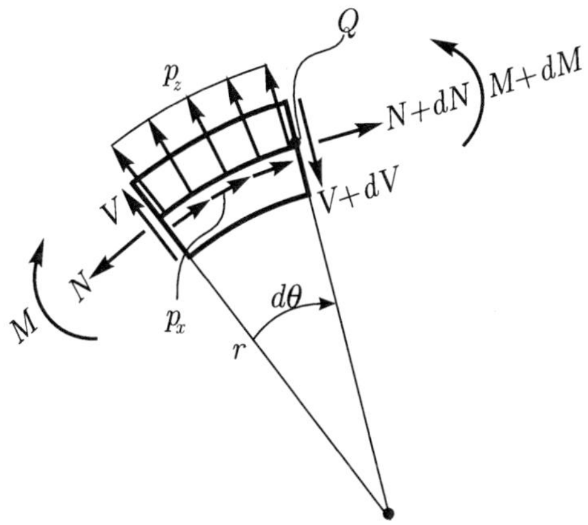

Fig. 4.87. Equilibrium of a differential element

leading to

$$\frac{dM}{ds} = V. \tag{4.228}$$

We note that due to the curved axis, equilibrium leads to coupled equations for the internal forces in contrast to the conditions for a straight bar for which the equilibrium of the axial forces is independent of that of the moment and shear forces.

Summarizing, equation (4.220) represents the strain compatibility conditions, equations (4.224) and (4.225) the constitutive relations, and equations (4.226) to (4.228) the equilibrium conditions. Therefore, when the boundary conditions at the end sections are introduced we have the complete formulation since all the requirements have been taken into account.

Of course, we could derive the stiffness matrix for a given curved bar solving the above differential formulation for unit end displacements/rotations as for the straight bar.

Our motivation to present the formulation of the curved bar problem was mainly to gain insight into how to tackle the analysis of curved bars including the basic assumptions used. Therefore we will not elaborate on the solution of the formulation, except for presenting the solution of a very simple problem as an illustration. Note that when the kinematic boundary conditions make the curved bar structure statically determinate, we can directly determine the internal forces and the solution becomes much simpler. In fact, substituting (4.217) and (4.221) into (4.224) and (4.225) respectively, we obtain

$$\frac{du}{ds} + \frac{w}{r} = \frac{N}{EA} \tag{4.229}$$

$$\frac{d^2w}{ds^2} - \frac{d}{ds}\left(\frac{u}{r}\right) = \frac{M}{EI} \tag{4.230}$$

with M and N known functions.

Example 4.16

Determine the displacement w for the circular cantilever beam shown in Figure 4.88.

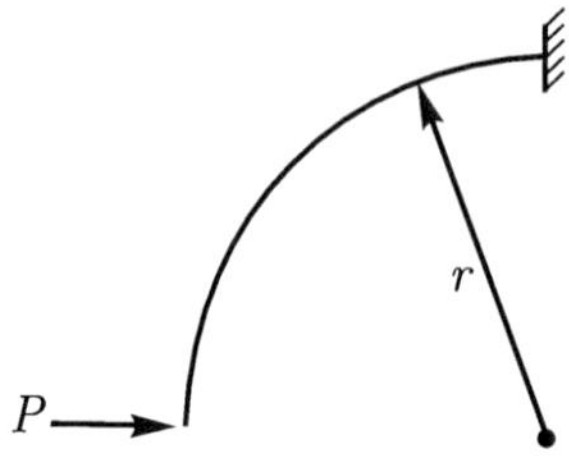

Fig. 4.88. Circular cantilever beam problem

Solution

We first determine the axial force and bending moment along the bar. Referring to Figure 4.89

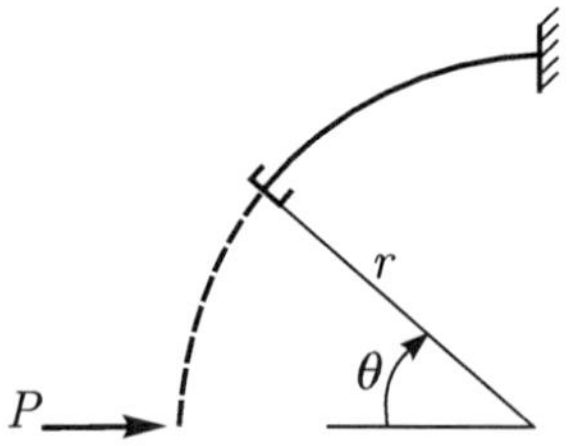

Fig. 4.89. Generic section of curved cantilever

$$N = -P\sin\theta \tag{4.231}$$

$$M = -Pr\sin\theta. \tag{4.232}$$

Since the radius of curvature is constant, we have that

$$s = \theta r$$

$$\frac{d(\cdot)}{ds} = \frac{d(\cdot)}{d\theta}\frac{d\theta}{ds} = \frac{1}{r}\frac{d(\cdot)}{d\theta}. \tag{4.233}$$

Hence, using (4.233) and substituting (4.229) into (4.230) yields

$$\frac{d^2w}{d\theta^2} + w = r^2\frac{M}{EI} + r\frac{N}{EA}$$

and introducing (4.231) and (4.232), we obtain

$$\frac{d^2w}{d\theta^2} + w = -\frac{Pr}{E}\left(\frac{r^2}{I} + \frac{1}{A}\right)\sin\theta. \tag{4.234}$$

A particular solution of equation (4.234) is given by

$$w_p = -\frac{k}{2}\theta\cos\theta$$

where $k = -\frac{Pr}{E}\left(\frac{r^2}{I} + \frac{1}{A}\right)$. The general solution of (4.234) can be written as

$$w = C_0\sin\theta + C_1\cos\theta - \frac{k}{2}\theta\cos\theta$$

where C_0 and C_1 are constants to be determined. The two kinematic boundary conditions at $\theta = \pi/2$ are

$$w\left(\frac{\pi}{2}\right) = 0, \qquad \frac{dw}{ds}\left(\frac{\pi}{2}\right) = 0$$

leading to

$$C_0 = 0 \qquad \text{and} \qquad C_1 = \frac{\pi k}{4}$$

and finally

$$w = \frac{k}{2}\cos\theta\left(\theta - \frac{\pi}{2}\right) - \frac{Pr}{2E}\left(\frac{r^2}{I} + \frac{1}{A}\right)\cos\theta\left(\theta - \frac{\pi}{2}\right).$$

Of course, we could now use the solution for w and equation (4.229) to evaluate u by a simple integration considering that $u(\pi/2) = 0$.

□

4.2.8 The Timoshenko beam model

We recall that the Bernoulli-Euler beam model was used to describe the behavior of bars subjected to transverse loading. The model considers the change of curvature of the beam axis induced by the bending moment. This deformation dominates the overall bar deformation as long as the bar is slender. As the height of the bar section h increases with respect to the characteristic bar length L the deformation induced by shear becomes no longer negligible and when h/L is about $\frac{1}{10}$ shear deformations need frequently to be included.

The Timoshenko beam model considers shear deformations. Its fundamental hypothesis is that the bar cross-sections which are initially orthogonal to the bar axis remain plane but not necessarily orthogonal to the deformed bar axis.

The loading and geometrical definitions used for the Bernoulli-Euler beam model are adopted here. The kinematic hypothesis is shown in Figure 4.90[11].

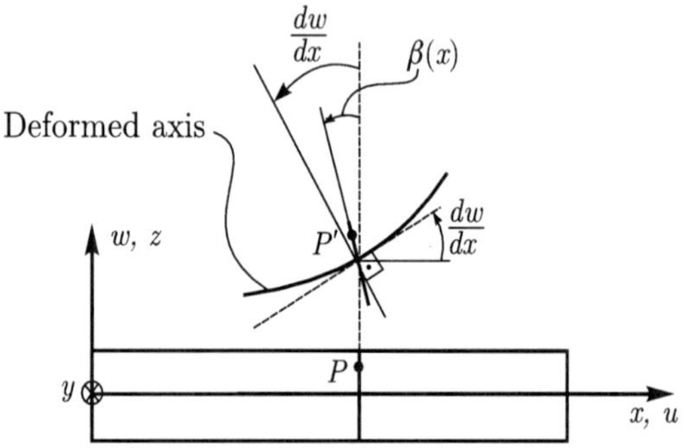

Fig. 4.90. Kinematics of beam sections for the Timoshenko beam model

We note that in addition to $w(x)$, which gives the transverse displacement of the beam axis, a new kinematic variable $\beta(x)$ which gives the section rotation with respect to the vertical direction is defined. Hence, we can write

$$u = -z\beta(x).$$

Hooke's law is also adopted for the longitudinal fibers leading to

$$\tau_{xx} = E\varepsilon_{xx} = -Ez\frac{d\beta}{dx}.$$

We can evaluate the transverse shear strain

$$\gamma_{xz} = \frac{\partial u}{\partial z} + \frac{\partial w}{\partial x} = \frac{dw}{dx} - \beta$$

and, therefore,

$$\tau_{xz} = G\gamma_{xz} = G\left(\frac{dw}{dx} - \beta\right).$$

According to these hypotheses the shear stresses are constant over the cross-section. However, the shear stresses must be zero at the top and bottom of the section, and we recall that, by static considerations alone, the Bernoulli-Euler model predicts a parabolic distribution of the transverse shear stresses for a rectangular cross-section.

[11] This same kinematic description was used in Figure 4.85

In order to retain the simple nature of the model while accounting for the deformations due to transverse shear strains, the value of γ_{xz} above is interpreted as a constant shear strain over a shear area A_s, where

$$k = \frac{A_s}{A}$$

and A is the actual cross-sectional area. Then the shear force is given by

$$V = \int_{A_s} -\tau_{xz}\, dA = -kGA\left(\frac{dw}{dx} - \beta\right). \tag{4.235}$$

The value of k depends on the stress/strain distributions over the cross-section and therefore depends on the cross-sectional shape. There are different procedures to evaluate k and we refer to Cowper, 1966 for a review of different methods[12]. In Bathe, 1996 a simple procedure is applied to obtain $k = 5/6$ for a rectangular cross-section.

The bending moment is given by

$$M = \int_A -Ez\frac{d\beta}{dx}(-z)\, dA = EI\frac{d\beta}{dx} \tag{4.236}$$

and the axial force is zero.

Considering the equilibrium in the transverse direction, we obtain

$$\frac{dV}{dx} = p$$

and from (4.235)

$$-kGA\frac{d}{dx}\left(\frac{dw}{dx} - \beta\right) = p.$$

Moment equilibrium leads to

$$\frac{dM}{dx} = V. \tag{4.237}$$

Using equation (4.235), (4.236)and (4.237), we arrive at

$$EI\frac{d^2\beta}{dx^2} = -kGA\left(\frac{dw}{dx} - \beta\right).$$

Introducing the boundary conditions, we can summarize the Timoshenko beam model formulation.

[12] A value obtained, which depends on the Poisson ratio, is $k = \frac{10+10\nu}{12+11\nu}$

Differential formulation of the Timoshenko beam model

Given the transversely distributed loading $p(x)$, find $\beta(x)$, $w(x)$ such that

$$-kGA\frac{d}{dx}\left(\frac{dw}{dx}-\beta\right)=p(x) \tag{4.238}$$

$$EI\frac{d^2\beta}{dx^2}=-kGA\left(\frac{dw}{dx}-\beta\right) \tag{4.239}$$

for all x. At $x=0$

$$\begin{aligned} \beta(0) &= \beta_0 \quad \text{or} \quad M(0)=-M_0 \\ w(0) &= w_0 \quad \text{or} \quad V(0)=Q_0 \end{aligned}$$

and at $x=L$

$$\begin{aligned} \beta(L) &= \beta_L \quad \text{or} \quad M(L)=M_L \\ w(L) &= w_L \quad \text{or} \quad V(L)=-Q_L \end{aligned}$$

where β_0, β_L are imposed rotations, w_0, w_L imposed transverse displacements, M_0, M_L prescribed moments and Q_0, Q_L prescribed transverse forces[13].

Example 4.17

Consider a bar with length L modeled as a Timoshenko beam. There is no transverse load and the following end displacements and rotations are imposed

$$w_0=\delta,\ w_L=0,\ \beta_0=0,\ \beta_L=0$$

Solution

Since $p(x)$ is zero, equation (4.238) becomes

$$\frac{dw}{dx}-\beta=C_1 \tag{4.240}$$

[13] The conventions for prescribed moments and forces are the same as those used for the Bernoulli-Euler beam model, see equations (4.136) to (4.140)

where C_1 is an integration constant. Substituting (4.240) into (4.239), we obtain

$$\frac{d^2\beta}{dx^2} = -\frac{kGA}{EI}C_1. \tag{4.241}$$

Integrating (4.241) twice and imposing $\beta(0) = 0$, we arrive at

$$\beta(x) = -\frac{kGA}{2EI}C_1x^2 + C_2x \tag{4.242}$$

where C_2 is also an integration constant. Now, substituting (4.242) into (4.240), integrating the resulting equation once and imposing the boundary condition $w(0) = \delta$, we obtain

$$w(x) = -\frac{kGA}{6EI}C_1x^3 + \frac{C_2}{2}x^2 + C_1x + \delta. \tag{4.243}$$

Considering the two remaining boundary conditions, $\beta(L) = 0$ and $w(L) = 0$, we obtain from (4.242) and (4.243)

$$-\frac{kGAL^2}{2EI}C_1 + LC_2 = 0$$

and

$$\left(-\frac{kGAL^3}{6EI} + L\right)C_1 + \frac{L^2}{2}C_2 = -\delta$$

which can be solved for C_1 and C_2 which substituted into (4.242) and (4.243) leads to the solution

$$\begin{aligned} w(x) &= \left(\frac{2x^3}{L^3(1+2g)} - \frac{3x^2}{L^2(1+2g)} - \frac{2xg}{L(1+2g)} + 1\right)\delta \\ \beta(x) &= \left(\frac{6x^2}{L^3(1+2g)} - \frac{6x}{L^2(1+2g)}\right)\delta \end{aligned}$$

where $g = \frac{6EI}{kGAL^2}$. In Example 4.7, this same problem was solved for the Bernoulli-Euler beam model. Comparing the solution for $w(x)$ given above with that of the Bernoulli-Euler beam model given in equation (4.167), we conclude that these solutions are consistent since when the shear rigidity given by GA is made large compared to the bending rigidity, given by EI, the solution based on Timoshenko beam theory approaches the solution based on Bernoulli-Euler beam theory and in the limit case, $GA \to \infty$, both solutions are the same.

□

As before, we can impose unit end displacements to derive a stiffness matrix for the Timoshenko beam. Using the bar nodal degrees of freedom convention of Figure 4.27, we can directly obtain the second column of the stiffness matrix from the solution of Example 4.7 imposing $\delta = 1$ and evaluating the end forces. In fact

$$\begin{aligned} M(0) &= EI\frac{d\beta}{dx}(0) = -\frac{6EI}{L^2\,(1+2g)} \\ M(L) &= EI\frac{d\beta}{dx}(L) = \frac{6EI}{L^2\,(1+2g)} \\ V(0) &= -kGA\left(\frac{dw}{dx} - \beta\right)\bigg|_{x=0} = \frac{12EI}{L^3\,(1+2g)} = V(L). \end{aligned}$$

Therefore

$$\tilde{\mathbf{k}}_{i2} = \begin{bmatrix} 0 \\ \frac{12EI}{L^3(1+2g)} \\ \frac{6EI}{L^2(1+2g)} \\ 0 \\ -\frac{12EI}{L^3(1+2g)} \\ \frac{6EI}{L^2(1+2g)} \end{bmatrix}.$$

If we proceed in an analogous manner, imposing unit end displacements/rotations for the remaining degrees of freedom and evaluating bar end forces and moments, we can construct the stiffness matrix for the Timoshenko beam

$$\tilde{\mathbf{k}} = \begin{bmatrix} \frac{EA}{L} & 0 & 0 & -\frac{EA}{L} & 0 & 0 \\ 0 & \frac{12EI}{L^3(1+2g)} & \frac{6EI}{L^2(1+2g)} & 0 & -\frac{12EI}{L^3(1+2g)} & \frac{6EI}{L^2(1+2g)} \\ 0 & \frac{6EI}{L^2(1+2g)} & \frac{2EI(2+g)}{L(1+2g)} & 0 & -\frac{6EI}{L^2(1+2g)} & \frac{2EI(1-g)}{L(1+2g)} \\ -\frac{EA}{L} & 0 & 0 & \frac{EA}{L} & 0 & 0 \\ 0 & -\frac{12EI}{L^3(1+2g)} & -\frac{6EI}{L^2(1+2g)} & 0 & \frac{12EI}{L^3(1+2g)} & -\frac{6EI}{L^2(1+2g)} \\ 0 & \frac{6EI}{L^2(1+2g)} & \frac{2EI(1-g)}{L(1+2g)} & 0 & -\frac{6EI}{L^2(1+2g)} & \frac{2EI(2+g)}{L(1+2g)} \end{bmatrix}.$$

Of course, the axial displacements are considered in the stiffness matrix above. We note that we can superpose the solution of the bar model under axial loading with either the Bernoulli-Euler or Timoshenko models since the sections do not rotate due to the axial loading.

Regarding comparisons in predictions obtained with the Bernoulli-Euler and Timoshenko models, we refer the reader to the modeling presented in Section 7.1.

4.3 Plates in bending

We first recall the plane stress model discussed in Section 4.1.2. The model was characterized geometrically as a thin plate and the mechanical loading had to act in the midsurface of the plate. These definitions were summarized in Figure 4.5.

The plate models we address in this section have the same geometries, however, the loading is acting transversely, *i.e.*, orthogonal to the midsurface of the plate, which induces bending, leading to a completely different structural behavior. While in the plane stress model the stresses are constant through the thickness, for the plate bending models, the stresses vary linearly through the thickness of the plate.

There is an interesting analogy between the behavior of bars and plates. The analogue of the bar subjected to axial loading is the plane stress model and that of the bar subjected to transverse loading is the plate bending model.

Insight into the behavior of a plate resisting transverse loading can be gained by interpreting the plate to act "like" beams in orthogonal directions as schematically shown in Figure 4.91. Of course, only the gross behavior is captured by this interpretation. We will return to this interpretation of the plate behavior later on in this section.

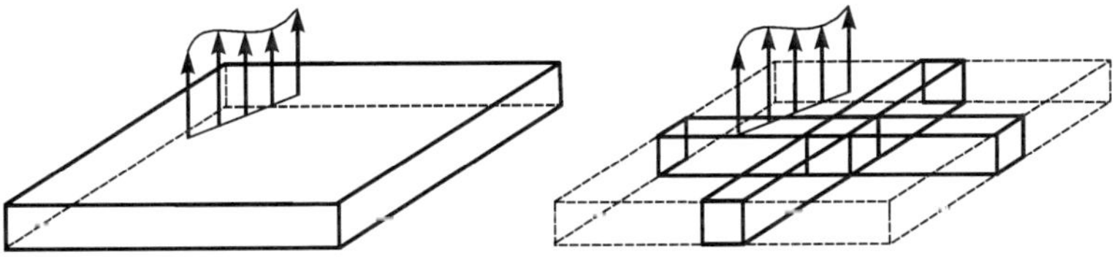

Fig. 4.91. Interpretation of bending behavior by beam action

The presentation of plate models is organized as follows. In the next section we discuss the Kirchhoff plate model which is the analogue of the Bernoulli-Euler beam model due to the similar kinematic hypothesis adopted. The emphasis is on model assumptions and on basic requirements. We derive the differential formulation of the model and, although solutions of the governing equations are not our primary objective, one classical solution is presented for illustrative purposes. A section is then dedicated to plate behavior. We end the presentation by briefly introducing the Reissner-Mindlin plate model which is the analogue of the Timoshenko beam model.

4.3.1 The Kirchhoff plate bending model

We restrict our discussion to the linear model. Therefore the displacements are assumed infinitesimally small and equilibrium is enforced in the undeformed configuration. The plate is supposed to be thin, *i.e.*, h, the thickness, is small compared to a characteristic geometric dimension L of the midsurface of the plate. Typically, Kirchhoff theory gives good predictions when $h/L < 1/20$.

Kinematics

The fundamental kinematic hypothesis of the model is that straight material lines which are initially orthogonal to the midsurface of the plate are also straight and orthogonal to the deformed midsurface. Additionally, the displacements in the transverse direction do not vary along the thickness of the plate.

We consider a generic plate described in Figure 4.92. Let P be a generic point on the midsurface and let P_z be a point on a straight line from P orthogonal to the midsurface. The coordinates of P are $(x, y, 0)$ and of P_z are (x, y, z). Suppose that the plate deforms under the action of the transversely distributed load $p(x, y)$ which is given per unit of midsurface area. In Figure 4.93a and 4.93b we show the deformed and undeformed configurations of the plate in the xz and yz planes, respectively.

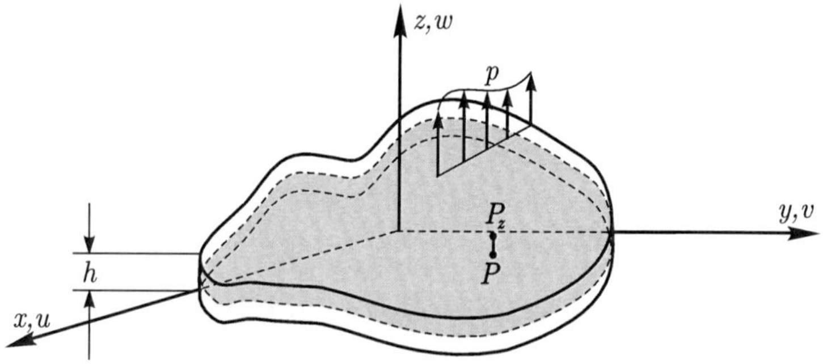

Fig. 4.92. Generic representation of a plate

Considering these definitions, the kinematic hypothesis translates into

$$w = w(x, y) \tag{4.244}$$

$$u = -z\frac{\partial w}{\partial x} \tag{4.245}$$

$$v = -z\frac{\partial w}{\partial y} \tag{4.246}$$

where we considered that the displacements are infinitesimally small.

Using the compatibility relations we obtain the strains

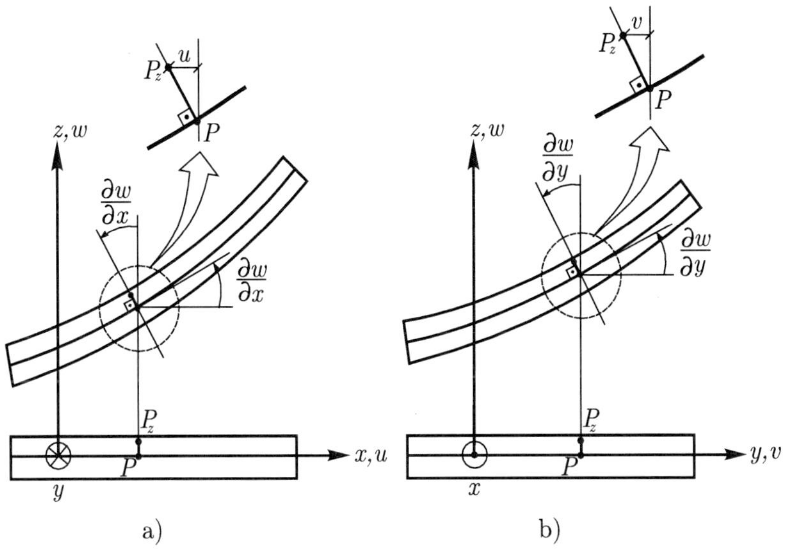

Fig. 4.93. Selected intersections of deformed and undeformed configurations of the plate with a vertical plane

$$\varepsilon_{xx} = \frac{\partial u}{\partial x} = -z\frac{\partial^2 w}{\partial x^2}$$

$$\varepsilon_{yy} = \frac{\partial v}{\partial y} = -z\frac{\partial^2 w}{\partial y^2}$$

$$\gamma_{xy} = \frac{\partial u}{\partial y} + \frac{\partial v}{\partial x} = -2z\frac{\partial^2 w}{\partial x \partial y}$$

and the remaining strain components are zero. We note that since we are considering infinitesimally small displacements the curvatures of the deformed midsurface are given by

$$\kappa_x = \frac{\partial^2 w}{\partial x^2}$$

$$\kappa_y = \frac{\partial^2 w}{\partial y^2}$$

$$\kappa_{xy} = \frac{\partial^2 w}{\partial x \partial y}$$

$$\kappa_{yx} = \kappa_{xy}.$$

Constitutive equation

The theory assumes that the plate is composed of a stack of laminae, as schematically shown in Figure 4.94 for a part of the plate, and that each lamina is in a state of plane stress. Therefore

$$\tau_{xx} = \frac{E}{(1-\nu^2)}(\varepsilon_{xx}+\nu\varepsilon_{yy}) = -\frac{Ez}{(1-\nu^2)}\left(\frac{\partial^2 w}{\partial x^2}+\nu\frac{\partial^2 w}{\partial y^2}\right) \quad (4.247)$$

$$\tau_{yy} = \frac{E}{(1-\nu^2)}(\varepsilon_{yy}+\nu\varepsilon_{xx}) = -\frac{Ez}{(1-\nu^2)}\left(\frac{\partial^2 w}{\partial y^2}+\nu\frac{\partial^2 w}{\partial x^2}\right) \quad (4.248)$$

$$\tau_{xy} = \frac{E}{2(1+\nu)}\gamma_{xy} = -\frac{Ez}{(1+\nu)}\frac{\partial^2 w}{\partial x\partial y}. \quad (4.249)$$

Equations (4.247) to (4.249) show that the stresses vary linearly in the thickness direction.

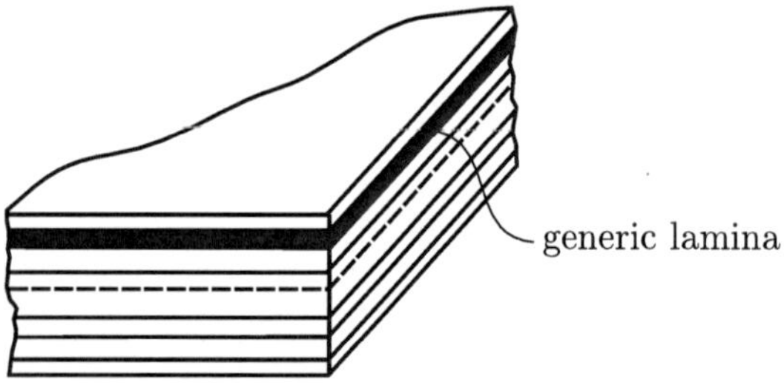

Fig. 4.94. Plate interpreted as a stack of laminae

Stress resultants

We extract a differential element from the plate and show in Figure 4.95a the stress distributions given by equations (4.247) to (4.249) when $\frac{\partial^2 w}{\partial x^2} > 0$, $\frac{\partial^2 w}{\partial y^2} > 0$ and $\frac{\partial^2 w}{\partial x\partial y} > 0$.

The moment resultant M_x per unit of length associated with the stress component τ_{xx} is

$$M_x = \int_{-h/2}^{+h/2} \tau_{xx}\,(-z)\;dz$$

where we use the same convention as for the beam: a positive value for the moment is associated with tension of the lower fibers, *i.e.*, below the midsurface of the plate. Using equation (4.247)

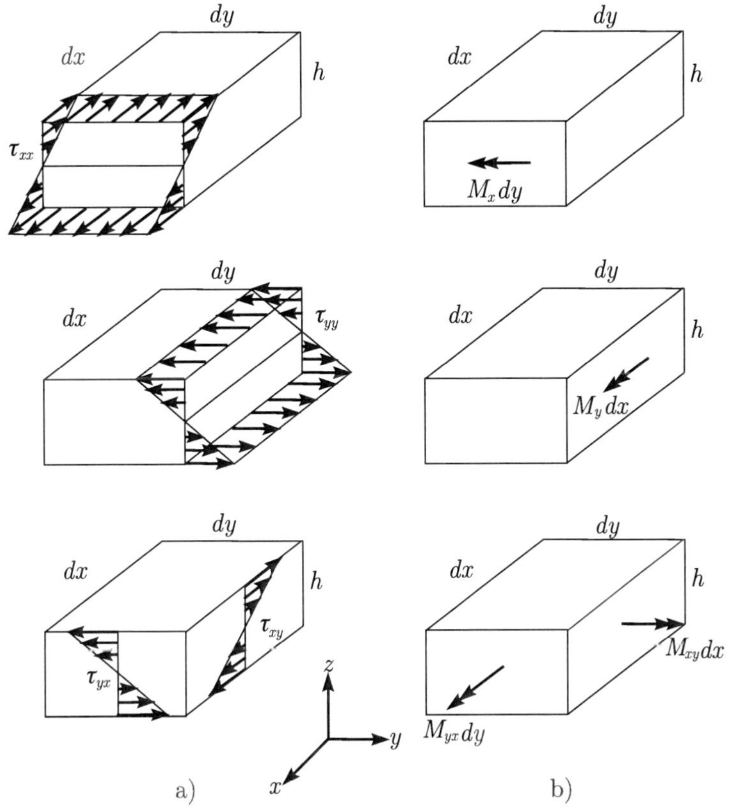

Fig. 4.95. Stress and stress resultant definitions; M_x, M_y and M_{yx} are positive whereas M_{xy} is negative for stresses shown

$$M_x = \frac{Eh^3}{12\,(1-\nu^2)}\left(\frac{\partial^2 w}{\partial x^2} + \nu\frac{\partial^2 w}{\partial y^2}\right) \tag{4.250}$$

and defining

$$D = \frac{Eh^3}{12\,(1-\nu^2)} \tag{4.251}$$

relation (4.250) becomes

$$M_x = D\left(\frac{\partial^2 w}{\partial x^2} + \nu\frac{\partial^2 w}{\partial y^2}\right). \tag{4.252}$$

Analogously, let M_y be the moment resultant per unit of length associated with τ_{yy} defined by

$$M_y = \int_{-h/2}^{+h/2} \tau_{yy}\,(-z)\;dz = D\left(\frac{\partial^2 w}{\partial y^2} + \nu\frac{\partial^2 w}{\partial x^2}\right). \tag{4.253}$$

Of course, both M_x and M_y are bending moments. Now let

$$M_{yx} = \int_{-h/2}^{+h/2} \tau_{yx}(-z)\ dz = D(1-\nu)\frac{\partial^2 w}{\partial x \partial y}. \tag{4.254}$$

We note that M_{yx} is a torsional moment per unit of length. This moment is also referred to as a twisting moment, and a positive value corresponds to the moment vector pointing out of the plate section on which the moment is acting. We also define

$$M_{xy} = \int_{-h/2}^{+h/2} \tau_{xy} z\ dz = -D(1-\nu)\frac{\partial^2 w}{\partial x \partial y} \tag{4.255}$$

where since $\tau_{xy} = \tau_{yx}$, we obtain

$$M_{yx} = -M_{xy}. \tag{4.256}$$

In Figure 4.95b we show the moments associated with the stress resultants defined above.

Equilibrium

The last requirement to be considered to complete the differential formulation is equilibrium. In Figure 4.96 we show the resultant forces and moments acting on a differential plate element which is shown twice in this figure merely for ease of visualization. We note that the shear resultants per unit of length, Q_x and Q_y, which are associated with the transverse shear stress components τ_{xz} and τ_{yz} respectively have been introduced. As for the shear force in the Bernoulli-Euler beam model, these shear forces do not enter the formulation through the constitutive relations since by the kinematic assumption the transverse shear strains are zero. However, they are required for equilibrium.

Imposing force equilibrium[14] in the z direction, we obtain

$$-Q_x dy + \left(Q_x + \frac{\partial Q_x}{\partial x}dx\right)dy - Q_y dx + \left(Q_y + \frac{\partial Q_y}{\partial y}dy\right)dx + p dx dy = 0.$$

Simplifying the equation above, we arrive at

$$\frac{\partial Q_x}{\partial x} + \frac{\partial Q_y}{\partial y} = -p. \tag{4.257}$$

[14] Note that the sign convention for the transverse shear force of plates and shells is here opposite to the convention used for beams. This sign convention for plates and shells is more natural considering the usual 3-D definition of strains, see Section 4.3.2

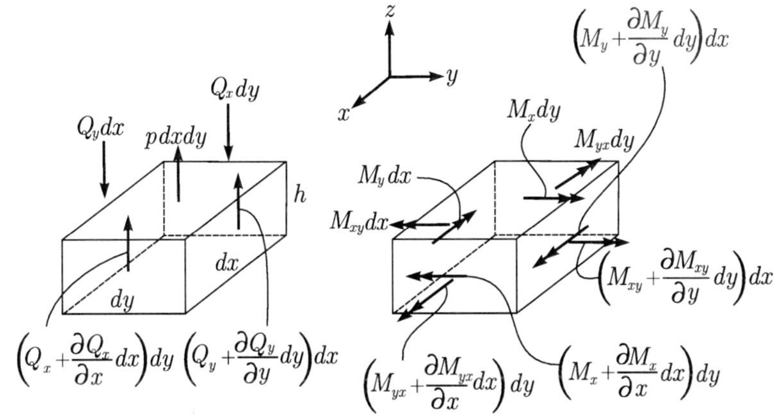

Fig. 4.96. Forces acting on a differential plate element.

Neglecting infinitesimals of higher-order, moment equilibrium about the x axis leads to

$$(Q_y dx)\, dy - M_y dx + \left(M_y + \frac{\partial M_y}{\partial y} dy \right) dx - M_{yx} dy + \left(M_{yx} + \frac{\partial M_{yx}}{\partial x} dx \right) dy = 0$$

which upon simplification gives

$$\frac{\partial M_y}{\partial y} + \frac{\partial M_{yx}}{\partial x} = -Q_y. \tag{4.258}$$

Finally, moment equilibrium in the y direction yields

$$-\,(Q_x dy)\, dx + M_x dy - \left(M_x + \frac{\partial M_x}{\partial x} dx \right) dy - M_{xy} dx + \left(M_{xy} + \frac{\partial M_{xy}}{\partial y} dy \right) dx = 0$$

$$\frac{\partial M_x}{\partial x} - \frac{\partial M_{xy}}{\partial y} = -Q_x. \tag{4.259}$$

Differential formulation

Summarizing, all requirements have been imposed, namely, compatibility (equations (4.244) to (4.246)), constitutive relations (equations (4.247) to (4.249)) and equilibrium (equations (4.257) to (4.259)). Therefore the differential equations of the Kirchhoff plate bending model were obtained.

It is usual to cast the complete formulation in terms of the transverse displacement $w(x, y)$ which is the only independent kinematic variable. For that we take derivatives of equation (4.258) with respect to y and of (4.259) with respect to x, sum them and use equation (4.257) to obtain

$$\frac{\partial^2 M_x}{\partial x^2} - 2\frac{\partial^2 M_{xy}}{\partial x \partial y} + \frac{\partial^2 M_y}{\partial y^2} = p. \tag{4.260}$$

We have also used that $M_{xy} = -M_{yx}$ to arrive at equation (4.260). Substituting (4.253), (4.254) and (4.255) into (4.260), we obtain

$$\frac{\partial^4 w}{\partial x^4} + 2\frac{\partial^4 w}{\partial x^2 \partial y^2} + \frac{\partial^4 w}{\partial y^4} = \frac{p}{D} \tag{4.261}$$

which is known as the Lagrange equation for the plate bending problem.

We note that the interpretation of D as the flexural rigidity of the plate is now obvious. We also remark that when equation (4.261) subjected to the appropriate boundary conditions is solved, *i.e.*, $w(x, y)$ is determined, the complete solution is known since we can obtain the stresses, strains and resultant forces from the transverse displacement field as shown in the equations above. Even the transverse shear forces which are not obtained from the constitutive relations can be calculated from the transverse displacements (as for the Bernoulli-Euler beam model). Namely, substituting the expressions for the moments in terms of the transverse displacements (equations (4.250), (4.252) and (4.254) into equations (4.258) and (4.259)), we obtain

$$Q_x = -D\frac{\partial}{\partial x}\left(\frac{\partial^2 w}{\partial x^2} + \frac{\partial^2 w}{\partial y^2}\right) \tag{4.262}$$

and

$$Q_y = -D\frac{\partial}{\partial y}\left(\frac{\partial^2 w}{\partial x^2} + \frac{\partial^2 w}{\partial y^2}\right). \tag{4.263}$$

Boundary conditions

Let us examine very briefly the boundary conditions for the Kirchhoff model. Consider an edge, parallel to the y axis and therefore given by $x = a$.

The *clamped* or *built-in* condition corresponds to imposing that the transverse displacements and the rotations about the y axis of material lines, such as AB shown in Figure 4.97, are zero. Therefore, the boundary conditions are given by

$$w|_{x=a} = 0 \quad \text{and} \quad \left.\frac{\partial w}{\partial x}\right|_{x=a} = 0.$$

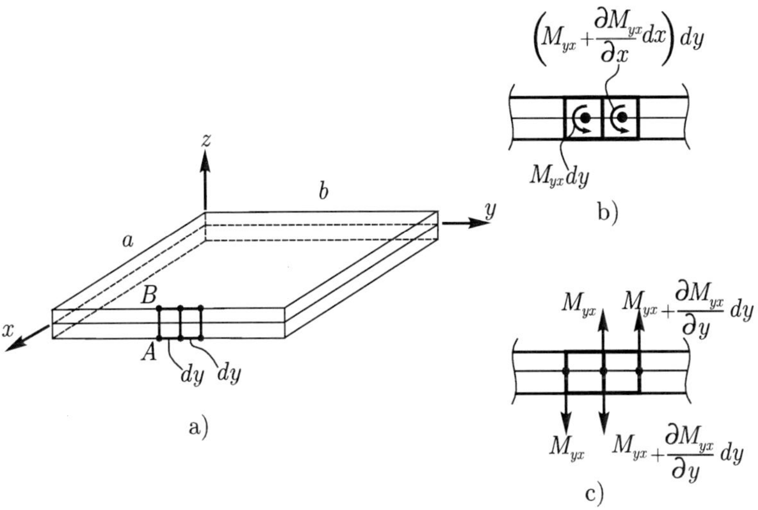

Fig. 4.97. Some definitions for a rectangular plate

Note that the condition on the rotation is imposed taking into account the kinematic assumption of the model.

Considering the *simply supported* condition, the transverse displacement should be restrained. For additional conditions, we observe that there are no kinematic restraints affecting the rotation about the y axis of material lines such as AB (see Figure 4.97), and there are no normal stresses τ_{xx} acting on the plate end section defined at $x = a$. Hence $M_x = 0$ at $x = a$. Summarizing, the two boundary conditions that represent the simply supported edge are

$$w\big|_{x=a} = 0 \tag{4.264}$$

and

$$M_x\big|_{x=a} = 0. \tag{4.265}$$

The last condition can also be expressed in terms of displacements when we consider equation (4.250) leading to

$$\left.\frac{\partial^2 w}{\partial x^2} + \nu\frac{\partial^2 w}{\partial y^2}\right|_{x=a} = 0$$

and because of (4.264) $\left.\frac{\partial^2 w}{\partial y^2}\right|_{x=a} = 0$, condition (4.265) becomes

$$\left.\frac{\partial^2 w}{\partial x^2}\right|_{x=a} = 0.$$

The last boundary condition to consider is the *free edge* condition. Since there are no kinematic restraints in this case, there are no stresses acting on the plate end section defined by $x = a$. The stress components which could possibly act on this edge section are τ_{xx}, τ_{yx} and τ_{zx}. These components are associated with the stress resultants M_x, M_{yx} and Q_x. If we would enforce the condition that these stress resultants are zero, we would obtain three boundary conditions instead of the two obtained in the case of the clamped and the simply supported edges. Historically, this apparent inconsistency was object of much controversy. Mathematically, considering the order of the differential equation (4.261), only two conditions are required per edge. It was shown that one condition is given by $M_x|_{x=a} = 0$ and the second involves a combination of the stress resultants M_{yx} and Q_x. Kelvin and Tait (see Timoshenko and Woinowsky-Krieger, 1959) gave a mechanical interpretation of this second condition that we discuss below.

Consider two generic differential elements of the edge section shown in Figure 4.97a. In Figure 4.97b, the twisting moment resultant acting on these two differential elements are shown. Each of these moments are represented in Figure 4.97c by a mechanically equivalent force couple. Therefore, we can see that the twisting moment distribution acting at the edge can be represented by a mechanically equivalent distribution of shear forces given by

$$Q'_x = \left. \frac{M_{yx} - \left(M_{yx} + \frac{\partial M_{yx}}{\partial y} dy\right)}{dy} \right|_{x=a} = - \left. \frac{\partial M_{yx}}{\partial y} \right|_{x=a}.$$

The net shear force distribution V_x considering both Q_x and Q'_x is then given by

$$V_x = Q_x|_{x=a} + Q'_x = \left. \left(Q_x - \frac{\partial M_{yx}}{\partial y}\right) \right|_{x=a}$$

and the boundary condition for the free edge which combines both effects of Q_x and M_{yx} is

$$V_x = \left. \left(Q_x - \frac{\partial M_{yx}}{\partial y}\right) \right|_{x=a} = 0.$$

This condition can be expressed in terms of displacements. Using relations (4.254) and (4.262), we obtain

$$\left. \left(\frac{\partial^3 w}{\partial x^3}\right) + (2 - \nu) \left(\frac{\partial^3 w}{\partial x \partial y^2}\right) \right|_{x=a} = 0. \tag{4.266}$$

Below we show a sample solution of a classical plate bending problem.

Example 4.18

Find the solution for the rectangular plate problem of Figure 4.97a when the plate is simply supported at the four edges and subjected to a distributed pressure $p = p(x, y)$. Particularize the solution for $p(x, y) = p_0$.

Solution

The boundary conditions can be derived generalizing those for the simply supported edge studied above when we consider the four plate edges. Then, we obtain

$$w\big|_{x=0} = 0, \quad \left.\frac{\partial^2 w}{\partial x^2}\right|_{x=0} = 0 \tag{4.267}$$

$$w\big|_{y=0} = 0, \quad \left.\frac{\partial^2 w}{\partial y^2}\right|_{y=0} = 0 \tag{4.268}$$

$$w\big|_{x=a} = 0, \quad \left.\frac{\partial^2 w}{\partial x^2}\right|_{x=a} = 0 \tag{4.269}$$

$$w\big|_{y=b} = 0, \quad \left.\frac{\partial^2 w}{\partial y^2}\right|_{y=b} = 0. \tag{4.270}$$

We use a classical approach to derive solutions for plate bending problems which is to use a Fourier series to construct solutions.

Assume

$$w(x,y) = \sum_{m=1}^{\infty}\sum_{n=1}^{\infty} w_{mn} \sin\frac{m\pi x}{a} \sin\frac{n\pi y}{b}. \tag{4.271}$$

We can write (4.271) since it is a mathematical fact that every smooth function defined on a 2-D domain can be expanded in a Fourier series as above. The fundamental property is that given this particular functional form for $w(x,y)$, there exist constant coefficients w_{mn} such that if the double sum is performed up to m and n high enough, such sum will be arbitrarily close[15] to $w(x,y)$.

Hence, our task is to determine the coefficients w_{mn} such that the sum given in (4.271) satisfies equation (4.261) and the boundary conditions given in equations (4.267) to (4.270). In fact, the Fourier series given in (4.271) was constructed such that it satisfies the boundary conditions independently of the values of w_{mn}. This property can be easily verified by checking that $w(x,y)$ written in the form of (4.271) satisfies conditions (4.267) to (4.270). In order to determine specific coefficients w_{mn} the load $p(x,y)$ should be characterized.

Given a smooth load distribution $p(x,y)$, it can also be expanded in a Fourier series as

$$p(x,y) = \sum_{m=1}^{\infty}\sum_{n=1}^{\infty} p_{mn} \sin\frac{m\pi x}{a} \sin\frac{n\pi y}{b} \tag{4.272}$$

[15] Of course, what we mean by arbitrarily close can be made mathematically precise, see Chapelle and Bathe, 2010*a*. However, for our present purposes the intuitive idea of this concept suffices

where

$$p_{mn} = \frac{4}{ab}\int_0^a\int_0^b p(x,y)\sin\frac{m\pi x}{a}\sin\frac{n\pi y}{b}\,dxdy. \tag{4.273}$$

Substituting (4.271) and (4.272) into (4.261), and, of course, performing the derivatives involved, we obtain

$$\sum_{m=1}^{\infty}\sum_{n=1}^{\infty}\left(\frac{m^4}{a^4}+\frac{2m^2n^2}{a^2b^2}+\frac{n^4}{b^4}\right)\pi^4 w_{mn}\sin\frac{m\pi x}{a}\sin\frac{n\pi y}{b}$$

$$= \frac{1}{D}\sum_{m=1}^{\infty}\sum_{n=1}^{\infty}p_{mn}\sin\frac{m\pi x}{a}\sin\frac{n\pi y}{b}.$$

Since the coefficients of the series on the left- and right-hand sides of the above equation should be the same, we obtain

$$w_{mn} = \frac{p_{mn}}{\pi^4 D\left(\frac{m^4}{a^4}+\frac{2m^2n^2}{a^2b^2}+\frac{n^4}{b^4}\right)}$$

and hence

$$w(x,y) = \frac{1}{\pi^4 D}\sum_{m=1}^{\infty}\sum_{n=1}^{\infty}\frac{p_{mn}}{\left(\frac{m^2}{a^2}+\frac{n^2}{b^2}\right)^2}\sin\frac{m\pi x}{a}\sin\frac{n\pi y}{b}. \tag{4.274}$$

When the load is uniform, *i.e.*, $p(x,y) = p_0$, we obtain from equation (4.273)

$$p_{mn} = \frac{16p_0}{\pi^2 mn} \qquad m,n = 1,3,5,...$$

and $p_{mn} = 0$ for m or n an even number. Hence, from (4.274) we can write the solution as

$$w(x,y) = \frac{16p_0}{\pi^6 D}\sum_{m=1}^{\infty}\sum_{n=1}^{\infty}\frac{\sin\frac{m\pi x}{a}\sin\frac{n\pi y}{b}}{mn\left(\frac{m^2}{a^2}+\frac{n^2}{b^2}\right)^2}$$

where $m = 1,3,5,...$ and $n = 1,3,5,...$

As a historical remark, we note that it was Navier who first proposed this Fourier series solution for plate bending problems (see Timoshenko and Woinowsky-Krieger, 1959).

Of course, having evaluated $w(x,y)$ all stresses and stress resultants can be obtained.

□

Plate bending behavior

In the introductory part of this section we interpreted the gross behavior of a rectangular plate as being captured by orthogonal bars in bending (see Figure 4.91). Our objective here is to obtain further insight into the plate resisting mechanisms.

Consider a rectangular simply supported plate. Let us examine the behavior of the plate when modeled by bars of rectangular cross-sections as those shown in Figure 4.9. Considering a bar with axis parallel to the global x axis, we see that besides the moment M_x and the shear force Q_x which are directly associated with the bending of this beam (refer to equations (4.147) and (4.142)), torsional moments M_{xy} and the shear forces Q_y acting on the lateral faces of this beam contribute to its equilibrium.

To obtain insight into how these effects stiffen the plate when equilibrium and compatibility are imposed, we consider the structure of Figure 4.98a subjected to a uniform transverse load.

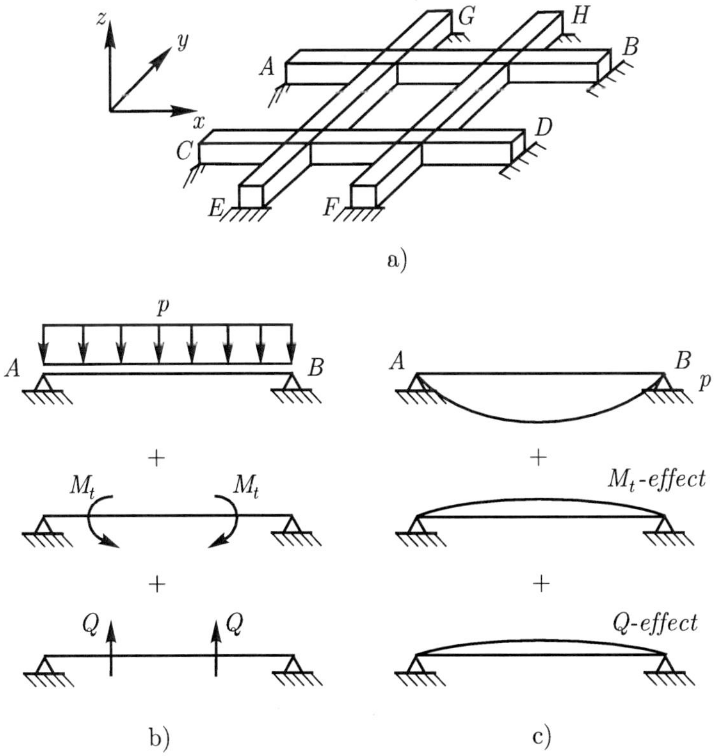

Fig. 4.98. Qualitative behavior of a rectangular plate based on a grid analogy

In Figure 4.98b, we show the actions onto bar AB. Besides the transverse load p, there are two additional contributions. The loads Q are a result of the compatibility of transverse displacements and represent the effect of the orthogonal beams working in bending. The moments M_t are a result of the compatibility of rotations. Namely, the section rotations of beam AB due to the transverse load induce rotations and torsion in the bars EG and FH. The sense of M_t indicated in Figure 4.98b reflects the fact that the bars EG and FH oppose the rotations due to the transverse load on the bar AB. Finally, in Figure 4.98c, we show, qualitatively, the transverse displacements induced by each load including the stiffening effect due to the M_t and Q effects.

Consider the rectangular plate in Figure 4.97. Let us examine the behavior of the plate as the relative size of the edges changes, *i.e.*, the relation a/b varies. Supposing that a/b is large, say $a/b > 5$, we show qualitatively in Figure 4.99 the deformations of two orthogonal slices of the midsurface. If we examine the contribution due to bending to equilibrate the transverse load, we conclude that the longer beam carries almost no loading, since for a simply supported beam the transverse stiffness is inversely proportional to L^3 where L is the beam length. Therefore, for a large portion of the plate, *i.e.*, the central part, away from the shorter edges, the resisting behavior corresponds to the bending along the short span and the plate behaves as a beam of large width. We note that the contribution of torsion for this part is also very small since $\frac{\partial^2 w}{\partial x \partial y}$ is close to zero and, hence, from equation (4.254), M_{xy} is very small. In fact, if we consider a plate infinitely long in the x direction, the deformation of the midsurface will be cylindrical (the end effects are at infinity), and $w = w(y)$. Hence $\frac{\partial^4 w}{\partial x^4} = 0$ and $\frac{\partial^4 w}{\partial x^2 \partial y^2} = 0$, and equation (4.261) becomes

$$\frac{d^4 w}{dy^4} = \frac{p}{D}$$

where, in essence, a beam of unit width is considered (see (4.136)). However, we use $D = E_* I$ with $E_* = E/\left(1 - \nu^2\right)$. Hence, a planar beam under plane strain conditions (refer to equation (4.58)) is solved (since anticlastic curvature is not allowed, see Figure 3.62). From a design perspective, it is relevant to note that for $a/b = 2$ the error incurred in assuming a large width beam behavior of the plate is of the order of 6.5%. Of course, this error decreases as a/b increases.

4.3.2 The Reissner-Mindlin plate bending model

As we mentioned, the Reissner-Mindlin plate model is the analogue of the Timoshenko beam model. This analogy is based on the kinematic hypothesis which includes modeling of transverse shear deformations. Hence, the Reissner-Mindlin model is adequate to model not only thin plates but also those which are moderately thick. All the assumptions used in the Kirchhoff model concerning linear analysis are also adopted here.

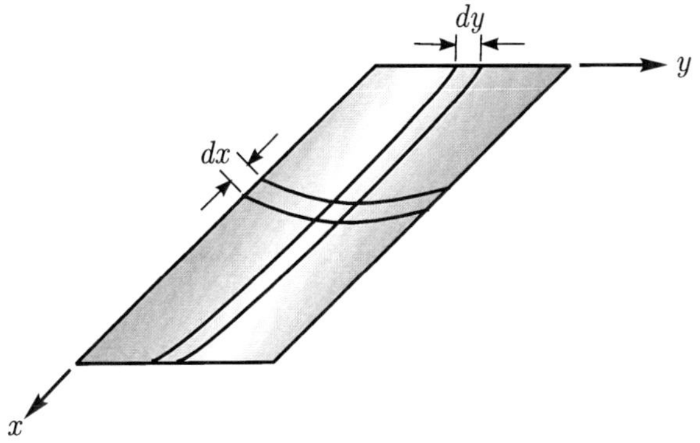

Fig. 4.99. Deformation of simply supported plate under uniform pressure for large a/b

Kinematics

The fundamental kinematic hypothesis of the model is that straight material lines which are initially orthogonal to the midsurface of the plate remain straight but not necessarily orthogonal to the deformed midsurface. Of course, we use the geometric and loading characterization of Figure 4.92, and Figure 4.100 is analogous to 4.93, but considering the kinematics of the Reissner-Mindlin model.

Note that $\beta_x(x, y)$ and $\beta_y(x, y)$ characterize the rotation of the material lines which are initially orthogonal to the midsurface. Again, we assume

$$w = w(x, y)$$

and referring to Figure 4.100, we can write

$$\begin{aligned} u &= -z\beta_x(x, y) \\ v &= -z\beta_y(x, y). \end{aligned}$$

Therefore $w(x, y)$, $\beta_x(x, y)$ and $\beta_y(x, y)$ – all referred to the midsurface – are the three independent degrees of freedom of the model and completely characterize the displacement field.

Considering the strain compatibility relations, we can write

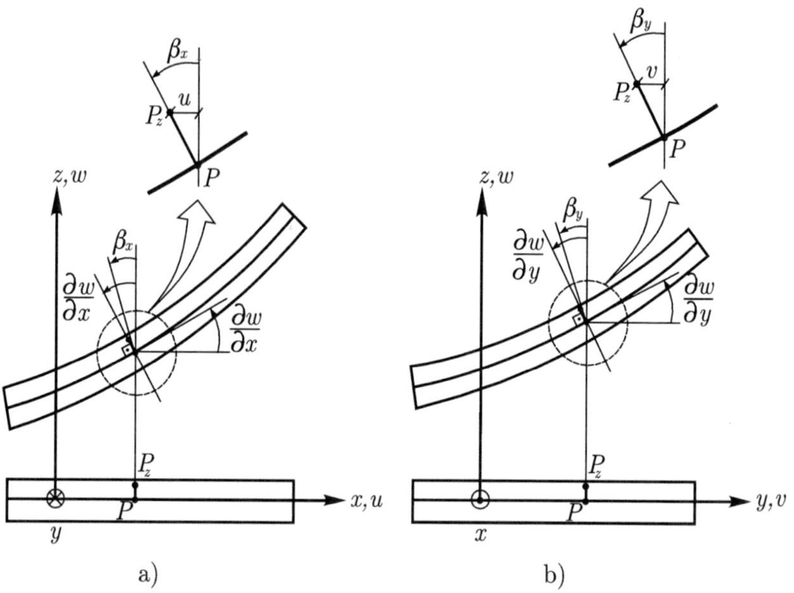

Fig. 4.100. Selected intersections of deformed and undeformed configurations of plate with a vertical plane. Reissner-Mindlin model

$$
\begin{aligned}
\varepsilon_{xx} &= \frac{\partial u}{\partial x} = -z\frac{\partial \beta_x}{\partial x} \\
\varepsilon_{yy} &= \frac{\partial v}{\partial y} = -z\frac{\partial \beta_y}{\partial y} \\
\gamma_{xy} &= \frac{\partial u}{\partial y} + \frac{\partial v}{\partial x} = -z\left(\frac{\partial \beta_x}{\partial y} + \frac{\partial \beta_y}{\partial x}\right) \\
\gamma_{xz} &= \frac{\partial u}{\partial z} + \frac{\partial w}{\partial x} = \frac{\partial w}{\partial x} - \beta_x \\
\gamma_{yz} &= \frac{\partial v}{\partial z} + \frac{\partial w}{\partial y} = \frac{\partial w}{\partial y} - \beta_y
\end{aligned}
$$

and $\varepsilon_{zz} = 0$.

Constitutive equation

The mechanical hypothesis of the Kirchhoff model assuming a plane stress condition of each lamina is used. Therefore

$$
\begin{aligned}
\tau_{xx} &= \frac{E}{(1-\nu^2)}(\varepsilon_{xx}+\nu\varepsilon_{yy}) = -\frac{Ez}{(1-\nu^2)}\left(\frac{\partial\beta_x}{\partial x}+\nu\frac{\partial\beta_y}{\partial y}\right)\\
\tau_{yy} &= \frac{E}{(1-\nu^2)}(\varepsilon_{yy}+\nu\varepsilon_{xx}) = -\frac{Ez}{(1-\nu^2)}\left(\frac{\partial\beta_y}{\partial y}+\nu\frac{\partial\beta_x}{\partial x}\right)\\
\tau_{xy} &= \frac{E}{2(1+\nu)}\gamma_{xy} = -\frac{Ez}{2(1+\nu)}\left(\frac{\partial\beta_x}{\partial y}+\frac{\partial\beta_y}{\partial x}\right).
\end{aligned}
$$

Additionally, we have the transverse shear relations which using the generalized Hooke's law can be written as

$$
\begin{aligned}
\tau_{xz} &= G\gamma_{xz} = G\left(\frac{\partial w}{\partial x}-\beta_x\right) = \frac{E}{2(1+\nu)}\left(\frac{\partial w}{\partial x}-\beta_x\right)\\
\tau_{yz} &= G\gamma_{yz} = G\left(\frac{\partial w}{\partial y}-\beta_y\right) = \frac{E}{2(1+\nu)}\left(\frac{\partial w}{\partial y}-\beta_y\right).
\end{aligned}
$$

Stress resultants

The stress resultants are the same as those of the Kirchhoff model and the relations between the stress resultants and the kinematic variables are

$$
\begin{aligned}
M_x &= D\left(\frac{\partial\beta_x}{\partial x}+\nu\frac{\partial\beta_y}{\partial y}\right)\\
M_y &= D\left(\frac{\partial\beta_y}{\partial y}+\nu\frac{\partial\beta_x}{\partial x}\right)\\
M_{yx} &= -M_{xy} = D\frac{(1-\nu)}{2}\left(\frac{\partial\beta_x}{\partial y}+\frac{\partial\beta_y}{\partial x}\right)\\
Q_x &= kGA\left(\frac{\partial w}{\partial x}-\beta_x\right)\\
Q_y &= kGA\left(\frac{\partial w}{\partial y}-\beta_y\right)
\end{aligned}
$$

where we have introduced the shear correction factor k which was defined for the Timoshenko beam model, see (4.235). Note that in the Reissner-Mindlin model the shear forces are obtained through the constitutive relations.

Equilibrium

The equilibrium conditions are those of the Kirchhoff model and are given by (4.257) to (4.259).

Boundary conditions

We note that due to the more general kinematic description in the Reissner-Mindlin model, we have three conditions to specify. As in the Kirchhoff model, let us examine the boundary conditions for an edge parallel to the y axis, *i.e.*, given by $x = a$.

For the *clamped* or *built-in* edge, we need to prevent the transverse displacements w and the rotation about the y axis of material lines such as AB of Figure 4.97. Therefore

$$w\big|_{x=a} = 0 \qquad \text{and} \qquad \beta_x\big|_{x=a} = 0.$$

Depending on the physical situation, these material lines such as AB may also be prevented to rotate about the x axis. Therefore, we either impose

$$\beta_y\big|_{x=a} = 0 \tag{4.275}$$

or

$$M_{yx}\big|_{x=a} = 0. \tag{4.276}$$

If we prevent the rotation β_y we have the "hard" condition and if we impose M_{yx} to be zero we have the "soft" condition.

For the *simply supported* edge, we would have

$$w\big|_{x=a} = 0 \qquad \text{and} \qquad M_x\big|_{x=a} = 0.$$

The third condition is also given by (4.275) and (4.276). Again (4.275) is called the "hard" condition while (4.276) is called the "soft" condition.

Finally, for the *free edge*, we have

$$M_x\big|_{x=a} = 0 \qquad \text{and} \qquad Q_x\big|_{x=a} = 0.$$

and either (4.275) or (4.276). Of course, if for our physical situation the plate section given by $x = a$ is stress free, then $M_{yx} = 0$ is the appropriate condition.

We note that, as we now have three independent kinematic variables, the free edge condition can be imposed in a more natural way than in the Kirchhoff model.

The differential formulation of the Reissner-Mindlin model is now complete since the compatibility, constitutive behavior, equilibrium and boundary conditions have been considered. Of course, the governing equations can be algebraically manipulated to arrive at a set of equations which are more convenient for the derivation of closed form solutions, but this is not our objective here.

The Reissner-Mindlin model is a hierarchically higher-order model when compared to the Kirchhoff model. One of the reasons is that it predicts transverse shear deformations. A less obvious improvement, which requires a more detailed study of the Reissner-Mindlin model (see Häggblad and Bathe, 1990), is given by the nature of the solution close to the boundaries. There is a region near the boundary – referred to as the boundary layer region – where the solution for the stress resultants may vary significantly.

In order to obtain more insight into this boundary layer behavior, while examining a situation of engineering interest, let us study the shear force at an edge of a simply supported plate subjected to a constant uniform pressure. Considering the Kirchhoff model, there are two contributions: the force Q_x and that due to the rate of change of the twisting moment, *i.e.*, $Q'_x = -\left.\frac{\partial M_{yx}}{\partial y}\right|_{x=a}$ as discussed above. The actual values along the edges can be obtained from the solution $w(x,y)$, discussed in Example 4.18, using relations (4.254) and (4.262). In Figure 4.101a, we show these shear force contributions.

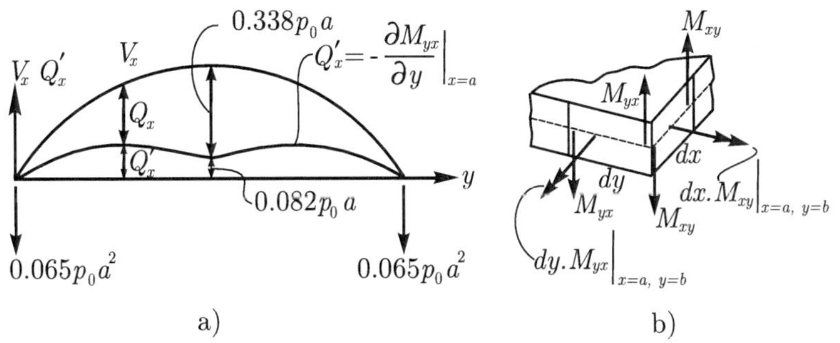

Fig. 4.101. Shear related quantities at edge of simply supported plate

If we refer to Figure 4.97 and, in particular, to Figure 4.97c we see that the shear force contribution given by $Q'_x = -\left.\frac{\partial M_{yx}}{\partial y}\right|_{x=a}$ is due to the balance of forces of two neighboring differential elements, not at a corner of the plate. That is, if we consider the differential element at the corner, such balance is no longer valid. In Figure 4.101b, we show the corner region. Here a concentrated reactive transverse force R is required for equilibrium

$$R = (M_{yx} - M_{xy})|_{x=a,y=b} = 2M_{yx}|_{x=a,y=b} = 2D(1-\nu)\left.\frac{\partial^2 w}{\partial x \partial y}\right|_{x=a,y=b}.$$

In Figure 4.101a these reactions are shown for both corners. We note that these reactions and also the shear forces are those compatible with $p(x,y) = -p_0$, *i.e.*, with a constant distributed load of intensity p_0 acting downwards. A physical interpretation of the concentrated corner reaction is that the simply supported plate subjected to $p(x,y) = -p_0$, unless held down, would lift up at its corners.

Now, considering the Reissner-Mindlin model, we need first to decide if we impose hard or soft conditions. If we admit that the kinematic restraints are such that the material lines orthogonal to the midsurface along the edge given by $x = a$ can not rotate around the x axis, *i.e.*, $\beta_y = 0$ (hard condition), then we would obtain a distribution of reactive twisting moments M_{yx} which is equivalent to that of Kirchhoff theory. Also, the Q_x would be that of Kirchhoff theory and there would be no reactions at the corner ($R = 0$).

If we impose soft conditions, $M_{yx}|_{x=a} = 0$, we obtain the distribution of shear forces Q_x shown in Figure 4.102. We can see that the distribution of shear forces near the center tends to that of Kirchhoff theory which includes the effect of the torsional moment. As we approach a corner, the Reissner-Mindlin shear force inverts sense. There is no concentrated force at the corner

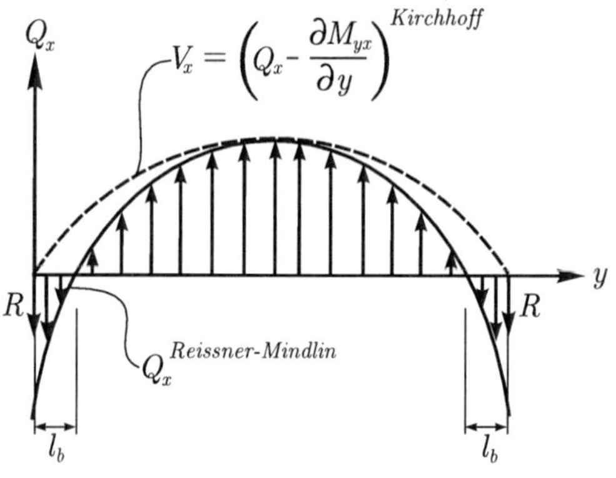

Fig. 4.102. Shear forces along simply supported edge for Kirchhoff and Reissner-Mindlin models

for the Reissner-Mindlin model and that shown in Figure 4.102 is from the Kirchhoff model.

Only from equilibrium considerations, we can conclude that V_x and R from the Kirchhoff model and Q_x from the Reissner-Mindlin model lead to the same resultant since they should equilibrate the same externally applied load. In other words, the corner concentrated reaction R of the Kirchhoff model is distributed along some distance from the corner as part of Q_x of the Reissner-Mindlin model. The magnitude of this distance depends on h/L. The situation shown in Figure 4.102 corresponds to a ratio h/L of the order of $1/10$. The quantity ℓ_b represents the length of the boundary layer. As the ratio h/L decreases ℓ_b also decreases and when h/L tends to zero, we approach the concentrated force situation predicted by the Kirchhoff model. We can clearly see, from the discussion above, that the Reissner-Mindlin model is a hierarchically higher-order model with respect to the Kirchhoff model regarding also the shear force predictions. Additional results regarding boundary layers in Reissner-Mindlin plates can be found in Häggblad and Bathe, 1990 and Arnold and Falk, 1990.

4.4 Shells

Shell structures comprise a very broad subject. On the one hand, there is a wealth of nature and man made shells. The reader may easily list a number of shell structures. Examples are many, covering a wide range of length scales. We could mention microscopic living cells, sea shells, egg shells, human skulls, biomedical devices, ship hulls, aircraft fuselages, car bodies, roofs, among many others. On the other hand, there is a vast literature on this subject ranging from shell structural behavior to the analysis and design of shells. Therefore, we first would like to mention our objectives in this section.

Our aim is to present an introductory discussion of shell structures focused on basic structural behavior. We would like to help the reader to acquire some elementary understanding of the issues in shell models and to gain insight into expected shell behavior. This knowledge is very valuable when modeling shells. Of course, we suppose that the shell model solutions will be obtained by using finite element methods. Therefore, no emphasis is given to obtaining analytical solutions to shell mathematical models. The very few solutions that will be presented are given to obtain some basic understanding.

A shell structure is geometrically characterized by a thin solid whose domain is defined by a curved midsurface and a thickness h. The shell is acted upon by surface tractions and body forces and is kinematically supported, usually along part or all of its periphery. A generic shell is schematically described in Figure 4.103.

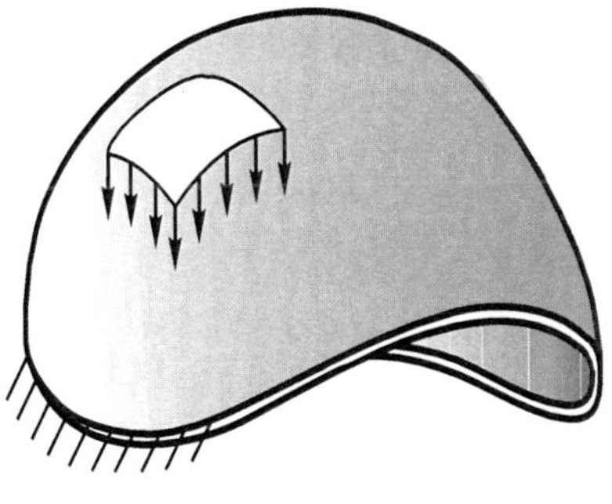

Fig. 4.103. A typical shell structure

The thickness h is supposed to be small. However, not too small that it would prevent the shell to sustain some level of compression and bending. In other words, we are not considering a membrane.

Below we discuss some basic geometrical facts required to introduce shell models. Next, we present an introductory discussion on shell mathematical models and on the formulation of the membrane-bending model. This formulation is then detailed for a class of problems – shells of revolution loaded axisymmetrically – and a few illustrative problems are solved.

4.4.1 Geometrical preliminaries

Let us consider a generic surface S. Let P be a point on the surface and $\mathbf{n}$ be the unit normal to S at P. Let π be a plane which contains $\mathbf{n}$. The intersection of this plane with the surface gives a curve C_π. These definitions are shown in Figure 4.104.

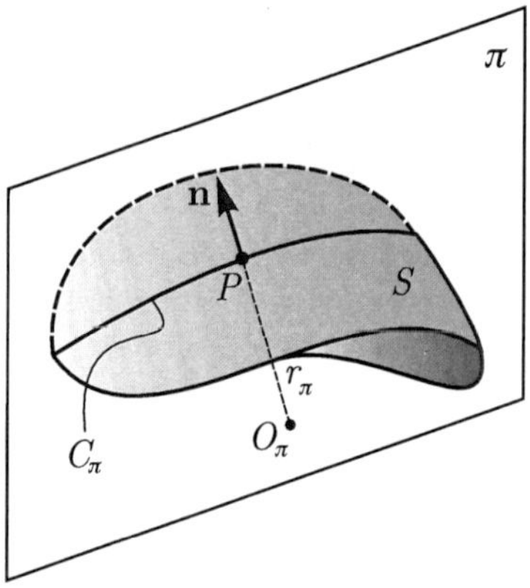

Fig. 4.104. Generic intersecting curve of surface S at point P

Considering the plane π, the curvature κ_π and the radius of curvature r_π at point P are well defined. We can also determine the center of curvature O_π for the curve C_π. Of course, there are infinitely many planes that contain the normal $\mathbf{n}$ and for each of these planes we can characterize the intersecting curve and the associated curvature definitions.

It is always possible to determine the maximum and minimum values for the curvatures and radii of curvatures at a point, κ_1, r_1 and κ_2, r_2 which are called the principal curvatures and principal radii of curvatures. The planes associated with these extreme values are orthogonal to each other (see Chapelle and Bathe, 2010*a*).

An important geometrical quantity is the Gaussian curvature κ_G defined at a point by

$$\kappa_G = \kappa_1 \kappa_2.$$

There are important geometric properties associated with the algebraic value of κ_G. When $\kappa_G > 0$ all the centers of curvatures of the intersecting curves are located on the same side of the tangent plane at that point, and the shell surface is called an *elliptic surface.* A dome like surface has this property for all points (see Figure 4.105).

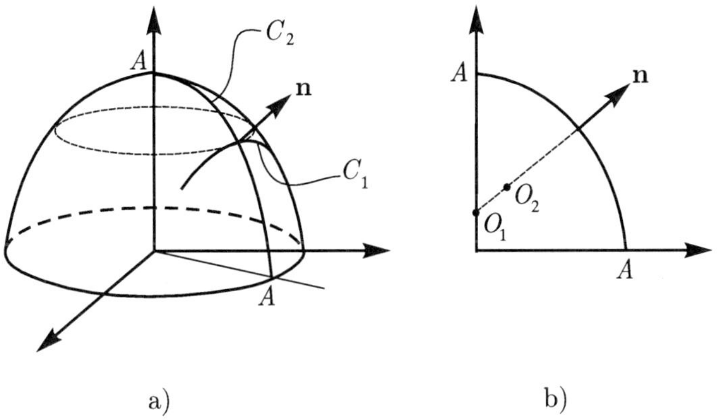

Fig. 4.105. Intersecting curves and centers of curvatures of the midsurface of a dome like structure; elliptic surface

When $\kappa_G < 0$ there are intersecting curves with centers of curvatures on opposite sides of the tangent plane at P, and the shell surface is called a *hyperbolic surface.* A saddle like surface is such an example as shown in Figure 4.106.

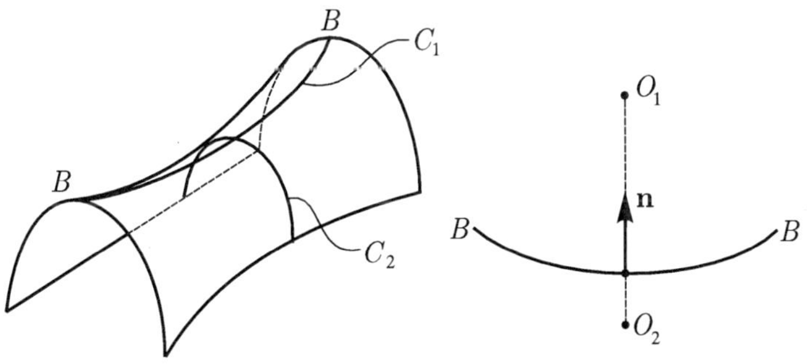

Fig. 4.106. Intersecting curves and centers of curvature for a saddle like surface; hyperbolic surface

Finally, $\kappa_G = 0$ corresponds to a surface with at least one curvature equal to zero (like for a cylinder or plate), and the shell surface is called a *parabolic surface.*

Therefore the algebraic value of κ_G helps to locally characterize a surface. Besides that, the algebraic value of κ_G of shell midsurfaces is one of the factors that greatly influences shell structural behavior.

4.4.2 Shell mathematical models

The particular choice of kinematic and mechanical hypotheses characterizes a given shell mathematical model. We adopt here the terminology used in Chapelle and Bathe, 2010*a* for the definitions of the mathematical models. There are two basic hypotheses which pertain to most shell mathematical models.

Kinematic hypothesis: Straight fibers initially orthogonal to the midsurface remain straight and unstretched during deformation. This kinematic assumption is called the Reissner-Mindlin kinematic assumption.

Mechanical hypothesis: The stress in the direction normal to the midsurface is zero.

The model characterized by these two assumptions is termed the *basic shell model.* While the displacement and rotation variables are referred to the shell midsurface, the strains and stresses of the *basic shell model* are given as for the 3-D continuum. When additional assumptions allow the analytical integration through the shell thickness, in a similar way as detailed for plates, the model is called the *shear-membrane-bending shell model* since these three behaviors can potentially arise.

When the kinematic hypothesis is stronger and it is further assumed that the straight lines initially orthogonal to the midsurface remain orthogonal to the midsurface after deformation, we have the *membrane-bending shell model* since transverse shear deformations are precluded. This kinematic hypothesis is known as the Kirchhoff-Love kinematic assumption, Love, 1934. We discuss below the membrane-bending shell model.

The membrane-bending shell model

Historically, this is an early shell model proposed; it is referred to as a classical shell model. There are many contributions associated with this model. Our presentation is closely based on the classical book of Timoshenko and Woinowsky-Krieger, 1959.

Stress resultants

Let us consider a generic point O on the midsurface of the shell. We choose a local Cartesian coordinate system such that z has the same direction as the normal at O and x and y are defined such that xz and yz are the planes associated with the principal curvatures at O. Let r_x and r_y be the principal

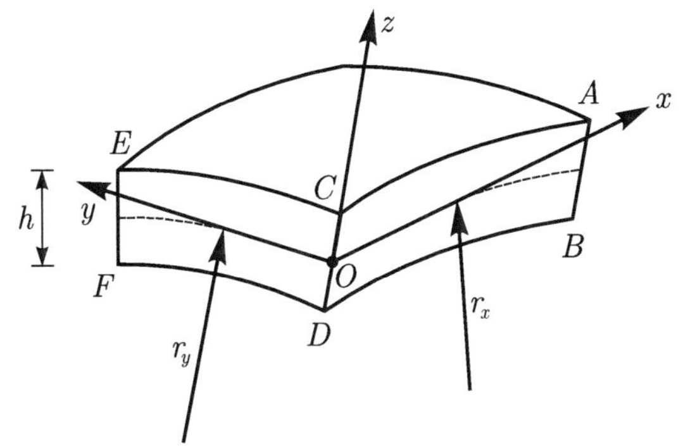

Fig. 4.107. A part extracted from the shell; r_x and r_y are radii of principal curvatures

radii of curvatures at O corresponding to the planes xz and yz respectively. In Figure 4.107, we show a differential element of the shell. The stress resultants per unit of length of coordinates along the midsurface acting on planes xz and yz are

$$N_x = \int_{-h/2}^{+h/2} \tau_{xx}\left(1+\frac{z}{r_y}\right) dz \tag{4.277}$$

$$N_y = \int_{-h/2}^{+h/2} \tau_{yy}\left(1+\frac{z}{r_x}\right) dz \tag{4.278}$$

$$N_{xy} = \int_{-h/2}^{+h/2} \tau_{xy}\left(1+\frac{z}{r_x}\right) dz \tag{4.279}$$

$$N_{yx} = \int_{-h/2}^{+h/2} \tau_{yx}\left(1+\frac{z}{r_y}\right) dz \tag{4.280}$$

$$Q_x = \int_{-h/2}^{+h/2} \tau_{zx}\left(1+\frac{z}{r_y}\right) dz \tag{4.281}$$

$$Q_y = \int_{-h/2}^{+h/2} \tau_{zy}\left(1+\frac{z}{r_x}\right) dz \tag{4.282}$$

$$M_x = \int_{-h/2}^{+h/2} \tau_{xx}(-z)\left(1+\frac{z}{r_y}\right) dz \tag{4.283}$$

$$M_y = \int_{-h/2}^{+h/2} \tau_{yy}(-z)\left(1+\frac{z}{r_x}\right) dz \tag{4.284}$$

$$M_{xy} = \int_{-h/2}^{+h/2} \tau_{xy} z \left(1+\frac{z}{r_x}\right) dz \tag{4.285}$$

$$M_{yx} = \int_{-h/2}^{+h/2} \tau_{yx}(-z)\left(1+\frac{z}{r_y}\right) dz. \tag{4.286}$$

In the resultant moment definitions the same sense conventions as for plates are used. Note that the stress resultants defined in equations (4.277) to (4.286) are the exact resultants associated with a given stress field. Since our presentation is restricted to thin shells, from now onwards we will neglect $\frac{z}{r_x}$ and $\frac{z}{r_y}$ with respect to 1, which implies obvious simplifications in the definitions of the stress resultants.

Kinematics

We adopt the Kirchhoff-Love kinematic assumption enunciated above and refer to Figure 4.83 where the deformation of a differential curved bar element was characterized. This same kinematics is used to describe the deformation of the shell sections $ABCD$ and $EFCD$ due to the analogous kinematic assumptions. In fact, Figure 4.83a can be used to characterize the undeformed configuration of either section $ABCD$ or $EFCD$ by considering instead of r, r_x or r_y respectively. The deformed configurations of these sections are given, in essence, by Figure 4.83b. Therefore, referring to equations (4.208) to (4.214), the shell strains are

$$\varepsilon_{xx} = \varepsilon_{xx_0} + z\left(\frac{1}{r'_x} - \frac{1}{r_x}\right) \tag{4.287}$$

$$\varepsilon_{yy} = \varepsilon_{yy_0} + z\left(\frac{1}{r'_y} - \frac{1}{r_y}\right) \tag{4.288}$$

where ε_{xx_0} and ε_{yy_0} are the strains at the midsurface. The approximations used for curved bars are employed here, *i.e.*, we are neglecting $\frac{z}{r_x}$, $\frac{z}{r_y}$, ε_{xx_0} and ε_{yy_0} with respect to 1. Note the complete analogy between (4.287), (4.288) and (4.214).

It is usual to define the changes of curvatures by

$$\chi_x = \frac{1}{r'_x} - \frac{1}{r_x}$$

$$\chi_y = \frac{1}{r'_y} - \frac{1}{r_y}.$$

Therefore equations (4.287) and (4.288) can be rewritten as

$$\varepsilon_{xx} = \varepsilon_{xx_0} + z\chi_x$$

$$\varepsilon_{yy} = \varepsilon_{yy_0} + z\chi_y.$$

In addition to these deformations of the shell, we still need to consider the deformations which induce shear strains in "planes" parallel to the midsurface. Consider that line AB rotates with respect to CD about the x axis. Let χ_{xy} give this rotation per unit of length. This quantity is actually the twist of the deformed midsurface and the induced shear strain is given by

$$\gamma_{xy} = \gamma_{xy_0} + 2z\chi_{xy}$$

where γ_{xy_0} is the shear strain at the midsurface.

Constitutive relations

Using the plane stress constitutive assumption we have

$$\tau_{xx} = \frac{E}{1-\nu^2}\left[\varepsilon_{xx_0} + \nu\varepsilon_{yy_0} + z\left(\chi_x + \nu\chi_y\right)\right] \tag{4.289}$$

$$\tau_{yy} = \frac{E}{1-\nu^2}\left[\varepsilon_{yy_0} + \nu\varepsilon_{xx_0} + z\left(\chi_y + \nu\chi_x\right)\right] \tag{4.290}$$

$$\tau_{xy} = \frac{E}{2\left(1+\nu\right)}\left[\gamma_{xy_0} + 2z\chi_{xy}\right]. \tag{4.291}$$

Substituting relations (4.289) to (4.291) into (4.277) to (4.286) and performing the integration through the thickness, we obtain

$$N_x = \frac{Eh}{1-\nu^2}\left(\varepsilon_{xx_0} + \nu\varepsilon_{yy_0}\right)$$

$$N_y = \frac{Eh}{1-\nu^2}\left(\varepsilon_{yy_0} + \nu\varepsilon_{xx_0}\right)$$

$$N_{xy} = N_{yx} = \frac{Eh\gamma_{xy_0}}{2\left(1+\nu\right)}$$

$$M_x = -D\left(\chi_x + \nu\chi_y\right) \tag{4.292}$$

$$M_y = -D\left(\chi_y + \nu\chi_x\right) \tag{4.293}$$

$$M_{xy} = -M_{yx} = D\left(1-\nu\right)\chi_{xy}$$

where D gives the flexural rigidity of the shell, and is given by (4.251), as for the plate.

If we were to follow the approach used to formulate the previous structural models, the next step would be to impose equilibrium to obtain the differential formulation of the model. Then, to solve problems, the specific shell midsurface geometry would need to be considered. We do not follow this approach due to the limited scope of our presentation. Instead, we present the formulation for a limited class of problems – shells of revolution loaded axisymmetrically.

Shell structures resist the external loads through membrane and bending internal actions. A special case is that of flat shells, that is, plates for which the transverse loads are resisted by bending while the in-plane loads are resisted by membrane internal actions. For curved midsurfaces this decoupling does not hold, see Section 4.2.7, and the loads are resisted by both membrane and bending actions. But, depending on the geometry of the midsurface, the boundary conditions and the loads, one of these internal actions – membrane or bending – may dominate.

Membrane actions dominate for example in shells of elliptic surfaces (for an example see Figure 4.105) when the loading can be resisted by membrane forces only and the restraints are applied all around the boundary and correspond to these internal membrane forces. In case the supports induce bending, these actions are local in the support regions. We discuss such a case below. However, if, for example, the elliptic shell is not supported all around the boundary and the external loading is inadmissible, then complex membrane and bending actions occur, see Bathe, Chapelle and Lee, 2003.

Considering hyperbolic surfaces (like the surface shown in Figure 4.106, which corresponds to half of a cooling tower), the boundary conditions and loading determine whether bending or membrane actions dominate. The same holds also for parabolic surfaces like cylinders. For a detailed discussion of the various cases that can arise, we refer to Chapelle and Bathe, 2010*a* and for numerical results see also Bathe and Lee, 2011, and Lee and Bathe, 2002.

Indeed, there are many different cases as to how a shell carries the externally applied loads – through bending, membrane or mixed stress state, varying over the shell surface, with possible boundary layers and internal layers of high stress gradients. Because of these various conditions that can arise, even just in linear analysis, the analysis of general shells considering linear and nonlinear behavior is a very challenging field in mechanics. We consider a relatively simple but practical and illustrative case of shell analyses below.

Cleary, from a structural design perspective resisting the loads primarily through membrane actions is most desirable and effective. Let us discuss one class of problems where this actually happens, namely, shells of revolution loaded axisymmetrically and supported on the whole boundary such that bending actions are small and mostly confined to the boundary only. This is a very special case of shell problems, but the discussion will illustrate some important general issues encountered in the analysis of shells. In our discussion we closely follow the work of Timoshenko and Woinowsky-Krieger, 1959.

4.4.3 Shells of revolution loaded axisymmetrically

In order to obtain insight into how transverse loads can be resisted by membrane actions in a shell of revolution loaded axisymmetrically consider a dome subjected to its own weight and to some surface loading as schematically described in Figure 4.108.

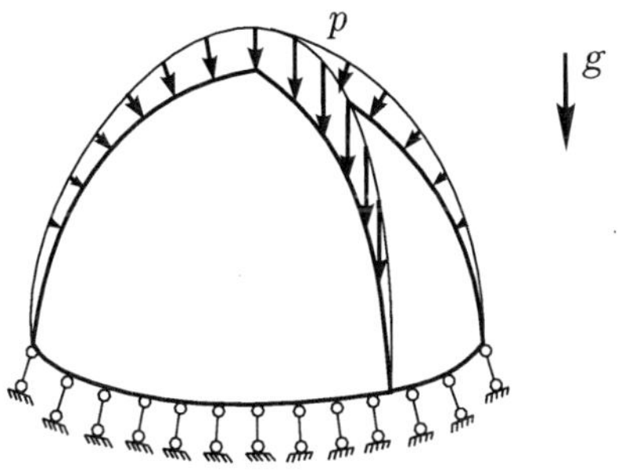

Fig. 4.108. A dome like structure

We compare the internal actions developed in an arch with those of a dome. In Figure 4.109a, we show a generic arch represented by its axis and in Figure 4.109c we describe a dome represented by its midsurface. This surface is obtained by revolving the curved axis of the arch around the vertical axis shown in Figure 4.109a.

The arch is acted on by the distributed load shown and since we assume that the arch axis does not correspond to the line of pressure for this load, bending will be developed in the arch as discussed in Section 4.2.7. The moment distribution is shown schematically in Figure 4.109b.

Consider now the distributed load acting on the shell, which is defined per unit of surface area, as schematically shown in Figure 4.109c. The spatial distribution of this load could be generated, for example, by revolving the load defined on the arch. We also show a force distribution at the lower shell boundary which equilibrates the applied load and whose evaluation is dealt with later on.

As we discuss in the sequel, as long as the applied distributed loading has a smooth variation, the shell "may" develop only internal forces tangential to its midsurface, that is, only membrane forces.

Let us examine two "slices" of the shell which are highlighted in Figure 4.109d. One is defined by the intersection of the shell with two meridian planes – planes which contain the vertical axis. The other is given by the intersection of the shell with two parallel planes – planes which are orthogonal to the vertical axis.

Assume now that the reactions on the periphery are such that only membrane forces develop in the shell, and consider the internal actions at the

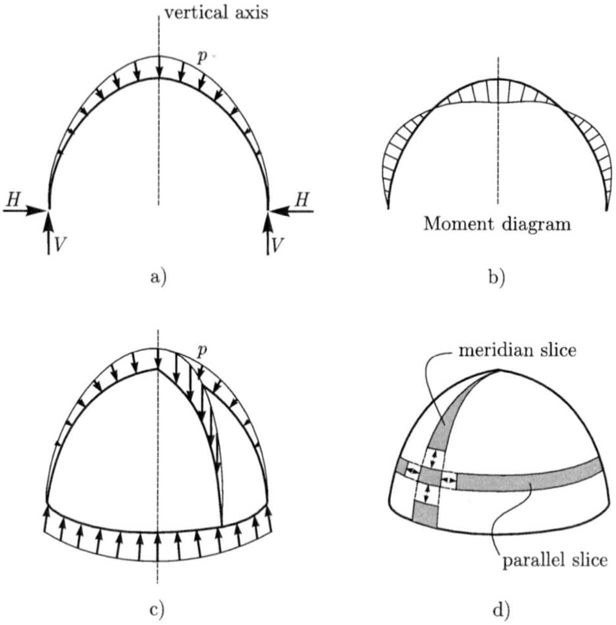

Fig. 4.109. Arch and shell actions

intersection of these two slices in Figure 4.109d. Then, an interesting interpretation for the shell resisting the load is as follows. The meridian slice behaves like an arch and the membrane forces on this slice due to the action of the parallel slices are such that the net loading – external loading superposed onto the membrane forces from the parallel slices – has as its line of pressure the line defined by the geometry of the meridian slice. Of course, under these conditions there will be no bending developed in the shell.

We can appreciate that a shell resisting loads through membrane forces only is a very efficient structure and such state is that one ideally sought by the structural designer. However, these conditions are very difficult to meet in practice and therefore some bending is generally induced even in this type of shell.

Membrane theory

We assume from the onset that there is no bending and torsion, *i.e.*,

$$M_x = M_y = M_{xy} = 0.$$

Therefore, we seek a field of membrane stress resultants, *i.e.*, N_x, N_y, N_{xy} which equilibrates the applied loading and which leads to deformations that are compatible with the kinematic boundary conditions.

A shell is called a shell of revolution when its midsurface is a surface of revolution, *i.e.*, it is obtained by revolving a planar curve around an axis, which lies in the plane of the curve, called the axis of revolution. The curve is called a meridian and this plane is referred to as a meridian plane. The load is axisymmetric when its distribution in all meridian planes is the same and it acts in these planes. We are considering surface tractions and body forces, but do not allow concentrated loads.

Under these geometrical and loading conditions the stress resultants and displacements also have an axial symmetry, *i.e.*, they are identical for each meridian plane. Further, if we cut the shell through a generic meridian plane there is no tendency of one part to slide with respect to the other and, hence, there are no shear stresses acting on any meridian plane.

Consider a differential element extracted from the shell as shown in Figure 4.110a and impose equilibrium. At the midsurface, the differential element

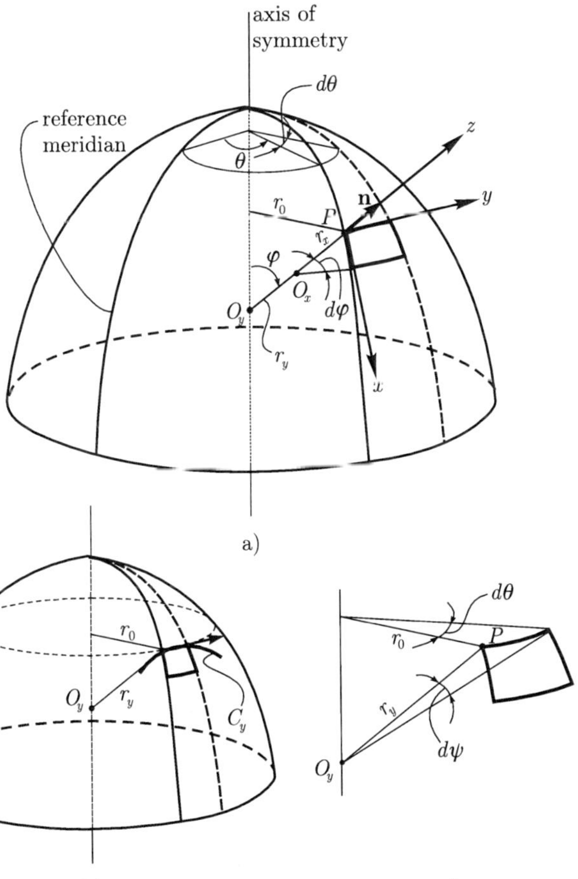

Fig. 4.110. Definitions for a shell of revolution

can be obtained by considering at point P a pair of meridian and a pair of parallel planes which are infinitesimally small distances apart.

It is possible to show that the meridian is always a principal curvature curve. We adopt a local coordinate system xyz with origin at P, z defined by the outward normal and x tangential to the meridian. The principal radius of curvature associated with the meridian is denoted by r_x with center O_x. In Figure 4.110b, the other principal normal section is shown. Its center of curvature O_y is located on the axis of symmetry. The parallel that passes through P is shown in Figure 4.110b and it is a geometrical fact that the parallel has the same unit tangent vector at P as the principal curve whose center is O_y. In Figure 4.110c a detail of 4.110a is shown which allows a better visualization of some quantities such as the differential angle increment $d\psi$ associated with the arc increment of the principal curve whose center is O_y. Also, r_0 is the radius of the parallel that contains the point P.

Let us first consider the equilibrium of the differential element in the z direction. Referring to Figures 4.111a to 4.111c, we can evaluate the contribution of the membrane forces. The contribution of N_x is given by

$$-\left(N_x + dN_x\right)\left(r_0 + dr_0\right) d\theta d\varphi$$

which, neglecting infinitesimals of higher-order, leads to

$$-N_x r_0 d\theta d\varphi. \qquad (4.294)$$

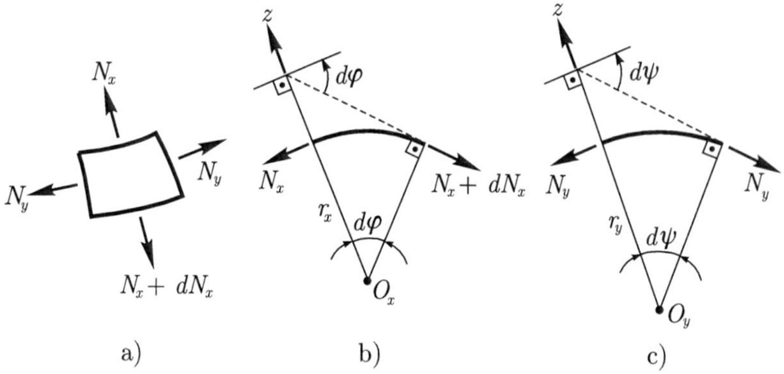

Fig. 4.111. Differential element and principal sections

We note that due to the axisymmetric nature of the solution, N_y is the same for every meridian plane. Its contribution is given by

$$-N_y r_x d\varphi d\psi. \qquad (4.295)$$

Next consider the contribution of the external load. Let p_z be the component of the surface load which is normal to the shell midsurface and taken positive when oriented as the z axis. Its contribution is given by

$$p_z r_x d\varphi r_y d\psi. \tag{4.296}$$

From Figure 4.110c, we can write the geometrical relation

$$r_0 d\theta = r_y d\psi. \tag{4.297}$$

Therefore using (4.294) to (4.297), equilibrium in the z direction leads to

$$N_x r_y d\psi d\varphi + N_y r_x d\varphi d\psi = p_z r_x r_y d\varphi d\psi$$

which yields

$$\frac{N_x}{r_x} + \frac{N_y}{r_y} = p_z. \tag{4.298}$$

We choose to obtain the additional equilibrium condition isolating a convenient part of the shell. This part is defined as the portion of the shell which lies above the parallel circumference defined by the angle φ as summarized in Figure 4.112, where the intersection of this part with a meridian plane is shown. Due to the axisymmetric nature of the external loading, a mechanically equivalent force system to the external loading acting on the selected part is given by its resultant R acting along the symmetry axis as shown in Figure 4.112. Equilibrium in this direction can be written as

$$N_x \sin\varphi \,(2\pi r_y \sin\varphi) = R$$

or

$$N_x = \frac{R}{2\pi r_y \,(\sin\varphi)^2}. \tag{4.299}$$

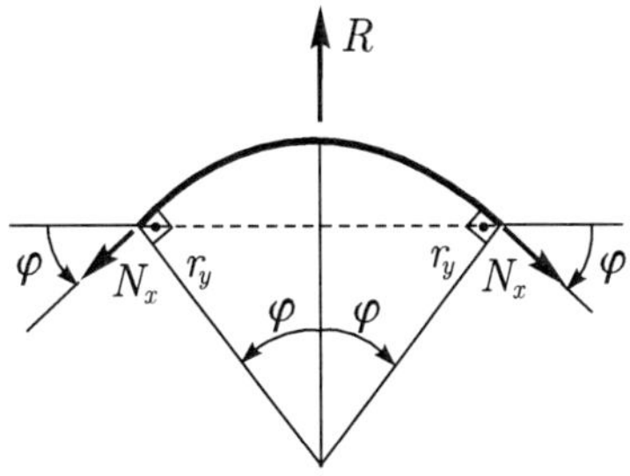

Fig. 4.112. Equilibrium of selected part of the shell

Equations (4.298) and (4.299) give the equilibrium conditions for the shell. Therefore, given the external load, these equations can be used to solve for the membrane forces. We demonstrate a typical solution in the following example.

Example 4.19[16]

Spherical domes are used extensively in engineering practice (*e.g.*, churches) and the major stresses developed are due to self-weight.

Consider the spherical dome in Figure 4.113a and find the membrane force distribution due to its own weight. Suppose that the required force distribution is applied at the periphery to guarantee equilibrium. The specific weight of the shell material is γ.

Find the membrane force distribution of the spherical dome characterized in Figure 4.113a subjected to its own weight. Suppose that the required force distribution is applied at the periphery to guarantee equilibrium. The specific weight of the shell material is γ.

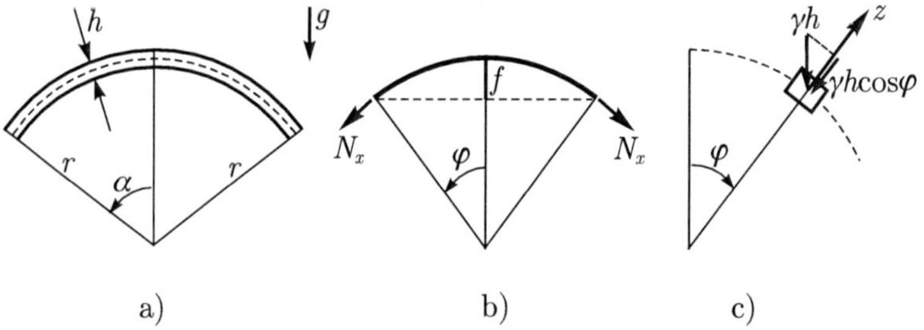

Fig. 4.113. a) Section of the dome; b) Selected part; c) Weight load of a differential element

Solution

We can evaluate the resultant force due to the dome weight for the part described in Figure 4.113b. The surface area of the midsurface of the spherical shell associated with this part is given by

$$S = 2\pi r f$$

and the resultant force, following the convention adopted in Figure 4.112, is given by

[16] This example is also presented in Timoshenko and Woinowsky-Krieger, 1959

$$R = -2\pi r f h\gamma = -2\pi r^2 \gamma h (1 - \cos\varphi)$$

and therefore from equation (4.299)

$$N_x = \frac{-2\pi r^2 \gamma h (1 - \cos\varphi)}{2\pi r (\sin\varphi)^2} = -\frac{\gamma r h}{1 + \cos\varphi}. \qquad (4.300)$$

Of course, for a spherical shell $r_x = r_y = r$. Referring to Figure 4.113c

$$p_z = -\gamma h \cos\varphi.$$

Considering equation (4.298), we obtain

$$\frac{-\gamma h}{1 + \cos\varphi} + \frac{N_y}{r} = -\gamma h \cos\varphi$$

and therefore

$$N_y = \gamma h r \frac{1 - \cos\varphi - \cos^2\varphi}{1 + \cos\varphi}. \qquad (4.301)$$

In Figure 4.114 we show the membrane force distributions obtained. We note

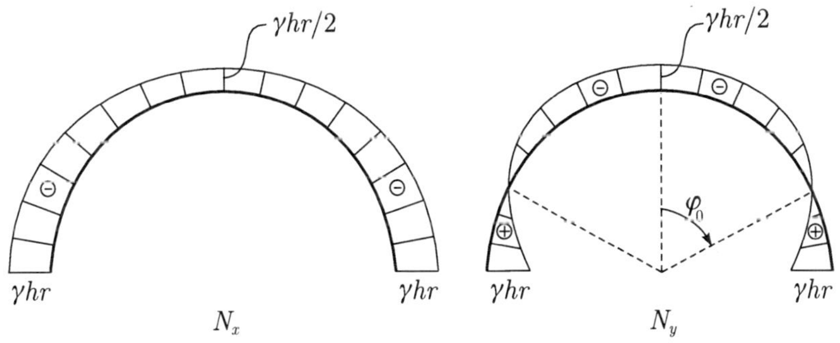

Fig. 4.114. Membrane forces for spherical shell problem

that N_x is always negative, *i.e.*, the meridian "slice" (refer to Figure 4.109d) works in compression. However, N_y changes from compression to tension when $\varphi = \varphi_0 \doteq 52°$. Therefore, the parallel "slices" defined by $\varphi < \varphi_0$ work in compression while those given by $\varphi > \varphi_0$ work in tension. Of course, when $\alpha < \varphi_0$ the whole shell is in compression.

□

We next consider the solution of the displacements. Again, due to the axisymmetric conditions, the displacements are completely characterized if they are known for a meridian plane.

In Figure 4.115, we show a meridian extracted from the midsurface of the shell described in Figure 4.110a. Let u be the displacement in the tangential direction x and w be that in the normal direction. The displacement increments associated with the increment $d\varphi$ are also shown in this figure. We note the complete analogy with the displacements of the axis of a curved bar which are described in Figure 4.84. Actually the axial strain given by equation (4.217) corresponds to the midsurface strain ε_{xx0} and, hence, it is given by

$$\varepsilon_{xx0} = \frac{1}{r_x}\frac{du}{d\varphi} + \frac{w}{r_x}. \tag{4.302}$$

The circumferential strain ε_{yy0} can be evaluated from the change in radius,

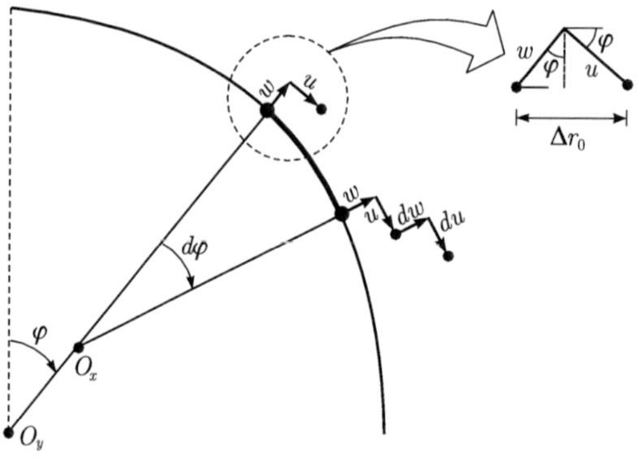

Fig. 4.115. Displacement conventions for a shell of revolution

Δr_0, of the parallel considering the undeformed and deformed configurations. This change in radius is given by

$$\Delta r_0 = u\cos\varphi + w\sin\varphi$$

and

$$\varepsilon_{yy0} = \frac{\Delta r_0}{r_0} = \frac{u\cos\varphi + w\sin\varphi}{r_y\sin\varphi}.$$

$$\varepsilon_{yy0} = \frac{u}{r_y}\cot\varphi + \frac{w}{r_y}. \tag{4.303}$$

We can eliminate w from equations (4.302) and (4.303) to arrive at

$$\frac{du}{d\varphi} - u\cot\varphi = r_x\varepsilon_{xx0} - r_y\varepsilon_{yy0}. \tag{4.304}$$

Using the plane stress constitutive relation (equation (4.44)) we have

$$\varepsilon_{xx_0} = \frac{1}{E}(\tau_{xx} - \nu\tau_{yy}) = \frac{1}{Eh}(N_x - \nu N_y) \tag{4.305}$$

$$\varepsilon_{yy_0} = \frac{1}{E}(\tau_{yy} - \nu\tau_{xx}) = \frac{1}{Eh}(N_y - \nu N_x)\,. \tag{4.306}$$

Substituting (4.305) and (4.306) into (4.304), we obtain

$$\frac{du}{d\varphi} - u\cot\varphi = \frac{1}{Eh}\left[N_x(r_x + \nu r_y) - N_y(r_y + \nu r_x)\right]. \tag{4.307}$$

We can denote the right-hand side of (4.307) by $f(\varphi)$, *i.e.*,

$$f(\varphi) = \frac{1}{Eh}\left[N_x(r_x + \nu r_y) - N_y(r_y + \nu r_x)\right] \tag{4.308}$$

which is a known function when we suppose that the membrane forces have been determined by equilibrium. Hence, the displacement u can be obtained by solving the following ordinary differential equation

$$\frac{du}{d\varphi} - u\cot\varphi = f(\varphi) \tag{4.309}$$

subject to appropriate kinematic boundary conditions, and from equation (4.303) we obtain

$$w = r_y\varepsilon_{yy_0} - u\cot\varphi. \tag{4.310}$$

Example 4.20

Consider the spherical dome subjected to its own weight studied in Example 4.19. Find the displacement fields when the shell is supported in the tangential direction as indicated in Figure 4.116.

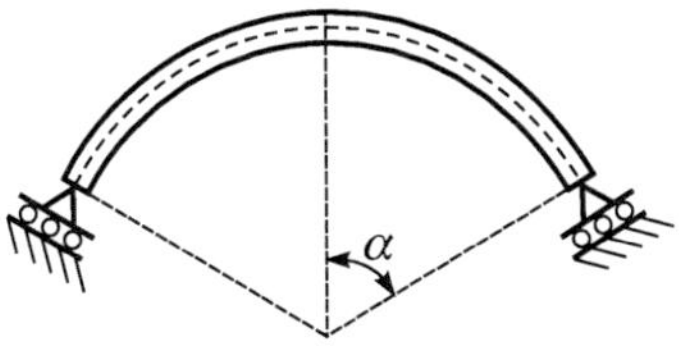

Fig. 4.116. Section of spherical dome with boundary conditions

Solution

Since the membrane forces have already been determined (see solution of Example 4.19), we can directly evaluate $f(\varphi)$ defined by equation (4.308) which leads to

$$f(\varphi) = \frac{(1+\nu)\,\gamma r^2}{E}\left(\cos\varphi - \frac{2}{1+\cos\varphi}\right) \tag{4.311}$$

and solving (4.309), we obtain

$$u = \frac{(1+\nu)\,\gamma r^2}{E}\left[\sin\varphi \ln(1+\cos\varphi) - \frac{\sin\varphi}{1+\cos\varphi}\right] + C\sin\varphi. \tag{4.312}$$

Imposing the boundary condition $u(\alpha) = 0$, the constant C can be determined and it is given by

$$C = \frac{(1+\nu)\,\gamma r^2}{E}\left[\frac{1}{1+\cos\alpha} - \ln(1+\cos\alpha)\right].$$

The displacement w is obtained from equation (4.310). Note that, as expected physically, the forces (expressed in (4.311)) and displacements u (given by (4.312)) and w (given by (4.310)) only vary with φ.

□

Using Examples 4.19 and 4.20, we have the solution of the spherical dome supported tangentially as described in Figure 4.116 within the membrane theory. However, if we change the supports such that the displacements are constrained in both directions, we can no longer obtain the solution using the membrane theory, since the solution obtained with this theory predicts a non-zero radial displacement at the supports. Of course, the transverse forces that would be developed at the supports associated with the restraint of the radial displacement w would induce bending and, hence, the membrane theory is no longer applicable since it neglects bending from the start. This observation prompts the discussion we present next.

Membrane-bending theory

In the membrane-bending model we need to consider, besides the membrane stress resultants N_x, N_y and N_{xy} which are already taken into account in the membrane theory, the moments M_x, M_y and M_{xy} and the transverse shear forces, Q_x and Q_y. Since in our simplified setting there are no shear stresses in any meridian plane the resultants to be considered are summarized in Figure 4.117a, which is a close up of the differential element of Figure 4.110. Equilibrium in the z direction is established as in the membrane theory but we need to include the shear force contribution. In Figure 4.117b, we repeat Figure 4.111b including the shear force. Therefore, equilibrium implies

$$-N_x d\varphi r_0 d\theta + (Q_x + dQ_x)(r_0 + dr_0)\, d\theta - Q_x r_0 d\theta$$

$$-N_y r_x d\varphi \sin\varphi d\theta + p_z r_0 d\theta r_x d\varphi = 0$$

which leads to

$$N_x r_0 + N_y r_x \sin\varphi - \frac{d}{d\varphi}(Q_x r_0) = p_z r_0 r_x. \tag{4.313}$$

The above equation, apart from the term $-\frac{d}{d\varphi}(Q_x r_0)$, is the same as equation

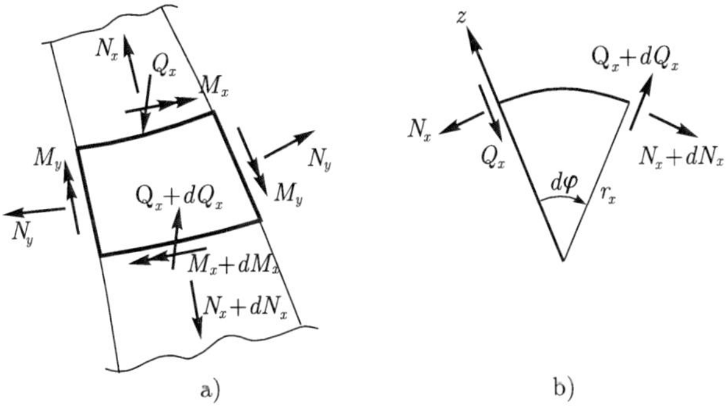

Fig. 4.117. a) Stress resultants acting in a differential element; b) Shear force appended to principal section defined by r_x

(4.298). Note, however, that the above equation was written in terms of r_0 instead of r_y because the additional term is more easily expressed considering r_0.

For the membrane theory, we did not consider the differential equilibrium in the x direction, since the second differential equilibrium condition was imposed in resultant form. Now, however, we need to consider the equilibrium in the x direction which referring to Figures 4.117 and 4.118 can be written as

$$-N_x r_0 d\theta + (N_x + dN_x)(r_0 + dr_0)\, d\theta - N_y r_x \cos\varphi d\varphi d\theta$$

$$+Q_x r_0 d\theta d\varphi + p_x r_0 r_x d\theta d\varphi = 0$$

where p_x is the x direction component of the surface load. Then, we obtain

$$\frac{d}{d\varphi}(N_x r_0) - N_y r_x \cos\varphi + Q_x r_0 = -p_x r_0 r_x \tag{4.314}$$

and referring to Figures 4.117 and 4.119 moment equilibrium about the y axis gives

$$-(M_x + dM_x)(r_0 + dr_0)\, d\theta + M_x r_0 d\theta + M_y r_x \cos\varphi d\varphi d\theta$$

$$-Q_x r_0 r_x d\varphi d\theta = 0$$

leading to

$$\frac{d}{d\varphi}(M_x r_0) - M_y r_x \cos\varphi + Q_x r_x r_0 = 0. \tag{4.315}$$

Therefore equations (4.313), (4.314) and (4.315) represent the equilibrium conditions for the differential element.

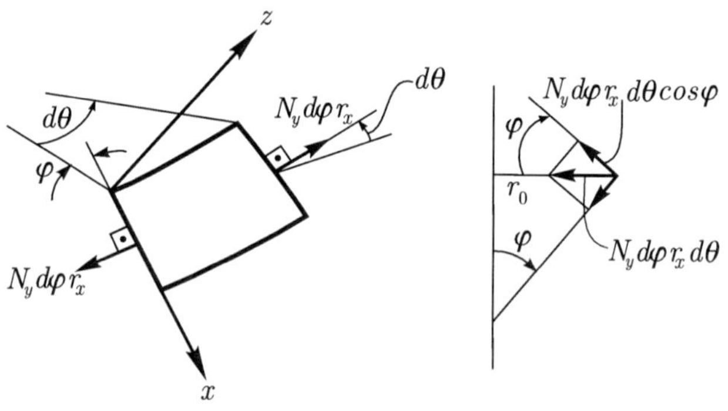

Fig. 4.118. Contribution of N_y to equilibrium in the x-direction, $N_y d\varphi r_x d\theta$ acts in the parallel plane and its $\cos\varphi$ component into the x-direction

The compatibility relations for the membrane strains have been already examined and are given by equations (4.302) and (4.303).

Using Hooke's law for the membrane components, we obtain

$$N_x = \frac{Eh}{1-\nu^2}\left[\frac{1}{r_x}\left(\frac{du}{d\varphi}+w\right)+\frac{\nu}{r_y}(u\cot\varphi+w)\right] \tag{4.316}$$

$$N_y = \frac{Eh}{1-\nu^2}\left[\frac{1}{r_y}(u\cot\varphi+w)+\frac{\nu}{r_x}\left(\frac{du}{d\varphi}+w\right)\right]. \tag{4.317}$$

The compatibility relations for the bending strains involve the changes in curvatures. Let us consider first the change in curvature of a meridian. We recognize that this change in curvature corresponds to the change in curvature of the axis of a curved bar. Adapting equation (4.221) to the shell quantities, we obtain

$$\chi_x = \frac{1}{r_x}\frac{d}{d\varphi}\left(\frac{u}{r_x}-\frac{dw}{r_x d\varphi}\right). \tag{4.318}$$

We note that in this evaluation of the change of curvature, the term $\left(\frac{u}{r_x}-\frac{dw}{r_x d\varphi}\right)$ corresponds to the angular change of the unit normal from the

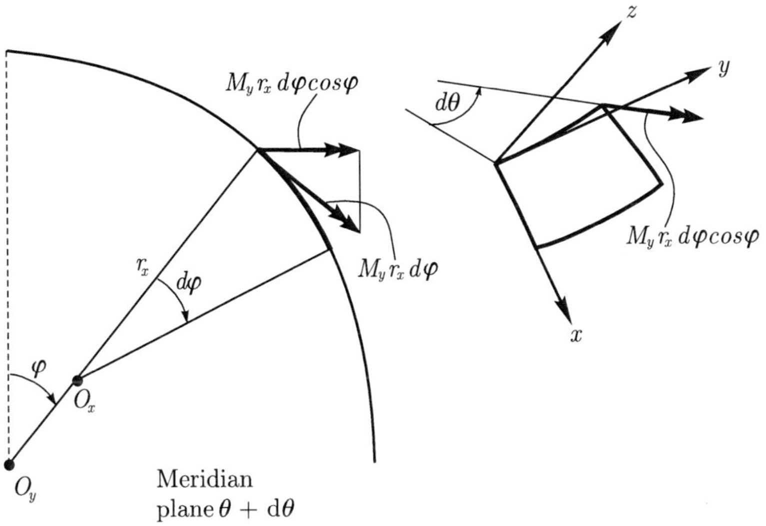

Fig. 4.119. Contribution of M_y to the moment equilibrium about the y axis

undeformed to deformed configuration. The rate of change in the x-direction gives χ_x. This same concept will be used for the evaluation of the change in curvature in the y direction, *i.e.*, we will compute the change in angle per unit of arc length along y.

Namely, referring to Figure 4.120, consider $\left(\frac{u}{r_x} - \frac{dw}{r_x d\varphi}\right)$ at point R; it corresponds to the change in angle of an infinitesimal meridian arc. Of course, due to the symmetry of the deformation, both, this normal and the arc remain in the meridian plane. Therefore, the rotation vector which characterizes this rotation is orthogonal to the meridian plane at R as shown in Figure 4.121. This rotation vector has a projection on the x axis which is different from zero. In Figure 4.121 all quantities used to evaluate this projection are shown and we obtain

$$-\left(\frac{u}{r_x} - \frac{dw}{r_x d\varphi}\right) \cos\varphi d\theta$$

which gives the change in angle of the normal around the x axis when we move from point P to R. Therefore, we have

$$\chi_y = \left(\frac{u}{r_x} - \frac{dw}{r_x d\varphi}\right) \frac{\cos\varphi d\theta}{r_0 d\theta} \tag{4.319}$$

leading to

$$\chi_y = \left(\frac{u}{r_x} - \frac{dw}{r_x d\varphi}\right) \frac{\cos\varphi}{r_0} = \left(\frac{u}{r_x} - \frac{dw}{r_x d\varphi}\right) \frac{\cot\varphi}{r_y}. \tag{4.320}$$

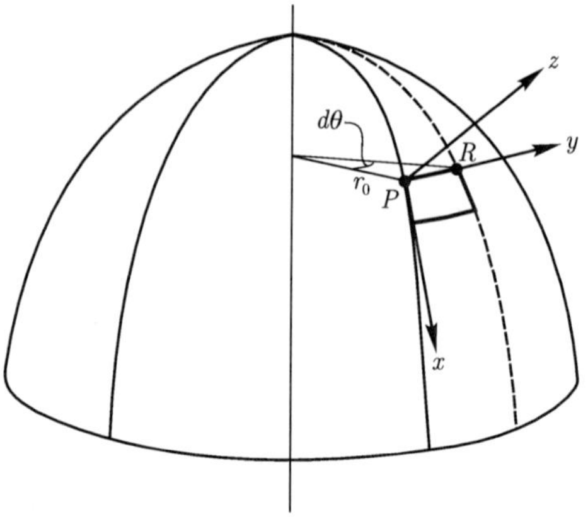

Fig. 4.120. Definition of points P and R for axisymmetrical shell

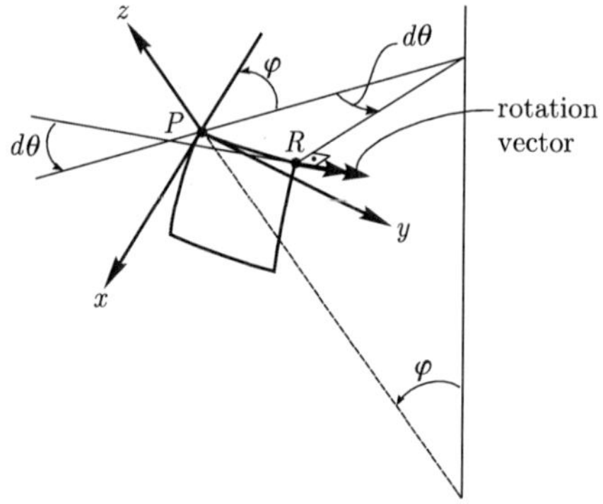

Fig. 4.121. Rotation vector at point R

Finally the bending moments can be evaluated using relation (4.292) and (4.293)

$$M_x = -D\left[\frac{1}{r_x}\frac{d}{d\varphi}\left(\frac{u}{r_x} - \frac{dw}{r_x d\varphi}\right) + \frac{\nu}{r_y}\left(\frac{u}{r_x} - \frac{dw}{r_x d\varphi}\right)\cot\varphi\right] \quad (4.321)$$

$$M_y = -D\left[\left(\frac{u}{r_x} - \frac{dw}{r_x d\varphi}\right)\frac{\cot\varphi}{r_y} + \frac{\nu}{r_x}\frac{d}{d\varphi}\left(\frac{u}{r_x} - \frac{dw}{r_x d\varphi}\right)\right]. \quad (4.322)$$

Now all conditions have been considered. Namely, equilibrium (equations (4.313), (4.314) and (4.315)), compatibility (equations (4.302), (4.303), (4.319) and (4.320)) and the constitutive (equations (4.316), (4.317), (4.321) and (4.322)). Therefore, these equations appended by appropriate boundary conditions represent the complete formulation.

We note that the substitution of equations (4.302), (4.303), (4.319), (4.320), (4.316), (4.317), (4.321) and (4.322) into (4.313), (4.314) and (4.315) reduces the problem to three equations in the variables u, w and Q_x. The solution of these equations has been addressed in many references, for example, in the classical work of Timoshenko and Woinowsky-Krieger, 1959. Of course, our objective is not to discuss these solutions. Actually, we presented the formulations only to gain some insight into the variables and into the resisting mechanisms of a shell in bending.

Membrane versus bending

We next want to mention some issues related to the accuracy of the membrane theory when compared to the bending theory which is, of course, a hierarchically higher-order theory. This discussion is also based on the book of Timoshenko and Woinowsky-Krieger, 1959.

Consider the membrane theory solution of the spherical shell subjected to its own weight for the displacements u and w (Example 4.20). These displacements can be substituted into equations (4.321) and (4.322) to obtain an estimate of the bending moments which are, of course, neglected in the membrane theory. These bending moments are given by

$$M_x = M_y = \frac{\gamma h^2}{12} \frac{2+\nu}{1-\nu} \cos\varphi. \tag{4.323}$$

If we use these moments to evaluate the ratio between the bending stresses and the compressive membrane stresses predicted by the membrane theory and compute the maximum value, we arrive at $3.29h/r$ for $\nu = 0.3$. Therefore, since for thin shells the ratio of the thickness h to the radius of curvature r is small, these bending stresses are negligible.

An improved estimate for the membrane forces N_x and N_y can be obtained if the moment estimates given in (4.323) are substituted into the equilibrium equations ((4.313), (4.314) and (4.315)). A comparison of these membrane forces with those of (4.300) and (4.301) shows that they differ by quantities that are proportional to $\frac{h^2}{r^2}$, a difference which becomes also negligible for thin shells. Therefore, as long as the boundary conditions are compatible with the membrane theory and the shell is thin the above discussion indicates that the membrane theory provides good predictions.

The effect on the solution of violating the boundary conditions which are compatible with the membrane theory can also be assessed.

Consider the spherical shell described in Figure 4.122. We report solutions for the problems summarized in Figures 4.123a and 4.123b. These solutions are based on further approximations of the differential formulation presented for the bending problem, which are increasingly appropriate as the shell becomes thinner.

These approximate solutions may be obtained as shown in Timoshenko and Woinowsky-Krieger, 1959 and are, for the problem of Figure 4.123a,

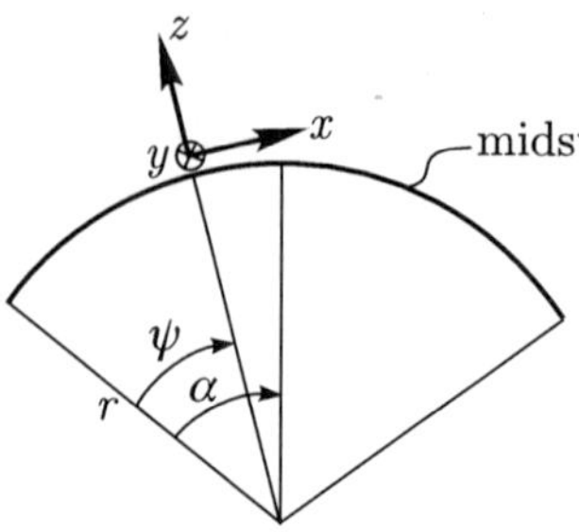

Fig. 4.122. Generic section of a spherical shell

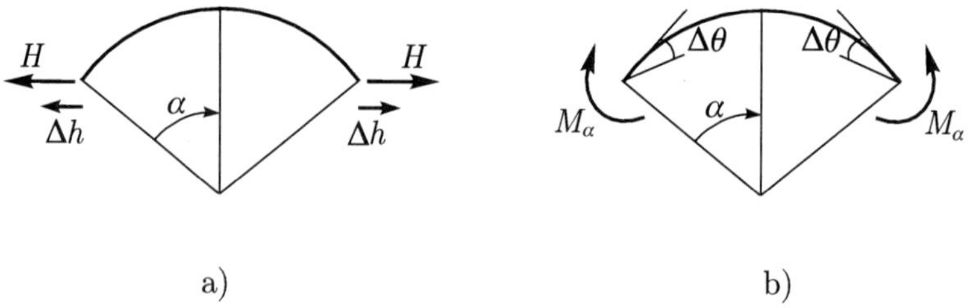

Fig. 4.123. End conditions for spherical shell problem: Δh is the displacement induced by H and $\Delta\theta$ is the rotation induced by M_α

$$
\begin{aligned}
N_x &= -\sqrt{2}\cot(\alpha-\psi)\sin\alpha e^{-\lambda\psi}\sin\left(\lambda\psi-\frac{\pi}{4}\right)H \\
N_y &= -2\lambda\sin\alpha e^{-\lambda\psi}\sin\left(\lambda\psi-\frac{\pi}{2}\right)H \\
M_x &= \frac{r}{\lambda}\sin\alpha e^{-\lambda\psi}\sin(\lambda\psi)H \\
\Delta h &= \frac{2r\lambda\sin^2\alpha}{Eh}H \\
\Delta\theta &= \frac{2\lambda^2\sin\alpha}{Eh}H
\end{aligned}
$$

and for the problem of Figure 4.123b

$$
\begin{aligned}
N_x &= -\frac{2\lambda}{r}\cot(\alpha-\psi)e^{-\lambda\psi}\sin(\lambda\psi)M_\alpha \\
N_y &= -\frac{2\sqrt{2}\lambda^2}{r}e^{-\lambda\psi}\sin\left(\lambda\psi-\frac{\pi}{4}\right)M_\alpha \\
M_x &= \sqrt{2}e^{-\lambda\psi}\sin\left(\lambda\psi+\frac{\pi}{4}\right)M_\alpha
\end{aligned}
$$

$$\Delta h = \frac{2\lambda^2 \sin\alpha}{Eh} M_\alpha$$

$$\Delta\theta = \frac{4\lambda^3 M_\alpha}{Erh}$$

where

$$\lambda^4 = 3\left(1-\nu^2\right)\left(\frac{r}{h}\right)^2. \tag{4.324}$$

Examining the functional form of these solutions, we note that the membrane forces and the bending moments damp out as we move away from the edge due to the factor $e^{-\lambda\psi}$. Of course, as the shell becomes thinner, λ becomes larger and the significant part of these solutions is closer to the boundary.

These solutions can be used to consider restraints at the boundary which are different from those compatible with the membrane theory in a similar way as solving statically indeterminate structures (refer, for example, to the discussion of Section 4.2.3 and to the example given below).

Example 4.21

Consider the spherical dome subjected to its own weight as described in Example 4.19 now clamped at the periphery. Obtain the solution for the bending moment M_x and the membrane force N_x using the membrane theory solution superimposed to the solutions for the problems given in Figure 4.123. To obtain numerical values use $r = 20$ m, $h = 0.2$ m, $\alpha = 90°$, $E = 2.1 \times 10^7$ kN/m^2, $\nu = 0.3$, $\gamma = 2.4g$ kN/m^3 where $g = 9.81$ m/s^2.

Solution

For the clamped condition both Δh and $\Delta\theta$ (refer to Figure 4.123 for the definitions of Δh and $\Delta\theta$) should be zero. Therefore

$$(\Delta h)_m + (\Delta h)_{horiz} + (\Delta h)_{moment} = 0 \tag{4.325}$$

$$(\Delta\theta)_m + (\Delta\theta)_{horiz} + (\Delta\theta)_{moment} = 0 \tag{4.326}$$

where the subscript m refer to the membrane solution, "*horiz*" to the contribution of the horizontal force H given per unit of length and "*moment*" to the contribution of the moment M_α also given per unit of length. Then

$$(\Delta h)_{horiz} = \frac{2r\lambda\sin^2\alpha}{Eh} H = F_{11} H$$

$$(\Delta h)_{moment} = \frac{2\lambda^2\sin\alpha}{Eh} M_\alpha = F_{12} M_\alpha$$

$$(\Delta\theta)_{horiz} = \frac{2\lambda^2 \sin\alpha}{Eh} H = F_{21}H = F_{12}H$$

$$(\Delta\theta)_{moment} = \frac{4\lambda^3}{Erh} M_\alpha = F_{22}M_\alpha$$

where λ is given in (4.324).

The values of H and M_α come from the solutions of (4.325) and (4.326) which can be re-written as

$$F_{11}H + F_{12}M_\alpha = -(\Delta h)_m \tag{4.327}$$

$$F_{21}H + F_{22}M_\alpha = -(\Delta\theta)_m. \tag{4.328}$$

Since the flexibility coefficients F_{11}, F_{12}, F_{22} are given above, we only need to determine $(\Delta h)_m$ and $(\Delta\theta)_m$. Referring to (4.310), we can write

$$(\Delta h)_m = r_0 \varepsilon_{yyo} = r_y \sin\alpha \varepsilon_{yyo}$$

since $u = 0$ and using (4.306), we obtain

$$(\Delta h)_m = \frac{r_y \sin\alpha}{Eh}(N_y - \nu N_x). \tag{4.329}$$

Introducing the numerical values we obtain

$$(\Delta h)_m = 5.830 \times 10^{-4} \text{ m}.$$

The change of angle $(\Delta\theta)_m$ can be evaluated as the change of the normal at the section given by α. Referring to equation (4.318) and considering the convention defined in Figure 4.123b, we obtain

$$(\Delta\theta)_m = -\frac{u}{r_x} + \frac{1}{r_x}\frac{dw}{d\psi}.$$

Considering that at the section given by α, $u = 0$ and using equation (4.310), we obtain

$$(\Delta\theta)_m = \frac{1}{r_x}\left(\frac{d}{d\psi}(r_y \varepsilon_{yyo}) - \frac{d}{d\psi}(u \cot\psi)\right)$$

leading to

$$(\Delta\theta)_m = \frac{1}{r_x}\left(\frac{d}{d\psi}\left(\frac{(\Delta h)_m}{\sin\psi}\right) - \frac{du}{d\psi}\cot\psi\right).$$

Using equation (4.307) we finally arrive at

$$(\Delta\theta)_m = \frac{1}{r_x}\left\{\frac{d}{d\psi}\left(\frac{(\Delta h)_m}{\sin\psi}\right) - \frac{\cot\psi}{Eh}\left[N_x(r_x + \nu r_y) - N_y(r_y + \nu r_x)\right]\right\}.$$

Since $(\Delta h)_m$ is given in (4.329), we obtain

$$(\Delta\theta)_m = \frac{r\gamma}{E}(2+\nu)\sin\alpha = 5.1572 \times 10^{-5} \text{ rad.}$$

The solution of (4.327) and (4.328) gives

$$H = -8.869 \text{ kN/m} \quad \text{and} \quad M_\alpha = 6.39 \text{ kN.}$$

If we plot N_x corresponding to the membrane solution and the value including the edge effects we would barely be able to distinguish both curves.

We obtain for M_x, which is identically zero in the membrane solution, the values shown in Figure 4.124. Note that the resulting moment has high

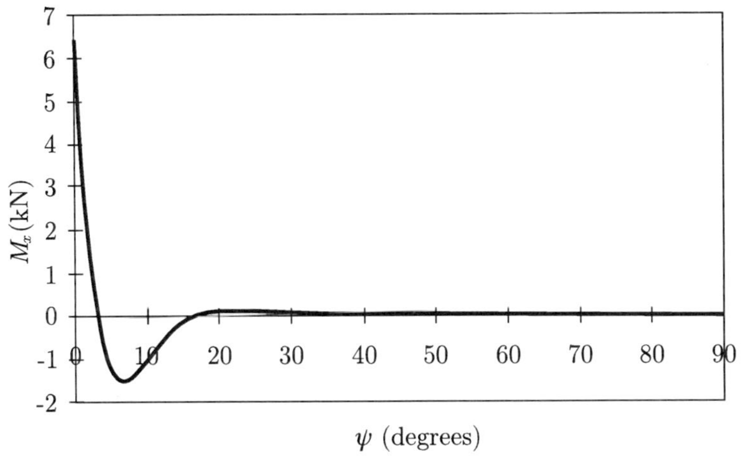

Fig. 4.124. Bending moment M_x for clamped shell

gradients near the edge but it is close to zero in the rest of the domain.

□

4.4.4 Remarks on shell modeling of engineering structures

In this introductory presentation on shells, the emphasis was placed on basic facts and behaviors. The formulation of the membrane-bending model was discussed for a very limited class of problems; namely, axisymmetric shells loaded axisymmetrically. Nevertheless, insight was gained into the relevant variables of the model, resisting mechanisms, membrane and bending behaviors, edge effects, among others.

As mentioned, the presentation was structured having in mind that the reader who is faced with the modeling of a shell structure will most probably be solving the shell models through finite element analysis. Therefore, the most important objective of the presentation was to focus on some basic understanding of shell structural behavior.

Of course, the formulations and results given above should also help the reader to understand and use other formulations, considering more general shell geometries, loadings and boundary conditions.

If a shell finite element analysis is undertaken – and not considering yet the task of making sure that the finite element solution is close enough to the solution of the mathematical model, a task which will be dealt with later on and which is very challenging for shells – the analyst needs to be aware that the response of shell structures is in general very sensitive to the shell geometry, loading and boundary conditions. Also, the behavior is not always easy to anticipate since the propagation of edge and perturbation effects may vary significantly depending on the conditions mentioned above, for detailed discussions see Chapelle and Bathe, 2010*a*. Hence, the hierarchical modeling concepts are clearly very valuable in shell analyses – as indeed already illustrated by the relatively simple shell solutions given above.

4.5 Summary of the mathematical models for structural mechanics

We end this long chapter with a summary of the mathematical models discussed. The objective is to synthesize the main aspects of each model for future reference. We list the basic model assumptions as well as the main variables, organized as kinematic, strain and stress type variables.

For all models, we define: the displacement, or generalized displacement, as the column matrix $\mathbf{u}$ which collects all independent kinematic variables of the model; the strain, or generalized strain, as the column matrix ε which collects the strain, or strain type, variables and the stress, or generalized stress, as the column matrix τ which collects the stress variables such as stresses and stress resultants.

Of course, as detailed in the presentation of each model, these stress type variables are used to enforce the equilibrium conditions. Also, the constitutive relations allow us to write

$$\tau = \mathbf{C}\varepsilon \tag{4.330}$$

and the strain compatibility is given by

$$\varepsilon = \partial_\varepsilon \mathbf{u}. \tag{4.331}$$

where we assume, as pointed out in Section 3.2.2, continuous displacements satisfying the displacement boundary conditions.

The specific forms of $\mathbf{C}$ and ∂_ε for the models studied are presented in Tables 4.3 and 4.4.

Plane strain model

Hypotheses

- Solid is prismatic
- Displacement assumptions

$u = u(x, y),\ v = v(x, y)$

$w = 0$

- Restrictions on loading

Volume

$f_x^B = f_x^B(x, y)$

$f_y^B = f_y^B(x, y)$

$f_z^B = 0$

Lateral surface

$f_x^S = f_x^S(x, y)$

$f_y^S = f_y^S(x, y)$

$f_z^S = 0$

Top and bottom surfaces

$f_x^S = 0,\ f_y^S = 0$

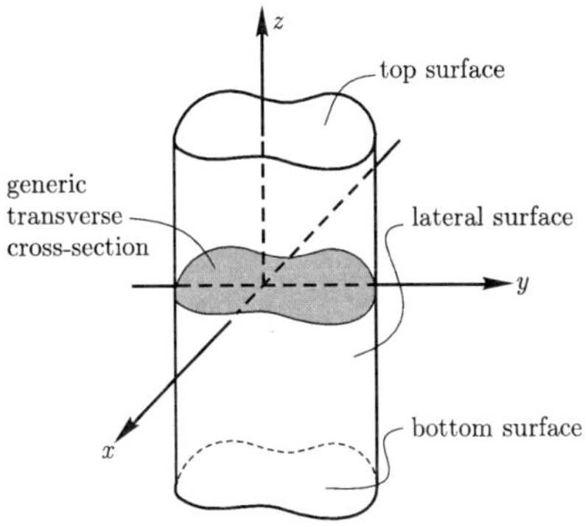

Primary variables

Displacements

$$\mathbf{u} = \begin{bmatrix} u(x, y) \\ v(x, y) \end{bmatrix}$$

Stresses

$$\tau = \begin{bmatrix} \tau_{xx} \\ \tau_{yy} \\ \tau_{xy} \end{bmatrix}$$

Strains

$$\varepsilon = \begin{bmatrix} \varepsilon_{xx} \\ \varepsilon_{yy} \\ \gamma_{xy} \end{bmatrix}$$

Remarks

Formulation is cast in the 2-D domain

Upon solution of $\mathbf{u}$, τ, ε, we can evaluate $\tau_{zz} = \nu(\tau_{xx} + \tau_{yy})$

Remaining variables are zero: $w = 0,\ \varepsilon_{zz} = \gamma_{xz} = \gamma_{yz} = 0,$

$$\tau_{xz} = \tau_{yz} = 0$$

Solution is exactly the same as for the 3-D model when restrictions are satisfied

Plane stress model

Hypotheses

- Solid is a plate
- Stress assumptions

$$\tau_{zz} = \tau_{xz} = \tau_{yz} = 0$$
$$\tau_{xx} = \tau_{xx}(x, y)$$
$$\tau_{yy} = \tau_{yy}(x, y)$$
$$\tau_{xy} = \tau_{xy}(x, y)$$

- Restrictions on loading

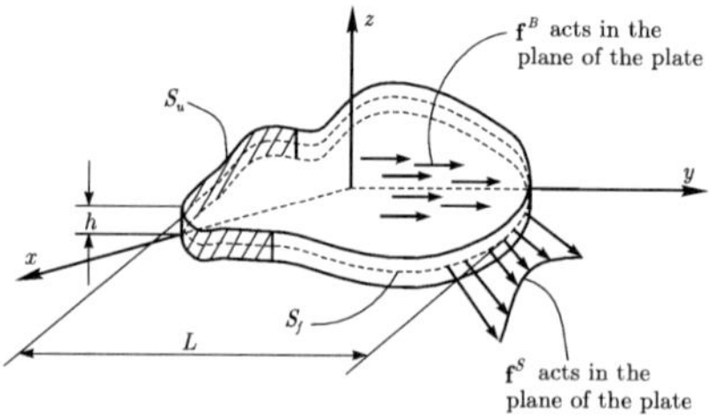

Volume	Lateral surface
$f_x^B = f_x^B(x, y)$	$f_x^S = f_x^S(x, y)$
$f_y^B = f_y^B(x, y)$	$f_y^S = f_y^S(x, y)$
$f_z^B = 0$	$f_z^S = 0$

Top and bottom surfaces

$f_x^S = 0,\ f_y^S = 0,\ f_z^S = 0$

Primary variables

Displacements	Stresses	Strains
$\mathbf{u} = \begin{bmatrix} u(x,y) \\ v(x,y) \end{bmatrix}$	$\tau = \begin{bmatrix} \tau_{xx} \\ \tau_{yy} \\ \tau_{xy} \end{bmatrix}$	$\varepsilon = \begin{bmatrix} \varepsilon_{xx} \\ \varepsilon_{yy} \\ \gamma_{xy} \end{bmatrix}$

Remarks

Formulation is cast in the 2-D domain

Upon solution of $\mathbf{u}$, τ, ε, we can evaluate: $\varepsilon_{zz} = -\frac{\nu}{E}(\tau_{xx} + \tau_{yy})$,

w by integration of ε_{zz}

Remaining variables are zero: $\tau_{zz} = \tau_{xz} = \tau_{yz} = 0$,

$\gamma_{xz} = \gamma_{yz} = 0$

Solution is only approximate compared with the solution of the 3-D model; however, it is close to the solution of the 3-D model as long as the restrictions are satisfied and the plate is thin.

Additional solution terms of the 3-D exact solution are proportional to z^2.

Axisymmetric model

Hypotheses

- Solid is of revolution
- Loading is axisymmetric

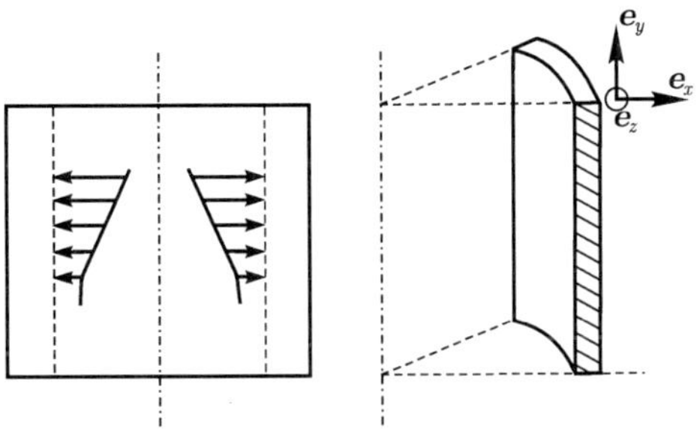

These hypotheses lead to

$$w = 0$$

$$\tau_{xz} = \tau_{yz} = 0$$

$$\gamma_{xz} = \gamma_{yz} = 0$$

$$\varepsilon_{zz} = \frac{u}{x}$$

Primary variables

Displacements Stresses Strains

$$\mathbf{u} = \begin{bmatrix} u(x,y) \\ v(x,y) \end{bmatrix} \quad \tau = \begin{bmatrix} \tau_{xx} \\ \tau_{yy} \\ \tau_{xy} \\ \tau_{zz} \end{bmatrix} \quad \varepsilon = \begin{bmatrix} \varepsilon_{xx} \\ \varepsilon_{yy} \\ \gamma_{xy} \\ \varepsilon_{zz} \end{bmatrix}$$

Remarks

Formulation is cast in the 2-D domain

The solution for $\mathbf{u}$, τ, ε appended by $w = 0$, $\tau_{xz} = \tau_{yz} = 0$, $\gamma_{xz} = \gamma_{yz} = 0$ is exactly the same as the solution of the 3-D model provided restrictions are satisfied

Bernoulli-Euler beam model

Hypotheses

- Solid is a straight bar
- Bar sections remain plane and orthogonal to the deformed axis
- Kinematics is fully described by:

  $u(x)$ displacement of the axis in the x direction

  $w(x)$ displacement of the axis in the z direction
- Loading

  $p(x)$ transverse distributed load

  $f(x)$ axial distributed load

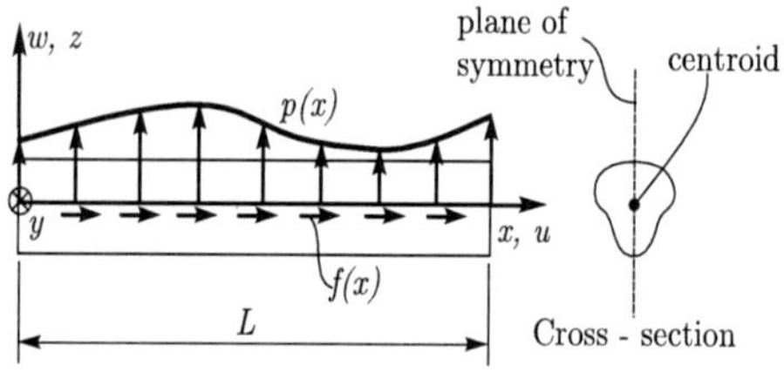

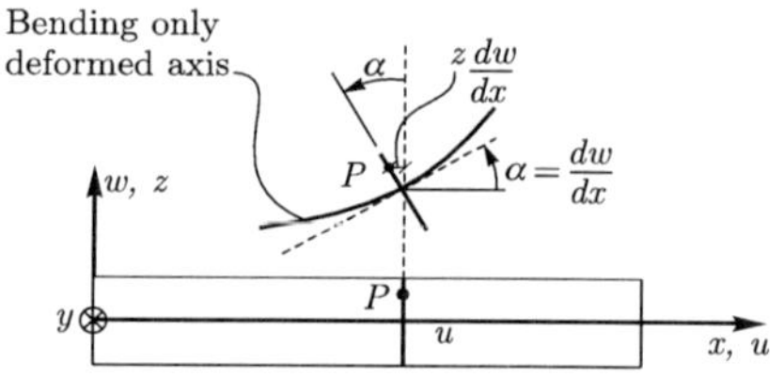

Primary variables

Displacements	Generalized stresses	Generalized strains
$\mathbf{u} = \begin{bmatrix} u(x) \\ w(x) \end{bmatrix}$	$\tau = \begin{bmatrix} N(x) \\ M(x) \end{bmatrix}$	$\varepsilon = \begin{bmatrix} \varepsilon \\ \kappa \end{bmatrix}$

Remarks

Formulation is cast in the 1-D domain

There is no transverse shear strain, *i.e.*, $\gamma_{xz} = 0$

Note that κ gives the change in curvature χ since the initial curvature is zero (bar is straight)

Curved bar model

Hypotheses

- Solid is a curved bar
- Bar sections remain plane and orthogonal to the deformed axis
- Kinematics is fully described by:

  $u(s)$ displacement of the axis in the tangential x direction

  $w(s)$ displacement of the axis in the normal z direction
- Loading

  $p_x(s)$ tangential distributed load

  $p_z(s)$ normal distributed load

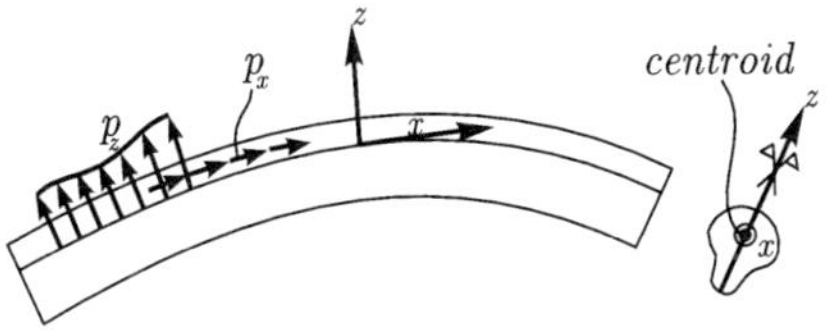

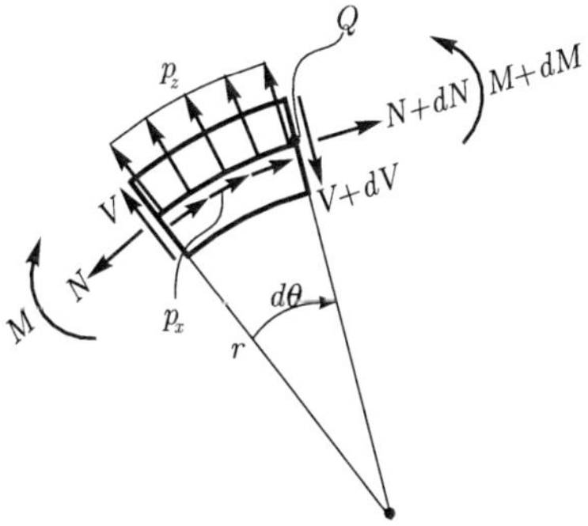

Primary variables

Displacements	Generalized stresses	Generalized strains
$\mathbf{u} = \begin{bmatrix} u(s) \\ w(s) \end{bmatrix}$	$\tau = \begin{bmatrix} N(s) \\ M(s) \end{bmatrix}$	$\varepsilon = \begin{bmatrix} \varepsilon_{xx_0} \\ -\chi \end{bmatrix}$

Remarks

Formulation is cast in a 1-D curved domain

There is no transverse shear strain, *i.e.*,$\gamma_{xz} = 0$

Kinematic assumptions of Bernoulli-Euler model

Timoshenko beam model

Hypotheses

- Solid is a straight bar
- Bar sections remain plane but not necessarily orthogonal to the deformed axis
- Kinematics is fully described by:

  $w(x)$ displacement of the axis in the z direction

  $\beta(x)$ rotation of the cross-section
- Loading

  $p(x)$ transverse distributed load

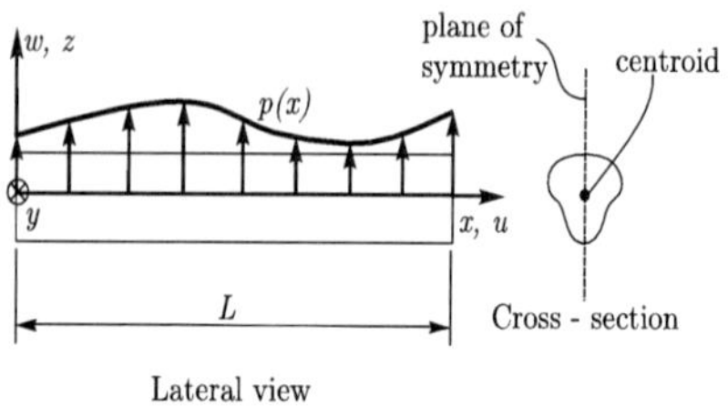

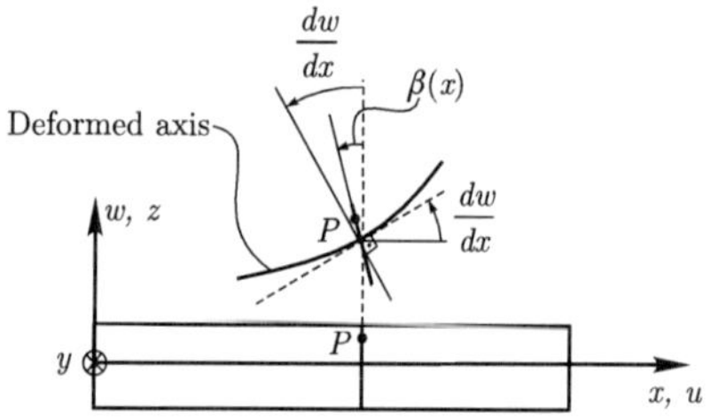

Primary variables

Displacements	Generalized stresses	Generalized strains
$\mathbf{u} = \begin{bmatrix} w(x) \\ \beta(x) \end{bmatrix}$	$\tau = \begin{bmatrix} V(x) \\ M(x) \end{bmatrix}$	$\varepsilon = \begin{bmatrix} \gamma_{xz} \\ \frac{d\beta}{dx} \end{bmatrix}$

Remarks

Formulation is cast in the 1-D domain

The model includes transverse shear deformations $\gamma_{xz} \neq 0$

The formulation can also be presented including the axial displacement $u(x)$ and the loading $f(x)$ as for the Bernoulli-Euler model

Kirchhoff plate model

Hypotheses

- Solid is a thin plate
- Straight material lines which are initially orthogonal to the midsurface of the plate remain straight and orthogonal to the deformed midsurface
- Displacements in the transverse direction do not vary along the thickness of the plate
- Each plate lamina is in a state of plane stress
- Kinematics is fully described by:

  $w = w(x, y)$
- Loading

  $p(x, y)$ transverse loading per unit of surface area

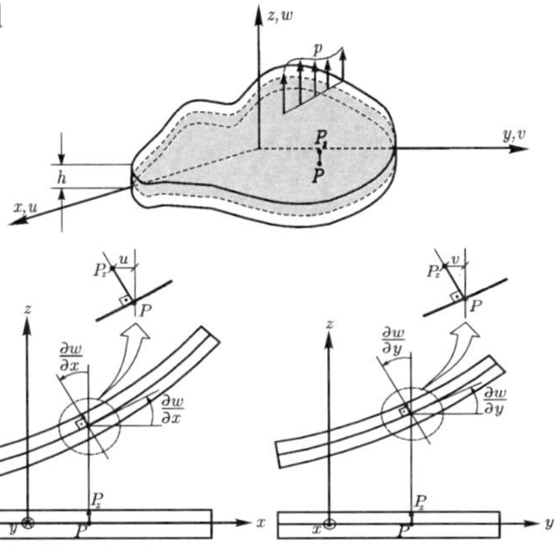

Primary variables

Displacements	Generalized stresses	Generalized strains
$\mathbf{u} = \begin{bmatrix} w(x,y) \end{bmatrix}$	$\tau = \begin{bmatrix} M_x \\ M_y \\ M_{yx} \end{bmatrix}$	$\varepsilon = \begin{bmatrix} \kappa_x \\ \kappa_y \\ \kappa_{yx} \end{bmatrix}$

Remarks

Formulation is cast in a 2-D domain, the midsurface of the plate

There are no transverse shear stresses: $\gamma_{xz} = \gamma_{yz} = 0$

Reissner-Mindlin plate model

Hypotheses

- Solid is a thin to moderately thick plate
- Straight material lines which are initially orthogonal to the midsurface of the plate remain straight but not necessarily orthogonal to the deformed midsurface
- For bending each plate lamina is in a state of plane stress
- Kinematics is fully described by:

$w = w(x, y)$

$\beta_x = \beta_x(x, y)$

$\beta_y = \beta_y(x, y)$

- Loading

$p(x, y)$ transverse loading per unit of surface area

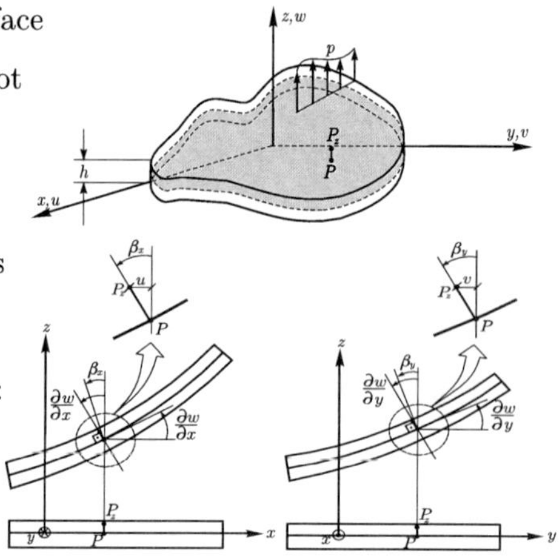

Primary variables

Generalized displacements	Generalized stresses	Generalized strains
$\mathbf{u} = \begin{bmatrix} w(x,y) \\ \beta_x(x,y) \\ \beta_y(x,y) \end{bmatrix}$	$\tau = \begin{bmatrix} M_x \\ M_y \\ M_{yx} \\ Q_x \\ Q_y \end{bmatrix}$	$\varepsilon = \begin{bmatrix} \frac{\partial \beta_x}{\partial x} \\ \frac{\partial \beta_y}{\partial y} \\ \frac{\partial \beta_x}{\partial y} + \frac{\partial \beta_y}{\partial x} \\ \gamma_{xz} \\ \gamma_{yz} \end{bmatrix}$

Remarks

Formulation is cast in a 2-D domain, the midsurface of the plate

Shell models

Hypotheses

Basic shell model

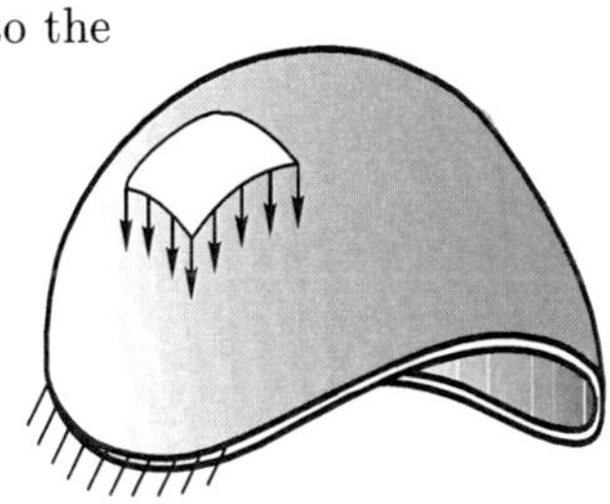

- Solid is a shell with a midsurface
- Straight material lines initially orthogonal to the midsurface remain straight and unstretched during deformations
- Kinematic variables are referred to the shell midsurface
- The stress in the direction orthogonal to the midsurface is zero

Membrane-shear-bending model

Additional hypotheses permit the analytical integration through the shell thickness

Membrane-bending model

- Additional hypothesis: straight material lines initially orthogonal to the midsurface remain orthogonal to the midsurface during deformations
- Hence transverse shear strains are zero

Remarks

This summary for shells is only giving some basic model hypotheses since much more discussion would be required before we could present for each shell model a summary analogous to those presented for the other models, see Chapelle and Bathe, 2010*a*

Table 4.3. Definitions of $\mathbf{C}$ and ∂_ε for linear elasticity mathematical models

Mathematical model (displacements)	Constitutive matrix $\mathbf{C}$	Strain-displacement matrix ∂_ε
3-D elasticity (u,v,w)	$\mathbf{C} = \frac{E(1-\nu)}{(1+\nu)(1-2\nu)} \begin{bmatrix} 1 & \frac{\nu}{1-\nu} & \frac{\nu}{1-\nu} & 0 & 0 & 0 \\ \frac{\nu}{1-\nu} & 1 & \frac{\nu}{1-\nu} & 0 & 0 & 0 \\ \frac{\nu}{1-\nu} & \frac{\nu}{1-\nu} & 1 & 0 & 0 & 0 \\ 0 & 0 & 0 & \frac{1-2\nu}{2(1-\nu)} & 0 & 0 \\ 0 & 0 & 0 & 0 & \frac{1-2\nu}{2(1-\nu)} & 0 \\ 0 & 0 & 0 & 0 & 0 & \frac{1-2\nu}{2(1-\nu)} \end{bmatrix}$	$\begin{bmatrix} \frac{\partial\cdot}{\partial x} & 0 & 0 \\ 0 & \frac{\partial\cdot}{\partial y} & 0 \\ 0 & 0 & \frac{\partial\cdot}{\partial z} \\ \frac{\partial\cdot}{\partial y} & \frac{\partial\cdot}{\partial x} & 0 \\ 0 & \frac{\partial\cdot}{\partial z} & \frac{\partial\cdot}{\partial y} \\ \frac{\partial\cdot}{\partial z} & 0 & \frac{\partial\cdot}{\partial x} \end{bmatrix}$
Plane strain (u,v)	$\frac{E(1-\nu)}{(1+\nu)(1-2\nu)} \begin{bmatrix} 1 & \frac{\nu}{1-\nu} & 0 \\ \frac{\nu}{1-\nu} & 1 & 0 \\ 0 & 0 & \frac{1-2\nu}{2(1-\nu)} \end{bmatrix}$	$\begin{bmatrix} \frac{\partial\cdot}{\partial x} & 0 \\ 0 & \frac{\partial\cdot}{\partial y} \\ \frac{\partial\cdot}{\partial y} & \frac{\partial\cdot}{\partial x} \end{bmatrix}$
Plane stress (u,v)	$\frac{E}{1-\nu^2} \begin{bmatrix} 1 & \nu & 0 \\ \nu & 1 & 0 \\ 0 & 0 & \frac{1-\nu}{2} \end{bmatrix}$	$\begin{bmatrix} \frac{\partial\cdot}{\partial x} & 0 \\ 0 & \frac{\partial\cdot}{\partial y} \\ \frac{\partial\cdot}{\partial y} & \frac{\partial\cdot}{\partial x} \end{bmatrix}$
Axisymmetric (u,v)	$\frac{E(1-\nu)}{(1+\nu)(1-2\nu)} \begin{bmatrix} 1 & \frac{\nu}{1-\nu} & 0 & \frac{\nu}{1-\nu} \\ \frac{\nu}{1-\nu} & 1 & 0 & \frac{\nu}{1-\nu} \\ 0 & 0 & \frac{1-2\nu}{2(1-\nu)} & 0 \\ \frac{\nu}{1-\nu} & \frac{\nu}{1-\nu} & 0 & 1 \end{bmatrix}$	$\begin{bmatrix} \frac{\partial\cdot}{\partial x} & 0 \\ 0 & \frac{\partial\cdot}{\partial y} \\ \frac{\partial\cdot}{\partial y} & \frac{\partial\cdot}{\partial x} \\ \frac{\cdot}{x} & 0 \end{bmatrix}$

Table 4.4. Definitions of $\mathbf{C}$ and ∂_ε for structural mathematical models

Mathematical model (generalized displacements)	Generalized constitutive matrix $\mathbf{C}$	Generalized strain-displacement matrix ∂_ε
Bernoulli-Euler beam (u,w)	$\begin{bmatrix} EA & 0 \\ 0 & EI \end{bmatrix}$	$\begin{bmatrix} \frac{d\,\cdot}{dx} & 0 \\ 0 & \frac{d^2\,\cdot}{dx^2} \end{bmatrix}$
Curved beam (u,w)	$\begin{bmatrix} EA & 0 \\ 0 & EI \end{bmatrix}$	$\begin{bmatrix} \frac{d\,\cdot}{ds} & \frac{\cdot}{r} \\ -\frac{d}{ds}\left(\frac{\cdot}{r}\right) & \frac{d^2\,\cdot}{ds^2} \end{bmatrix}$
Timoshenko beam (w,β)	$\begin{bmatrix} kGA & 0 \\ 0 & EI \end{bmatrix}$	$\begin{bmatrix} \frac{d\,\cdot}{dx} & -1 \\ 0 & \frac{d\,\cdot}{dx} \end{bmatrix}$
Kirchhoff plate (w)	$D\begin{bmatrix} 1 & \nu & 0 \\ \nu & 1 & 0 \\ 0 & 0 & (1-\nu) \end{bmatrix}$	$\begin{bmatrix} \frac{\partial^2\,\cdot}{\partial x^2} \\ \frac{\partial^2\,\cdot}{\partial y^2} \\ \frac{\partial^2\,\cdot}{\partial x \partial y} \end{bmatrix}$
Reissner-Mindlin plate (w,β_x,β_y)	$\begin{bmatrix} D & \nu D & 0 & 0 & 0 \\ \nu D & D & 0 & 0 & 0 \\ 0 & 0 & \frac{D(1-\nu)}{2} & 0 & 0 \\ 0 & 0 & 0 & kGA & 0 \\ 0 & 0 & 0 & 0 & kGA \end{bmatrix}$	$\begin{bmatrix} 0 & \frac{\partial\,\cdot}{\partial x} & 0 \\ 0 & 0 & \frac{\partial\,\cdot}{\partial y} \\ 0 & \frac{\partial\,\cdot}{\partial y} & \frac{\partial\,\cdot}{\partial x} \\ \frac{\partial\,\cdot}{\partial x} & -1 & 0 \\ \frac{\partial\,\cdot}{\partial y} & 0 & -1 \end{bmatrix}$

5. The principle of virtual work

The principle of virtual work is the starting point for the formulation of the finite element method for solids and structures and we dedicate this chapter to its study.

We start by introducing the principle of virtual work for a 1-D model and by exploring the fundamental facts. Then, the principle is given for the 3-D elasticity model and for the remaining mathematical models discussed in Chapter 4.

In the last section, we introduce some energy concepts in 3-D analysis which provide alternative ways to formulate the mathematical models of Chapters 3 and 4. These energy based formulations will be useful in Chapters 6 and 8.

5.1 The principle of virtual work for the bar problem

We introduce in this section the principle of virtual work considering a one-dimensional problem, namely a bar subjected to distributed axial loading $f(x)$ and a concentrated load R at its end as shown in Figure 5.1a.

The mathematical model for this problem was discussed in Section 4.2.1. However, for completeness, we recall below the differential formulation and stress the key steps.

Equilibrium

As usual, we consider a differential element. The equilibrium requirement gives (see Figure 5.1b)

$$-\tau A + (\tau + d\tau) A + f dx = 0 \qquad \text{within the bar} \tag{5.1}$$

and so

$$A\frac{d\tau}{dx} + f = 0. \tag{5.2}$$

Also, at the right end of the bar we have the equilibrium condition

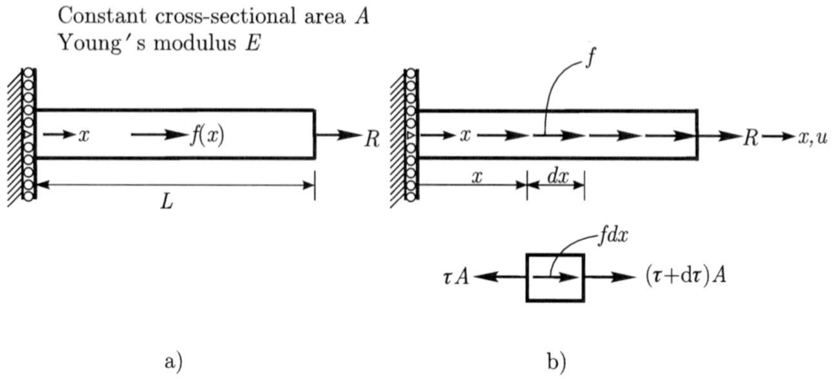

Fig. 5.1. a)Bar problem; b) Differential equilibrium for bar problem (plane sections remain plane)

$$\tau(L)A = R.$$

Compatibility

The relevant strain for this model is $\varepsilon = \varepsilon_{xx}$. The strain compatibility condition is

$$\varepsilon = \frac{du}{dx} \tag{5.3}$$

where u is the displacement field[1] with the condition $u(0) = 0$.

Constitutive equation

We are considering a linear elastic material. Therefore

$$\tau = E\varepsilon \tag{5.4}$$

where E is the Young's is modulus.

Boundary conditions

The boundary conditions at the ends of the bar are listed as

$$u(0) = 0 \tag{5.5}$$

[1] In this chapter we suppose that the functions such as $u(x)$ are sufficiently smooth to allow the evaluation of all required integrations and differentiations. In Section 6.2 we will address this issue more rigorously

and

$$\tau(L)A = R. \tag{5.6}$$

Considering equations (5.2) to (5.6), which represent the basic equilibrium, compatibility and constitutive behavior requirements (including the boundary conditions at both ends), we can write the differential formulation for the problem in terms of the displacement field only.

Differential formulation of the bar problem

$$EA\frac{d^2u}{dx^2} + f = 0 \qquad \text{within the bar} \tag{5.7}$$

$$u(0) = 0 \tag{5.8}$$

$$EA\frac{du}{dx}(L) = R \tag{5.9}$$

Given a particular functional form for $f(x)$, an analytical solution can be calculated by integrating equation (5.7) and imposing the boundary conditions (5.8) and (5.9).

We would like to derive an alternative form, yet equivalent, to the differential formulation presented above. Let us consider the equilibrium equation given in (5.2). We recognize that this equation is valid for any section within the bar. Therefore we can write

$$\left(A\frac{d\tau}{dx} + f\right)\delta u = 0 \tag{5.10}$$

where $\delta u(x)$ is a continuous function defined in the interval $0 \leqslant x \leqslant L$ and zero at the point where the boundary condition in displacement of the differential formulation is prescribed, *i.e.*, $\delta u(0) = 0$. The function $\delta u(x)$ is called the virtual displacement field.

Hence, we also have

$$\int_0^L \left(A\frac{d\tau}{dx} + f\right)\delta u \; dx = 0 \tag{5.11}$$

or equivalently

$$-\int_0^L A\frac{d\tau}{dx}\delta u \; dx = \int_0^L f\delta u \; dx. \tag{5.12}$$

Integrating the left-hand side of equation (5.12) by parts we arrive at

$$-\int_0^L A\frac{d\tau}{dx}\delta u\ dx = -\ (A\tau\delta u)|_0^L + \int_0^L \frac{d\delta u}{dx}A\tau\ dx$$

and substituting into equation (5.12), we obtain

$$\int_0^L \frac{d\delta u}{dx}A\tau\ dx = \int_0^L f\delta u\ dx + (A\tau\delta u)|_0^L \tag{5.13}$$

Calling $\delta\varepsilon = \frac{d\delta u}{dx}$ the virtual strain field, since it is the strain field associated with the virtual displacement field, and using the force boundary condition $\tau(L)A = R$ we can re-write (5.13) as

$$\int_0^L \delta\varepsilon\,\tau A\ dx = \int_0^L f\delta u\ dx + R\,\delta u|_{x=L} \tag{5.14}$$

where we employed $\delta u|_{x=0} = 0$.

Using the notion of work from basic mechanics let us interpret the terms of equation (5.14):

- $R\delta u|_{x=L}$ is the work done by the force R for the virtual displacement at $x = L$.
- $\int_0^L f\delta u\ dx$ is the work done by the distributed force f for the virtual displacement field δu.
- $\int_0^L \delta\varepsilon\,\tau A\ dx$ is the internal work done by the stress field τ, solution of the differential formulation, for the virtual strain field $\delta\varepsilon$ which – it is important to note – corresponds to δu (see also Section 2.3.8).

We denote by δW_i the *internal virtual work* given by

$$\delta W_i = \int_0^L \delta\varepsilon\,\tau A\ dx$$

and by δW_e the *external virtual work* given by

$$\delta W_e = \int_0^L f\delta u\ dx + R\,\delta u|_{x=L}\,.$$

Of course,

$$\delta W_i = \delta W_e. \tag{5.15}$$

Equation (5.14) or equivalently (5.15) is the mathematical statement of the "Principle of Virtual Work" or the "Principle of Virtual Displacements" for the bar problem. We note that, for the stress field which satisfies equilibrium within the bar and at $x = L$, the external virtual work is equal to the internal virtual work for any virtual displacement field that is zero at $x = 0$ (where the displacement boundary condition is imposed).

Hence, we have shown that equilibrium implies the virtual work equation, but not yet the vice-versa. In order to prove the equivalence, we also need to show that the virtual work statement, equation (5.14), implies equilibrium, *i.e.*, equations (5.2) and (5.6).

Using integration by parts we have the following identity

$$\int_0^L \delta\varepsilon \tau A dx = \int_0^L \frac{d\delta u}{dx}\tau A dx = \delta u \tau A|_0^L - \int_0^L \delta u \frac{d\tau}{dx} A dx. \tag{5.16}$$

Substituting (5.16) into (5.14) we obtain

$$\int_0^L \left(A\frac{d\tau}{dx} + f\right)\delta u dx - \delta u \tau A|_0^L + R\delta u|_{x=L} = 0$$

and using that $\delta u|_{x=0} = 0$, we arrive at

$$\int_0^L \left(A\frac{d\tau}{dx} + f\right)\delta u\ dx + \left[(R - \tau A)\,\delta u\right]|_{x=L} = 0. \tag{5.17}$$

Equation (5.17) has to be valid for any $\delta u\,(x)$ with $\delta u\,(0) = 0$. Let us take

$$\delta u\,(x) = x\left(1 - \frac{x}{L}\right)\left(A\frac{d\tau}{dx} + f\right). \tag{5.18}$$

Although we do not explicitly know the functional form of $(A\frac{d\tau}{dx} + f)$, the above choice for $\delta u\,(x)$ is certainly allowed, since this function satisfies $\delta u\,(0) = 0$. Since for this choice $\delta u\,(L) = 0$, equation (5.17) becomes

$$\int_0^L \left(A\frac{d\tau}{dx} + f\right)^2 x\left(1 - \frac{x}{L}\right) dx = 0.$$

Considering that for $(0, L)$, $x > 0$, $\left(1 - \frac{x}{L}\right) > 0$ and $\left(A\frac{d\tau}{dx} + f\right)^2 \geq 0$, the only possibility for the above integral to vanish is to have

$$A\frac{d\tau}{dx} + f = 0 \qquad \text{within the bar} \tag{5.19}$$

i.e., to satisfy the equilibrium equation (5.2).

Using (5.19), we can now return to equation (5.17) which then reads

$$\left[(R - \tau A)\,\delta u\right]|_{x=L} = 0.$$

Since the value of $\delta u\,(x)|_L$ is arbitrary, we conclude that

$$R = \tau A|_{x=L}\,.$$

In other words, the equilibrium at the boundary, *i.e.*, the force boundary condition is also implicitly contained in the virtual work statement.

Hence, we proved the equivalence of the equilibrium equations (5.2), (5.6) and the principle of virtual work equation (5.14). This is a fundamental result that we emphasize in Figure 5.2. Although we proved this equivalence for the 1-D bar problem, this is a general result which is valid for *every* mathematical model in solid and structural mechanics.

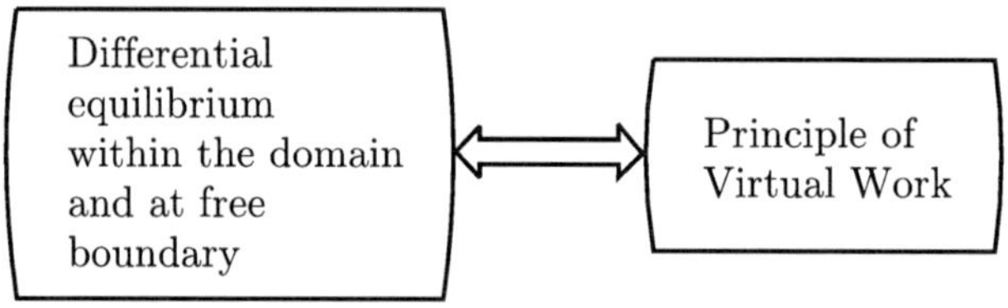

Fig. 5.2. Equivalence between equilibrium and the principle of virtual work

The above discussion shows that we may interpret the principle of virtual work to be an integral form of the equilibrium equations, and it is also referred to as a variational formulation because the virtual displacements can be thought of as a variation of the real displacements.

We summarize what we have discussed so far in Table 5.1. The repetition shown in this table is intentional, because we want to emphasize which equations are the same and which are not the same for the two formulations.

In equations (5.7), (5.8) and (5.9) we wrote the differential formulation in terms of the displacement field only. We write below the variational formulation in terms of displacements only.

Variational formulation of the bar problem

Find $u\,(x)$, $u(0) = 0$ such that

$$\int_0^L EA\frac{d\delta u}{dx}\frac{du}{dx}dx = \int_0^L f\delta u dx + R\delta u\big|_{x=L} \tag{5.20}$$

for any $\delta u\,(x)$, $\delta u\,(0) = 0$.

The relation (5.20) is the principle of virtual work in terms of displacements only, and also contains the compatibility and constitutive requirements.

Table 5.1. Summary of differential and variational formulations for bar problem

Differential Formulation	**Variational Formulation**
Find $\tau(x), \varepsilon(x)$ and continuous $u(x)$ such that the following holds:	Find $\tau(x), \varepsilon(x)$ and continuous $u(x)$ such that the following holds:
Displacement boundary condition $u(0) = 0$	**Displacement boundary condition** $u(0) = 0$
Differential equilibrium $A\frac{d\tau}{dx} + f = 0$ **Equilibrium at the boundary** $\tau A = R$ for $x = L$	**Principle of virtual work** $\int_0^L \delta\varepsilon\tau A\, dx = \int_0^L f\delta u\, dx + R\delta u\vert_{x=L}$ for any $\delta u(x)$, $\delta u(0) = 0$ and $\delta\varepsilon = \frac{d\delta u}{dx}$
Compatibility $\varepsilon = \frac{du}{dx}$	**Compatibility** $\varepsilon = \frac{du}{dx}$
Constitutive equation $\tau = E\varepsilon$	**Constitutive equation** $\tau = E\varepsilon$

Example 5.1

Considering the bar problem of Figure 5.1, show that the principle of virtual work contains the condition of global equilibrium of the bar.

Solution

In Figure 5.3, we show the reaction at the support represented by F, which is given by

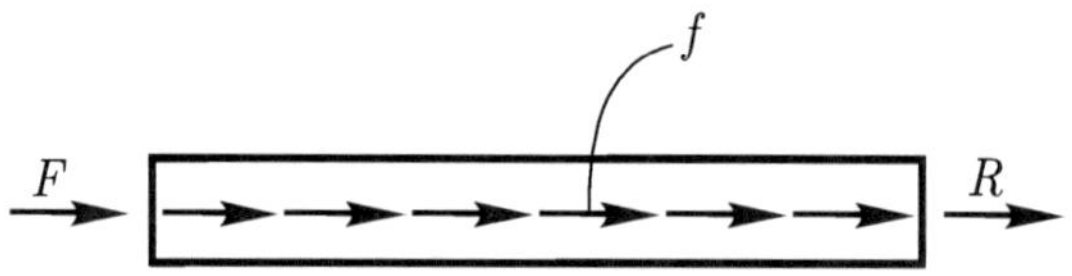

Fig. 5.3. Bar problem with reaction explicitly represented

$$F = -\tau(0)\,A.$$

In this case, the reaction can be calculated directly by imposing equilibrium in the x direction to obtain

$$F = -\left(\int_0^L f\,dx + R\right).$$

Consider the problem described in Figure 5.3 with no support and let us apply the principle of virtual work taking as a virtual displacement field a rigid body translation in the x direction of magnitude Δ. Since it is a rigid body motion, the virtual strain field associated with it is identically zero (in fact, $\varepsilon = \frac{d\delta u}{dx} = \frac{d}{dx}\Delta = 0$). Therefore, the right-hand side of the virtual work statement is also zero. Then, the principle of virtual work yields

$$\int_0^L \delta\varepsilon\,\tau\,A\,dx = 0 = \int_0^L f\Delta dx + F\Delta + R\Delta$$

$$0 = \Delta\left(\int_0^L f\,dx + F + R\right)$$

Since Δ is arbitrary

$$\int_0^L f\,dx + F + R = 0$$

and therefore

$$F = -\left(\int_0^L f dx + R \right)$$

which is the global equilibrium equation of the bar.

□

Hence, we note that the principle of virtual work contains not only the local, *i.e.,* the differential equilibrium written in integral form as previously shown, but also the global equilibrium. Although we explored this fact in this very particular setting, this result is general and it will be detailed further for a 2-D case (see Example 5.6).

In other words, if we introduce the support reactions as external loads, suppress the kinematic restraints provided by the supports, and use as virtual displacement fields rigid body modes, the principle of virtual work yields the global equilibrium equations.

The condition that the equilibrium of every part of a solid implies global equilibrium was also exemplified for a truss structure (see Section 2.2). Hence, the above result that the principle of virtual work contains the global equilibrium condition must be expected, since this principle is equivalent to differential equilibrium, that is, equilibrium of every part of the bar.

Example 5.2

Consider the problem of a steel bar which supports a weight as described in Figure 5.4.

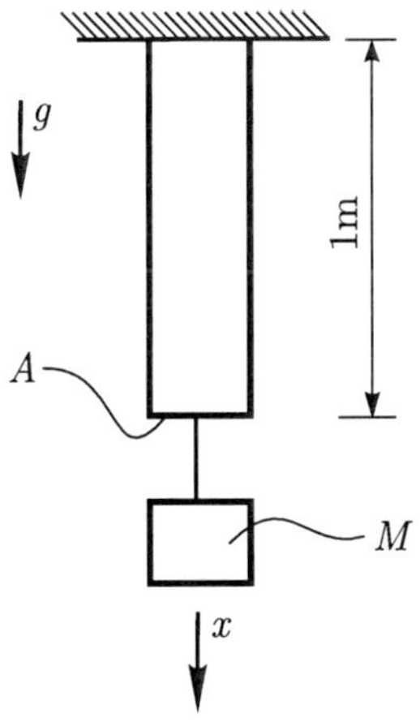

Fig. 5.4. Steel bar subjected to gravity. The density $\rho[\text{kg/m}^3]$, the area $A[\text{m}^2]$ and the mass $M[\text{kg}]$

(i) State and solve the differential formulation of this problem.
(ii) Show explicitly, considering the exact solution derived in (i), that the principle of virtual work is satisfied for the following virtual displacement patterns: $\delta u(x) = ax$ and $\delta u(x) = ax^2$, where a is a constant.

Solution

This problem is clearly a particular case of the bar problem studied above and its differential formulation can be written as

$$EA\frac{d^2u}{dx^2} + \rho g A = 0 \tag{5.21}$$

$$u(0) = 0 \tag{5.22}$$

$$EA\left.\frac{du}{dx}\right|_{x=L} = Mg \tag{5.23}$$

where $L = 1$ m is the bar length.

The solution can be found by substitution and integration and is given by

$$u(x) = \frac{g}{E}\left[-\frac{\rho x^2}{2} + \left(\frac{M}{A} + \rho L\right)x\right].$$

(ii) The exact solution for the stress field is

$$\tau(x) = E\frac{du}{dx} = g\left[-\rho x + \left(\frac{M}{A} + \rho L\right)\right].$$

Considering the virtual displacement $\delta u(x) = ax$, the associated virtual strain field is

$$\delta\varepsilon(x) = \frac{d\delta u(x)}{dx} = a.$$

For the principle of virtual work to be satisfied for this virtual displacement field, we need to have that

$$\int_0^L ag\left[-\rho x + \left(\frac{M}{A} + \rho L\right)\right] Adx = \int_0^L \rho g A a x dx + MgaL$$

which indeed is satisfied.

For the second virtual displacement field $\delta u(x) = ax^2$, we obtain

$$\delta\varepsilon(x) = 2ax$$

and the equation to be verified is

$$\int_0^L 2axg\left[-\rho x + \left(\frac{M}{A} + \rho L\right)\right] A dx = \int_0^L \rho g A a x^2 dx + MgaL^2$$

which also holds.

□

Example 5.3

Consider the bar problem with varying cross-section described in Figure 5.5.

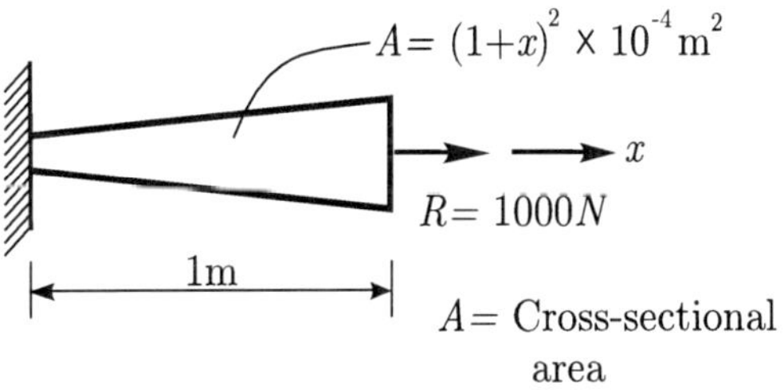

Fig. 5.5. Bar problem with varying bar cross-sectional area

(i) Establish the differential formulation and the principle of virtual work for this problem.

(ii) Find the exact solution for this problem.

(iii) Show that the principle of virtual work is satisfied for the exact stress field and a virtual displacement field $\delta u(x) = ax$.

Solution

(i) Since the cross-sectional area is not constant along the bar, we need to generalize equation (5.1) as,

$$-\tau A + (\tau + d\tau)(A + dA) + f dx = 0$$

$$d\tau A + \tau dA + d\tau dA + f dx = 0$$

and hence obtain

$$\frac{d}{dx}(\tau A) + f = 0. \tag{5.24}$$

Substituting the strain and constitutive equations into (5.24) we obtain the differential formulation for this problem.

Find $u(x)$ such that

$$\frac{d}{dx}\left(EA\frac{du}{dx}\right) + f = 0 \quad \text{within the bar} \tag{5.25}$$

$$u(0) = 0 \tag{5.26}$$

$$EA\frac{du}{dx}|_{x=L} = R. \tag{5.27}$$

(ii) For this example, $f = 0$, and so (5.25) becomes

$$\begin{aligned} \frac{E}{10^4}\frac{d}{dx}\left[(1+x)^2\frac{du}{dx}\right] &= 0 \\ (1+x)^2\frac{d^2u}{dx^2} + 2(1+x)\frac{du}{dx} &= 0 \end{aligned}$$

or

$$\frac{d^2u}{dx^2} + \frac{2}{(1+x)}\frac{du}{dx} = 0$$

leading to

$$u(x) = \frac{R \cdot 10^4}{E}\left(\frac{x}{1+x}\right).$$

(iii) To verify that the principle of virtual work holds for $\delta u(x) = ax$ we use

$$\delta\varepsilon(x) = a.$$

The stress field is given by

$$\tau(x) = E\frac{du}{dx} = R \cdot 10^4 \frac{1}{(1+x)^2}$$

and the principle obviously holds

$$\int_0^L a\frac{R \cdot 10^4}{(1+x)^2}(1+x)^2 \cdot 10^{-4} dx = RaL.$$

□

Example 5.4

Consider the problem described in Figure 5.6. The Young's modulus is constant and equal to E_1 for $0 < x < L/2$ and constant and equal to E_2 for $L/2 < x < L$ with $E_1 \neq E_2$.

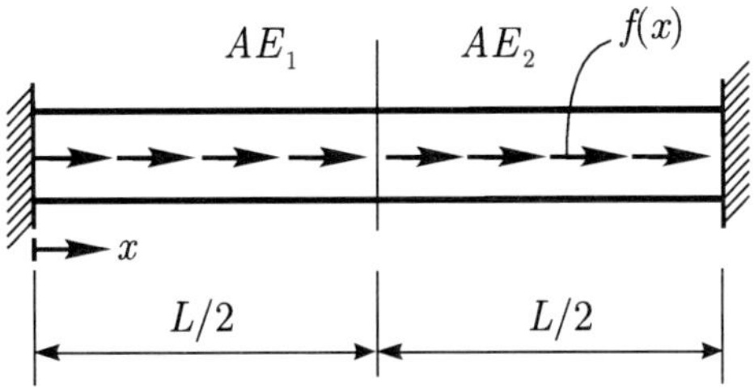

Fig. 5.6. Bar problem with varying material properties

(i) Establish the differential formulation for this problem.

(ii) Establish the principle of virtual work formulation for this problem.

Solution

(i) The differential equilibrium equation of the bar in terms of displacements when both E and A may vary was derived in Example 5.3

$$\frac{d}{dx}\left(EA\frac{du}{dx}\right) + f = 0$$

However in this problem, E changes discontinuously at $x = \frac{L}{2}$ therefore we need to use the differential formulation for subdomain 1

$$E_1 A\frac{d^2u_1}{dx^2} + f = 0 \qquad 0 < x < \frac{L}{2}$$

and for subdomain 2

$$E_2 A\frac{d^2u_2}{dx^2} + f = 0 \qquad \frac{L}{2} < x < L.$$

where $u_1(x)$ and $u_2(x)$ represent the axial displacement for subdomains 1 and 2 respectively. The displacement boundary conditions are

$$u_1(0) = 0$$

$$u_2(L) = 0$$
$$u_1|_{(\frac{L}{2})} = u_2|_{(\frac{L}{2})}.$$

Also, the normal force N at $x = \frac{L}{2}$ should be the same when we approach $x = \frac{L}{2}$ from the right side or from the left side of the section. From the left side

$$N|_{(\frac{L}{2})} = E_1 A \frac{du_1}{dx}|_{(\frac{L}{2})}$$

and from the right side

$$N|_{(\frac{L}{2})} = E_2 A \frac{du_2}{dx}|_{(\frac{L}{2})}.$$

Hence we have the force boundary condition

$$E_1 A \frac{du_1}{dx}|_{(\frac{L}{2})} = E_2 A \frac{du_2}{dx}|_{(\frac{L}{2})}.$$

(ii) The principle of virtual work can however be directly written for the whole bar

$$\int_0^L \delta\varepsilon\tau A dx = \int_0^L \delta u f dx$$

or

$$\int_0^{L/2} \frac{d\delta u}{dx} E_1 \frac{du}{dx} A dx + \int_{L/2}^{L} \frac{d\delta u}{dx} E_2 \frac{du}{dx} A dx = \int_0^L \delta u f dx$$

because the discontinuous change of E does not introduce a difficulty in this equation.

□

5.2 The principle of virtual work in 2-D and 3-D analyses

In the previous section we introduced the principle of virtual work for the one-dimensional bar problem. That setting, due to its simplicity, allowed us to concentrate on the fundamental facts.

We now plan to state the principle for the other mathematical models discussed in Chapters 3 and 4. We choose to first present the principle of virtual work for the three-dimensional elasticity mathematical model and then discuss its specialization to the remaining mathematical models.

Recalling the discussion of the principle of virtual work for the bar problem, we concluded that satisfying the principle of virtual work is equivalent to satisfying local (and therefore also global) equilibrium:

(I) If a stress field τ for which the virtual work of the stresses is equal to the virtual work of the externally applied forces for any virtual displacement field δu, zero at the fixed boundary, then this stress field also satisfies the differential equilibrium (equation (5.2)) and the stress equilibrium at the free boundary, *i.e.*, the force boundary condition (equation (5.6)).

(II) If a stress field τ satisfies differential equilibrium (equation (5.2)) and the force boundary condition (equation (5.6)) then for any virtual displacements δu, zero at the fixed boundary, the internal virtual work of this stress field is equal to the virtual work of the externally applied forces (the reverse of **(I)**).

5.2.1 The principle of virtual work for 3-D elasticity

In Figure 3.59 of Chapter 3, we described the 3-D elasticity problem. For a given choice of body forces $\mathbf{f}^B$, surface tractions $\mathbf{f}^S$ on S_f and displacement boundary conditions $\mathbf{u} = \hat{\mathbf{u}}$ on S_u, let $\tau_{xx}, \tau_{yy}, \tau_{zz}, \tau_{xy}, \tau_{xz}, \tau_{yz}$ be the stress field that corresponds to the solution of this elasticity problem. Therefore, this stress field satisfies differential equilibrium in the volume of the body and at the boundary, *i.e.*, the force boundary conditions on S_f.

The above statements **(I)** and **(II)**, which were proved in the context of the one-dimensional bar problem, are also valid for 2-D and 3-D problems. However, we need to precisely define, in each case, the internal virtual work.

The internal virtual work for the 3-D problem is a generalization of the internal virtual work for the 1-D bar problem for which the stress and strain states were fully represented by only one stress (τ) and one strain component (ε). The virtual work of the 3-D stress field $\tau_{xx}, \tau_{yy}, \tau_{zz}, \tau_{xy}, \tau_{xz}, \tau_{yz}$ considering a virtual strain field $\delta\varepsilon_{xx}, \delta\varepsilon_{yy}, \delta\varepsilon_{zz}, \delta\gamma_{xy}, \delta\gamma_{xz}, \delta\gamma_{yz}$ (derived from the virtual displacements) is given by

$$\delta W_i = \int (\delta\varepsilon_{xx}\tau_{xx} + \delta\varepsilon_{yy}\tau_{yy} +$$

$$+\delta\varepsilon_{zz}\tau_{zz} + \delta\gamma_{xy}\tau_{xy} + \delta\gamma_{xz}\tau_{xz} + \delta\gamma_{yz}\tau_{yz})\; dV \tag{5.28}$$

which can be re-written using the stress and strain column matrices as

$$\delta W_i = \int_V \delta\varepsilon^T \tau \; dV = \int_V \bar{\varepsilon}^T \tau \; dV.$$

As in the expression above, we now denote the virtual quantities by an overbar.

The external virtual work is given by

$$\delta W_e = \int_V \bar{\mathbf{u}}^T \mathbf{f}^B \; dV \; + \int_{S_f} \bar{\mathbf{u}}^T \; \mathbf{f}^S \; dS$$

where

$$\bar{\mathbf{u}} = \begin{bmatrix} \bar{u}(x,y,z) \\ \bar{v}(x,y,z) \\ \bar{w}(x,y,z) \end{bmatrix}$$

is the column matrix which represents the virtual displacement field. Note that by the definition of the virtual displacement field $\bar{\mathbf{u}} = \mathbf{0}$ on S_u.

Therefore the virtual work statement reads[2]

$$\int_V \bar{\varepsilon}^T \tau \, dV = \int_V \bar{\mathbf{u}}^T \mathbf{f}^B \, dV + \int_{S_f} \bar{\mathbf{u}}^T \, \mathbf{f}^S \, dS. \tag{5.29}$$

In the following example we demonstrate an important result which was already shown for the 1-D bar problem in Section 5.1.

Example 5.5

In the context of a 3-D body, show that if a stress field satisfies differential equilibrium and the force boundary conditions then the virtual work statement is satisfied by this stress field.

Solution

The differential equilibrium equations for a 3-D body are given by equations (3.114) and the force boundary conditions by equations (3.117) which are assumed to be satisfied by the stress field τ_{xx}, τ_{yy}, τ_{zz}, τ_{xy}, τ_{xz}, τ_{yz} considered.

Let $\bar{u}(x,y,z), \bar{v}(x,y,z)$ and $\bar{w}(x,y,z)$ be an arbitrarily chosen displacement field such that

$$\bar{u} = 0,\ \bar{v} = 0,\ \bar{w} = 0 \quad \text{on } S_u. \tag{5.30}$$

Hence $\bar{u}$, $\bar{v}$, $\bar{w}$ are a virtual displacement field.

Using the equilibrium equations, we write

$$\left(\frac{\partial \tau_{xx}}{\partial x} + \frac{\partial \tau_{xy}}{\partial y} + \frac{\partial \tau_{xz}}{\partial z} + f_x^B \right) \bar{u} = 0 \tag{5.31}$$

$$\left(\frac{\partial \tau_{yx}}{\partial x} + \frac{\partial \tau_{yy}}{\partial y} + \frac{\partial \tau_{yz}}{\partial z} + f_y^B \right) \bar{v} = 0 \tag{5.32}$$

$$\left(\frac{\partial \tau_{zx}}{\partial x} + \frac{\partial \tau_{zy}}{\partial y} + \frac{\partial \tau_{zz}}{\partial z} + f_z^B \right) \bar{w} = 0. \tag{5.33}$$

[2] It is implicitly understood that for the integral over S_f the virtual displacements $\bar{\mathbf{u}}$ in the integrand represent the virtual displacements evaluated on S_f. We could have used $\bar{\mathbf{u}}^{S_f}$ to emphasize this fact

Using the derivative product rule, we have

$$\begin{aligned}\frac{\partial}{\partial x}\left(\tau_{xx}\bar{u}+\tau_{yx}\bar{v}+\tau_{zx}\bar{w}\right) &= \frac{\partial \tau_{xx}}{\partial x}\bar{u}+\frac{\partial \tau_{yx}}{\partial x}\bar{v}+\frac{\partial \tau_{zx}}{\partial x}\bar{w}\\ &+\tau_{xx}\frac{\partial \bar{u}}{\partial x}+\tau_{yx}\frac{\partial \bar{v}}{\partial x}+\tau_{zx}\frac{\partial \bar{w}}{\partial x}\end{aligned}$$

and re-arranging terms, we obtain

$$\begin{aligned}\frac{\partial \tau_{xx}}{\partial x}\bar{u}+\frac{\partial \tau_{yx}}{\partial x}\bar{v}+\frac{\partial \tau_{zx}}{\partial x}\bar{w} &= \frac{\partial}{\partial x}\left(\tau_{xx}\bar{u}+\tau_{yx}\bar{v}+\tau_{zx}\bar{w}\right)\\ &-\left(\tau_{xx}\frac{\partial \bar{u}}{\partial x}+\tau_{yx}\frac{\partial \bar{v}}{\partial x}+\tau_{zx}\frac{\partial \bar{w}}{\partial x}\right).\end{aligned}$$

Now, using analogous expressions for the derivatives with respect to y and z, and equations (5.31) to (5.33) we can write

$$\begin{aligned}&\int_V \left[\frac{\partial}{\partial x}\left(\tau_{xx}\bar{u}+\tau_{yx}\bar{v}+\tau_{zx}\bar{w}\right)+\frac{\partial}{\partial y}\left(\tau_{xy}\bar{u}+\tau_{yy}\bar{v}+\tau_{zy}\bar{w}\right)+\right.\\ &+\frac{\partial}{\partial z}\left(\tau_{xz}\bar{u}+\tau_{yz}\bar{v}+\tau_{zz}\bar{w}\right)+\\ &-\left(\tau_{xx}\frac{\partial \bar{u}}{\partial x}+\tau_{yy}\frac{\partial \bar{v}}{\partial y}+\tau_{zz}\frac{\partial \bar{w}}{\partial z}+\tau_{yx}\frac{\partial \bar{v}}{\partial x}+\tau_{xy}\frac{\partial \bar{u}}{\partial y}+\right.\\ &\left.\tau_{zx}\frac{\partial \bar{w}}{\partial x}+\tau_{xz}\frac{\partial \bar{u}}{\partial z}+\tau_{zy}\frac{\partial \bar{w}}{\partial y}+\tau_{yz}\frac{\partial \bar{v}}{\partial z}\right)\\ &\left.+f_x^B\bar{u}+f_y^B\bar{v}+f_z^B\bar{w}\;\right]\,dV=0.\end{aligned}\tag{5.34}$$

The following mathematical identity follows from the use of the divergence theorem[3]

$$\begin{aligned}&\int_V \left[\frac{\partial}{\partial x}\left(\tau_{xx}\bar{u}+\tau_{yx}\bar{v}+\tau_{zx}\bar{w}\right)+\frac{\partial}{\partial y}\left(\tau_{xy}\bar{u}+\tau_{yy}\bar{v}+\tau_{zy}\bar{w}\right)+\right.\\ &\left.+\frac{\partial}{\partial z}\left(\tau_{xz}\bar{u}+\tau_{yz}\bar{v}+\tau_{zz}\bar{w}\right)\right]\,dV\\ &=\int_S \left[\left(\tau_{xx}\bar{u}+\tau_{yx}\bar{v}+\tau_{zx}\bar{w}\right)n_x+\left(\tau_{xy}\bar{u}+\tau_{yy}\bar{v}+\tau_{zy}\bar{w}\right)n_y+\right.\\ &\left.\left(\tau_{xz}\bar{u}+\tau_{yz}\bar{v}+\tau_{zz}\bar{w}\right)n_z\right]\,dS\end{aligned}\tag{5.35}$$

where n_x, n_y and n_z are the components of the outward unit normal to the surface S. But

[3] The divergence theorem states: Let $\mathbf{F}$ be a vector field in the volume V; then $\int_V \left(\frac{\partial F_x}{\partial x}+\frac{\partial F_y}{\partial y}+\frac{\partial F_z}{\partial z}\right)dV=\int_S \mathbf{F}\cdot\mathbf{n}\,dS$ where $\mathbf{n}$ is the unit outward normal on the surface S of V

$$\int_S [(\tau_{xx}n_x + \tau_{xy}n_y + \tau_{xz}n_z)\,\bar{u} + (\tau_{yx}n_x + \tau_{yy}n_y + \tau_{yz}n_z)\,\bar{v} +$$

$$(\tau_{zx}n_x + \tau_{zy}n_y + \tau_{zz}n_z)\,\bar{w}]\ dS$$

$$= \int_{S_f} \left[\bar{u}f_x^S + \bar{v}f_y^S + \bar{w}f_z^S\right]\ dS \tag{5.36}$$

where we used (3.117) and (5.30).

Next, using the symmetry of the stress tensor and substituting (5.35) and (5.36) into (5.34), we arrive at

$$\int_V (\bar{\varepsilon}_{xx}\tau_{xx} + \bar{\varepsilon}_{yy}\tau_{yy} + \bar{\varepsilon}_{zz}\tau_{zz} + \bar{\gamma}_{xy}\tau_{xy} + \bar{\gamma}_{yz}\tau_{yz} + \bar{\gamma}_{xz}\tau_{xz})\ dV =$$

$$\int_V \left[\bar{u}f_x^B + \bar{v}f_y^B + \bar{w}f_z^B\right]\ dV + \int_{S_f} \left[\bar{u}f_x^S + \bar{v}f_y^S + \bar{w}f_z^S\right]\ dS. \tag{5.37}$$

We note that the strains with the overbar are obtained from the virtual displacement field using the strain-displacement relations (3.115) and therefore are virtual strains. Re-writing (5.37) using matrix quantities gives

$$\int_V \bar{\varepsilon}^T \tau\ dV = \int_V \bar{\mathbf{u}}^T \mathbf{f}^B\ dV + \int_{S_f} \bar{\mathbf{u}}^T \mathbf{f}^S\ dS \tag{5.38}$$

which is the virtual work statement (5.29).

□

Hence we proved statement **(II)** for the 3-D problem. The proof of statement **(I)** is obtained in the same way as for the 1-D case, by starting from equation (5.38) – the virtual work statement – and perform the mathematical operations backwards.

Regarding the principle of virtual work equation, we emphasize once more:

- The virtual strains $\bar{\varepsilon}$ are calculated by the differentiations given in (3.115) from the assumed virtual displacements $\bar{\mathbf{u}}$.
- The virtual displacements $\bar{\mathbf{u}}$ must be smooth enough to allow the evaluation of the integral in (5.29), with $\bar{\mathbf{u}}$ equal to zero on S_u.
- All integrations are performed over the original volume and surface area of the body, unaffected by the imposed virtual displacements.

The principle of virtual work in terms of the displacement variables only is the variational formulation which contains all requirements of equilibrium, compatibility and constitutive behavior.

Variational formulation of the 3-D elasticity model

Find $\mathbf{u}$, $\mathbf{u} = \hat{\mathbf{u}}$ on S_u such that[4]

$$\int_V (\partial_\epsilon \bar{\mathbf{u}})^T \mathbf{C} \, \partial_\epsilon \bar{\mathbf{u}} \, dV = \int_V \bar{\mathbf{u}}^T \, \mathbf{f}^B \, dV + \int_{S_f} \bar{\mathbf{u}}^T \, \mathbf{f}^S \, dS \tag{5.39}$$

for any $\bar{\mathbf{u}}$ which is zero on S_u

Note that we have used equations (4.330) and (4.331) to enforce the strain-displacement and constitutive requirements.

Since all field and boundary conditions of the differential formulation are represented in the above variational formulation and vice-versa, these formulations are equivalent.

Of course, we can similarly derive the principle of virtual work for all differential formulations considered in Chapter 4 and all the results above are valid for these models. We detail below the derivation of the principle of virtual work for some of these models.

5.2.2 The principle of virtual work for the plane stress model

Figure 4.5 describes a plane stress situation. The non-zero stress components are τ_{xx}, τ_{yy} and τ_{xy}, and the non-zero strain components are $\varepsilon_{xx}, \varepsilon_{yy}, \gamma_{xy}, \varepsilon_{zz}$. Starting from the general expression of the internal virtual work of the 3-D elasticity problem and introducing the plane stress assumptions we have

$$\int_V \bar{\varepsilon}^T \tau \; dV = \int_A \int_{-h/2}^{+h/2} \left(\bar{\varepsilon}_{xx}\tau_{xx} + \bar{\varepsilon}_{yy}\tau_{yy} + \bar{\varepsilon}_{zz} \cdot 0 + \bar{\gamma}_{xy}\tau_{xy}\right) \; dz \; dA$$

$$= h \int_A \left(\bar{\varepsilon}_{xx}\tau_{xx} + \bar{\varepsilon}_{yy}\tau_{yy} + \bar{\gamma}_{xy}\tau_{xy}\right) \; dA$$

where we used h to be constant.

Since both the body forces and surface tractions are assumed to have non-zero components only on the xy plane and do not depend on z, the external virtual work reads

$$\int_V \mathbf{u}^T \mathbf{f}^B \; dV + \int_{S_f} \mathbf{u}^T \; \mathbf{f}^S \; dS =$$

[4] Throughout the discussion, we assume that $\mathbf{u}$ and $\bar{\mathbf{u}}$ are continuous

$$= \int_A \int_{-h/2}^{+h/2} \left(\bar{u}\ f_x^B + \bar{v}\ f_y^B\right)\ dz\ dA + \int_{L_f} \int_{-h/2}^{+h/2} \left(\bar{u}\ f_x^S + \bar{v}\ f_y^S\right)\ dz\ dL$$

$$= h \int_A \left(\bar{u}\ f_x^B + \bar{v}\ f_y^B\right) dA + h \int_{L_f} \left(\bar{u}\ f_x^S + \bar{v}\ f_y^S\right) dL.$$

with $V = h \times A$ and $S_f = h \times L_f$.

Using the column matrix quantities defined in Chapter 4 for the plane stress and plane strain models, we are led to the principle of virtual work for plane stress

$$h \int_A \bar{\varepsilon}^T \tau\ dA = h \int_A \bar{\mathbf{u}}^T\ \mathbf{f}^B dA + h \int_{L_f} \bar{\mathbf{u}}^T\ \mathbf{f}^S\ dL \tag{5.40}$$

for all $\bar{\mathbf{u}}$, $\bar{\mathbf{u}} = \mathbf{0}$ on L_u and $\bar{\varepsilon} = \partial_\varepsilon \bar{\mathbf{u}}$.

Formally, the same kind of expression for the virtual work principle as for the 3-D case is obtained, that is,

$$\int_A \bar{\varepsilon}^T \tau\ dA = \int_A \bar{\mathbf{u}}^T\ \mathbf{f}^B dA + \int_{L_f} \bar{\mathbf{u}}^T\ \mathbf{f}^S\ dL \tag{5.41}$$

for all $\bar{\mathbf{u}}$, $\bar{\mathbf{u}} = \mathbf{0}$ on S_u, with $S_u = h \times L_u$ and $\bar{\varepsilon} = \partial_\varepsilon \bar{\mathbf{u}}$.

We note that in the above expressions $\mathbf{f}^B$ is given per unit of volume and $\mathbf{f}^S$ is given per unit of surface area.

Sometimes, it is more convenient to use (5.40) instead of (5.41). For example, when considering concentrated loads applied at the boundary (see the example below).

Example 5.6

Consider the plane stress problem defined in Figure 5.7.

Calculate the reactions by simple statics, and calculate the reactions using the principle of virtual work.

Solution

Referring to Figure 5.7, the equilibrium equations of statics are

$$\sum F_x = 0,\ \sum F_y = 0,\ \sum M_A = 0$$

which leads to

$$X_A = 0,\ Y_A + Y_B + t \int_A f_y^B dA + t \int_0^L f_y^S dx = 0$$

and

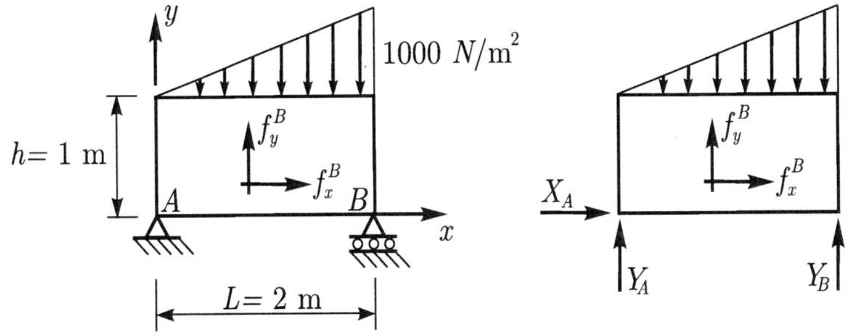

Fig. 5.7. a) Definition of the plane stress problem, $f_x^B = 0$, $f_y^B = -2400\ N/m^3$, $t = 0.1\,m$ (thickness); b) Equilibrium of the plate including reactions

$$Y_B L + t\int_A f_y^B x dA + t\int_0^L f_y^S x dx = 0$$

from which we obtain $Y_A = 273.33\,N$, $Y_B = 306.67\,N$.

We use equation (5.40) since we have an extra term on the right-hand side corresponding to the virtual work of the concentrated forces given by

$$\sum_{i=1}^{N_c} F_i\,\delta d_i$$

where F_i, $i = 1, \cdots, N_C$ are the magnitudes of the N_C concentrated forces and δd_i are the virtual displacements in the direction and the same orientation as the concentrated loads F_i.

Note that if we considered equation (5.41), we would need to divide the magnitude of the concentrated load by the thickness t since for this equation we divided by the constant thickness.

To calculate the reactions we use 3 independent virtual displacement fields corresponding to virtual rigid body mode displacements. Then, the internal virtual work is zero.

Choosing first, as virtual displacement, a translation in the direction x of magnitude Δ_x, we obtain

$$\begin{aligned} 0 &= t\int_A \begin{bmatrix} \Delta_x & 0 \end{bmatrix} \begin{bmatrix} 0 \\ f_y^B \end{bmatrix} dA + t\int_0^L \begin{bmatrix} \Delta_x & 0 \end{bmatrix} \begin{bmatrix} 0 \\ f_y^S \end{bmatrix} dL + X_A\Delta_x \\ 0 &= X_A \end{aligned}$$

which is the equilibrium equation in the x direction.

Considering now a translation in the y direction of magnitude Δ_y, we obtain

$$0 = t\int_A \begin{bmatrix} 0 & \Delta_y \end{bmatrix} \begin{bmatrix} 0 \\ f_y^B \end{bmatrix} dA + t\int_0^L \begin{bmatrix} 0 & \Delta_y \end{bmatrix} \begin{bmatrix} 0 \\ f_y^S \end{bmatrix} dL + Y_A\Delta_y + Y_B\Delta_y$$

or

$$0 = Y_A + Y_B + t\int_A f_y^B dA + t\int_0^L f_y^S dx$$

which is the equilibrium equation in the y direction.

As the last virtual displacement field, we choose an infinitesimal rigid rotation about point A which is given by

$$\bar{\mathbf{u}} = \begin{bmatrix} -d\theta\ y \\ d\theta\ x \end{bmatrix}.$$

The virtual work expression becomes

$$\begin{aligned} 0 &= Y_B(Ld\theta) + t\int_A [-d\theta\, y \quad d\theta\, x] \begin{bmatrix} 0 \\ f_y^B \end{bmatrix} dA + t\int_0^L [-d\theta\, y \quad d\theta\, x] \begin{bmatrix} 0 \\ f_y^S \end{bmatrix} dx \\ 0 &= Y_B L + t\int_A f_y^B x dA + t\int_0^L f_y^S x dx \end{aligned}$$

which corresponds to the moment equilibrium condition. Since we obtained exactly the same equilibrium equations, the reactions are those already calculated. □

5.2.3 The principle of virtual work for the plane strain model

The plane strain model was discussed in Section 4.1.1. Due to the kinematic assumptions the strains $\varepsilon_{zz} = \gamma_{yz} = \gamma_{zx} = 0$ and the non-zero components are, therefore, ε_{xx}, ε_{yy} and γ_{xy}. The stresses τ_{xx}, τ_{yy}, τ_{xy} and τ_{zz} are, in general, non-zero.

Since in plane strain we have for every section the same stress and strain fields, and loading, we can take a unit thickness to write the principle of virtual work. Analogously to what we did for the plane stress model, we can use the matrix quantities defined in Chapter 4. Following the derivations of the plane stress case and considering a unit thickness, the principle of virtual work reads

$$\int_A \bar{\varepsilon}^T \tau \, dA = \int_A \bar{\mathbf{u}}^T \mathbf{f}^B dA + \int_{L_f} \bar{\mathbf{u}}^T \mathbf{f}^S \, dL. \tag{5.42}$$

for all $\bar{\mathbf{u}}$, $\bar{\mathbf{u}} = \mathbf{0}$ in S_u (*i.e.* L_u) and $\bar{\varepsilon} = \partial_\varepsilon \bar{\mathbf{u}}$.

5.2.4 The principle of virtual work for the axisymmetric model

Considering the discussion of the axisymmetric model in Section 4.1.3, besides the usual stress and strain components $\tau_{xx}, \tau_{yy}, \tau_{xy}$ and $\varepsilon_{xx}, \varepsilon_{yy}, \gamma_{xy}$ for the plane models, we have the hoop stress and strain components τ_{zz} and ε_{zz} which are also non-zero.

In Figure 4.12 a generic cross-section of the solid of revolution is shown. The y axis is the symmetry axis and the x axis corresponds to the radial direction. We need to define a volume to apply the principle of virtual work. It is usual to take the volume corresponding to an angle of one radian centered at the symmetry axis. Then, for the evaluation of the internal virtual work, we have

$$\int_V \bar{\varepsilon}^T \tau \, dV = \int_A \left(\int_0^1 \bar{\varepsilon}^T \tau \; x \, d\theta \right) dA = \int_A \bar{\varepsilon}^T \tau \; x \, dA$$

where θ is the circumferential angle coordinate of the cylindrical system used (refer to Section 4.1.3 for details) and we adopt the column matrices of the axisymmetric model. Using the same integration procedure for the external work terms, we arrive at the principle of virtual work

$$\int_A \bar{\varepsilon}^T \tau \; x \, dA = \int_A \bar{\mathbf{u}}^T \mathbf{f}^B \, x \, dA + \int_{L_f} \bar{\mathbf{u}}^T \mathbf{f}^S x \, dL.$$

for all $\bar{\mathbf{u}}$, $\bar{\mathbf{u}} = \mathbf{0}$ in S_u (*i.e.* L_u) and $\bar{\varepsilon} = \partial_\varepsilon \bar{\mathbf{u}}$.

5.2.5 The principle of virtual work for the Bernoulli-Euler beam model

Let us derive the principle of virtual work for a structural mathematical model. We choose as a typical model, the Bernoulli-Euler beam model which we discussed in Section 4.2.2. Recognizing that the stress and strain components that do work are τ_{xx} and ε_{xx}, the internal virtual work reads

$$\delta W_i = \int_V \bar{\varepsilon}_{xx} \tau_{xx} \, dV.$$

Referring to equation (4.145) re-written below

$$\tau_{xx} = -\frac{M}{I} z$$

and to equations (4.124) and (4.141) leading to

$$\varepsilon_{xx} = -z \frac{d^2 w}{dx^2}$$

we obtain

$$\delta W_i = \int_0^L \int_A \left(-z\frac{d^2\bar{w}}{dx^2}\right)\left(-\frac{M}{I}z\right)\, dA\, dx = \int_0^L \frac{d^2\bar{w}}{dx^2} M\, dx$$

Recall that $\kappa = \frac{d^2 w}{dx^2}$ is the curvature of the axis of the beam for infinitesimally small displacements. Considering only a transverse distributed force p as the external load, the principle of virtual work for a beam, for example, clamped at both ends reads

$$\int_0^L \bar{\kappa} M\, dx = \int_0^L \bar{w} p\, dx$$

for all $\bar{w}$, $\bar{w}(0) = \bar{w}(L) = 0 = \frac{d\bar{w}}{dx}(0) = \frac{d\bar{w}}{dx}(L)$ and $\kappa = \frac{d^2\bar{w}}{dx^2}$.

This result was anticipated since for this model $\tau = [M]$ and $\varepsilon = [\kappa]$ (note that, if one end of the beam is free and subjected to an externally applied force or moment, their virtual work contribution should be included in δW_e).

5.3 Strain and potential energy in 3-D

In this section we discuss the concept of strain and potential energy for 3-D elastic solids.

5.3.1 Strain energy

Let us extend, to a 3-D linear elastic solid, the concept of *strain energy* introduced in Chapter 2 for a truss bar, *i.e.*, for a one-dimensional state of stress and strain.

Recall that the internal work per unit of volume in a 1-D state of stress and strain for a linear elastic material is given by

$$W(\varepsilon) = \int_0^T \tau(\varepsilon)\, d\varepsilon = \frac{1}{2}\tau\varepsilon = \frac{1}{2}E\varepsilon^2$$

which gives the total work per unit of volume at a point performed by the stress throughout the deformation, that is, from a stress-free state to the current state. Then, the total strain energy is defined as

$$\mathcal{U}(\varepsilon) = \int_V W(\varepsilon)\, dV.$$

Also recall that the strain energy depends only on the current state of deformation and gives the energy stored as elastic deformation.

Considering a 3-D elastic solid, we can generalize the above concepts defining the internal work per unit of volume or the strain energy density by

$$W(\varepsilon_{xx}, \varepsilon_{yy}, \varepsilon_{zz}, \gamma_{xy}, \gamma_{xz}, \gamma_{yz}) = \frac{1}{2}(\tau_{xx}\varepsilon_{xx} + \tau_{yy}\varepsilon_{yy}$$

$$+\tau_{zz}\varepsilon_{zz} + \tau_{xy}\gamma_{xy} + \tau_{yz}\gamma_{yz} + \tau_{zx}\gamma_{zx}) = \frac{1}{2}\varepsilon^T\tau$$

$$= \frac{1}{2}\varepsilon^T\mathbf{C}\varepsilon$$

and the total strain energy by

$$\mathcal{U}(\varepsilon) = \int_V W(\varepsilon)\, dV = \frac{1}{2}\int_V \varepsilon^T\mathbf{C}\varepsilon\, dV.$$

Sometimes we use the notation

$$a(\mathbf{v}, \mathbf{v}) = 2\mathcal{U}(\varepsilon(\mathbf{v})) = \int_V (\partial_\varepsilon \mathbf{v})^T \mathbf{C}\partial_\varepsilon \mathbf{v}\; dV \tag{5.43}$$

which gives twice the strain energy associated with the displacement field $\mathbf{v}$.

5.3.2 The total potential energy

The *total potential* of an elastic solid subjected to a field of body forces $\mathbf{f}^B$ in V and a field of surface tractions $\mathbf{f}^S$ on S_f is defined as

$$\Pi(\mathbf{v}) = \mathcal{U}(\varepsilon(\mathbf{v})) - P(\mathbf{v})$$

where $\mathbf{v}$ is a *compatible displacement field*, that is, a sufficiently smooth displacement field (the strains $\varepsilon(\mathbf{v}) = \partial_\varepsilon\mathbf{v}$ can be evaluated) which satisfies the kinematic boundary conditions on S_u, and $P(\mathbf{v})$ is the potential of the external loads, that is,

$$P(\mathbf{v}) = \int_V \mathbf{v}^T\mathbf{f}^B\, dV + \int_{S_f} \mathbf{v}^T\mathbf{f}^S\, dS.$$

There are two important results regarding the total potential energy which we discuss below.

(i) *The stationarity condition $\delta\Pi = 0$ is equivalent to the principle of virtual work and the displacement field* $\mathbf{u}$ *which makes Π stationary is the solution of the elasticity problem.*

Proof

Let us consider a variation $\delta\mathbf{u}$ from a compatible displacement field $\mathbf{u}$. Let $\delta\varepsilon$ be the strain corresponding to $\delta\mathbf{u}$. Then

$$\delta\Pi = \frac{1}{2}\int_V \left[\delta\varepsilon^T\mathbf{C}\,\varepsilon(\mathbf{u}) + \varepsilon^T(\mathbf{u})\,\mathbf{C}^T\delta\varepsilon\right] dV$$

$$- \int_V \delta \mathbf{u}^T \mathbf{f}^B \, dV - \int_{S_f} \delta \mathbf{u}^T \mathbf{f}^S \, dS = 0.$$

Since $\mathbf{C}$ is symmetric, we obtain from the above equation

$$\int_V \delta\varepsilon^T \mathbf{C} \, \varepsilon \, (\mathbf{u}) \, dV = \int_V \delta \mathbf{u}^T \mathbf{f}^B \, dV + \int_{S_f} \delta \mathbf{u}^T \mathbf{f}^S \, dS. \tag{5.44}$$

The variation $\delta\mathbf{u}$ is arbitrary provided that $\mathbf{u}+\delta\mathbf{u}$ is a compatible displacement field. Hence, $\delta\mathbf{u} = \mathbf{0}$ on S_u, since $\mathbf{u}$ satisfies the displacement boundary conditions on S_u.
The variation $\delta\mathbf{u}$ can be interpreted as a virtual displacement field since it is an arbitrary displacement field which is zero on S_u. Since (5.44) holds for an arbitrary virtual displacement, it corresponds to the principle of virtual work in terms of displacements and, hence, the displacement $\mathbf{u}$ is the solution of the elasticity problem.

(ii) *The total potential energy attains its minimum value for the solution* $\mathbf{u}$ *of the elasticity problem.*

Proof

Let us evaluate the potential energy for the displacement field $\mathbf{u}+\delta\mathbf{u}$ where $\mathbf{u}$ is the solution of the elasticity problem and $\delta\mathbf{u}$ an arbitrary variation. Also, let ε and $\delta\varepsilon$ be the strain corresponding to the displacements $\mathbf{u}$ and $\delta\mathbf{u}$, respectively.

$$\Pi(\mathbf{u}+\delta\mathbf{u}) = \frac{1}{2} \int_V (\varepsilon+\delta\varepsilon)^T \, \mathbf{C} \, (\varepsilon+\delta\varepsilon) \, dV$$

$$- \int_V \left(\mathbf{u}^T + \delta\mathbf{u}^T\right) \mathbf{f}^B \, dV - \int_{S_f} \left(\mathbf{u}^T + \delta\mathbf{u}^T\right) \mathbf{f}^S \, dS$$

$$= \frac{1}{2} \int_V \varepsilon^T \mathbf{C} \, \varepsilon \, dV - \int_V \mathbf{u}^T \mathbf{f}^B \, dV - \int_{S_f} \mathbf{u}^T \mathbf{f}^S \, dS \tag{5.45}$$

$$+ \int_V \delta\varepsilon^T \mathbf{C} \, \varepsilon \, dV - \int_V \delta\mathbf{u}^T \mathbf{f}^B \, dV - \int_{S_f} \delta\mathbf{u}^T \mathbf{f}^S \, dS$$

$$+ \frac{1}{2} \int_V \delta\varepsilon^T \mathbf{C} \, \delta\varepsilon \, dV$$

Considering that $\delta\Pi = 0$, equation (5.45) becomes

$$\Pi(\mathbf{u}+\delta\mathbf{u}) = \Pi\,(\mathbf{u}) + \frac{1}{2} \int_V \delta\varepsilon^T \mathbf{C} \, \delta\varepsilon \, dV.$$

Note that

$$\mathcal{U}(\delta\mathbf{u}) = \frac{1}{2}\int_V \delta\varepsilon^T \mathbf{C}\,\delta\varepsilon\,dV$$

is the strain energy corresponding to the displacement variation. The strain energy is always greater than or equal to zero and when the body is properly supported the strain energy is strictly positive for any compatible displacement field. Therefore, we can conclude that the total potential energy attains its minimum at the solution of the elasticity problem.

□

We note that the definition of the total potential energy

$$\Pi(\mathbf{u}) = \frac{1}{2}\int_V (\partial_\varepsilon \mathbf{u})^T \mathbf{C}\partial_\varepsilon \mathbf{u}\,dV - \int_V \mathbf{u}^T\mathbf{f}^B\,dV - \int_{S_f} \mathbf{u}^T\mathbf{f}^S\,dS$$

is quite general and is valid for all mathematical models discussed in Chapters 3 and 4 provided, of course, that the variables are those defined in Tables 4.3 and 4.4. Hence the above results (i) and (ii) are also valid for all these models.

6. The finite element process of solution

At this point it should be clear that the principle of virtual work is an integral form of expressing equilibrium, and when we introduce the other two basic requirements to be satisfied, namely, compatibility and the constitutive behavior into the virtual work statement, we have the mathematical model formulated in integral form. It is then called the variational formulation of the problem.

The question that might be asked is: what is the practical value of the principle of virtual work formulation? For the differential formulation of problems, there are methods of finding analytical solutions. However, for the virtual work formulation no method of solution has been presented so far.

As an example, let us recall our model problem – the 1-D bar problem. Examining the formulation given by equation (5.20), we recognize that even for this problem – the simplest problem – there is no systematic method of finding a displacement field $u(x)$ which would satisfy equation (5.20) for every virtual displacement $\delta u(x)$. On the other hand, we know how to find the solution $u(x)$ for the differential formulation. Therefore, if our objective were to only solve the bar problem, there would be no point in exploring the principle of virtual work as a basis of solution. However, as we step up our ladder of hierarchical mathematical models, *i.e.*, as we consider increasingly more complex models, we realize that to find the analytical solution of the differential formulation of these models becomes increasingly more difficult and indeed impossible, for example, for 3-D geometrically complex problems.

Of course, the principle of virtual work formulations of such problems would not offer an easier route for an exact analytical solution. However, as it will be seen shortly, the principle of virtual work formulation provides an effective basis for a numerical solution – namely, the finite element solution. We introduce next the finite element formulation starting with the solution of the 1-D bar problem.

6.1 Finite element formulation of 1-D bar problem

Let us consider the principle of virtual work formulation for the bar problem as given in equation (5.20), with the notation $\bar{u}(x) = \delta u(x)$. Let the exact solution be $u(X)$ where X is the global longitudinal coordinate. We use this principle to establish a finite element solution $u_h(X)$ which, if properly constructed, can be as close as we wish to the exact solution $u(X)$. The governing equation to find $u_h(X)$ is

$$\int_0^L EA \frac{d\bar{u}_h}{dX}\frac{du_h}{dX}dX = \int_0^L f\bar{u}_h dX + R\bar{u}_h \,|_{X=L} \tag{6.1}$$

for all $\bar{u}_h(X)$ with $\bar{u}_h(0) = 0$.

To proceed further we need to assume a spatial variation for $u_h(X)$ and $\bar{u}_h(X)$. Let us assume the functional form of $u_h(X)$ and $\bar{u}_h(X)$ shown in Figure 6.1.

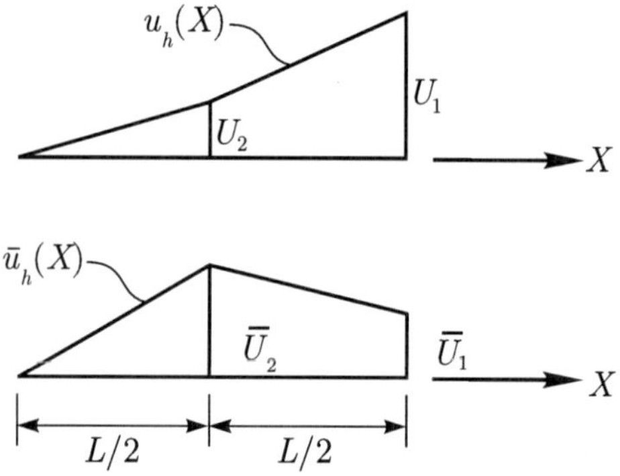

Fig. 6.1. Definition of $u_h(X)$ and $\bar{u}_h(X)$

As implicitly defined in the Figure 6.1, $u_h(X)$ and $\bar{u}_h(X)$ have the following properties:

- They are zero at $X = 0$.
- They assume the values of U_2, $\bar{U}_2$ at $X = L/2$ and U_1, $\bar{U}_1$ at $X = L$.
- They are linear for $0 \le X \le L/2$ and $L/2 \le X \le L$.

Hence the values U_1, U_2 and $\bar{U}_1$, $\bar{U}_2$ completely define $u_h(X)$ and $\bar{u}_h(X)$.

Before further developing the discussion, let us analytically characterize $u_h(X)$ and $\bar{u}_h(X)$. In Figure 6.1, we consider two subdomains: subdomain (1) corresponding to $0 \le X \le L/2$ and subdomain (2) corresponding to $L/2 \le X \le L$. Figure 6.2 shows these subdomains and also defines domain local coordinate systems represented by x.

Consider $u^{(1)}(x)$ as the $u_h(X)$ for subdomain (1) and $u^{(2)}(x)$ as the $u_h(X)$ for subdomain (2). Likewise $\bar{u}^{(1)}(x)$, $\bar{u}^{(2)}(x)$ represent $\bar{u}_h(X)$ for subdomains (1) and (2). From this point onwards, we refer to the subdomain (m) as element (m) and use the superscript (m). We have

$$u^{(1)} = \begin{bmatrix} 0 & \frac{x}{(L/2)} \end{bmatrix} \begin{bmatrix} U_1 \\ U_2 \end{bmatrix} = \mathbf{H}^{(1)}\mathbf{U} \tag{6.2}$$

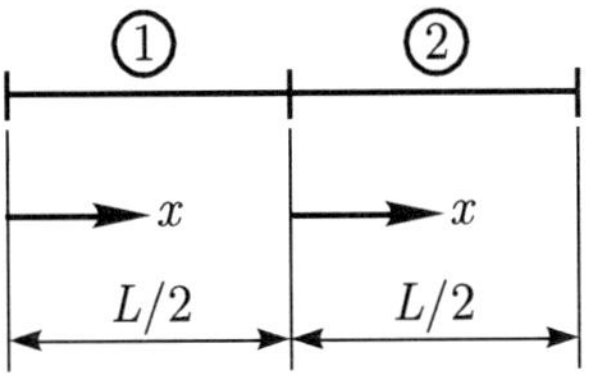

Fig. 6.2. Subdivision of the domain for bar problem

$$u^{(2)} = \begin{bmatrix} \frac{x}{(L/2)} & 1 - \frac{x}{(L/2)} \end{bmatrix} \begin{bmatrix} U_1 \\ U_2 \end{bmatrix} = \mathbf{H}^{(2)}\mathbf{U} \tag{6.3}$$

where $\mathbf{U}^T = \begin{bmatrix} U_1 & U_2 \end{bmatrix}$ and the $\mathbf{H}^{(1)}$ and $\mathbf{H}^{(2)}$ are the displacement interpolation matrices. Similarly

$$\bar{u}^{(1)} = \mathbf{H}^{(1)}\bar{\mathbf{U}} \qquad \bar{u}^{(2)} = \mathbf{H}^{(2)}\bar{\mathbf{U}} \tag{6.4}$$

where $\bar{\mathbf{U}}^T = \begin{bmatrix} \bar{U}_1 & \bar{U}_2 \end{bmatrix}$.

The strains in each element are

$$\varepsilon^{(1)}(x) = \frac{du^{(1)}(x)}{dx} = \begin{bmatrix} 0 & \frac{1}{L/2} \end{bmatrix} \begin{bmatrix} U_1 \\ U_2 \end{bmatrix} = \mathbf{B}^{(1)}\mathbf{U} \tag{6.5}$$

$$\varepsilon^{(2)}(x) = \frac{du^{(2)}(x)}{dx} = \begin{bmatrix} \frac{1}{L/2} & -\frac{1}{L/2} \end{bmatrix} \begin{bmatrix} U_1 \\ U_2 \end{bmatrix} = \mathbf{B}^{(2)}\mathbf{U}. \tag{6.6}$$

These expressions give the strain interpolation matrices $\mathbf{B}^{(m)}$, $m - 1, 2$. The virtual strains are given by

$$\bar{\varepsilon}^{(1)} = \mathbf{B}^{(1)}\bar{\mathbf{U}} \tag{6.7}$$

$$\bar{\varepsilon}^{(2)} = \mathbf{B}^{(2)}\bar{\mathbf{U}}. \tag{6.8}$$

Substituting (6.2) to (6.8) into (6.1) yields

$$\int_0^{L/2} EA\bar{\varepsilon}^{(1)}\varepsilon^{(1)}dx + \int_0^{L/2} EA\bar{\varepsilon}^{(2)}\varepsilon^{(2)}dx$$

$$= \int_0^{L/2} \bar{u}^{(1)} f^{(1)}\, dx + \int_0^{L/2} \bar{u}^{(2)} f^{(2)}\, dx + R\bar{u}^{(2)}\big|_{x=L/2}$$

or

$$EA\int_0^{L/2} \bar{\mathbf{U}}^T {\mathbf{B}^{(1)}}^T \mathbf{B}^{(1)}\,\mathbf{U}\; dx + EA\int_0^{L/2} \bar{\mathbf{U}}^T {\mathbf{B}^{(2)}}^T \mathbf{B}^{(2)}\,\mathbf{U}\; dx$$

$$= \int_0^{L/2} \bar{\mathbf{U}}^T \mathbf{H}^{(1)^T} f^{(1)} dx + \int_0^{L/2} \bar{\mathbf{U}}^T \mathbf{H}^{(2)^T} f^{(2)} dx + \bar{\mathbf{U}}^T R \mathbf{H}^{(2)^T}\Big|_{x=L/2} \tag{6.9}$$

where we used $f^{(m)}$ to express the function $f(X)$ in the element (m) local coordinate x. Defining

$$\mathbf{K}^{(1)} = EA \int_0^{L/2} \mathbf{B}^{(1)^T} \mathbf{B}^{(1)}\, dx \tag{6.10}$$

$$\mathbf{K}^{(2)} = EA \int_0^{L/2} \mathbf{B}^{(2)^T} \mathbf{B}^{(2)}\, dx \tag{6.11}$$

$$\mathbf{R}_B^{(1)} = \int_0^{L/2} \mathbf{H}^{(1)^T} f^{(1)}\, dx \tag{6.12}$$

$$\mathbf{R}_B^{(2)} = \int_0^{L/2} \mathbf{H}^{(2)^T} f^{(2)}\, dx \tag{6.13}$$

$$\mathbf{R}_C = R\mathbf{H}^{(2)^T}\Big|_{x=L/2} = R \begin{bmatrix} 1 \\ 0 \end{bmatrix} = \begin{bmatrix} R \\ 0 \end{bmatrix} \tag{6.14}$$

equation (6.9) can be re-written as

$$\bar{\mathbf{U}}^T \left[\left(\mathbf{K}^{(1)} + \mathbf{K}^{(2)} \right) \mathbf{U} - \left(\mathbf{R}_B^{(1)} + \mathbf{R}_B^{(2)} \right) - \mathbf{R}_C \right] = 0. \tag{6.15}$$

Defining

$$\mathbf{K} = \mathbf{K}^{(1)} + \mathbf{K}^{(2)} = \frac{EA}{(L/2)} \begin{bmatrix} 1 & -1 \\ -1 & 2 \end{bmatrix} \tag{6.16}$$

$$\mathbf{R}_B = \mathbf{R}_B^{(1)} + \mathbf{R}_B^{(2)} = \begin{bmatrix} \int_0^{L/2} \frac{x}{L/2} f^{(2)}\, dx \\ \int_0^{L/2} \frac{x}{L/2} f^{(1)}\, dx + \int_0^{L/2} \left(1 - \frac{x}{L/2}\right) f^{(2)}\, dx \end{bmatrix} \tag{6.17}$$

(6.15) becomes

$$\bar{\mathbf{U}}^T \left[\mathbf{K}\mathbf{U} - \mathbf{R}_B - \mathbf{R}_C \right] = 0. \tag{6.18}$$

The fact that $\bar{u}_h(X)$ is arbitrary, provided it obeys the functional form given in Figure 6.1, is guaranteed by letting $\bar{\mathbf{U}}$ be an arbitrary vector. The only way for the product of the row matrix $\bar{\mathbf{U}}^T$ by the column matrix $[\mathbf{KU} - \mathbf{R}_B - \mathbf{R}_C]$ to be zero, with $\bar{\mathbf{U}}$ arbitrary is to have

$$\mathbf{K}\mathbf{U} = \mathbf{R}_B + \mathbf{R}_C. \tag{6.19}$$

Equation (6.19) represents two linear algebraic equations to be solved for U_1 and U_2, which give $u_h(X)$ by equations (6.2) and (6.3).

We have shown that by choosing a functional form for both the solution sought and the virtual displacements, we can find a solution that satisfies the virtual work statement (equation (6.1) within the assumed displacement field).

Considering equation (6.19), we realize that these equations correspond to those of the truss model shown in Figure 6.3 when the nodal forces are applied at the truss nodes according to (6.14) and (6.17). The $\mathbf{K}$ matrix

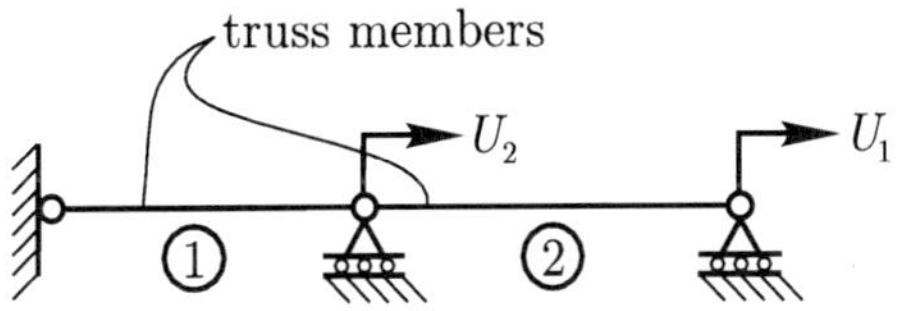

Fig. 6.3. Truss model

of equation (6.16) has the natural interpretation of a stiffness matrix. The column matrices $\mathbf{R}_B$ and $\mathbf{R}_C$ collect nodal loads. In fact, $\mathbf{R}_B$ is lumping the force per unit length to the nodes.

Of course, the principle of virtual work in (6.1) also holds when the bar is unrestrained at the left end. Hence, the problem can be solved by using this principle for the unrestrained bar and then imposing the restraint prior to solving the matrix equations. In this case, we would consider u_h and $\bar{u}_h$ as defined in Figure 6.4.

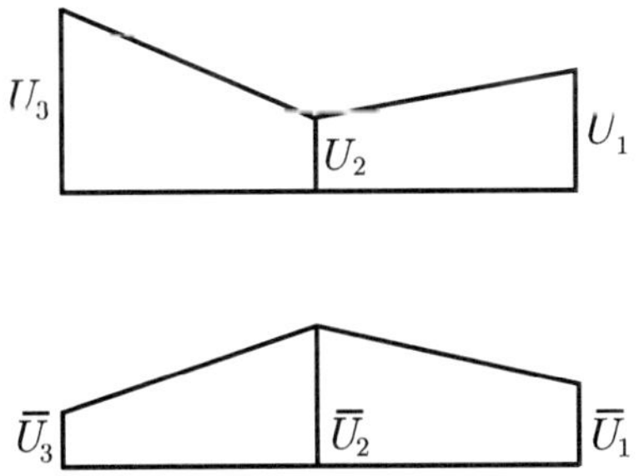

Fig. 6.4. Alternative definition of $u_h(x)$ and $\bar{u}_h(x)$

The derivations would be as above, but using $\mathbf{U}^T = \begin{bmatrix} U_1 & U_2 & U_3 \end{bmatrix}$, $\bar{\mathbf{U}}^T = \begin{bmatrix} \bar{U}_1 & \bar{U}_2 & \bar{U}_3 \end{bmatrix}$ and

$$\mathbf{H}^{(1)} = \begin{bmatrix} 0 & \frac{x}{(L/2)} & 1 - \frac{x}{(L/2)} \end{bmatrix}$$

$$\mathbf{H}^{(2)} = \begin{bmatrix} \frac{x}{(L/2)} & 1 - \frac{x}{(L/2)} & 0 \end{bmatrix}$$

$$\mathbf{B}^{(1)} = \begin{bmatrix} 0 & \frac{1}{(L/2)} & -\frac{1}{(L/2)} \end{bmatrix}$$

$$\mathbf{B}^{(2)} = \begin{bmatrix} \frac{1}{(L/2)} & -\frac{1}{(L/2)} & 0 \end{bmatrix}$$

leading to

$$\mathbf{K}^{(1)} = \frac{EA}{(L/2)} \begin{bmatrix} 0 & 0 & 0 \\ 0 & 1 & -1 \\ 0 & -1 & 1 \end{bmatrix} \tag{6.20}$$

$$\mathbf{K}^{(2)} = \frac{EA}{(L/2)} \begin{bmatrix} 1 & -1 & 0 \\ -1 & 1 & 0 \\ 0 & 0 & 0 \end{bmatrix} \tag{6.21}$$

and the matrices of equation (6.19) would be

$$\mathbf{K} = \frac{EA}{(L/2)} \begin{bmatrix} 1 & -1 & 0 \\ -1 & 2 & -1 \\ 0 & -1 & 1 \end{bmatrix} \tag{6.22}$$

$$\mathbf{R}_B = \begin{bmatrix} \int_0^{L/2} \frac{x}{L/2} f^{(2)}\, dx \\ \int_0^{L/2} \frac{x}{L/2} f^{(1)}\, dx + \int_0^{L/2} \left(1 - \frac{x}{L/2}\right) f^{(2)}\, dx \\ \int_0^{L/2} \left(1 - \frac{x}{L/2}\right) f^{(1)}\, dx \end{bmatrix} \tag{6.23}$$

and

$$\mathbf{R}_C = \begin{bmatrix} R \\ 0 \\ R_b \end{bmatrix}. \tag{6.24}$$

Here R_b is the reaction at $X = 0$. We would now impose $U_3 = 0$ and solve the equations for U_1 and U_2.

We note that if $f = 0$, the solutions based on the virtual work approach and that of the truss model of Figure 6.3 are the same. In fact, the matrices $\mathbf{K}$ and $\mathbf{R}_C$ are exactly the same for both approaches and $\mathbf{R}_B = \mathbf{0}$. Since the truss model solution is exact for forces applied to the nodes only, the solution of the virtual work approach is also exact in such a case. Of course, this happens since the functional assumption for the displacements (see Figure 6.1) is the functional form of the analytical solution and, in fact, only one element gives the exact solution when $f = 0$.

When $f(x) \neq 0$ the integration of the differential equation (5.7) gives a $u(X)$ which is not linear in X. Hence, since $u_h(X)$ is piecewise linear, as described in Figure 6.1, the method can not give the exact displacement and stress solution for all points in the bar (unless the procedure described in Section 6.4.2 is used as well, see also Bathe, 1996). We address this issue in more detail later on.

When we anticipate that the exact solution $u(X)$ is not piecewise linear, it is intuitive to subdivide the original domain into more than two subdomains, *i.e.*, to consider more elements. In such way, the $u_h(X)$ – although still piecewise linear – may be closer to $u(X)$.

Let us consider the subdivision of the original domain into more than two subdomains (elements) as schematically represented in Figure 6.5. In

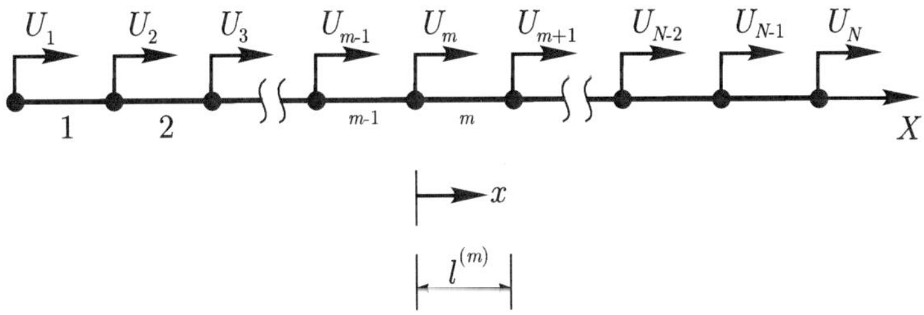

Fig. 6.5. Bar problem discretized by several elements

this figure we also give the nodal points and nodal degrees of freedom. The function $u_h(X)$ will be defined to be linear in each element and at nodal point i of coordinate X_i, $u_h(X_i) = U_i$. Therefore

$$u^{(m)}(x) = U_m \left(1 - \frac{x}{l^{(m)}}\right) + U_{m+1}\frac{x}{l^{(m)}} \tag{6.25}$$

where $l^{(m)}$ is the length of element (m). Note that this displacement interpolation can represent the rigid body motion of the element, which is indeed a fundamental requirement for displacement interpolations, see Bathe, 1996. Letting

$$\mathbf{U}^T = \begin{bmatrix} U_1 & U_2 & \dots & U_N \end{bmatrix}$$

we can define $\mathbf{H}^{(m)}$

$$u^{(m)} = \mathbf{H}^{(m)}\mathbf{U} \tag{6.26}$$

and $\mathbf{B}^{(m)}$

$$\varepsilon^{(m)} = \mathbf{B}^{(m)}\mathbf{U} \tag{6.27}$$

where $\mathbf{H}^{(m)}$ and $\mathbf{B}^{(m)}$ are

$$\mathbf{H}^{(m)} = \begin{bmatrix} 0 & \dots & 0 & \left(1 - \frac{x}{l^{(m)}}\right) & \frac{x}{l^{(m)}} & 0 & \dots & 0 \end{bmatrix}$$

$$\mathbf{B}^{(m)} = \begin{bmatrix} 0 & \dots & 0 & -\frac{1}{l^{(m)}} & \frac{1}{l^{(m)}} & 0 & \dots & 0 \end{bmatrix}$$

with only two non-zero entries. The virtual quantities are also defined by

$$\bar{u}^{(m)} = \mathbf{H}^{(m)}\bar{\mathbf{U}} \tag{6.28}$$

$$\bar{\varepsilon}^{(m)} = \mathbf{B}^{(m)}\bar{\mathbf{U}} \tag{6.29}$$

where

$$\bar{\mathbf{U}}^T = \begin{bmatrix} \bar{U}_1 & \bar{U}_2 & \dots & \bar{U}_N \end{bmatrix}$$

is the column matrix of virtual nodal displacements.

Let $\mathbf{R}_C$ collect the nodal concentrated forces corresponding to $\mathbf{U}$

$$\mathbf{R}_C^T = \begin{bmatrix} R_{C_1} & R_{C_2} & \cdots & R_{C_N} \end{bmatrix}. \tag{6.30}$$

Of course, many entries might be zero. Also some entries might correspond to unknown nodal reactions associated with the degrees of freedom for which the displacements are prescribed.

Substituting (6.26) to (6.30) into the virtual work expression (6.1), we obtain

$$\sum_{m=1}^{n_e} \int_0^{l^{(m)}} \bar{\mathbf{U}}^T \mathbf{B}^{(m)^T} E^{(m)} A^{(m)} \mathbf{B}^{(m)} \mathbf{U} \, dx$$
$$= \sum_{m=1}^{n_e} \int_0^{l^{(m)}} \bar{\mathbf{U}}^T \mathbf{H}^{(m)^T} f^{(m)} \, dx + \bar{\mathbf{U}}^T \mathbf{R}_C$$

where n_e is the number of elements, and $E^{(m)}$ and $A^{(m)}$ are the Young's modulus and the cross-sectional area of element (m). This equation can be re-written as

$$\bar{\mathbf{U}}^T \left(\sum_{m=1}^{n_e} \int_0^{l^{(m)}} \mathbf{B}^{(m)^T} E^{(m)} A^{(m)} \mathbf{B}^{(m)} \, dx \right) \mathbf{U}$$
$$= \bar{\mathbf{U}}^T \left(\sum_{m=1}^{n_e} \int_0^{l^{(m)}} \mathbf{H}^{(m)^T} f^{(m)} \, dx + \mathbf{R}_C \right).$$

Defining

$$\mathbf{K}^{(m)} = \int_0^{l^{(m)}} \mathbf{B}^{(m)^T} E^{(m)} A^{(m)} \mathbf{B}^{(m)} \, dx \tag{6.31}$$

$$\mathbf{R}_B^{(m)} = \int_0^{l^{(m)}} \mathbf{H}^{(m)^T} f^{(m)} \, dx \tag{6.32}$$

$$\mathbf{K} = \sum_{m=1}^{n_e} \mathbf{K}^{(m)} \tag{6.33}$$

$$\mathbf{R}_B = \sum_{m=1}^{n_e} \mathbf{R}_B^{(m)} \tag{6.34}$$

we obtain

$$\bar{\mathbf{U}}^T \left[\mathbf{K}\mathbf{U} - \mathbf{R}_B - \mathbf{R}_C\right] = 0.$$

We should impose that $\bar{\mathbf{U}}^T$ is arbitrary to reflect the condition that $\bar{u}_h(X)$ is any function which obeys the specified functional form (linear in each element). Using the same arguments as before, this condition implies that

$$\mathbf{K}\mathbf{U} = \mathbf{R} \tag{6.35}$$

with $\mathbf{R} = \mathbf{R}_B + \mathbf{R}_C$.

Equation (6.35) is a system of N algebraic equations to be solved for the nodal displacements. We note that at least one nodal displacement should be prescribed to prevent rigid body motion.

The matrix $\mathbf{K}^{(m)}$, which can be interpreted as the element (m) stiffness matrix, can be evaluated in closed form due to the simplicity of the element used. In fact, from equation (6.31), we can write

$$\mathbf{K}^{(m)} = \frac{E^{(m)} A^{(m)}}{l^{(m)}} \begin{bmatrix} \ddots & & & & \\ & 1 & -1 & & \\ & -1 & 1 & & \\ & & & \ddots & \\ & & & & \ddots \end{bmatrix}_{(N \times N)} \begin{matrix} \\ m \\ m+1 \\ \\ \\ \end{matrix}$$

(columns m, $m+1$)

where only the non-zero entries are shown. We observe that this stiffness matrix is exactly the stiffness matrix we would obtain for a truss element if it were placed linking nodes m and $m+1$ and if only the axial degrees of freedom were considered. The assemblage of the **K** matrix is given by equation (6.33) which is exactly equal to equation (2.49). Recall that equation (2.49), which gives the assemblage of the **K** matrix for trusses, was derived imposing equilibrium and displacement compatibility. In the finite element formulation equation (6.33) is obtained directly from the principle of virtual work. Also

$$\mathbf{R}_B^{(m)} = \begin{bmatrix} \cdot \\ \cdot \\ \cdot \\ \int_0^{l^{(m)}} \left(1 - \frac{x}{l^{(m)}}\right) f^{(m)}\, dx \\ \int_0^{l^{(m)}} \frac{x}{l^{(m)}} f^{(m)}\, dx \\ \cdot \\ \cdot \\ \cdot \end{bmatrix} \begin{matrix} \\ \\ \\ m \\ m+1 \\ \\ \\ \\ \end{matrix}$$

where again only the non-zero entries are shown.

In the finite element assemblage, we can define an element matrix containing only the stiffness coefficients associated with the element degrees of freedom in a local numbering as shown in Figure 6.6. Considering element (m), the element matrix is given by

$$\mathbf{k}^{(m)} = \frac{EA}{l} \begin{bmatrix} 1 & -1 \\ -1 & 1 \end{bmatrix}$$

where E, A and l are those for element (m).

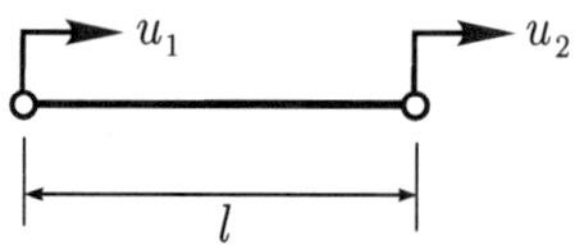

Fig. 6.6. Generic two-node element

The assemblage process implicitly given by equation (6.33) is efficiently performed by using the element matrices corresponding to the local numbering and adding their contributions to the global matrix using the correspondence between local and global degrees of freedom. This assemblage procedure has been discussed in detail in Chapters 2 and 4.

Example 6.1

Consider the problem described in Figure 5.1 with $f(X) = aX$, $L = 2$ m, $a = 10$ kN/m^2, $R = 20$ kN, $A = 2 \times 10^{-4}$ m^2 and $E = 2.1 \times 10^8$ kN/m^2.

(i) Derive the analytical solution.
(ii) Detail the solution for a uniform discretization of 2 elements.
(iii) Formulate and solve the problem for uniform discretizations of 4, 8 and 16 elements.
(iv) Discuss the stress solutions.

Solution

(i) The analytical solution can be derived by solving the differential formulation which for this case reads (see equations (5.7) to (5.9))

$$EA\frac{d^2u}{dX^2} + aX = 0$$

$$u(0) = 0$$

$$EA\frac{du}{dX}(L) = R.$$

The solution is given by

$$u(X) = -\frac{a}{6EA}X^3 + \left(\frac{R}{EA} + \frac{aL^2}{2EA}\right)X. \tag{6.36}$$

(ii) Using equation (6.23), we have

$$\mathbf{R}_B = \begin{bmatrix} \int_0^{L/2} \frac{x}{(L/2)} a\left(x + \frac{L}{2}\right)\, dx \\ \int_0^{L/2} \frac{x}{(L/2)} ax\, dx + \int_0^{L/2}\left(1 - \frac{x}{(L/2)}\right) a\left(x + \frac{L}{2}\right)\, dx \\ \int_0^{L/2}\left(1 - \frac{x}{(L/2)}\right) ax\, dx \end{bmatrix}$$

where we used the degree of freedom numbering shown in Figure 6.4. Therefore

$$\mathbf{R}_B = \frac{aL^2}{24}\begin{bmatrix} 5 \\ 6 \\ 1 \end{bmatrix}$$

and using (6.22) and (6.24), equation (6.35) reads

$$\frac{EA}{(L/2)}\begin{bmatrix} 1 & -1 & 0 \\ -1 & 2 & -1 \\ 0 & -1 & 1 \end{bmatrix}\begin{bmatrix} U_1 \\ U_2 \\ U_3 \end{bmatrix} = \frac{aL^2}{24}\begin{bmatrix} 5 \\ 6 \\ 1 \end{bmatrix} + \begin{bmatrix} R \\ 0 \\ R_b \end{bmatrix}.$$

Imposing $U_3 = 0$, we obtain

$$U_1 = 1.587 \times 10^{-3} \text{ m}, \qquad U_2 = 9.127 \times 10^{-4} \text{ m}.$$

The reaction at node 1 is

$$R_b = -40 \text{ kN}$$

and the solution for each element is

$$u^{(1)}(x) = \frac{x}{(L/2)}U_2$$

and

$$u^{(2)}(x) = \frac{x}{(L/2)}U_1 + \left(1 - \frac{x}{(L/2)}\right)U_2.$$

The analytical and finite element solutions are plotted in Figure 6.7.

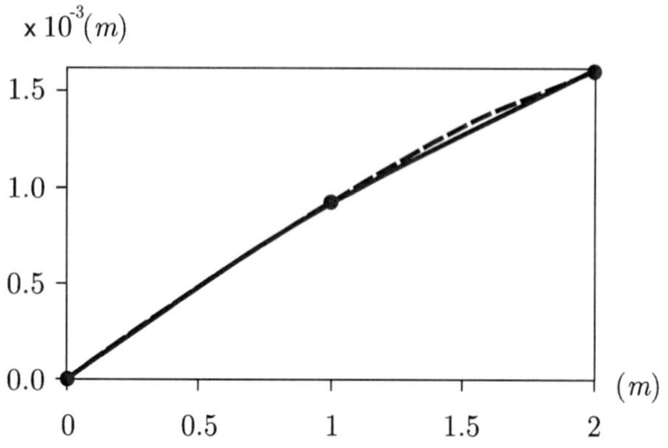

Fig. 6.7. Solutions for displacements are shown for the 2 element mesh (solid line) and the exact solution (dashed line)

(iii) The matrix equation is generically given by

$$\frac{EA}{\ell}\begin{bmatrix} 1 & -1 & & & & & \\ -1 & 2 & -1 & & & & \\ & -1 & 2 & -1 & & & \\ & & \cdot & \cdot & \cdot & & \\ & & & \cdot & \cdot & \cdot & \\ & & & & -1 & 2 & -1 \\ & & & & & -1 & 1 \end{bmatrix}\begin{bmatrix} U_1 \\ U_2 \\ \cdot \\ \cdot \\ \cdot \\ \cdot \\ U_N \end{bmatrix} = \begin{bmatrix} R_B^1 \\ R_B^2 \\ \cdot \\ \cdot \\ \cdot \\ \cdot \\ R_B^N \end{bmatrix} + \begin{bmatrix} R \\ 0 \\ \cdot \\ \cdot \\ \cdot \\ 0 \\ R_b \end{bmatrix}$$

where $\ell = L/n_e$ and $N = n_e + 1$ (n_e is the number of elements). The $\mathbf{R}_B^{(m)}$ and $\mathbf{R}_B$ are obtained using (6.32) and (6.34) respectively. We do not show graphically the displacement solutions for the refined meshes since they are all very close to the analytical solution.

(iv) The stress field can be evaluated in each element by

$$\tau^{(m)} = E\varepsilon^{(m)} = E\frac{du^{(m)}}{dX} = E\mathbf{B}^{(m)}\mathbf{U}.$$

Since the displacements are piecewise linear, the stress in each element is constant. In Figure 6.8 we report the stress distributions obtained. Of course,

$$A\frac{d\tau^{(m)}}{dX} + f \neq 0$$

since $f(X)$ is linear and $\tau^{(m)}$ is constant. The analytical solution for the stress can be derived from (6.36)

$$\tau(X) = -\frac{aX^2}{2EA} + \frac{aL^2}{2EA}.$$

The finite element solutions for the stresses display discontinuities at the element interfaces which are referred to as stress jumps. Of course, the exact solution is a continuous stress field since no concentrated forces are applied along the bar. As more elements are used the magnitude of the stress jumps is decreasing and an acceptable level of stress jumps can be reached. More details on the solution of this problem are given in Bathe, 1996.

□

Equilibrium properties of finite element solutions in 1-D

As exemplified in Example 6.1, the differential equilibrium is in general not satisfied by the finite element solution. However, two equilibrium properties are always satisfied by a finite element solution no matter how coarse the mesh might be. These properties are:

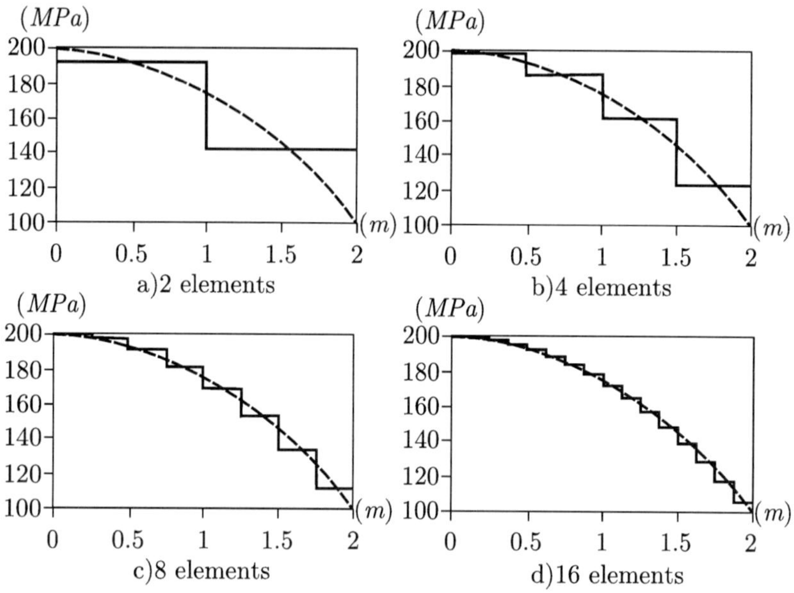

Fig. 6.8. Stress solutions. The stepped full line represents the finite element solutions and the dashed line represents the analytical solution

1. *Nodal point equilibrium*
2. *Element equilibrium*

We detail below these properties following Bathe, 1996. First, we define the column matrix $\mathbf{F}^{(m)}$ by

$$\mathbf{F}^{(m)} = \mathbf{K}^{(m)}\mathbf{U} \tag{6.37}$$

which gives the nodal point forces for element (m). The nodal displacements given by $\mathbf{U}$ correspond to the solution of the problem in consideration and, hence, these forces are *acting onto the element* at the element nodal points.

We re-write (6.37) as

$$\mathbf{F}^{(m)} = \mathbf{K}^{(m)}\mathbf{U} = \int_{l^{(m)}} \mathbf{B}^{(m)T}\tau^{(m)}A\,dx$$

where $\tau^{(m)}$ is the solution for the stress field for element (m), *i.e.*, $\tau^{(m)} = E\mathbf{B}^{(m)}\mathbf{U}$. From this expression $\mathbf{F}^{(m)}$ can also be interpreted to be equivalent in the virtual work sense to the element stresses, since we can left multiply the equation above by any finite element virtual displacement field and obtain

$$\bar{\mathbf{U}}^T\mathbf{F}^{(m)} = \int_{l^{(m)}} \bar{\varepsilon}^{(m)}\tau^{(m)}A\,dx. \tag{6.38}$$

Property 1 follows from

$$\sum_{m=1}^{n_e} \mathbf{F}^{(m)} = \sum_{m=1}^{n_e} \mathbf{K}^{(m)}\mathbf{U} = \mathbf{R}$$

$$\sum_{m=1}^{n_e} \mathbf{F}^{(m)} = \mathbf{R}. \tag{6.39}$$

Hence, at each node the sum of the element nodal point forces is in equilibrium with the externally applied nodal loads (which include the lumped distributed forces ($\mathbf{R}_B$)). Here, $\mathbf{F}^{(m)}$ are the forces that give the stress contribution of element (m) to the nodal point equilibrium.

To show Property 2 we use equation (6.38) with the nodal virtual displacements

$$\bar{\mathbf{U}}^T = \begin{bmatrix} \Delta & \cdots & \Delta & \Delta & \cdots & \Delta \end{bmatrix}$$

where Δ is a constant displacement applied to the element nodes, *i.e.*, we are considering a rigid element translation as the virtual displacement field. We obtain

$$\bar{\mathbf{U}}^T\mathbf{F}^{(m)} = 0 \tag{6.40}$$

since $\bar{\varepsilon}^{(m)} = 0$. Also

$$\mathbf{F}^{(m)^T} = \begin{bmatrix} 0 & \cdots & 0 & F_1 & F_2 & 0 & \cdots & 0 \end{bmatrix}$$

and hence (6.40) yields

$$F_1 = -F_2 \tag{6.41}$$

which proves that element (m) is in equilibrium under its nodal point forces.

In Figure 6.9 we summarize the above results for the two element mesh of Example 6.1. We observe that for all nodes equilibrium is satisfied considering: the external loads; the lumped distributed loads and the element nodal point forces. Actually this is the "pictorial" form of equation (6.39). We also show the element nodal point forces acting on the elements which self-equilibrate (see equation (6.41)).

Summary of the finite element procedure for 1-D linear elements

Although the bar problem and the finite element solutions using linear elements are a special and simple application, the solution process already displays some key characteristics of the finite element method for structural mechanics:

- The finite element solution is based on the use of the principle of virtual work.
- The problem domain is partitioned into subdomains called elements.

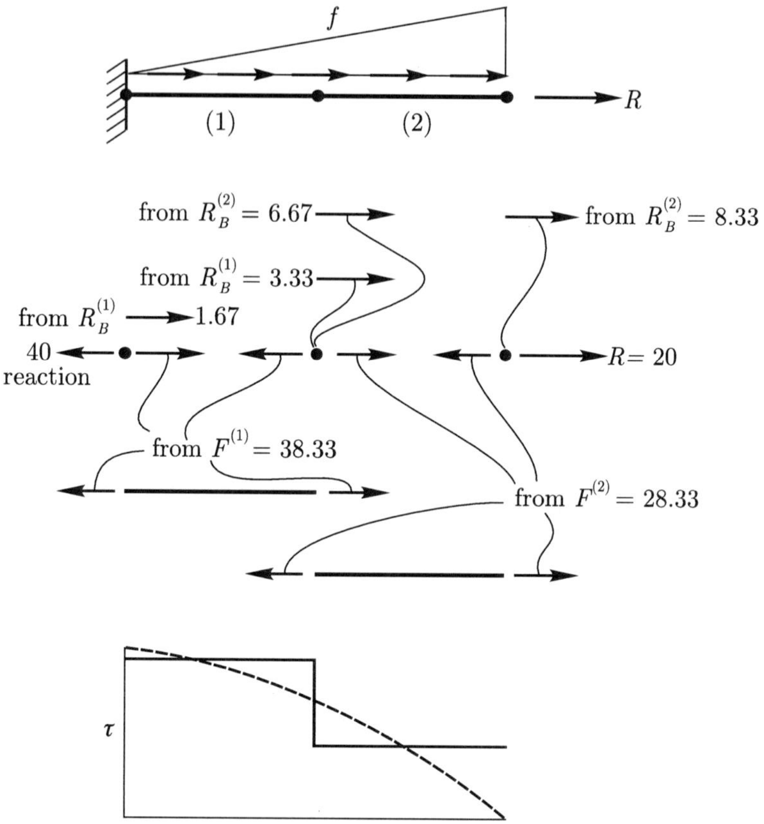

Fig. 6.9. Schematic representation of forces/stress variables for bar problem. Values shown are obtained considering $L = 2$ m, $A = 2 \times 10^{-4}$ m^2, $E = 2.1 \times 10^8$ kN/m^2, $f = 10X$ kN/m, $R = 20$ kN

- A set of points, called nodal points, are defined in each element (the nodal points are at the element end points for the above formulation).
- The unknown displacements (the displacement of the bar sections $u(X)$ for the above problem) are assumed to have a given functional form through the interpolation functions within each element (linear for the above problem).
- The values of the displacements at every point within the elements are uniquely defined by these interpolation functions and their nodal point values (see equation (6.25) for the above problem).
- Given the values at the nodal points, $u_h(X)$ is defined for all X, $0 \leq X \leq L$.
- The principle of virtual work is imposed considering the assumptions for the unknown displacements. The virtual quantities are assumed to have

the same functional form as the real displacements (see equation (6.25) and (6.28) for the above problem).
- The application of the principle of virtual work under these conditions leads to a system of algebraic equations (equation (6.35)).
- The solution of the algebraic equations gives the values of the displacements at the nodes $(\mathbf{U})$.
- Using the displacement nodal values, the displacements throughout the domain are given (use again equation (6.25) for the above problem to obtain $u_h(X)$).
- The finite element solution is an approximation for the exact solution of the problem.
- The finite element solution satisfies: exactly *compatibility*, since the *strain-displacement relation* (equation (6.27) for this case) and the displacement boundary conditions are imposed; exactly the *constitutive relation*, since the stress-strain law ($\tau^{(m)} = E^{(m)}\varepsilon^{(m)}$ for the above problem) is imposed; approximately equilibrium (see the discussion of "nodal equilibrium" and "element equilibrium").
- As we consider a finer partition of the domain, *i.e*, more and smaller elements, the conditions of "nodal equilibrium" and "element equilibrium" (as detailed for the two element bar problem) represent closer and closer differential equilibrium. Therefore, it is intuitive that we can satisfy differential equilibrium as precisely as required, by considering sufficiently small elements. We can therefore arrive at finite element solutions that are as close to the exact solution of the mathematical model as desired.
- A given partition of the domain into elements is referred to as a finite element mesh. The process of further subdividing the domain into more elements is referred to as mesh refinement.

Quadratic 1-D finite elements

In the previous example we used linear elements, *i.e*, the displacement interpolation was defined by a linear function in each element. We can define a quadratic element, *i.e*, an element in which the displacement interpolation is parabolic, but then need to use three-nodes per element.

In Figure 6.10 we show a generic three-node element and also the local numbering of the element degrees of freedom. The interpolation can be defined by

$$u^{(m)}(x) = u_1 h_1(x) + u_2 h_2(x) + u_3 h_3(x). \tag{6.42}$$

Of course, $U_{k-1} = u_1$, $U_{k+1} = u_2$ and $U_k = u_3$. The interpolation functions can be easily derived by observing that each $h_i(x)$ should be parabolic and should be 1 at the position of the node i and 0 at the remaining nodal positions, that is

$$u^{(m)}(x = 0) = u_1 = U_{k-1}$$

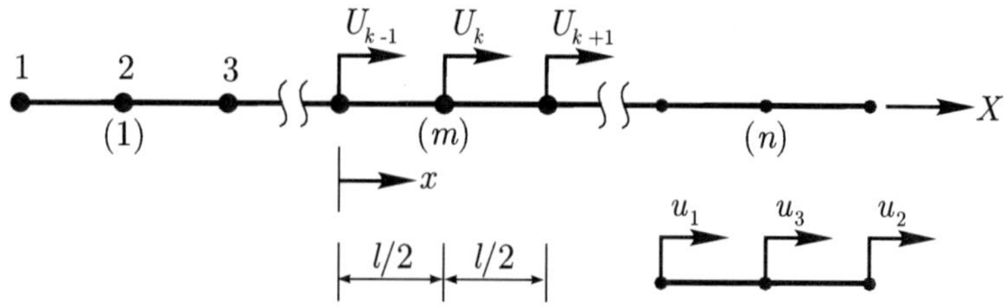

Fig. 6.10. Generic three-node element

$$u^{(m)}(x = l/2) = u_3 = U_k$$
$$u^{(m)}(x = l) = u_2 = U_{k+1}.$$

If these conditions are satisfied, we have that $u^{(m)}(x)$ is parabolic in x and we directly obtain

$$h_1(x) = \frac{\left(x - \frac{l}{2}\right)(x - l)}{\left(0 - \frac{l}{2}\right)(0 - l)} = \frac{2x^2 - 3xl + l^2}{l^2}$$

$$h_2(x) = \frac{(x - 0)\left(x - \frac{l}{2}\right)}{(l - 0)\left(l - \frac{l}{2}\right)} = \frac{2x^2 - xl}{l^2}$$

$$h_3(x) = \frac{(x - 0)(x - l)}{\left(\frac{l}{2} - 0\right)\left(\frac{l}{2} - l\right)} = -\frac{4\left(x^2 - xl\right)}{l^2}$$

Of course, l varies from element to element and could have been represented by $l^{(m)}$.

The interpolation matrix $\mathbf{H}^{(m)}$ is given by

$$u^{(m)} = \mathbf{H}^{(m)}\mathbf{U}$$

where

$$\mathbf{H}^{(m)} = \begin{bmatrix} 0 & \dots & 0 & h_1 & h_3 & h_2 & 0 & \dots & 0 \end{bmatrix}$$

where h_1 is placed in column $k-1$, h_3 in k and h_2 in $k+1$. We note that $\sum_{i=1}^{3} h_i = 1$ which is the requirement that the element can represent a rigid body motion. The derivations leading to $\mathbf{KU} = \mathbf{R}$ would follow equations (6.26) to (6.35).

Example 6.2

Solve the bar problem of Figure 5.1 for $f(x) = a$ (constant) and $f(x) = ax$ (linear) with only one three-node element.

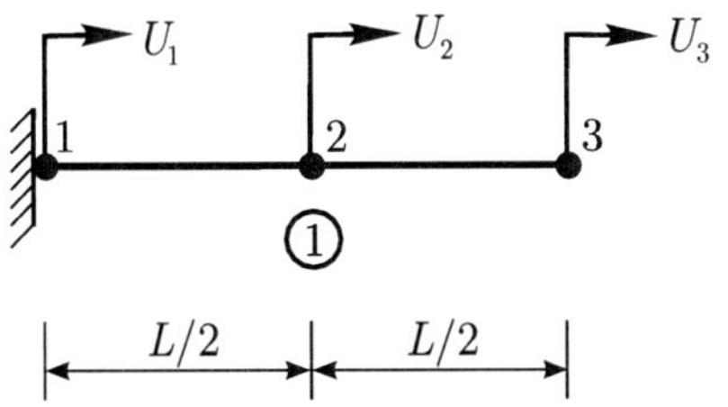

Fig. 6.11. Bar model with one three-node element

Solution

Let us consider the finite element model as described in Figure 6.11. Since we have only one element

$$\mathbf{K} = \mathbf{K}^{(1)} = \int_0^{L/2} \mathbf{B}^{(1)^T} E^{(1)} A^{(1)} \mathbf{B}^{(1)} \, dx$$

where

$$\mathbf{B}^{(1)} = \left[\begin{array}{ccc} \frac{4x}{L^2} - \frac{3}{L} & \frac{-8x}{L^2} + \frac{4}{L} & \frac{4x}{L^2} - \frac{1}{L} \end{array} \right]$$

and therefore

$$\mathbf{K} = \frac{EA}{3L} \begin{bmatrix} 7 & -8 & 1 \\ -8 & 16 & -8 \\ 1 & -8 & 7 \end{bmatrix}$$

Considering first $f = a$

$$\mathbf{R}_B = \mathbf{R}_B^{(1)} = \int_0^{L/2} \mathbf{H}^{(1)^T} f^{\,(1)} dx$$

$$\mathbf{R}_B = \frac{aL}{3} \begin{bmatrix} \frac{1}{2} \\ 2 \\ \frac{1}{2} \end{bmatrix}$$

and

$$\mathbf{R}_C = R \mathbf{H}^{(1)^T} |_{x=l} = R \begin{bmatrix} 0 \\ 0 \\ 1 \end{bmatrix}$$

Considering $R = 1$ the algebraic system is

$$\frac{EA}{3L}\begin{bmatrix} 7 & -8 & 1 \\ -8 & 16 & -8 \\ 1 & -8 & 7 \end{bmatrix}\begin{bmatrix} U_1 \\ U_2 \\ U_3 \end{bmatrix} = \begin{bmatrix} \frac{aL}{6} \\ \frac{2aL}{3} \\ \frac{aL}{6}+1 \end{bmatrix}$$

Since $U_1 = 0$ we are lead to the reduced system

$$\frac{EA}{3L}\begin{bmatrix} 16 & -8 \\ -8 & 7 \end{bmatrix}\begin{bmatrix} U_2 \\ U_3 \end{bmatrix} = \begin{bmatrix} \frac{2aL}{3} \\ \frac{aL}{6}+1 \end{bmatrix}. \tag{6.43}$$

The solution is

$$\mathbf{U}^T = \begin{bmatrix} 0 & \frac{3L^2a+4L}{8EA} & \frac{L^2a+2L}{2EA} \end{bmatrix}$$

which leads to

$$u^{(1)}(x) = \mathbf{H}^{(1)}\mathbf{U} = \frac{-ax^2+2Lax+2x}{2EA}$$

since we have only one element

$$u_h(X) = \frac{-aX^2+2LaX+2X}{2EA} \tag{6.44}$$

Since the interpolation is quadratic and the analytical solution is also quadratic in displacements for a constant f, we should expect the finite element solution to be exact. In fact, substituting (6.44) in the differential equation we have

$$EA\frac{d^2u_h}{dX^2} + f = -a + a = 0.$$

In order to obtain the solution for $f = ax$ we need only to calculate the corresponding $\mathbf{R}_B$

$$\mathbf{R}_B = \frac{aL^2}{3}\begin{bmatrix} 0 \\ 1 \\ \frac{1}{2} \end{bmatrix}$$

and solving (6.43) using this $\mathbf{R}_B$, we find

$$\mathbf{U}^T = \begin{bmatrix} 0 & \frac{11aL^3+24L}{48EA} & \frac{aL^3+3L}{3EA} \end{bmatrix}.$$

Therefore

$$u_h(X) = \frac{-3X^2aL+7aL^2X+12X}{12EA}$$

and substituting into the differential equation

$$EA\frac{d^2u_h}{dx^2} + f = \frac{-La}{2} + aX \neq 0$$

which shows, as expected, that the differential equation is not exactly satisfied.

□

6.2 Convergence properties of 1-D finite element solutions

We introduced the finite element formulation for the 1-D problem in Section 6.1. The fundamental idea was to choose a functional form for the axial displacements and invoke the principle of virtual work. Let us focus, for the moment, on the piecewise linear interpolation of the displacements.

We derived how to obtain the finite element solution $u_h(x)$ for this problem and pointed out that this solution is in general only an approximation to the exact solution. However, taking into account the compatibility, constitutive and equilibrium conditions that the finite element solution satisfies, we discussed that as we refine the mesh, considering more and more elements, the finite element solution $u_h(x)$ "converges" to the exact solution, *i.e.*, the finite element solution becomes successively closer to the exact solution, both for displacements and associated stresses. The results of Example 6.1 are an illustration. The objective of this section is to discuss why and how convergence is obtained for 1-D finite element solutions. In Section 6.5.1 we briefly discuss convergence for 2-D and 3-D displacement-based finite element solutions.

The discussion of convergence requires that we measure the "distance" between the finite element and exact solutions and requires some mathematical tools which we shall introduce at the minimum level needed. For a deeper discussion see Bathe, 1996, Chapelle and Bathe, 2010*a* and Grätsch and Bathe, 2005.

Let us define the bilinear form

$$a(g, w) = \int_0^L EA\frac{dg}{dx}\frac{dw}{dx}\, dx$$

and the linear form

$$(g, w) = \int_0^L gw\, dx$$

where $g(x)$ and $w(x)$ are real valued functions defined on the interval $[0, L]$. The constant value L represents the bar length, E represents the Young's modulus and A the bar cross-sectional area.

Then the principle of virtual work (see equation (5.20)) can be written as:

Find $u(x)$, $u(0) = 0$, such that

$$a(u, v) = (f, v) + Rv(L) \tag{6.45}$$

for any $v(x)$, $v(0) = 0$.

Until now, the displacement $u(x)$ and the virtual displacement $v(x)$ were assumed to be functions defined on the interval $[0, L]$. However, we note that not all real functions $g(x)$ and $w(x)$ lead to a finite value of $a(g, w)$, for example, $g(x) = x^p$ and $w(x) = x^q$ with $p+q \leq 1$ do not. Surely, any possible solution $u(x)$ should correspond to a finite value of the strain energy given by (see Chapter 2)

$$\mathcal{U}\left(u(x)\right) = \frac{1}{2}\int_0^L EA\frac{du}{dx}\frac{du}{dx}\,dx = \frac{1}{2}a(u, u)$$

where $u(x)$ is the axial displacement field. Therefore $a(u, u)$ gives twice the strain energy associated with the displacement field u. Now we can restate the problem given by equation (6.45) to include this requirement as:

Find $u(x) \in V$ such that

$$a(u, v) = (f, v) + Rv(L) \tag{6.46}$$

for any $v(x) \in V$, where the space V is defined as

$$V = \{v/\ v : [0, L] \to \mathbb{R};\ a(v, v) < \infty;\ v(0) = 0\} \tag{6.47}$$

which reflects the fact that we require all functions, that we work with, to correspond to a finite strain energy.

We referred to the set of functions V defined above as a space. Of course, the reader is familiar with the space of Euclidean vectors for which the addition of vectors and the multiplication of a vector by a scalar (real number) are well defined. A fundamental property is that

$$\alpha\mathbf{x} + \beta\mathbf{y} \quad \text{for any vectors } \mathbf{x} \text{ and } \mathbf{y} \text{ and scalars } \alpha,\ \beta \tag{6.48}$$

is also a vector.

A set for which we can define, both, addition and multiplications by scalar operations and for which the equivalent of (6.48) also gives an element of the set is called a *linear vector space*[1].

Considering the usual addition and multiplication operations, it is not difficult to show that V is a linear vector space.

[1] There are other properties that are important to hold in order to work with a linear space (see Bathe, 1996)

The finite element solutions considered in Section 6.2 are then obtained by solving:

Find $u_h(x) \in V_h$ such that

$$a(u_h, v_h) = (f, v_h) + Rv_h(L) \tag{6.49}$$

for any $v_h(x) \in V_h$. Here V_h is the space of all possible displacement functions $v_h(x)$, $v_h(x) = 0$, that can be obtained with the specific mesh used. Of course, we have $V_h \subset V$ and (6.49) is a restatement of (6.1) in the notation introduced here.

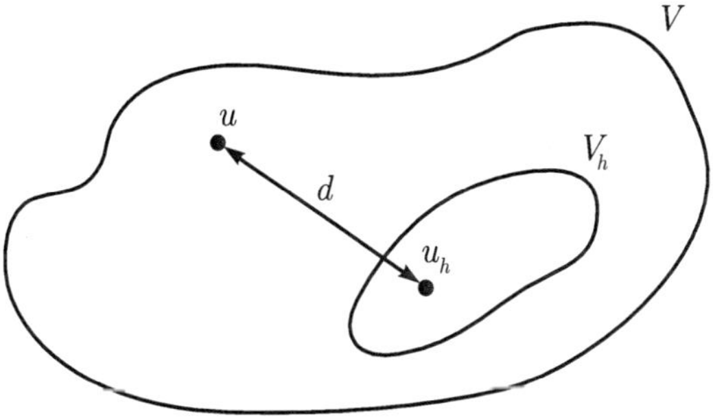

Fig. 6.12. Pictorial representation of the exact and finite element solutions

In Figure 6.12 we summarize pictorially the solutions u and u_h in the spaces V and V_h, with $V_h \subset V$. The space V_h is smaller than V since V represents all possible displacement functions and not just the finite element functions contained in V_h. The exact solution u lies in general outside V_h. We also show schematically the "distance" d between u and u_h, still to be defined. Of course, if we choose a coarse mesh, V_h is "small" and u_h might be far from the exact solution. As the mesh is refined and V_h is made larger, u_h may get closer to u (*i.e.*, the distance d may become smaller). The fact that u_h, indeed, approaches u, and under what conditions, is discussed below.

In order to introduce the concept of convergence more precisely, let us consider a sequence of meshes, using the same kind of finite elements, starting with mesh 1, which characterizes V_h^1, and refining sequentially to obtain meshes 2, $\cdots$, i, $\cdots$, n such that

$$V_h^1 \subset V_h^2 \subset \cdots \subset V_h^i \subset \cdots \subset V_h^n. \tag{6.50}$$

Here, when refining mesh k to obtain mesh $k+1$ the nodal positions of mesh k shall be preserved which guarantees that $V_h^k \subset V_h^{k+1}$.

In Table 6.1 we show some meshes with two-node elements which define finite element spaces that obey condition (6.50). Note that in the left column uniform discretizations are shown, *i.e.*, for each mesh the elements are all

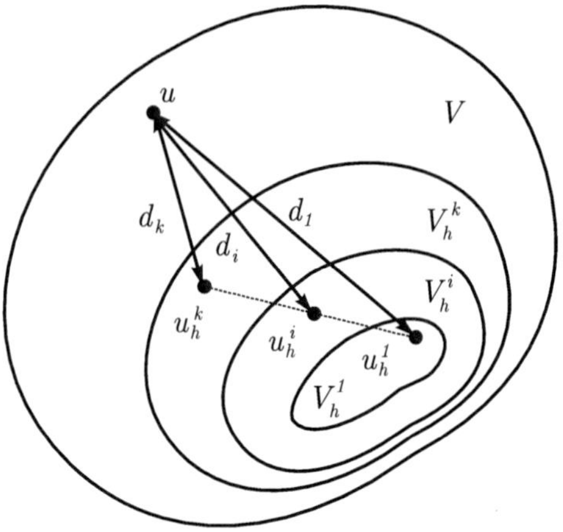

Fig. 6.13. Desirable convergence behavior as meshes are refined

Table 6.1. Typical mesh refinements for 1-D problem

Mesh 1 corresponding to V_h^1	Mesh 1 corresponding to V_h^1
Mesh 2 corresponding to V_h^2	Mesh 2′ corresponding to $V_h^{2'}$ 16 equal elements
Mesh 4 corresponding to V_h^4 32 equal elements	Mesh 2″ corresponding to $V_h^{2''}$

of equal length and the next finer mesh is obtained by subdividing every element into two new equal length elements.

In the right column of Table 6.1 we show two possible mesh refinements. Note that the mesh corresponding to $V_h^{2'}$ is obtained by further subdividing the right half of the domain into 16 equal elements, while to obtain the mesh corresponding to $V_h^{2''}$, the left half of the domain is represented by 16 elements of graded lengths.

In Figure 6.13 we reproduce Figure 6.12 considering a sequence of finite element spaces. This figure represents what would be the desirable behavior of the finite element solutions, *i.e.*, as we refine the mesh, the finite element solution becomes "closer" to the exact solution. It is also implied that we could obtain a finite element solution as "close" to the exact solution as desired if we sufficiently refine the mesh.

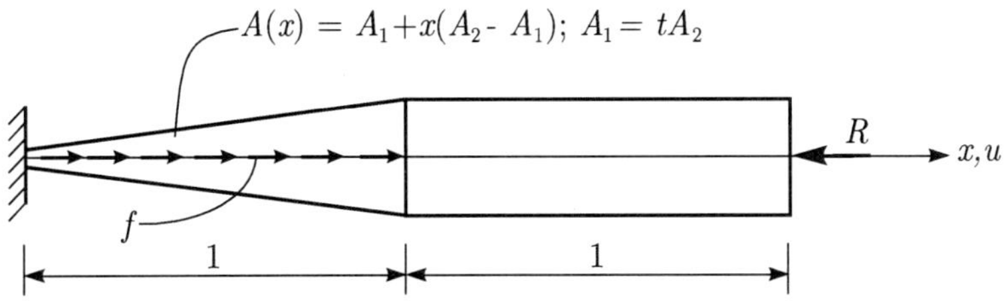

Fig. 6.14. Bar problem definition. $E = 2.0 \times 10^7$ kN/m^2, $A_2 = 1$ m^2, $t = 0.01$, $R = 15859.075383$ kN. The distributed force $f = 2.0 \times 10^4$ kN/m is acting only for $0 \le x \le 1$

To illustrate some convergence behaviors, consider the problem definition given in Figure 6.14. The exact solution and the finite element solution for the mesh 1 of Table 6.1 are shown in Figure 6.15a. We see that there is a large error.

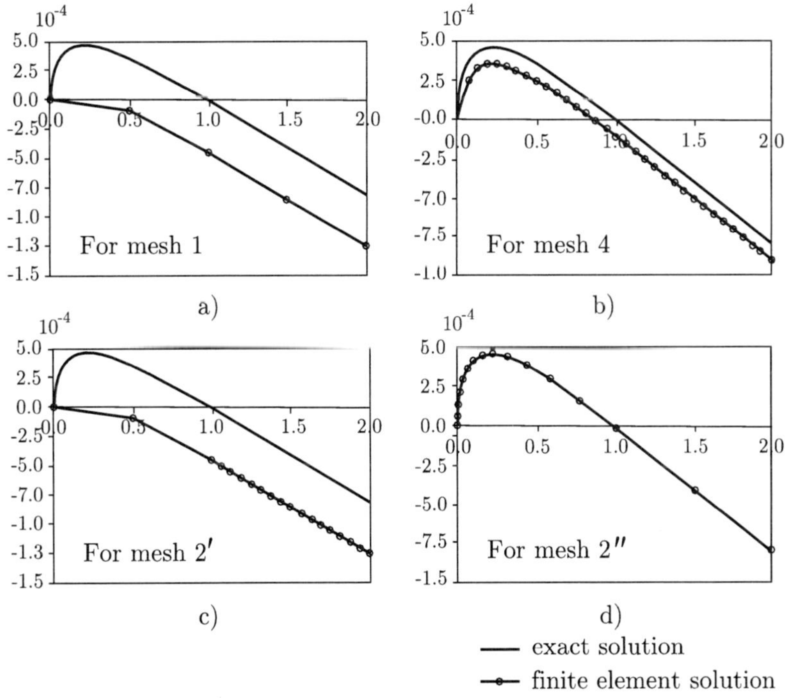

Fig. 6.15. Finite element and exact solutions for problem of Figure 6.14. The vertical axis shows the axial displacement and the horizontal axis gives the section coordinate, both in meters

In Figure 6.15b we show the finite element solution obtained with mesh 4, *i.e.* with 32 equally spaced elements, *i.e.*, following the refinement of the left column of Table 6.1. Of course, we note a significant improvement in the finite element solution accuracy. But, still, there is a considerable "distance" between the exact and finite element solutions.

In Figure 6.15c we report the finite element solution obtained with mesh $2'$. Although we refined the right half of the domain, there is no improvement of the finite element solution, that is, $u_h^1(x) = u_h^{2'}(x)$.

In Figure 6.15d we show the finite element solution for the mesh $2''$ which is "close" to the exact solution, and in the scale of the figure, we can barely distinguish between the finite element and exact solutions.

Note that the number of degrees of freedom of meshes $2'$ and $2''$ are the same. Nevertheless, the finite element solution accuracies are quite different. Note also that, although the number of degrees of freedom of mesh $2''$ is smaller than that of mesh 4 (18 versus 32) the solution $u_h^{2''}(x)$ is much more accurate than the solution $u_h^4(x)$. These numerical results bring up some convergence related issues that we examine next.

6.2.1 Convergence conditions

There are two fundamental conditions which should be met to assure that we obtain the desired convergence behavior.

Condition 1

The finite element interpolation should be such that it allows to obtain sequences of finite element spaces as defined in (6.50) *for which the space* V_h^n *becomes arbitrarily close to any element of* V*, i.e., any possible solution.*

By *finite element interpolation* we mean the definition of a generic finite element space V_h. For example, piecewise continuous linear functions correspond to the finite element interpolation associated with 2-node 1-D elements.

We also note that it is not any strategy of mesh refinement which obeys (6.50) that leads to a space V_h^n arbitrarily close to any element of V even when $n \to \infty$. For example, referring to the problem of Figure 6.14, considering the mesh refinement strategy in which only the right half of the domain is subdivided – specifically we use mesh 1, mesh 2', etc. – we can not find a function $v_h^n(x) \in V_h^n$ arbitrarily close to $u(x)$ even if we were allowed to freely choose the nodal displacement values. The reason, of course, is that the $v_h^n(x)$ is piecewise linear in the first half of the domain as given by only 2 elements that are not subdivided in the refinement process, and the analytical solution is far from that assumption.

Of course, it is quite intuitive that if we choose a uniform refinement strategy, we would obtain $v_h^n(x) \in V_h^n$ arbitrarily close to $u(x)$. It would suffice to make the element small enough and choose as nodal values, the values of $u(x)$ at the nodal points.

Condition 2

The finite element solution u_h is the closest, in some distance measure, to the exact solution among all elements of V_h.

It is easy to see that if Conditions 1 and 2 are met, convergence is guaranteed. In fact, Condition 1 assures that we can construct finite element spaces which are arbitrarily close to the exact solution. If we do so, then Condition 2 assures that the finite element solution will be arbitrarily close to the exact solution (since it is the function closest to the exact solution considering all functions in the finite element space).

Our next task is to further discuss Conditions 1 and 2 making them mathematically precise, and for that we first need to introduce some basic definitions.

6.2.2 Distances and norms of functions

Of course, the reader is familiar with the concept of the norm of a vector which gives the length of the vector. Let us introduce the concept of the norm of a function because we seek a scalar quantity which gives a measure of the distance between two functions, *i.e.*, how far apart they are. In Figure 6.16a we show two real valued functions $g(x)$ and $w(x)$ defined on the domain $[a, b]$. It is natural to take the difference of these two functions, $g(x) - w(x)$,

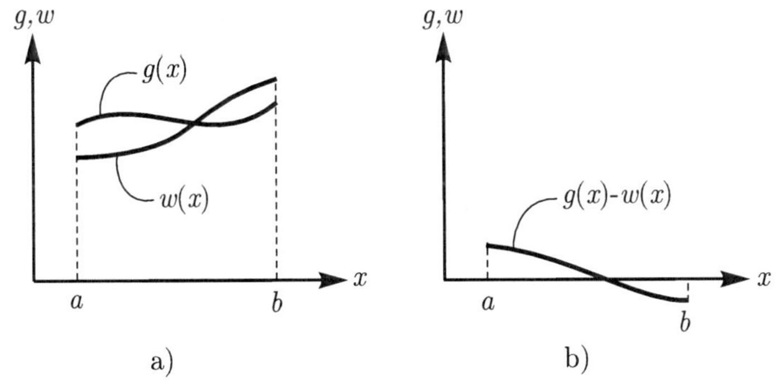

Fig. 6.16. Two generic real valued functions

which is plotted in Figure 6.16b. To obtain a scalar, we could integrate this difference function. However, we must avoid that algebraic positive differences are compensated by those which are negative. Therefore, we square the difference function and then take the square root of the result to arrive at a norm on $g(x) - w(x)$

$$\|g(x) - w(x)\|_0 = \left(\int_a^b (g(x) - w(x))^2 \, dx \right)^{1/2} . \tag{6.51}$$

Namely, a norm defined on a linear vector space W is a function $\|\cdot\| : W \rightarrow \mathbb{R}^+$ which obeys the following properties

$$\|v\| = 0 \Leftrightarrow v = 0 \tag{6.52}$$

$$\|\alpha v\| = |\alpha| \; \|v\| \quad \text{for every real number } \alpha \tag{6.53}$$

$$\|v + w\| \leq \|v\| + \|w\| \quad \text{for every } v, \, w \text{ in } W. \tag{6.54}$$

These properties have a clear geometrical interpretation for Euclidean vectors. The property given in (6.54) is called the triangle inequality since when v and w are linearly independent Euclidean vectors it corresponds to the geometrical property that for any triangle the length of a side is smaller than the sum of the other two side lengths.

Based on equation (6.51), we can define a norm of a function $f(x)$ given in a domain $[a, b]$ in a linear vector space of functions such as V (see (6.47)) by

$$\|f(x)\|_0 = \left(\int_a^b (f(x))^2 \, dx \right)^{1/2} .$$

It can be shown that this norm definition verifies properties (6.52) to (6.54) and, hence, leads to a valid norm. There are several different norms for spaces of functions. The norm above is referred to as the L^2-norm or the the 0-norm as the sub-index indicates. The 1-norm can also be defined and it takes into account not only the values of the function but also those of its first derivative leading to

$$\|f(x)\|_1 = \left(\int_a^b \left[(f(x))^2 + \left(\frac{df(x)}{dx} \right)^2 \right] dx \right)^{1/2} . \tag{6.55}$$

In fact, the p-norm is defined by

$$\|f(x)\|_p = \left(\int_a^b \left[(f(x))^2 + \left(\frac{df(x)}{dx} \right)^2 + \left(\frac{d^2 f(x)}{dx^2} \right)^2 + \cdots + \left(\frac{d^p f(x)}{dx^p} \right)^2 \right] dx \right)^{1/2}$$

where p is an integer greater than zero. A useful norm in mechanics is the strain energy norm which for a displacement function $v(x)$ can be defined by

$$\|v(x)\|_E = (a(v, v))^{1/2} .$$

These norms will be sufficient for the convergence study in 1-D. Of course, these concepts can be generalized to obtain norms for spaces of functions defined in several dimensions, see Bathe, 1996.

We can define the distance measured in the norm V for two real valued functions $w(x)$ and $g(x)$ defined in the interval $[a, b]$ by

$$d(w, v)_V = \|w - v\|_V \,.$$

Norm V can be 0, 1, E or any other norm defined on the vector space considered.

6.2.3 Convergence properties

Let us present Condition 1 in a mathematically precise manner. The distance of the exact solution u of any 1-D bar problem to the finite element space V_h^i in the norm V is defined by[2]

$$d(u, V_h^i)_V = \min\left\{\left\|u - v_h^i\right\|_V \text{ for all } v_h^i \in V_h^i\right\}. \tag{6.56}$$

We emphasize that norm V could be the 0-norm, the 1-norm or the energy norm. Note also that (6.56) is the mathematical statement that $d(u, V_h^i)_V$ is the distance from u to the closest element of V_h^i measured in the norm V.

The Condition 1 would be given by:
There exist sequences of finite element spaces as defined in (6.50) such that

$$d(u, V_h^i)_V \to 0 \quad \text{when } i \to \infty \text{ for all } u \in V.$$

Note that Condition 1 is not dependent on the finite element solution. It represents an approximation property that the finite element spaces should satisfy.

The Condition 2 can be defined as:

$$\left\|u - u_h^i\right\|_E \leq \left\|u - v_h^i\right\|_E \quad \text{for all } v_h^i \in V_h^i \tag{6.57}$$

i.e. the finite element solution u_h^i is the best solution considering all elements of V_h^i when the distance is measured in the energy norm.

Note that when Condition 1 is satisfied for the energy norm, *Condition 2* gives

$$\left\|u - u_h^i\right\|_E \to 0 \quad \text{when } i \to \infty$$

which shows that the finite element solution converges in the energy norm to the exact solution.

To obtain the above convergence result for the 1-norm, we consider the norm equivalence given by

---

[2] Actually, if we would like to be mathematically precise, we should have used the infimum of the set

$$C_1 \left\| v \right\|_1 \leq \left\| v \right\|_E \leq C_2 \left\| v \right\|_1 \quad \text{for all } v \in V \tag{6.58}$$

for positive constants C_1 and C_2. From (6.57) and (6.58), we can write

$$C_1 \left\| u - u_h^i \right\|_1 \leq \left\| u - u_h^i \right\|_E \leq \left\| u - v_h^i \right\|_E \leq C_2 \left\| u - v_h^i \right\|_1 \quad \text{for all } v_h^i \in V_h^i$$

which yields

$$\left\| u - u_h^i \right\|_1 \leq C \left\| u - v_h^i \right\|_1 \quad \text{for all } v_h^i \in V_h^i \tag{6.59}$$

where $C = C_2/C_1$.

Equation (6.59) can be interpreted as Condition 2 in the 1-norm and when Condition 1 is satisfied for the 1-norm, we obtain

$$\left\| u - u_h^i \right\|_1 \to 0 \quad \text{when } i \to \infty \tag{6.60}$$

which shows that the finite element solution converges in the 1-norm to the exact solution.

Let us examine Conditions 1 and 2 for linear finite elements in 1-D for the 1-norm.

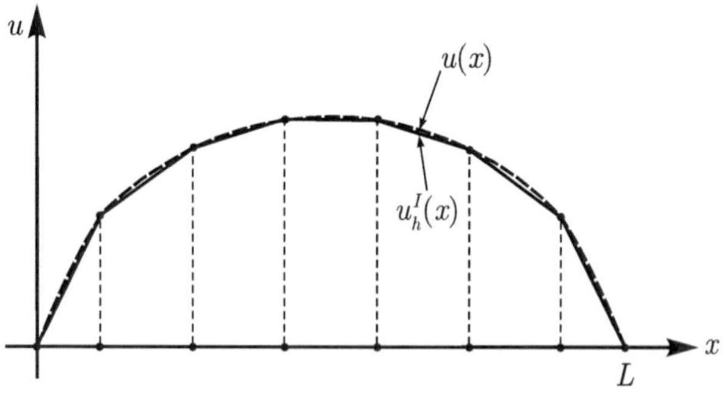

Fig. 6.17. Typical representation of $u_h^I(x)$

Let $u(x)$ be the exact solution of an arbitrary problem given by (6.46) with EA varying. Of course, u is a function in V. Let V_h be a finite element space implicitly defined by the choice of a finite element mesh. In Figure 6.17 a typical situation is represented where we also show the function $u_h^I(x)$. This function is called the interpolant and it is the function in V_h for which

$$u(x_i) = u_h^I(x_i) \qquad i = 1, \cdots, n_p$$

where the x_i are the nodal point coordinates and n_p is the number of nodal points. Note that the values of the interpolant coincide with the value of $u(x)$ at the nodal positions. It can be shown (see, for example, Bathe, 1996) that

$$\left\| u - u_h^I \right\|_1 \leq C^* h \left| u \right|_2$$

where h is the maximum element length, C^* is a positive constant independent of h and $u(x)$, and

$$\left| u \right|_2 = \left(\int_0^L \left(\frac{d^2 u}{dx^2} \right)^2 dx \right)^{1/2} .$$

Of course, we are assuming that $u(x)$ is sufficiently smooth such that the above equation is well defined. Since $u_h^I \in V_h$, we can write

$$d(u, V_h)_1 \leq \left\| u - u_h^I \right\|_1 \leq C^* h \left| u \right|_2 . \tag{6.61}$$

If we define our sequence of finite element spaces given in (6.50) such that the largest element of mesh n tends to zero when $n \rightarrow \infty$, (6.61) would lead to

$$d(u, V_h)_1 \rightarrow 0 \text{ as } n \rightarrow \infty$$

since $C \left| u \right|_2$ is a positive constant. Therefore, Condition 1 is satisfied.

Related to Condition 2, it can be shown (see, for example, Bathe, 1996) that

$$a(u - u_h, u - u_h) \leq a(u - v_h, u - v_h) \text{ for all } v_h \in V_h$$

where u_h is the finite element solution for V_h. Hence

$$\left\| u - u_h \right\|_E \leq \left\| u - v_h \right\|_E \text{ for all } v_h \in V_h. \tag{6.62}$$

The above equation yields Condition 2 for the energy norm. The property in (6.62) has an important consequence; namely, that when the exact solution is contained in the finite element space, *i.e.*, $u \in V_h$, then the finite element solution will be the exact solution, *i.e.*, $u = u_h$.

Of course, due to the norm equivalence (6.58) Condition 2 holds for the 1-norm as given in (6.59). Therefore, convergence is assured as given in (6.60).

Since we are considering a specific interpolation, we can obtain an error estimate. In fact, from (6.59)

$$\left\| u - u_h \right\|_1 \leq C \left\| u - u_h^I \right\|_1 \leq C C^* h \left| u_2 \right| .$$

Defining $\tilde{C} = C^* C \left| u \right|_2$, we obtain

$$\left\| u - u_h \right\|_1 \leq \tilde{C} h \tag{6.63}$$

which is a convergence result for the 1-norm for linear finite element approximations of the problem given by (6.46).

Note that (6.63) shows that the error, *i.e.* the distance between the exact solution and the finite element solution, measured in the 1-norm can be made arbitrarily small. It suffices to refine the mesh until h, the size of the largest element, is small enough. Additionally (6.63) guarantees that not only the function $u_h(x)$ is close to $u(x)$ but also its derivative. Since when solving (6.46) we are interested in both displacements and stresses, which are obtained from the derivative of the displacement function, (6.63) is referred to as *the* convergence result. The convergence for linear elements is said to be linear since the error decreases with the first power of h. If the order of the displacement interpolation is increased, say from linear to quadratic, the convergence is controlled by the second power of h (in the equivalent of (6.63)) and is then referred to as "quadratic convergence". The general result is that for piecewise polynomials of order p the convergence (provided the exact solution u is sufficiently smooth) is controlled by the p_{th} power of h, *i.e.*, the convergence rate is of order p.

Example 6.3

Study the convergence of the finite element solutions to the exact solution for the problem described in Figure 6.18. Consider uniform mesh refinements.

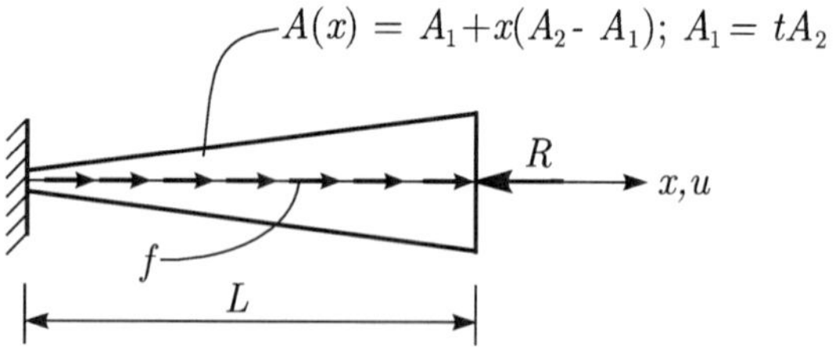

Fig. 6.18. Bar problem definition. $E = 2.0 \times 10^7$ kN/m^2, $A_2 = 1$ m^2, $t = 0.01$, $L = 1$ m, $R = 15859.075383$ kN, $f = 2.0 \times 10^4$ kN/m

Solution

The differential formulation for this problem is given in equations (5.25) to (5.27). The exact solution is

$$u(x) = -\frac{1}{1980}\frac{2x\ln(2) + 2x\ln(5) - \ln(1+99x)}{\ln(2)+\ln(5)}. \tag{6.64}$$

The sequence of finite element spaces is obtained by starting with a mesh of two 2-node elements of equal length and subdividing in each mesh refinement every element into two new elements of equal length. Our most refined mesh has 1024 elements. We measure the error in the 1-norm, that is,

$$E_h = \|u - u_h\|_1 \, .$$

In Figure 6.19 we show the error as the mesh is refined and in Figure 6.20 we show the exact solution together with the finite element solution for our most refined mesh. To use the error estimate given in (6.63) we need first to

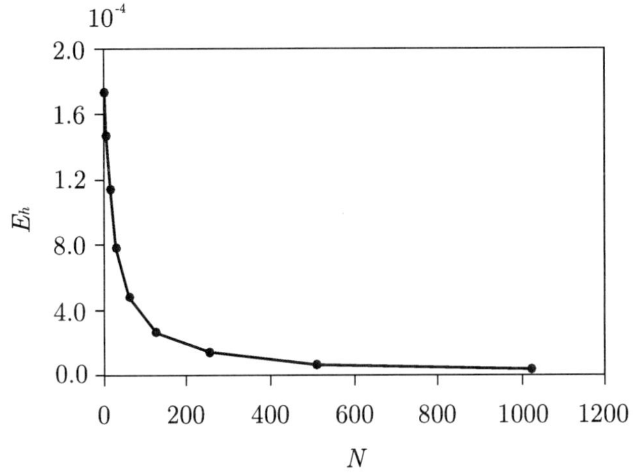

Fig. 6.19. Convergence diagram. N is the number of free degrees of freedom

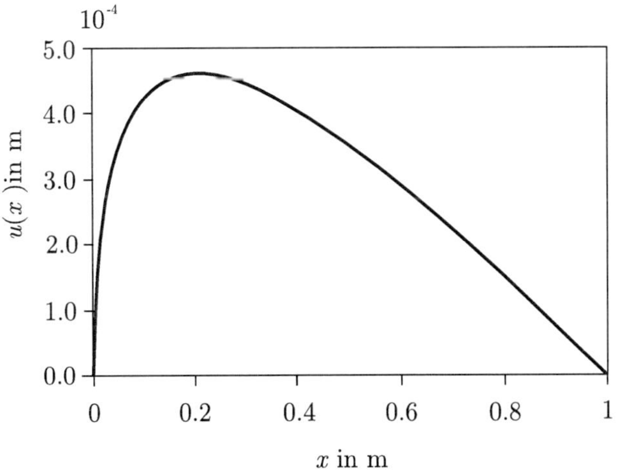

Fig. 6.20. Finite element solution for the 1024 element mesh and exact solution. No difference can be seen in the scale of the graph

verify that

$$|u(x)|_2 = \left(\int_0^L \left(\frac{d^2 u}{dx^2} \right) dx \right)^{1/2} < \infty$$

which, indeed is true, and can be evaluated using the exact solution (6.64).

Considering the error estimate (6.63), we have

$$\log \|u - u_h\|_1 \leq \log \left(\tilde{C} h \right)$$

$$\log E_h \leq \log \tilde{C} + \log h.$$

Using this inequality we can construct the diagram of Figure 6.21. Therefore, for a given mesh the ordered pair $(\log E_h, \log h)$ has to lie either on the straight line or in the highlighted region and in practice we find that the numerical results are frequently close to the straight line.

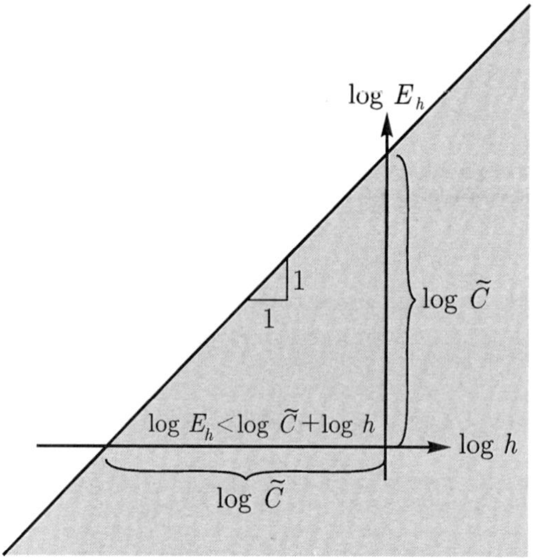

Fig. 6.21. Log-log diagram used for convergence study

If the solution is known it is convenient to use the relative error which is defined as

$$E_h^r = \frac{\|u - u_h\|_1}{\|u\|_1} = \frac{E_h}{\|u\|_1}$$

since E_h^r gives the error measured with respect to the 1-norm value of the exact solution. Also, note that from (6.63), we can write

$$\frac{\|u - u_h\|_1}{\|u\|_1} \leq \frac{\tilde{C}}{\|u\|_1} h$$

leading to

$$\log E_h^r \leq \log \widehat{C} + \log h.$$

In Figure 6.22 we show the pairs $(\log E_h^r, \log h)$ obtained in the solution of the problem described in Figure 6.18. The convergence rate evaluated based on the first two meshes is 0.244 and it increases as we consider pairs of more refined meshes. The convergence rate for the two most refined meshes is 0.9958 which is close to 1.

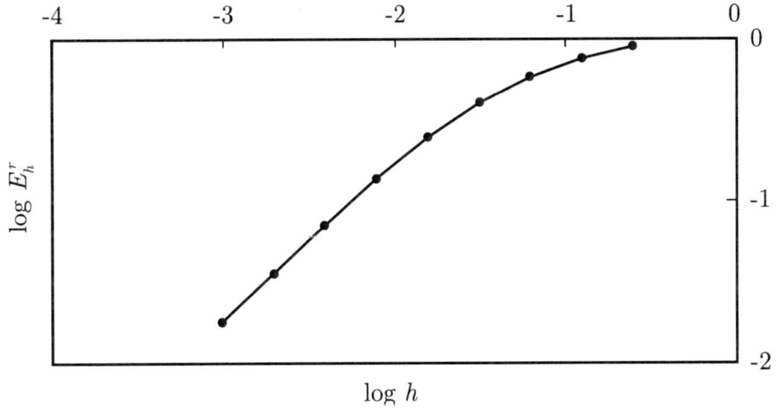

Fig. 6.22. Convergence diagram for problem of Figure 6.18 for uniform meshes

It is usual to measure convergence with respect to the number of degrees of freedom. The number of free degrees of freedom N corresponding to the 2-node element meshes used to solve the problem of Figure 6.18 is equal to the number of elements, that is,

$$N = n_e = \frac{L}{h} \quad \text{or} \quad h = N^{-1}L$$

and, therefore the error estimate (6.63) can be re-written in terms of N as

$$\|u - u_h\|_1 \leq \tilde{C}LN^{-1}$$

or

$$\|u - u_h\|_1 \leq C'N^{-1} \tag{6.65}$$

which leads to

$$\log E_h \leq \log C' - \log N$$

and a diagram, analogous to that shown in Figure 6.21, is reported in Figure 6.23. Of course, we can use E_h^r instead, which leads to

$$\log E_h^r \leq \log \overline{C} - \log N.$$

The convergence results shown in Figure 6.22 are reported in terms of N in Figure 6.24.

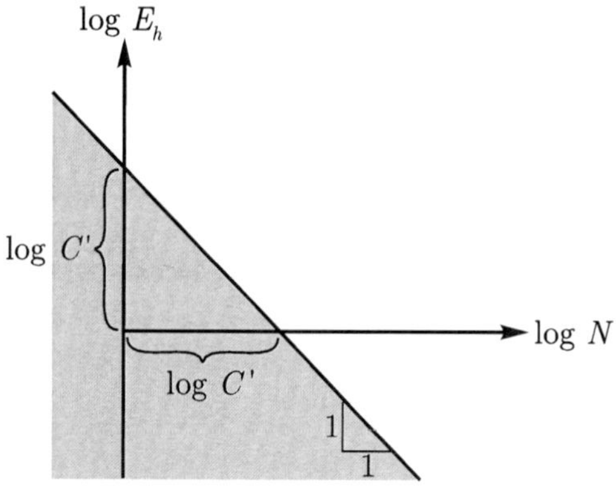

Fig. 6.23. Log-log diagram in terms of N (number of free degrees of freedom)

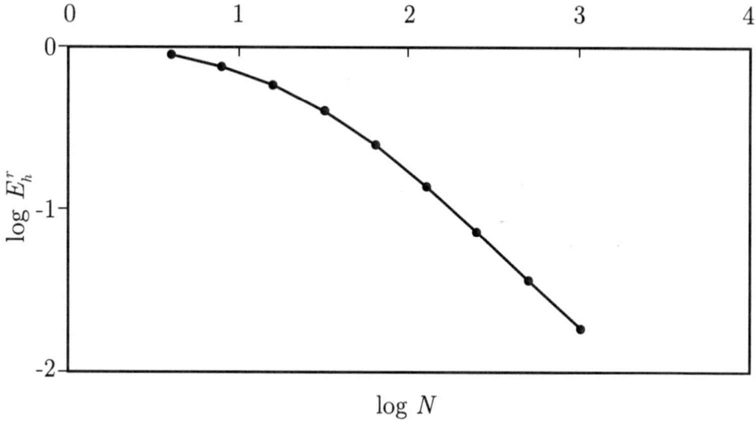

Fig. 6.24. Convergence diagram for problem of Figure 6.18 for uniform meshes in terms of N

Referring to the exact solution $u(x)$ of the problem described in Figure 6.18, which is reported in Figure 6.20, we see that in the region near the support $u(x)$ varies with highly varying gradients, and that for $x > 0.5$, the gradient of $u(x)$ is almost constant. Therefore, it is quite intuitive that we can potentially improve the finite element predictions with respect to using a uniform mesh if we use smaller elements in the region of highly varying gradients and larger elements in the rest of the region. This kind of refinement strategy is known as *mesh grading* and we exemplify its potential benefits by means of the next example.

□

Example 6.4

For the problem of Figure 6.18, study the convergence behavior of the finite element solutions using a mesh grading strategy.

Solution

Let us define the mesh grading by specifying the nodal positions as

$$x_i = \left(\frac{i}{n_e}\right)^\beta \quad , \quad i = 0, 1, 2, \cdots, n_e \quad ; \quad n_e \geq 2$$

where $x_0, x_1, \cdots, x_{n_e}$ are the nodal coordinates, n_e is the number of 2-node elements and β is a positive real constant which defines the grading. Note that $\beta = 1$ gives, as a particular case, the uniform mesh. Since we would like smaller elements as we approach $x = 0$, we consider $\beta > 1.0$.

In Figure 6.25 we show the convergence behavior of the finite element solutions for $\beta = 4.0$. The rate of convergence evaluated using the first two meshes ($n_e = 4$ and $n_e = 8$) is 0.9928 and using the last two meshes ($n_e = 512$ and $n_e = 1024$) is 0.99999926.

In Table 6.2 we show some selected results for some β values. We note that any grading refinement strategy leads to superior performance when compared to the uniform refinement ($\beta = 1$) which can be measured by the error of the most refined mesh. Note that a rate of convergence very close to 1 is attained for the last two meshes for all values of β.

We can now revisit the finite element solutions reported in Figure 6.15 for the problem described in Figure 6.14. We note that for the left half of the domain, *i.e.*, for $0 \leq x \leq 1$, this problem is identical to the problem discussed in Examples 6.3 and 6.4. Therefore, the discussion presented in these examples is directly applicable and explains why the grading leading to the results shown in Figure 6.15d is efficient.

Regarding the right half of the domain, of course the exact displacements in this region varies linearly and only one element predicts the exact strain. However, the displacements are in error as long as the displacement at $x = 1$ is not the exact displacement. □

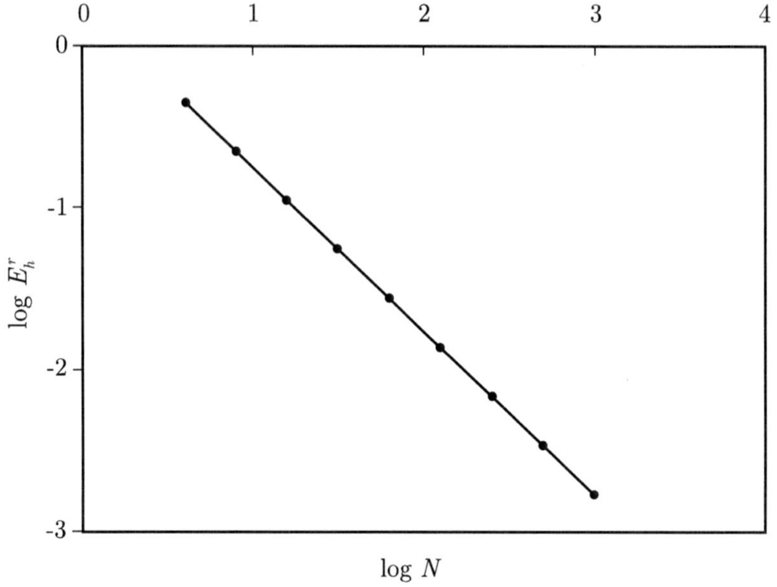

Fig. 6.25. Convergence diagram for problem of Figure 6.18 for graded meshes

Table 6.2. Selected results for various β values

β	Length of smallest elem. (1024 elem.)	Length of largest elem (1024 elem.)	Relative error (E_h^r) (1024 elem.)
1.0	0.976×10^{-3}	0.976×10^{-3}	3.5×10^{-5}
2.0	0.954×10^{-3}	1.95×10^{-3}	5.0×10^{-6}
3.0	0.931×10^{-3}	2.93×10^{-3}	3.6×10^{-6}
4.0	0.909×10^{-3}	3.90×10^{-3}	3.4×10^{-6}

β	Convergence rate for first two meshes	Convergence rate for last two meshes
1.0	0.24	0.995486
2.0	0.99	0.999974
3.0	0.76	0.999992
4.0	0.99	0.999999

Finally, we should note that we discussed here only the convergence of displacement-based elements. The behavior of mixed interpolated elements referred to in Section 6.5 can be much more difficult to assess, see *e.g.* Brezzi and Fortin, 1991, Chapelle and Bathe, 2010*a* and Hiller and Bathe, 2003.

6.2.4 Smoothness of the solution

We mentioned earlier that to derive the error estimate (6.63), we must have

$$|u|_2 < \infty \tag{6.66}$$

and if higher-order elements are used even stricter conditions may be applicable, that is the exact solution need be sufficiently smooth.

Hence if (6.66) is not satisfied, the estimate (6.63) can not directly be used and further mathematical analysis is required to obtain valid estimates for specific situations. In fact, it can be shown that the same rate of convergence as for the smooth solution case can still be achieved if the mesh is properly graded (Ciarlet, 1978 and Bathe, 1996 and Chapelle and Bathe, 2010*a*). Considering the full realm of elasticity, solutions with poor smoothness properties are encountered in problems with sharp corners and cracks and special error estimates are then required.

6.2.5 The h, p and $h - p$ finite element methods

The strategy used so far to refine finite element meshes is called the *h method*: the elements of size h are simply subdivided. In the p method, we start with a mesh and obtain larger finite element spaces by increasing the order of the interpolation functions in each element. For example, we start with a mesh of linear elements and then consider quadratic, cubic ... elements. The element interpolations of order p are constructed in a special way to render the procedure as efficient as possible.

In the $h - p$ method, elements are subdivided and the polynomial order of the elements is increased. This method combines the advantages of the h and p methods.

Of course, there are several issues concerning the implementation, general applicability and efficiency of these methods specially when we consider 2-D, 3-D solids, and plate and shell elements.

In engineering applications of the finite element method mostly linear elements (2-node elements in 1-D, 4-node elements in 2-D and 8-node elements in 3-D) or quadratic elements are used. Then the h method is employed to refine the meshes. For these analyses computer programs are available to generate and refine the meshes for complex geometries of engineering interest.

6.3 Displacement-based finite element formulation for solids

In the previous sections we introduced the fundamental concepts of a finite element formulation and solution considering one-dimensional problems.

In this section, we shall extend the finite element formulation to 3-D elastic bodies. Of course, some complexity is added, since the problem domain is multi-dimensional. However, as seen shortly, the concepts involved are the same as for the one-dimensional case.

In 2-D solutions the finite element interpolation functions can be pictorially well presented and then the generalization for 3-D solutions is apparent and straight forward. Hence, we start with the 2-D formulation and in Section 6.3.5 we discuss the 3-D formulation.

Our starting point is, as for the 1-D case, the principle of virtual work for 3-D

$$\int_V \bar{\varepsilon}^T \tau dV = \int_V \bar{\mathbf{u}}^T \mathbf{f}^B dV + \int_{S_f} \bar{\mathbf{u}}^T \mathbf{f}^S dS. \tag{6.67}$$

This equation is applicable for plane stress, plane strain and axisymmetric solids (see Section 5.2.1). We shall consider the plane stress case (see Section 5.2.2) but all concepts are also directly applicable to the other 2-D cases.

We consider (5.40) which gives the principle of virtual work for plane stress conditions when the thickness t (here assumed constant) is kept in the equation. For the finite element solution we express this principle in terms of displacements only. Therefore, we introduce the constitutive equation $\tau = \mathbf{C}\varepsilon$ and the compatibility relations, *i.e.*, the strain-displacement relations $\varepsilon = \partial_\varepsilon \mathbf{u}$ given in Table 4.3. Substituting, we obtain:

Find $\mathbf{u}$ such that

$$\begin{aligned} t\int_A (\partial_\varepsilon \bar{\mathbf{u}})^T \; \mathbf{C}\, \partial_\varepsilon \mathbf{u}\, dA &= t\int_A \bar{\mathbf{u}}^T \mathbf{f}^B dA + t\int_{L_f} \bar{\mathbf{u}}^T \mathbf{f}^S dL \\ \forall\, \bar{\mathbf{u}},\; \bar{\mathbf{u}} &= \mathbf{0} \;\text{ in }\; L_u \end{aligned} \tag{6.68}$$

In the finite element analysis we assume $\mathbf{u}$ and $\bar{\mathbf{u}}$ to have a particular functional form given by the interpolation functions and we proceed as discussed below.

6.3.1 Discretization methodology

Consider a typical finite element discretization as shown in Figure 6.26 where a generic element (m) is highlighted. For each element in the discretization a set of nodal points is defined, see Figures 6.27 and 6.28. We are taking four nodes per element, *i.e.*, we are considering 4-node elements, and note that a nodal point is usually shared by several elements. For instance, we can see in Figure 6.28 that node k belongs to elements p, l, m and q. In order to define a typical finite element approximation for the exact unknown displacements, we assume all element displacements to vary according to interpolation functions. As in the 1-D case, the displacements inside an element depend only on the displacement values at the element nodal points.

Consider the rectangular element (m) as shown in Figure 6.29 with the global nodal degrees of freedom. It is effective to use a local coordinate

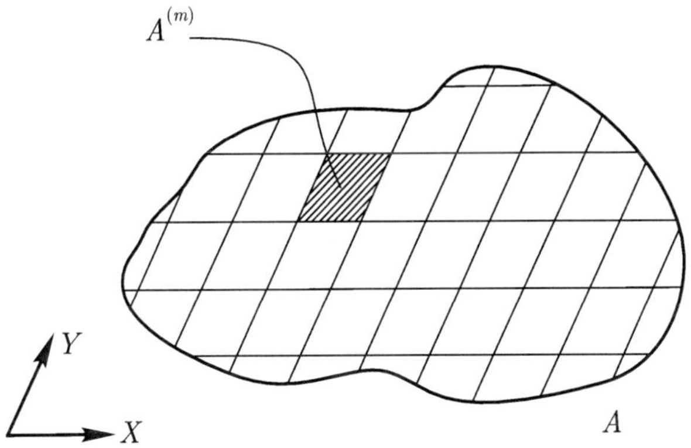

Fig. 6.26. Partition of the 2-D domain into subdomains, *i.e.*, elements. Only part of the domain is shown

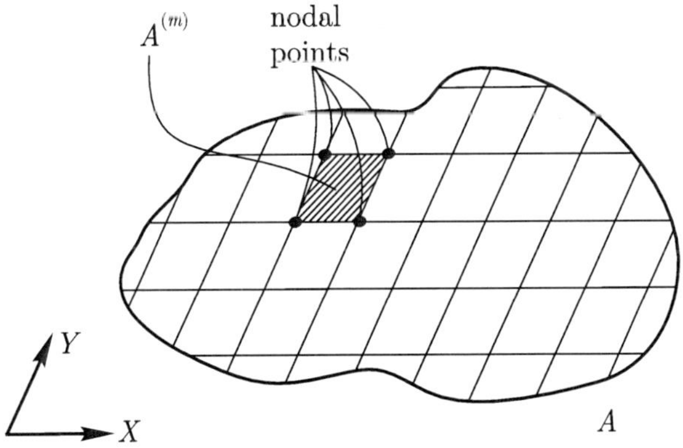

Fig. 6.27. Nodal points for element (m)

system and a local node numbering as shown in Figure 6.30 to define the interpolation functions. Then, the interpolation function associated with node 1 is given by

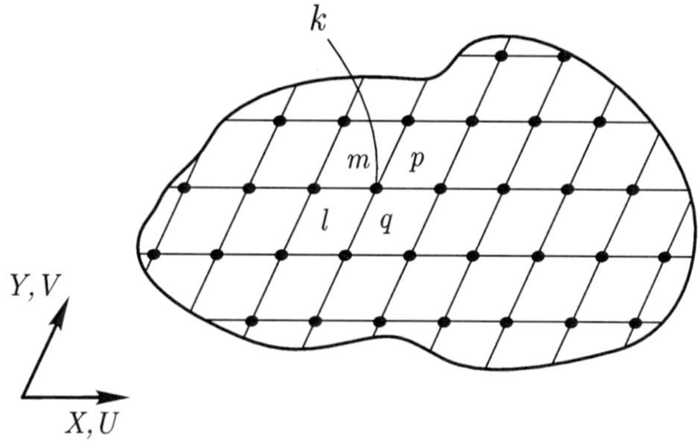

Fig. 6.28. Nodal points considering all elements in the part of the domain shown

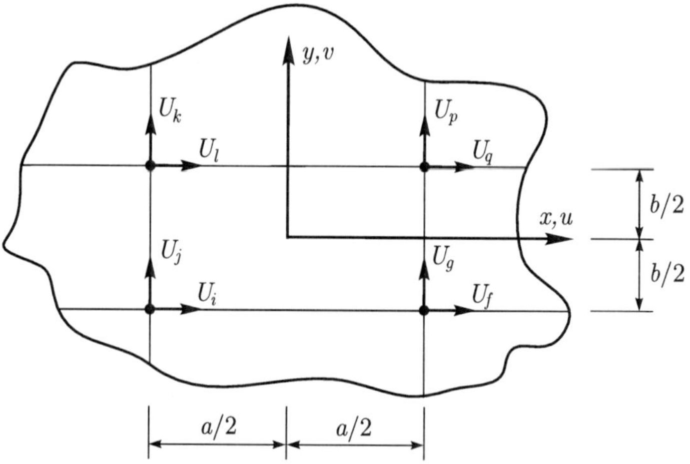

Fig. 6.29. Detail of a generic element (m) indicating the global nodal degrees of freedom

$$h_1(x, y) = \frac{1}{4ab}(a + 2x)(b + 2y). \tag{6.69}$$

This function is pictorially shown in Figure 6.31. Note that $h_1(x, y)$ is equal to 1 for the nodal coordinates of node 1, *i.e.*, at $x = a/2$, $y = b/2$, and is equal to 0 for the nodal coordinates of nodes 2, 3 and 4. The function $h_1(x, y)$ is linear in x for a fixed y and linear in y for a fixed x. We can define $h_i(x, y)$, $i = 2, 3, 4$ with analogous properties to h_1, *i.e.*, being 1 for the coordinates of node i and 0 for the coordinates of node j, $j \neq i$, by

$$h_2(x, y) = \frac{1}{4ab}(a - 2x)(b + 2y) \tag{6.70}$$

$$h_3(x,y) = \frac{1}{4ab}(a-2x)(b-2y) \tag{6.71}$$

$$h_4(x,y) = \frac{1}{4ab}(a+2x)(b-2y)\,. \tag{6.72}$$

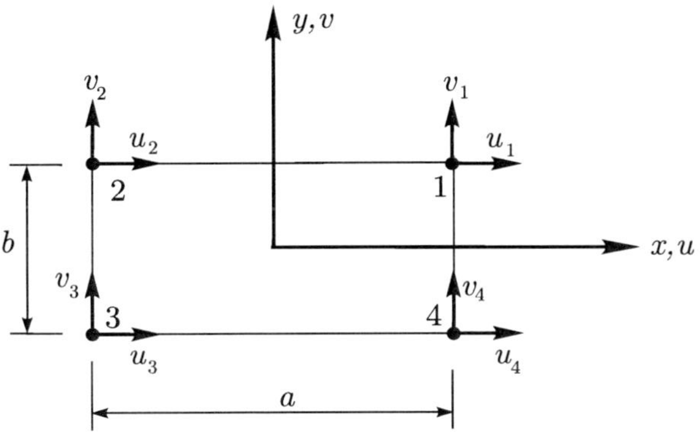

Fig. 6.30. Local degrees of freedom; the local degrees of freedom act into the same directions as the global degrees of freedom

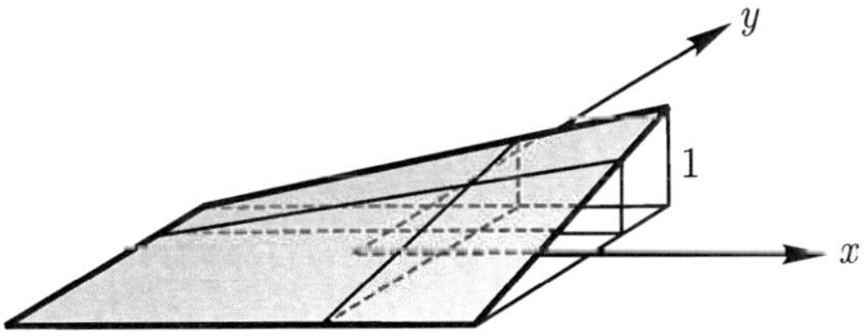

Fig. 6.31. Schematic description of function $h_1(x,y)$

Using these interpolation functions we can define the assumed displacements within element (m) at any point (x,y) from their values at the nodal points by

$$u^{(m)}(x,y) = \sum_{i=1}^{4} h_i(x,y)\; u_i$$

and

$$v^{(m)}(x,y) = \sum_{i=1}^{4} h_i(x,y)\; v_i.$$

Denoting the coordinate of node i by (x_i, y_i), it is easy to verify that

$$u^{(m)}(x_i, y_i) = u_i$$

$$v^{(m)}(x_i, y_i) = v_i.$$

Using matrix notation, we have

$$\mathbf{u}^{(m)} = \begin{bmatrix} u^{(m)}(x,y) \\ v^{(m)}(x,y) \end{bmatrix} = \mathbf{H}\,\hat{\mathbf{u}} \tag{6.73}$$

where

$$\mathbf{H} = \begin{bmatrix} h_1 & 0 & h_2 & 0 & h_3 & 0 & h_4 & 0 \\ 0 & h_1 & 0 & h_2 & 0 & h_3 & 0 & h_4 \end{bmatrix}. \tag{6.74}$$

and

$$\hat{\mathbf{u}}^T = \begin{bmatrix} u_1 & v_1 & u_2 & v_2 & u_3 & v_3 & u_4 & v_4 \end{bmatrix}$$

We shall use the hat symbol ($\hat{}$) to denote nodal quantities, such as nodal displacements in equation (6.73), whenever the distinction between nodal and continuum quantities is not obviously implied.

For the derivation below, it is convenient to define a column matrix which collects all N global nodal displacement degrees of freedom

$$\mathbf{U}^T = [U_1 \; U_2 \; \dots \; U_N]$$

and also define $\mathbf{H}^{(m)}$ by

$$\mathbf{u}^{(m)} = \mathbf{H}^{(m)}\mathbf{U} \tag{6.75}$$

In equations (6.73) and (6.75) $\mathbf{H}$ corresponds to the local nodal displacement degrees of freedom ((m) being implied), whereas $\mathbf{H}^{(m)}$ corresponds to the global nodal displacement degrees of freedom. We assume that the local degrees of freedom act into the same directions as the global degrees of freedom, in which case the columns in $\mathbf{H}^{(m)}$ are either zero or given by the columns defined in $\mathbf{H}$. For example, considering the element in Figure 6.29, we have

$$\mathbf{u}^{(m)} = \begin{bmatrix} \dots & \overset{f}{h_4} & \overset{g}{0} & . & \overset{i}{h_3} & \overset{j}{0} & \overset{k}{0} & \overset{l}{h_2} & \dots & \overset{p}{0} & \overset{q}{h_1} & \dots \\ \dots & 0 & h_4 & . & 0 & h_3 & h_2 & 0 & \dots & h_1 & 0 & \dots \end{bmatrix} \mathbf{U}. \tag{6.76}$$

Only the columns of $\mathbf{H}^{(m)}$ with non-zero entries are shown.

We can now evaluate the strains from $\mathbf{u}^{(m)}$ using the strain-displacement relations

$$\varepsilon^{(m)} = \mathbf{B}^{(m)}\mathbf{U} \tag{6.77}$$

where

$$\mathbf{B}^{(m)} = \partial_{\varepsilon}\mathbf{H}^{(m)} \tag{6.78}$$

and the stresses are obtained using the constitutive equation for element (m)

$$\tau^{(m)} = \mathbf{C}^{(m)}\,\varepsilon^{(m)}. \tag{6.79}$$

Since the virtual displacements are interpolated in the same way as the real displacements we have

$$\bar{\mathbf{u}}^{(m)} = \mathbf{H}^{(m)}\bar{\mathbf{U}} \tag{6.80}$$

and

$$\bar{\varepsilon}^{(m)} = \mathbf{B}^{(m)}\bar{\mathbf{U}}. \tag{6.81}$$

Therefore all quantities involved in the principle of virtual displacements are defined elementwise. Since we can write the integral over the domain A as a sum of integrals performed over the elements, we obtain

$$\begin{aligned}\sum_{m=1}^{n_e} t\int_{A^{(m)}} \bar{\varepsilon}^T\tau^{(m)}dA^{(m)} &= \sum_{m=1}^{n_e} t\int_{A^{(m)}} \bar{\mathbf{u}}^{(m)^T}\mathbf{f}^{B^{(m)}}dA^{(m)} + \\ &\quad \sum_{m=1}^{n_e} t\int_{L_1^{(m)}\dots\, L_q^{(m)}} \bar{\mathbf{u}}^{(m)^T}\mathbf{f}^{S^{(m)}}dL^{(m)}\end{aligned} \tag{6.82}$$

where n_e is the number of elements in the mesh and $L_1^{(m)} \dots\, L_q^{(m)}$ represent the sides of element (m) that belong to the boundary L_f. Note that an element in the interior of the domain does not have any sides that are on L_f, and therefore there is no contribution from such an element to the last term in (6.82).

Substituting from (6.75) to (6.81) into (6.82) we arrive at

$$\begin{aligned}&\bar{\mathbf{U}}^T\left[\sum_{m=1}^{n_e} t\int_{A^{(m)}} \mathbf{B}^{(m)T}\mathbf{C}^{(m)}\mathbf{B}^{(m)}\,dA^{(m)}\right]\mathbf{U} = \\ &\bar{\mathbf{U}}^T\left[\sum_{m=1}^{n_e} t\left(\int_{A^{(m)}} \mathbf{H}^{(m)T}\mathbf{f}^{B^{(m)}}dA^{(m)}\right)\right. \\ &\left. + \sum_{m=1}^{n_e} t\left(\int_{L_1^{(m)},\dots,L_q^{(m)}} \mathbf{H}^{(m)T}\mathbf{f}^{S^{(m)}}dL^{(m)}\right)\right]\end{aligned} \tag{6.83}$$

Note that $\bar{\mathbf{U}}$ and $\mathbf{U}$ were taken out of the sums and integrals since they are independent of (m). Defining

$$\mathbf{K}^{(m)} = t\int_{A^{(m)}} \mathbf{B}^{(m)^T}\mathbf{C}^{(m)}\mathbf{B}^{(m)}dA^{(m)} \tag{6.84}$$

$$\mathbf{R}_B^{(m)} = t\int_{A^{(m)}} \mathbf{H}^{(m)^T}\mathbf{f}^{B^{(m)}}dA^{(m)} \tag{6.85}$$

$$\mathbf{R}_S^{(m)} = t\int_{L_1^{(m)},\dots,L_q^{(m)}} \mathbf{H}^{(m)^T}\mathbf{f}^{S^{(m)}}dL^{(m)} \tag{6.86}$$

and

$$\mathbf{K} = \sum_{m=1}^{n_e}\mathbf{K}^{(m)} \tag{6.87}$$

$$\mathbf{R}_B = \sum_{m=1}^{n_e}\mathbf{R}_B^{(m)} \tag{6.88}$$

$$\mathbf{R}_S = \sum_{m=1}^{n_e}\mathbf{R}_S^{(m)} \tag{6.89}$$

$$\mathbf{R} = \mathbf{R}_B + \mathbf{R}_S$$

we can write (6.83) as

$$\bar{\mathbf{U}}^T\left(\mathbf{KU}-\mathbf{R}\right) = 0.$$

Since the virtual displacements are arbitrary, $\bar{\mathbf{U}}$ should be made arbitrary leading to

$$\mathbf{KU} = \mathbf{R}. \tag{6.90}$$

Example 6.5

Consider the problem shown in Figure 6.32. For the given finite element discretization calculate:

(i) The nodal displacements.
(ii) The displacements and stresses at the material point P which has coordinates $X = l/2, Y = h$.

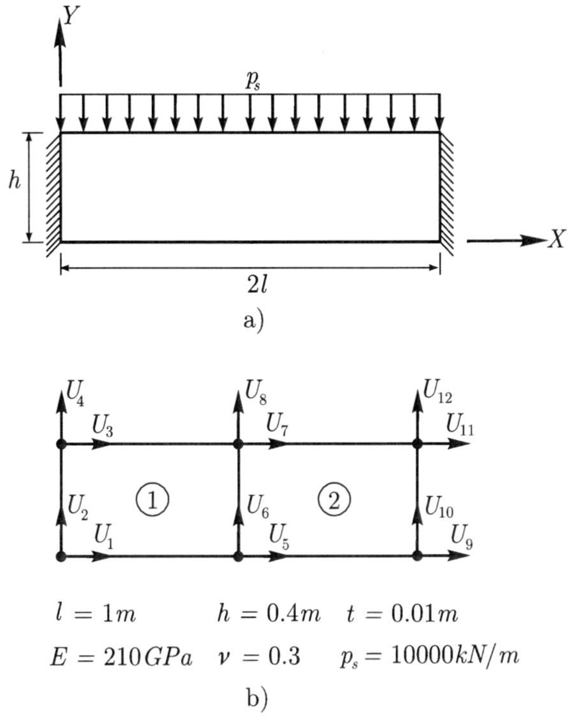

Fig. 6.32. Two-dimensional plane stress problem

Solution

We can write (6.90) as

$$\begin{bmatrix} \mathbf{K}_{aa} & \mathbf{K}_{ab} \\ \mathbf{K}_{ba} & \mathbf{K}_{bb} \end{bmatrix} \begin{bmatrix} \mathbf{U}_a \\ \mathbf{U}_b \end{bmatrix} = \begin{bmatrix} \mathbf{R}_a \\ \mathbf{R}_b \end{bmatrix}$$

where $\mathbf{U}_a$ and $\mathbf{U}_b$ list respectively the unknown and known nodal point displacements (see also (2.53) and (2.54)). The equations to be solved are

$$\mathbf{K}_{aa}\mathbf{U}_a = \mathbf{R}_a - \mathbf{K}_{ab}\mathbf{U}_b \tag{6.91}$$

where then

$$\mathbf{R}_b = \mathbf{K}_{ba}\mathbf{U}_a + \mathbf{K}_{bb}\mathbf{U}_b.$$

Considering the problem in Figure 6.32, we recognize that due to symmetry we only need to solve the problem given in Figure 6.33 with only U_6 and U_8 as unknowns. Therefore, using equation (6.91)

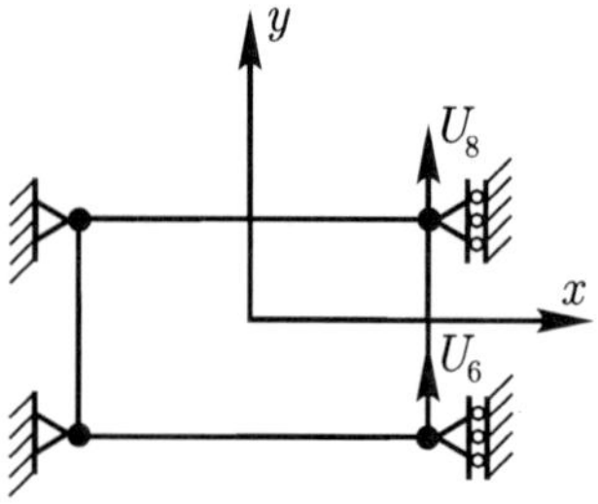

Fig. 6.33. Finite element problem to be solved considering symmetry

$$\begin{bmatrix} K_{66} & K_{68} \\ K_{86} & K_{88} \end{bmatrix} \begin{bmatrix} U_6 \\ U_8 \end{bmatrix} = \begin{bmatrix} R_6 \\ R_8 \end{bmatrix}. \tag{6.92}$$

Let us evaluate the required terms. Denoting by $\mathbf{B}_j^{(m)}$ the column matrix corresponding to the jth column of $\mathbf{B}^{(m)}$ and considering equation (6.84), we have

$$K_{ij}^{(m)} = t \int_{A^{(m)}} \mathbf{B}_i^{(m)^T} \mathbf{C}^{(m)} \mathbf{B}_j^{(m)} dA^{(m)}.$$

The evaluation of $\mathbf{B}_6^{(1)}$ and $\mathbf{B}_8^{(1)}$ is therefore required. Since in the local element numbering U_6 corresponds to v_4 and U_8 to v_1

$$\mathbf{B}_6^{(1)^T} = \begin{bmatrix} 0 & \frac{\partial h_4}{\partial y} & \frac{\partial h_4}{\partial x} \end{bmatrix}, \quad \mathbf{B}_8^{(1)^T} = \begin{bmatrix} 0 & \frac{\partial h_1}{\partial y} & \frac{\partial h_1}{\partial x} \end{bmatrix}.$$

From equations (6.69) to (6.72), with $a = l, b = h$

$$\frac{\partial h_1}{\partial x} = \frac{1}{2lh}(h + 2y), \quad \frac{\partial h_1}{\partial y} = \frac{1}{2lh}(l + 2x)$$

$$\frac{\partial h_4}{\partial x} = \frac{1}{2lh}(h - 2y), \quad \frac{\partial h_4}{\partial y} = -\frac{1}{2lh}(l + 2x).$$

Hence,

$$K_{66} = K_{66}^{(1)} = \frac{Et}{(1-\nu^2)} \int_{-l/2}^{l/2} \int_{-h/2}^{h/2} \begin{bmatrix} 0 & -\frac{1}{2lh}(l+2x) & \frac{1}{2lh}(h-2y) \end{bmatrix}$$

$$\begin{bmatrix} 1 & \nu & 0 \\ \nu & 1 & 0 \\ 0 & 0 & \frac{1-\nu}{2} \end{bmatrix} \begin{bmatrix} 0 \\ -\frac{1}{2lh}(l+2x) \\ \frac{1}{2lh}(h-2y) \end{bmatrix} dydx =$$

$$= \frac{Et}{(1-\nu^2)} \frac{h^2(1-\nu)+2l^2}{6hl}$$

Analogously, we obtain

$$K_{88} = K_{88}^{(1)} = \frac{Et}{(1-\nu^2)} \int_{-l/2}^{l/2} \int_{-h/2}^{h/2} \begin{bmatrix} 0 & \frac{1}{2lh}(l+2x) & \frac{1}{2lh}(h+2y) \end{bmatrix}$$

$$\begin{bmatrix} 1 & \nu & 0 \\ \nu & 1 & 0 \\ 0 & 0 & \frac{1-\nu}{2} \end{bmatrix} \begin{bmatrix} 0 \\ \frac{1}{2lh}(l+2x) \\ \frac{1}{2lh}(h+2y) \end{bmatrix} dydx =$$

$$= \frac{Et}{(1-\nu^2)} \frac{h^2(1-\nu)+2l^2}{6hl}$$

$$K_{68} = K_{86} = \frac{Et}{(1-\nu^2)} \int_{-l/2}^{l/2} \int_{-h/2}^{h/2} \begin{bmatrix} 0 & \frac{1}{2lh}(l+2x) & \frac{1}{2lh}(h+2y) \end{bmatrix}$$

$$\begin{bmatrix} 1 & \nu & 0 \\ \nu & 1 & 0 \\ 0 & 0 & \frac{1-\nu}{2} \end{bmatrix} \begin{bmatrix} 0 \\ -\frac{1}{2lh}(l+2x) \\ \frac{1}{2lh}(h-2y) \end{bmatrix} dydx =$$

$$= \frac{Et}{(1-\nu^2)} \frac{h^2(1-\nu)-4l^2}{12hl}.$$

For the load vector, denoting the jth column of $\mathbf{H}^{(m)}$ by $\mathbf{H}_j^{(m)}$ and the jth line of the column matrix $\mathbf{R}_S^{(m)}$ by $R_{S_j}^{(m)}$, we can write

$$R_{S_j}^{(m)} = t \int_{L_1^{(m)},\ldots,L_q^{(m)}} \mathbf{H}_j^{(m)^T} \mathbf{f}^S dL^{(m)}.$$

Hence

$$R_{S_6} = R_{S_6}^{(1)} = 0$$

$$\begin{aligned} R_{S_8} &= R_{S_8}^{(1)} = t \int_{-l/2}^{l/2} \begin{bmatrix} 0 & h_1(x, y = h/2) \end{bmatrix} \begin{bmatrix} 0 \\ -p_s/t \end{bmatrix} dx = \\ &= \int_{-l/2}^{l/2} \frac{1}{2l}(l + 2x)(-p_s)\, dx = -\frac{p_s l}{2}. \end{aligned} \quad (6.93)$$

Since p_s is a line load, we divided by t to obtain the equivalent pressure load corresponding to $\mathbf{f}^S$. Of course, if we bring t inside the integral in equation (6.93) we can use p_s directly. Note that in the evaluation of $\mathbf{R}_S^{(1)}$ the only element sides that are in L_f are those given by $y = h/2, y = -h/2$. Since h_4 which enters in the evaluation of $\mathbf{R}_{S_6}^{(1)}$ is zero for $y = +h/2$ and $\mathbf{f}^S$ is zero for $y = -h/2$, it is immediate to conclude that $\mathbf{R}_{S_6}^{(1)} = 0$. Also the integral for the evaluation of $\mathbf{R}_{S_8}^{(1)}$ is performed only for $y = +h/2$ since $\mathbf{f}^S = \mathbf{0}$ for $y = -h/2$. Therefore

$$\frac{Et}{(1-\nu^2)} \begin{bmatrix} \frac{h^2(1-\nu)+2l^2}{6hl} & \frac{h^2(1-\nu)-4l^2}{12hl} \\ \frac{h^2(1-\nu)-4l^2}{12hl} & \frac{h^2(1-\nu)+2l^2}{6hl} \end{bmatrix} \begin{bmatrix} U_6 \\ U_8 \end{bmatrix} = \begin{bmatrix} 0 \\ -\frac{p_s l}{2} \end{bmatrix}. \quad (6.94)$$

Substituting the numerical values and solving

$$\begin{aligned} U_6 &= -0.0148352 \text{ m} \\ U_8 &= -0.0161172 \text{ m}. \end{aligned}$$

In order to evaluate the stresses at P we use

$$\tau|_P = \mathbf{C}^{(1)} \left. \mathbf{B}^{(1)} \right|_P \mathbf{U}.$$

The local coordinates of P are $x = 0, y = h/2$ which are used in the evaluation of $\mathbf{B}^{(1)}|_P$. Therefore

$$\tau|_P = \frac{E}{1-\nu^2} \begin{bmatrix} 1 & \nu & 0 \\ \nu & 1 & 0 \\ 0 & 0 & \frac{1-\nu}{2} \end{bmatrix} \begin{bmatrix} 0 \\ \frac{\partial h_4}{\partial y} U_6 + \frac{\partial h_1}{\partial y} U_8 \\ \frac{\partial h_4}{\partial x} U_6 + \frac{\partial h_1}{\partial x} U_8 \end{bmatrix}_{x=0,\ y=h/2}$$

resulting in

$$\tau|_P = \begin{bmatrix} \tau_{xx} \\ \tau_{yy} \\ \tau_{xy} \end{bmatrix} = \begin{bmatrix} -0.11094 \text{ GPa} \\ -0.36981 \text{ GPa} \\ -1.30177 \text{ GPa} \end{bmatrix}.$$

□

Of course, this example was chosen to merely demonstrate the finite element analysis process. A much finer mesh (of many more elements) would be needed to solve the problem in Figure 6.32a accurately.

We showed in (6.74) to (6.76) how the matrices $\mathbf{H}$ and $\mathbf{H}^{(m)}$ are related.

Analogously, we can define the matrices $\mathbf{B}$ and $\mathbf{k}^{(m)}$ which would correspond to $\mathbf{B}^{(m)}$ and $\mathbf{K}^{(m)}$ when we consider only the entries associated with the local element degrees of freedom. Then $\mathbf{B}$ is defined by

$$\varepsilon^{(m)} = \mathbf{B}\hat{\mathbf{u}}. \tag{6.95}$$

For instance, for the four-node element in Figure 6.30

$$\mathbf{B} = \begin{bmatrix} \frac{\partial h_1}{\partial x} & 0 & \frac{\partial h_2}{\partial x} & 0 & \frac{\partial h_3}{\partial x} & 0 & \frac{\partial h_4}{\partial x} & 0 \\ 0 & \frac{\partial h_1}{\partial y} & 0 & \frac{\partial h_2}{\partial y} & 0 & \frac{\partial h_3}{\partial y} & 0 & \frac{\partial h_4}{\partial y} \\ \frac{\partial h_1}{\partial y} & \frac{\partial h_1}{\partial x} & \frac{\partial h_2}{\partial y} & \frac{\partial h_2}{\partial x} & \frac{\partial h_3}{\partial y} & \frac{\partial h_3}{\partial x} & \frac{\partial h_4}{\partial y} & \frac{\partial h_4}{\partial x} \end{bmatrix}. \tag{6.96}$$

The matrix $\mathbf{k}^{(m)}$ can be defined by

$$\mathbf{k}^{(m)} = t \int_A \mathbf{B}^T \, \mathbf{C} \, \mathbf{B} \, dA \tag{6.97}$$

where the domain A, and the matrices $\mathbf{B}$ and $\mathbf{C}$ correspond to the element considered.

A generic entry in the matrix $\mathbf{K}^{(m)}$ is given by (see also Example 6.5)

$$K_{ij}^{(m)} = t \int_{A^{(m)}} \mathbf{B}_i^{(m)^T} \, \mathbf{C} \, \mathbf{B}_j^{(m)} dA^{(m)} \tag{6.98}$$

where $K_{ij}^{(m)}$ is different from zero only when $\mathbf{B}_i^{(m)}$ and $\mathbf{B}_j^{(m)}$ have entries that are non-zero. This happens only when the degrees of freedom U_i and U_j belong to element (m). Therefore all entries of $\mathbf{K}^{(m)}$ are also in $\mathbf{k}^{(m)}$ since all columns of $\mathbf{B}^{(m)}$ which have non-zero entries are in $\mathbf{B}$. Hence, we can obtain $\mathbf{K}^{(m)}$ from $\mathbf{k}^{(m)}$ by properly assigning the entries in $\mathbf{k}^{(m)}$ to entries in $\mathbf{K}^{(m)}$. This is achieved by using the element connectivity array $\mathbf{LM}^{(m)}$ which lists the global degrees of freedom numbers ordered according to the local numbering. As already detailed for matrix structural analysis (see Sections 2.3 and 4.2.4), considering the element described in Figure 6.29, we would have

$$\mathbf{LM}^{(m)T} = \begin{array}{c} \\ [\end{array} \begin{array}{cccccccc} u_1 & v_1 & u_2 & v_2 & u_3 & v_3 & u_4 & v_4 \\ q & p & l & k & i & j & f & g \end{array} \begin{array}{c} \\ ]. \end{array} \tag{6.99}$$

The contributions of element (m) to the global stiffness matrix $\mathbf{K}$ can be obtained by summing

$$\begin{aligned}
&k_{11}^{(m)} \text{ to the entry } qq \text{ of the array;} \\
&k_{12}^{(m)} \text{ to the entry } qp; \\
&\quad\vdots \\
&k_{18}^{(m)} \text{ to the entry } qg; \\
&k_{22}^{(m)} \text{ to the entry } pp; \\
&\quad\vdots \\
&k_{28}^{(m)} \text{ to the entry } pg;
\end{aligned}$$

and so on.

Here by taking advantage of the symmetry of $\mathbf{K}$ we assemble only the upper part of $\mathbf{K}$ including the diagonal entries. This is why in the above procedure after adding the contribution of k_{18} we did not start from k_{21}. The complete process for all finite elements corresponds to the summation of stiffness values in (6.87), where we start with a global stiffness matrix containing only zero entries and successively add the element stiffness matrices. If a degree of freedom is restrained to a zero value, a zero is placed in the corresponding position of $\mathbf{LM}$ for each element, and the stiffness contribution is not added to the stiffness matrix. In other words, we are only constructing the stiffness matrix corresponding to the unknown degrees of freedom ($\mathbf{K}_{aa}$ in (2.57)).

We used the rectangular element to introduce the basic ideas of the finite element procedure for 2-D analysis. The formulation of other types of elements follows the same approach and rests on the definition of the interpolation functions. The evaluation of the stiffness matrix and load vectors follows the same steps as described for rectangular elements.

Example 6.6

Consider the mesh of triangular elements shown in Figure 6.34. Define the interpolation functions for a generic element of the mesh. Indicate how to obtain the stiffness matrix of this element.

Solution

The interpolation functions can be constructed obeying the same property used in the formulation of the rectangular elements, *i.e.*, the interpolation functions h_i assumes the value of one at node i and zero at node j, $j \neq i$.

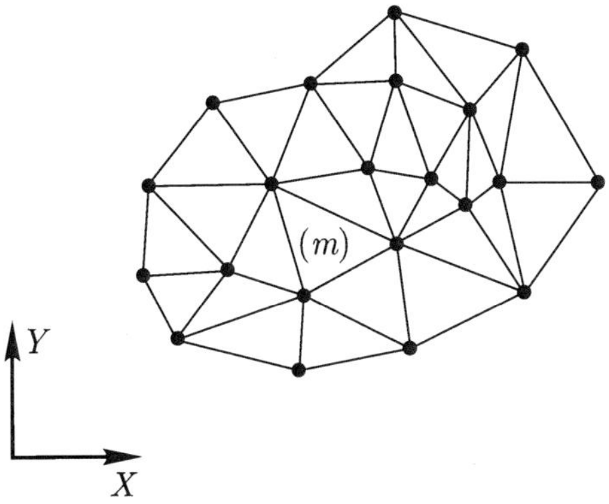

Fig. 6.34. Mesh of three-node triangular elements

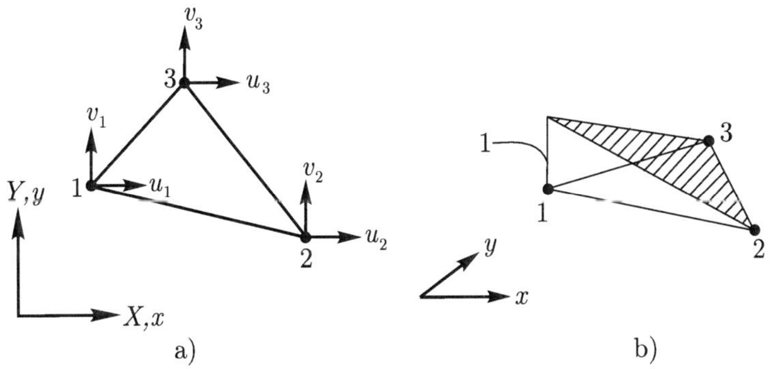

Fig. 6.35. a) Generic three-node triangular element; b) Interpolation function h_1 geometrically defined

Therefore, using the property that three non-colinear points determine a unique plane, we construct the interpolation functions. For instance, h_1 is given by the unique plane which passes through the point that has out-of-plane value of one at the coordinates of node 1 and through nodes 2 and 3. This is pictorially shown in Figure 6.35b. The analytical expression of h_1 is

$$h_1(x, y) = \frac{1}{2A_t}(a_1 + b_1 x + c_1 y) \tag{6.100}$$

where A_t is the area of the triangular element which can be calculated by

$$2A_t = x_1 y_2 + x_2 y_3 + x_3 y_1 - y_1 x_2 - y_2 x_3 - y_3 x_1$$

where (x_1, y_1), (x_2, y_2) and (x_3, y_3) are the coordinates of nodes 1, 2 and 3 respectively. The remaining constants are

$$\begin{aligned} a_1 &= x_2 y_3 - x_3 y_2 \\ b_1 &= y_2 - y_3 \\ c_1 &= x_3 - x_2. \end{aligned}$$

It is immediate to verify that equation (6.100) is, in fact, the analytical form of the interpolation function sought since, by substitution,

$$h_1(x_1, y_1) = 1, \qquad h_1(x_2, y_2) = 0, \qquad h_1(x_3, y_3) = 0$$

and since the functional form of $h_1(x, y)$ corresponds to a plane, all requirements are satisfied. Analogously,

$$h_2(x, y) = \frac{1}{2A_t}(a_2 + b_2 x + c_2 y)$$

and

$$h_3(x, y) = \frac{1}{2A_t}(a_3 + b_3 x + c_3 y)$$

with

$$\begin{aligned} a_2 &= x_3 y_1 - x_1 y_3, \quad a_3 = x_1 y_2 - x_2 y_1 \\ b_2 &= y_3 - y_1, \quad b_3 = y_1 - y_2 \\ c_2 &= x_1 - x_3, \quad c_3 = x_2 - x_1. \end{aligned}$$

Having the interpolation functions defined, we can evaluate the matrices $\mathbf{H}$, $\mathbf{B}$ and $\mathbf{k}^{(m)}$. We note that from the functional form of the interpolation functions $h_i(x, y)$, the strains ε_{xx}, ε_{yy} and γ_{xy}, and consequently the stresses τ_{xx}, τ_{yy} and τ_{xy} are constant over the element. This element is usually referred to as the constant strain triangle.

□

6.3.2 Equilibrium properties of the finite element solutions for 2-D

As discussed for the 1-D finite element formulation, when we introduce the finite element formulation assumptions into the principle of virtual work statement expressed only in terms of displacements, we are satisfying exactly compatibility and the stress-strain law. However, equilibrium is satisfied only in an approximate manner. We also showed that there are two equilibrium properties that are satisfied for any finite element discretization, no matter how coarse the mesh might be (see also Bathe, 1996). Namely:

1. *Nodal equilibrium.* The external applied nodal forces are equilibrated by the element nodal point forces.

2. *Element equilibrium.* The element nodal point forces are always in equilibrium for any given element.

These same properties can be generalized for the 2-D finite element discretization.

We define the element nodal point forces for a generic element (m) by

$$\mathbf{F}^{(m)} = \int_{A^{(m)}} \mathbf{B}^{(m)T} \tau^{(m)} \, dA^{(m)} = \mathbf{K}^{(m)} \mathbf{U} \tag{6.101}$$

where $\mathbf{U}$ is the finite element nodal displacement solution[3]. Hence, these nodal point forces give the forces acting onto the element at the element nodal points.

To show Property 1, part of a generic mesh of triangular elements (actually, the results that are demonstrated do not depend on the mesh being of triangular elements) and the nodal point forces of each element are shown in Figure 6.36.

For every node, the element nodal forces are in equilibrium with the externally applied nodal forces. In fact

$$\sum_{m=1}^{n_e} \mathbf{F}^{(m)} = \sum_{m=1}^{n_e} \mathbf{K}^{(m)} \mathbf{U} = \mathbf{K}\mathbf{U} = \mathbf{R}.$$

In Figure 6.36 we give a pictorial representation of this property for node k.

To illustrate Property 2 consider the nodal point forces of a generic element (m) as shown in Figure 6.37 which are given by

$$\begin{aligned} \mathbf{F}^{(m)T} = \{0 \cdots \overset{e}{F_e} \cdots \overset{f}{F_f} \cdots \overset{g}{F_g} \cdots \overset{h}{F_h} \cdots \\ \overset{k}{F_k} \cdots \overset{\ell}{F_\ell} \cdots 0\} \end{aligned}$$

where only the non-zero entries are shown and consider a vector of nodal displacements corresponding to a rigid translation of intensity Δ in the x direction, *i.e.*,

$$\bar{\mathbf{U}}^T = \{0 \cdots \overset{e}{\Delta} \cdots \overset{g}{\Delta} \cdots \overset{k}{\Delta} \cdots 0\}.$$

We can write

$$\bar{\mathbf{U}}^T \mathbf{F}^{(m)} = \Delta.F_e + \Delta.F_g + \Delta.F_k \tag{6.102}$$

and using the definition of $\mathbf{F}^{(m)}$ given in (6.101), we obtain

---

[3] The equations in this section are presented for the plane strain model of unit thickness. For the plane stress and axisymmetric models the integrations would have to be performed accordingly (see Sections 5.2.2 and 5.2.4)

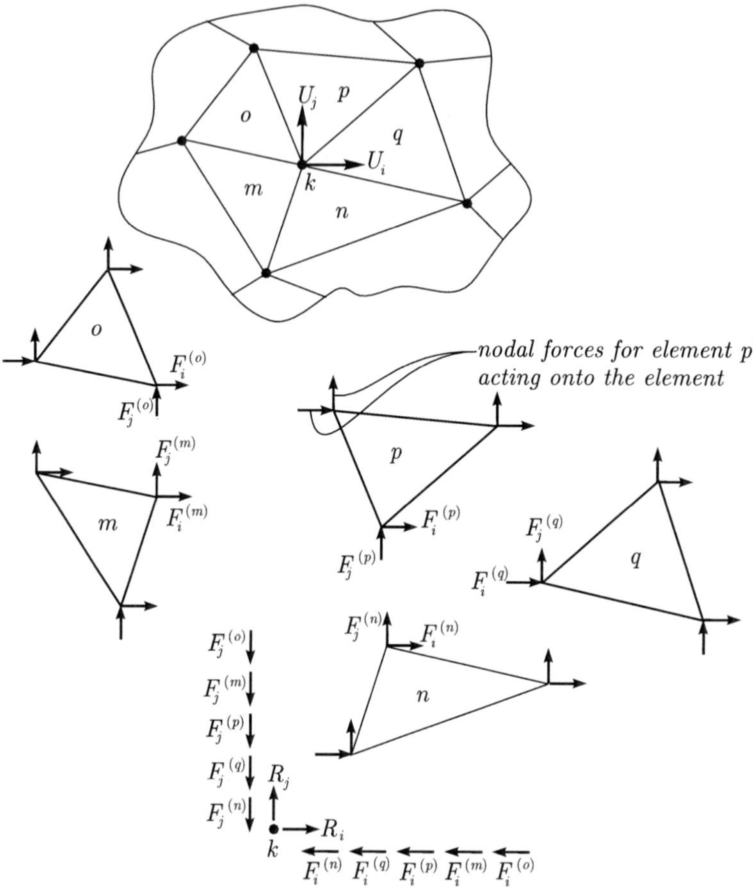

Fig. 6.36. Nodal forces for some elements in a mesh

$$
\begin{aligned}
\bar{\mathbf{U}}^T\mathbf{F}^{(m)} &= \bar{\mathbf{U}}^T\mathbf{K}^{(m)}\mathbf{U} \\
&= \bar{\mathbf{U}}^T \int_{A^{(m)}} \mathbf{B}^{(m)T}\mathbf{C}\mathbf{B}^{(m)}\ dA^{(m)}\ \mathbf{U} \\
&= \int_{A^{(m)}} \bar{\mathbf{U}}^T\mathbf{B}^{(m)T}\mathbf{C}\mathbf{B}^{(m)}\ dA^{(m)}\ \mathbf{U}
\end{aligned}
$$

Note that

$$
\bar{\mathbf{U}}^T\mathbf{B}^{(m)T} = \bar{\epsilon}^{(m)T}
$$

are the strains associated with the displacements $\bar{\mathbf{u}}^{(m)} = \mathbf{H}^{(m)}\bar{\mathbf{U}}$. As mentioned already, it is a fundamental requirement that the element displacement interpolations must contain the rigid body mode displacements (see see Bathe, 1996). Since, of course, this element represents exactly rigid body

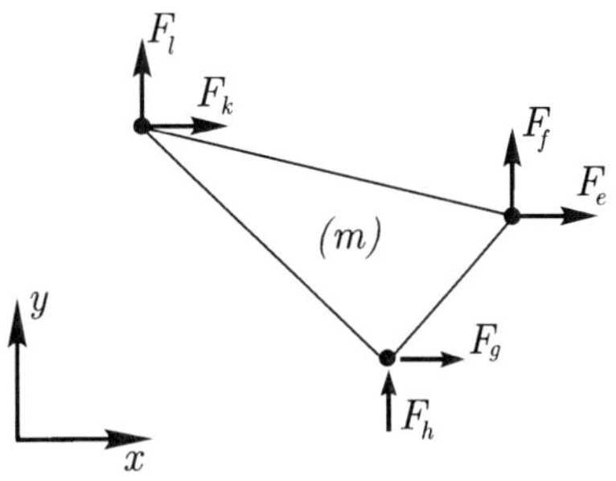

Fig. 6.37. Nodal forces for a generic element (m).

modes – that is, for nodal displacements corresponding to a rigid body mode, the interpolated displacements inside the element correspond to the same rigid body mode – we have $\bar{\varepsilon}^{(m)} = 0$. Therefore

$$\bar{\mathbf{U}}^T \mathbf{F}^{(m)} = 0. \qquad (6.103)$$

Then, from equations (6.102) and (6.103), we obtain

$$F_e + F_g + F_k = 0$$

which is the equilibrium condition in the x direction. Following an analogous derivation, considering a rigid translation in the y direction, we arrive at the equilibrium condition in the y direction, *i.e.*,

$$F_f + F_h + F_\ell = 0.$$

Finally, if we consider a rigid body rotation we arrive at the moment equilibrium condition for the element (m) nodal forces.

6.3.3 Isoparametric finite elements

In engineering practice we need to discretize, in general, very complex geometric domains requiring general quadrilateral and triangular elements. The isoparametric element concept provides a very powerful methodology to construct such general elements. In particular, also elements with more than 3 or 4 nodes can be constructed allowing curved element sides.

Consider the finite element mesh shown in Figure 6.38 in which we have 4-node quadrilateral and 3-node triangular elements. In order to introduce the concept of isoparametric element formulations consider the generic quadrilateral element (m) shown in Figure 6.38. To formulate the element matrices, we have to define the interpolation of the displacements inside the element from its nodal point displacements. For this purpose we now use an auxiliary reference domain of coordinates r and s as shown in Figure 6.39.

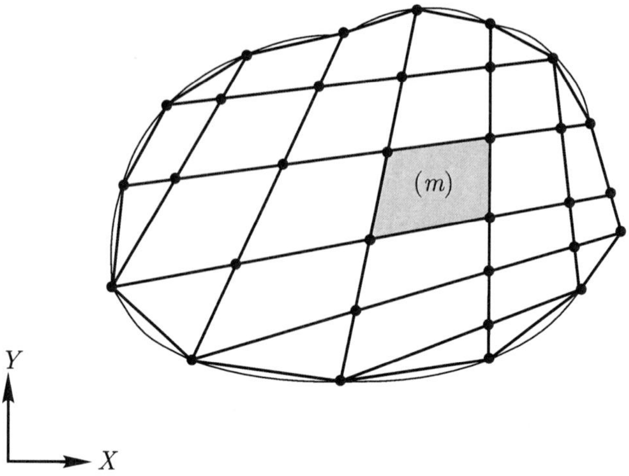

Fig. 6.38. Mesh with quadrilateral and triangular elements

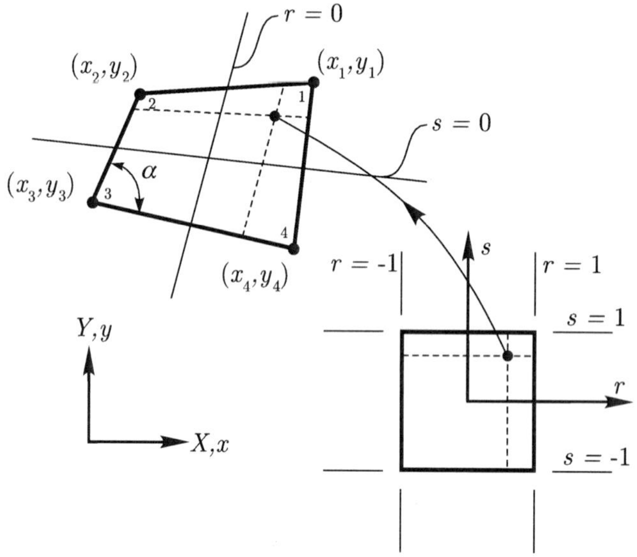

Fig. 6.39. Schematic representation of the domain and the mapping; interior angle $\alpha < 180^0$

The mapping we use, which relates every point (r, s) of the reference domain to a point (x, y) in the physical element domain, is

$$x(r, s) = \sum_{i=1}^{4} h_i(r, s)\, x_i \tag{6.104}$$

$$y(r,s) = \sum_{i=1}^{4} h_i(r,s)\, y_i \tag{6.105}$$

where the (x_i, y_i) are the nodal point coordinates of element (m), see Figure 6.39. We assume that this mapping is defined uniquely for every (r, s) and (x, y) and hence invertible at all points. This requires that all interior angles α, see Figure 6.39, be smaller than 180°, see Bathe, 1996. The interpolation functions $h_i(r,s)$ are given by

$$\begin{aligned} h_1 &= \frac{1}{4}(1+r)(1+s) \\ h_2 &= \frac{1}{4}(1-r)(1+s) \\ h_3 &= \frac{1}{4}(1-r)(1-s) \\ h_4 &= \frac{1}{4}(1+r)(1-s) \end{aligned} \tag{6.106}$$

Of course, these functions are the interpolation functions of a 2 by 2 square element, defined in the local coordinates r and s. A straight line given in the reference domain by a constant r is mapped to a straight line in the element physical domain, and the same holds for a straight line given by a constant s. Therefore, the sides of the reference domain are mapped to the sides of the actual physical element and the vertices of the reference domain are mapped to the element nodes.

In Figure 6.40 we show the element nodal point displacements following the local node numbering. The element displacement interpolation can now be defined by

$$u(r,s) = \sum_{i=1}^{4} h_i(r,s)\, u_i \tag{6.107}$$

and

$$v(r,s) = \sum_{i=1}^{4} h_i(r,s)\, v_i. \tag{6.108}$$

The term "isoparametric" refers to the fact that the same interpolations are used for the geometry and for the displacements.

Note that, since the h_i are defined as functions of r and s, the displacements (u, v) for an interior point of the element in the directions of x and y, respectively, are given in terms of r and s by equations (6.107) and (6.108). However, considering that the mapping defined in (6.104) and (6.105) is invertible, we can write

$$r = r(x,y) \tag{6.109}$$

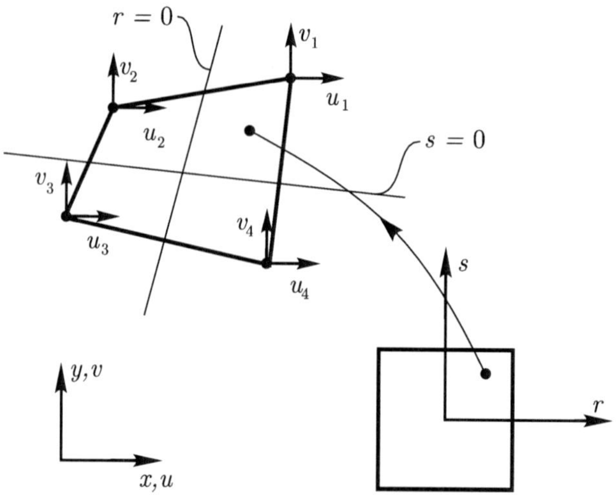

Fig. 6.40. Nodal displacements of element in local numbering

$$s = s(x, y). \tag{6.110}$$

Hence, given (x, y) in the physical element, (6.109) and (6.110) give the corresponding (r, s) in the reference domain. Therefore, using equations (6.109) and (6.110), we can re-write (6.107) and (6.108) as

$$\begin{aligned} u(r(x,y), s(x,y)) &= \sum_{i=1}^{4} h_i(r(x,y), s(x,y))\, u_i \\ &= \sum_{i=1}^{4} \tilde{h}_i(x,y)\, u_i \end{aligned} \tag{6.111}$$

$$\begin{aligned} v(r(x,y), s(x,y)) &= \sum_{i=1}^{4} h_i(r(x,y), s(x,y))\, v_i \\ &= \sum_{i=1}^{4} \tilde{h}_i(x,y)\, v_i . \end{aligned} \tag{6.112}$$

Hence, the displacement interpolation is well defined which allows the evaluation of $\mathbf{H}$, $\mathbf{B}$ and $\mathbf{k}^{(m)}$. In practice, the inversion used in equations (6.109) and (6.110) is not explicitly performed, because the evaluation of the matrices $\mathbf{H}$, $\mathbf{B}$ and $\mathbf{k}^{(m)}$ do not require such an inversion. However, it is important to realize that in theory, such an inversion must be possible. The Jacobian $\mathbf{J}$ of the isoparametric mapping defined in equations (6.104) and (6.105) is given by

$$\mathbf{J} = \begin{bmatrix} \frac{\partial x}{\partial r} & \frac{\partial y}{\partial r} \\ \frac{\partial x}{\partial s} & \frac{\partial y}{\partial s} \end{bmatrix} \tag{6.113}$$

with $dxdy = det\mathbf{J}\, drds$, see Bathe, 1996. Hence the condition for uniqueness between (x, y) and (r, s) and a proper formulation is that $det\mathbf{J} > 0$ for $-1 \leq r \leq 1$, and $-1 \leq s \leq 1$. Hence, when constructing the mesh, the nodal points should be placed in such way that $det\mathbf{J} > 0$. This is achieved when all interior angles are smaller than 180 degrees, see Figure 6.39.

The procedures given above are directly applicable to higher-order elements. Figure 6.41 shows a generic 9-node element, which is much used in practice. The interpolation function h_i should be 1 at (x_i, y_i) and zero at every other node. Of course, the interpolation functions will now involve quadratic monomials. Then, the mapping is given by

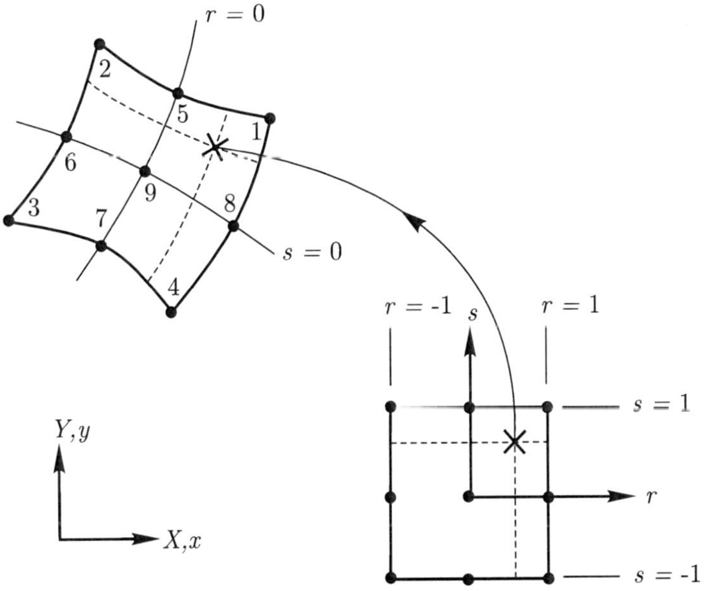

Fig. 6.41. Generic 9-node element and associated mapping

$$x = \sum_{i=1}^{9} h_i(r, s)\, x_i \tag{6.114}$$

$$y = \sum_{i=1}^{9} h_i(r, s)\, y_i \tag{6.115}$$

and the interpolation of the displacement is given by

$$u = \sum_{i=1}^{9} h_i\,(r,s)\,u_i \tag{6.116}$$

$$v = \sum_{i=1}^{9} h_i\,(r,s)\,v_i. \tag{6.117}$$

The objective of our discussion on isoparametric finite elements was only to present basic equations of this class of elements which are used in virtually all general purpose finite programs due to their versatility and good approximation properties. Of course, there are many details regarding these element formulations and we refer to Bathe, 1996 for a more complete discussion on isoparametric finite elements.

6.3.4 Numerical integration

In order to complete our presentation of finite element formulations, we want to briefly introduce the numerical integration procedures which are widely used to evaluate finite element matrices.

In finite element analysis, we are required to obtain the numerical value of

$$\int_V g(x,y,z)\,dV, \quad \int_S w(x,y)\,dS, \quad \int_L f(x)\,dL$$

where the real valued functions $g(x,y,z)$, $w(x,y)$ and $f(x)$ are known. Since the integral over 2-D and 3-D domains can be obtained as successive integrations in 1-D, we want a procedure to evaluate

$$\int_a^b f(x)\,dx \quad \text{for } f(x) \text{ known.}$$

Since $f(x)$ may be a complicated function, which is problem dependent, it is not effective and general (and may not be possible) to perform an analytical integration. Instead, we integrate a function $\phi(x)$ for which $\phi(x_i) = f(x_i)$ at selected points x_i of the function domain – sampling points – and take

$$\int_a^b f(x)\,dx \doteq \int_a^b \phi(x)\,dx.$$

Of course, the evaluation of the integral of $\phi(x)$ should be systematic, computationally effective and, if sufficient points x_i are used, should lead to an accurate approximation of the integral of $f(x)$.

Consider the function $f(x)$ shown in Figure 6.42a and the selected points x_i, $i = 1, \cdots, 4$. In Figure 6.42b we show $\phi(x)$ which is defined as the polynomial function for which $\phi(x_i) = f(x_i)$, $i = 1, \cdots, 4$. The integral of $\phi(x)$

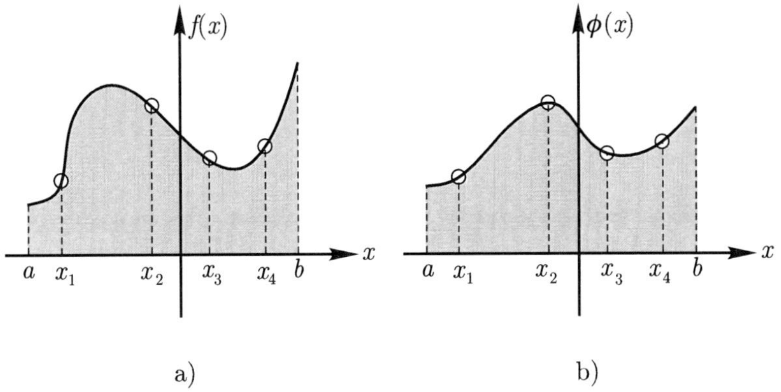

Fig. 6.42. a) Function $f(x)$ to be integrated and sampling points; b) Polynomial function that assumes the values of $f(x)$ at given points

which is given by the shaded area in Figure 6.42b is an approximation of the integral of $f(x)$ given by the shaded area in Figure 6.42a.

There are two main integration schemes used in finite element analysis. In the Newton-Cotes integration scheme the evaluation points are defined by subdividing the domain into equal parts and also include the end points. In this case, if $f(x)$ is a polynomial of order n, we need $n+1$ points to exactly integrate $f(x)$, since the unique polynomial $\phi(x)$ of order n passes through the $n+1$ points and $\phi(x) \equiv f(x)$. On the other hand, in the Gauss integration scheme, specific locations of the evaluation points are used in order to maximize the precision of the numerical integration procedure and these do not include the end points. As shown in Bathe, 1996, using the Gauss scheme only n points are required to integrate exactly a polynomial function of order $2n-1$.

Let us consider, without loss of generality, our domain of integration to be $[-1, 1]$. We can write

$$\int_{-1}^{+1} f(r)dr \doteq \int_{-1}^{+1} \phi(r)dr$$

where $\phi(r)$ is the polynomial function which assumes the values of $f(r)$ at the n sampling points, *i.e.*, $\phi(r_i) = f(r_i)$ $i = 1, 2, ..., n$. It can be shown that

$$\int_{-1}^{+1} \phi(r)dr = \sum_{i=1}^{n} \alpha_i f(r_i)$$

where $\alpha_i, r_i, i = 1, ..., n$ are available for a given numerical integration order. The α_i are called the weights, r_i the sampling points and n the integration order. In Table 6.3 we show these data for Gauss integration up to order 3.

The procedures used in 1-D integrations can directly be applied in 2-D and 3-D integrations. Consider the two 2-D domains shown in Figure 6.43

Table 6.3. Coordinates and weights for the Gauss points in one-dimension

Integration order	Gauss sampling points (r coordinate)	Weights
1	0.000000000000000	2.000000000000000
2	±0.577350269189626	1.000000000000000
3	±0.774596669241483	0.555555555555556
	0.000000000000000	0.888888888888889

which are related by the mapping $x(r,s), y(r,s)$. Let $f(x,y)$ be a function defined in the domain A. We have

$$\int_A f(x,y)dxdy = \int_{-1}^{+1}\int_{-1}^{+1} f\left(x\left(r,s\right), y(r,s)\right) \det \mathbf{J}\ drds \tag{6.118}$$

where $\mathbf{J}$ is the Jacobian of the mapping. For example, the stiffness matrix of

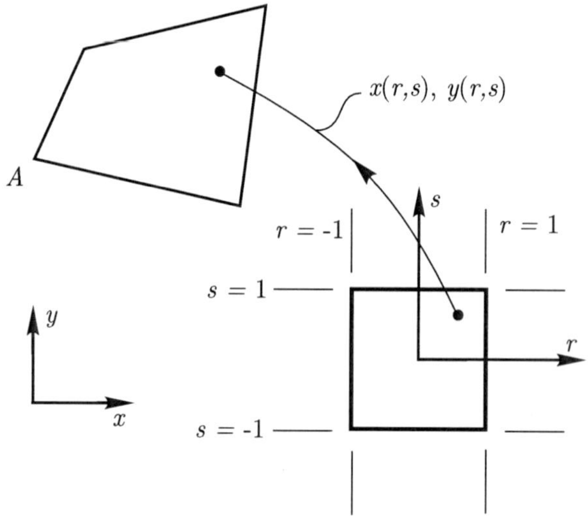

Fig. 6.43. Two-dimensional domains related by the isoparametric mapping

a 2-D isoparametric element is, for a unit thickness,

$$\mathbf{k} = \int_{-1}^{+1}\int_{-1}^{+1} \mathbf{B}^T\mathbf{C}\mathbf{B} \det \mathbf{J} drds \tag{6.119}$$

where, for a 4-node element, $x(r,s)$ and $y(r,s)$ are given by (6.107) and (6.108). Here each entry k_{ij} is of the form (6.118).

The rule of numerical integration is given by

$$\int_{-1}^{+1}\int_{-1}^{+1} g(r,s)drds \cong \sum_{i=1}^{n_{intr}}\sum_{j=1}^{n_{ints}} \alpha_i\alpha_j g(r_i,s_j) \tag{6.120}$$

where $g(r,s) = f(x(r,s), y(r,s)) \det \mathbf{J}$. In (6.120) n_{intr} and n_{ints} represent the number of integration points in the r and s directions, respectively, α_i and α_j are the weights and r_i and s_j the sampling points, each corresponding to one-dimensional integration, see Table 6.3. In Figure 6.44 we show the locations of the sampling points of the most common integration rules used for 2-D isoparametric elements.

An important question is "What order of numerical integration should be used for a specific element?" This question and other considerations are discussed in Bathe, 1996.

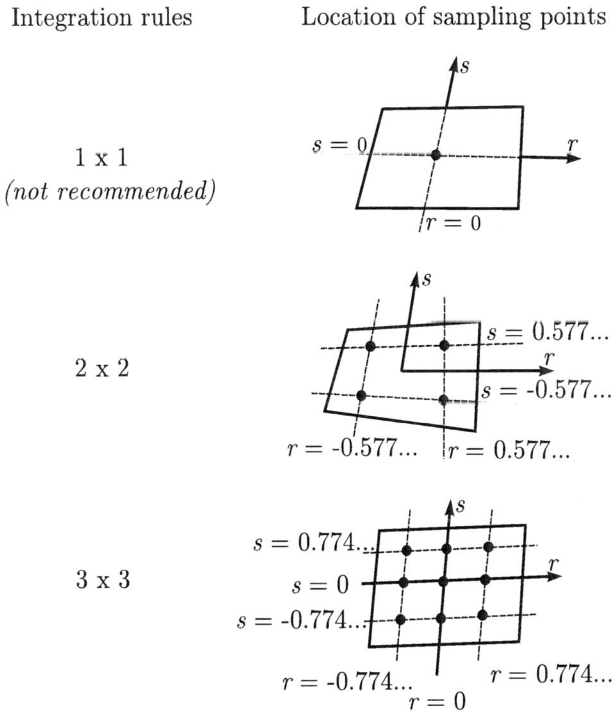

Fig. 6.44. Location of the sampling points for quadrilateral elements

6.3.5 Displacement-based finite element formulations for 3-D solids

The starting point is the principle of virtual work for 3-D elasticity written in terms of displacements given in (6.67).

We need to define the displacement interpolations. In Figure 6.45, we show a typical 3-D element inside the solid being discretized. Our task is to define $u^{(m)}$, $v^{(m)}$ and $w^{(m)}$, *i.e.*, the displacements for element (m) from those at the nodes. A typical node k and its nodal degrees of freedom are also shown in Figure 6.45. A convenient way to define the interpolation is to use the isoparametric formulation. In the 3-D case the reference domain is a cube (see Figure 6.45).

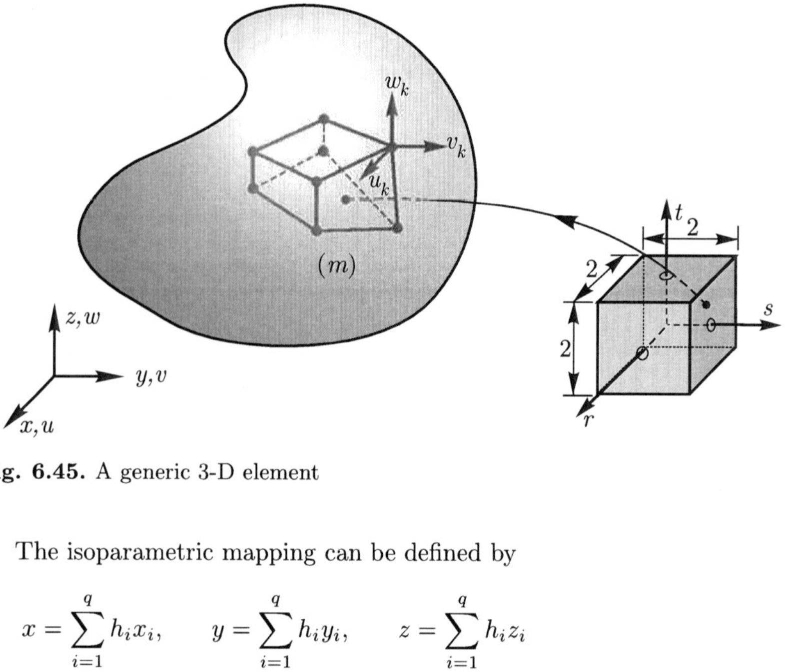

Fig. 6.45. A generic 3-D element

The isoparametric mapping can be defined by

$$x = \sum_{i=1}^{q} h_i x_i, \qquad y = \sum_{i=1}^{q} h_i y_i, \qquad z = \sum_{i=1}^{q} h_i z_i$$

where the x_i, y_i, z_i are the element nodal coordinates and the $h_i(r, s, t)$ are the interpolation functions defined in the reference domain as for the 2-D case. The number of nodes, which is 8 for the element shown in Figure 6.45, is generically given by q. The displacement interpolations are given by

$$u = \sum_{i=1}^{q} h_i u_i, \qquad v = \sum_{i=1}^{q} h_i v_i, \qquad w = \sum_{i=1}^{q} h_i w_i$$

where u_i, v_i, w_i are the nodal displacements of node i. Therefore

$$
\mathbf{H} = \left[\begin{array}{ccc|c|ccc|c|ccc} h_1 & 0 & 0 & & h_i & 0 & 0 & & h_q & 0 & 0 \\ 0 & h_1 & 0 & \cdots\cdots & 0 & h_i & 0 & \cdots\cdots & 0 & h_q & 0 \\ 0 & 0 & h_1 & & 0 & 0 & h_i & & 0 & 0 & h_q \end{array} \right] .
$$

Of course, when $\mathbf{H}$ is defined we can evaluate $\mathbf{B}$ and $\mathbf{k}^{(m)}$ (see equations (6.122) and (6.123) below). The interpolation functions are given, for example, in Bathe, 1996.

6.3.6 Framework to formulate displacement-based finite elements

In Chapter 5 we presented a generalized form of the principle of virtual work which is applicable to the mathematical models of Chapter 4 and to the 3-D elasticity model presented in Chapter 3. Finite element formulations can then be obtained defining finite element interpolations for the kinematic variables. In fact, to arrive at a specific element formulation we would define the nodal points and the displacement interpolation, that is,

$$
\mathbf{u} = \mathbf{H}\hat{\mathbf{u}} \tag{6.121}
$$

where $\mathbf{u}$ is the generalized displacement column matrix (see Section 4.5) and $\hat{\mathbf{u}}$ is the specific element nodal displacement column matrix. The generalized strain is given by

$$
\varepsilon = \partial_\varepsilon \mathbf{u} = \partial_\varepsilon \mathbf{H}\hat{\mathbf{u}} = \mathbf{B}\hat{\mathbf{u}}. \tag{6.122}
$$

Then, the element stiffness matrix can be obtained by

$$
\mathbf{k}^{(m)} = \int_V \mathbf{B}^T \mathbf{C} \mathbf{B} \, dV. \tag{6.123}
$$

In the above expression $\mathbf{C}$ is the generalized constitutive matrix. We presented above the definitions of the matrices $\mathbf{H}$, $\mathbf{B}$ and $\mathbf{k}^{(m)}$ at the element level. Of course, it is implied that $\mathbf{H}^{(m)}$, $\mathbf{B}^{(m)}$ can be defined and the finite element equations would then be established, in principle, as above (see Section 6.3.1).

6.4 Finite elements for beams, plates and shells

In this section we will discuss finite elements which together with those for 2-D and 3-D solids can be used to undertake comprehensive modeling of engineering structures. Of course, the mathematical models of Chapter 4 are the basis for the formulations to be presented.

In Section 6.3.6, we presented a framework to formulate displacement-based finite elements which is applicable to all mathematical models discussed in this book. However, we will not detail the finite element formulation for all these models. Instead, we concentrate on fewer models which capture the relevant structural behaviors rendering the associated finite elements widely applicable and very attractive for engineering structural analysis.

In Section 6.4.1, we present the Bernoulli-Euler beam finite element, in Section 6.4.2 we compare the finite element and matrix structural analysis formulations in the modeling of frame structures, in Section 6.4.3 we briefly introduce plate and shell finite elements, and in Section 6.4.4 we present more general beam elements.

6.4.1 Bernoulli-Euler beam finite element

Let us consider first the Bernoulli-Euler model for transverse loading only as detailed in Section 4.2.2.

Following the framework of Section 6.3.6, we recall that

$$\begin{aligned} \tau &= [M(x)]\,, \quad \varepsilon = [\kappa(x)]\,, \quad \mathbf{C} = [EI]\,, \\ \mathbf{u} &= [w(x)]\,, \quad \partial_\varepsilon = \left[\frac{d^2}{dx^2}\right]. \end{aligned}$$

We note that we need to interpolate the transverse displacement $w(x)$. The curvature (the generalized strain)

$$\kappa = \frac{d^2 w}{dx^2}$$

involves the second derivative of $w(x)$. So far, we detailed finite element formulations for which the strains involved only the first derivatives of the displacement functions.

As discussed in Section 6.2, for the principle of virtual work formulation to be well defined both the displacement solution and the virtual displacement should belong to a given space of functions V. This space should be such that the strain energy is finite for every displacement function in V. We recall that the strain energy for the Bernoulli-Euler beam model is given by

$$a(w,w) = \frac{1}{2}\int_0^L \varepsilon^T \mathbf{C} \varepsilon \, dV = \frac{1}{2}\int_0^L \kappa E I \kappa \, dx = \frac{1}{2}\int_0^L EI \left(\frac{d^2 w}{dx^2}\right)^2 dx.$$

It can be shown that for $w(x)$ and $\frac{dw}{dx}(x)$ piecewise continuous the strain energy is finite. Hence, when we choose a finite element mesh, both $w_h(x)$ and $\frac{dw_h(x)}{dx}$ should be continuous at the element interfaces, *i.e.*, at the nodes. To have this property, we use the so-called Hermitian interpolation functions, which we have already used. In fact, we can refer to Table 4.1 where these interpolation functions are defined.

Elements for which continuity of the kinematic variables and of their first derivatives is satisfied are referred to as C^1 elements since a C^1 function is continuous with continuous first derivatives. The elements for which only the continuity of the kinematic variables is satisfied are referred to as C^0 elements since a C^0 function is continuous.

To obtain physical insight regarding these mathematical conditions, we note that for elastic solids the displacements should be continuous throughout the domain since the material particles should not separate from each other, that is, the solid should not "break". This same restriction physically justifies why $\frac{dw}{dx}$ should also be continuous. Namely, since $\theta = \frac{dw}{dx}$ gives the beam section rotation, a discontinuity would lead to a section "breakage" (a hinge). Figure 6.46 summarizes these considerations.

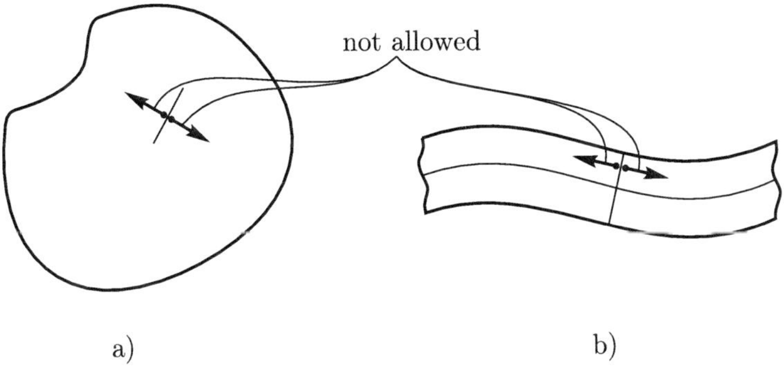

Fig. 6.46. Requirements for kinematic variables

Let (m) be a generic element of length l as summarized in Figure 6.47a, in which the local coordinate system given by the x axis and the element nodal displacement and rotation degrees of freedom w_1, $\theta_1 = \left.\frac{dw}{dx}\right|_{x=0}$, w_2, $\theta_2 = \left.\frac{dw}{dx}\right|_{x=l}$ are shown. Then, we can write

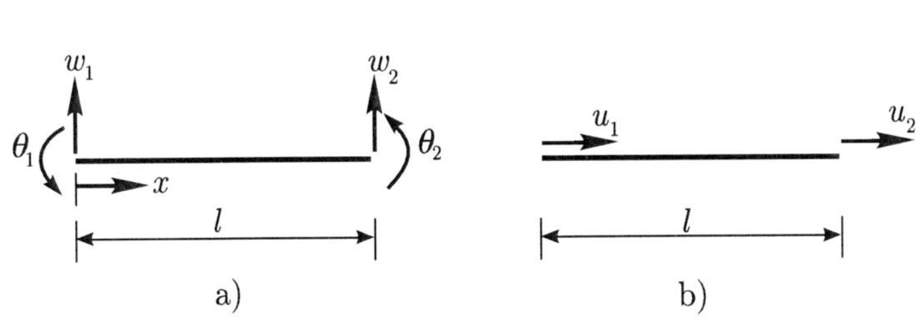

Fig. 6.47. a) Two-node beam element; b) Axial degrees of freedom for two-node beam element

$$w_h(x) = h_2(x)w_1 + h_5(x)w_2 + h_3(x)\theta_1 + h_6(x)\theta_2$$

where the interpolation functions are those given in Table 4.1 when we take $L = l$. Continuity of the derivatives is assured as long as we take the same nodal rotation degrees of freedom at element interfaces. The matrix $\mathbf{B}$ is implicitly given by

$$\kappa = \mathbf{B}\mathbf{u} = \begin{bmatrix} \frac{d^2h_2}{dx^2} & \frac{d^2h_3}{dx^2} & \frac{d^2h_5}{dx^2} & \frac{d^2h_6}{dx^2} \end{bmatrix} \begin{bmatrix} w_1 \\ \theta_1 \\ w_2 \\ \theta_2 \end{bmatrix}$$

and the stiffness matrix is given by

$$\mathbf{k} = \int_0^l \mathbf{B}^T EI \mathbf{B}\, dx = \begin{bmatrix} \frac{12EI}{l^3} & \frac{6EI}{l^2} & -\frac{12EI}{l^3} & \frac{6EI}{l^2} \\ \frac{6EI}{l^2} & \frac{4EI}{l} & -\frac{6EI}{l^2} & \frac{2EI}{l} \\ -\frac{12EI}{l^3} & -\frac{6EI}{l^2} & \frac{12EI}{l^3} & -\frac{6EI}{l^2} \\ \frac{6EI}{l^2} & \frac{2EI}{l} & -\frac{6EI}{l^2} & \frac{4EI}{l} \end{bmatrix}.$$

Note that we can also consider the axial deformation introducing the degrees of freedom shown in Figure 6.47b. The interpolation of the axial displacements is linear in each element as detailed before, and if we define

$$\mathbf{u}^T = \begin{bmatrix} u_1 & w_1 & \theta_1 & u_2 & w_2 & \theta_2 \end{bmatrix}$$

we obtain exactly the stiffness matrix associated with the matrix formulation of the Bernoulli-Euler beam theory given in (4.176).

6.4.2 Matrix and finite element structural analysis for frames

We would like to briefly comment on the similarities and differences between the matrix and the finite element formulation for bars.

We recall that the matrix formulation for a single bar gives the exact solution as long as the bar is subjected to end forces only. When deriving the matrix formulation for an assemblage of bars, equilibrium and compatibility were enforced at every node and, therefore, the exact solution is also obtained as long as the loads are applied to the nodes. When there are distributed forces along the members, we can still obtain the exact solution if we use superposition of effects as detailed in Section 4.2.4.

We note that the finite element formulation for the Bernoulli-Euler beam detailed in the previous section gives the exact solution of the mathematical model whenever the loads are applied to the nodes. The reason is that the interpolations used for the transverse displacements are those of the exact solution when there is no distributed loading applied to the bars, and, as discussed in Section 6.2, when the exact solution is contained in the finite element space, the finite element solution is the exact solution. Hence, of course, the same stiffness matrix is obtained for both formulations.

In the finite element formulation when there are distributed loads applied to the bars, they are transferred to the nodes as consistent nodal forces. Differential equilibrium is no longer satisfied since the interpolation functions correspond to the exact solution when there is no distributed loading. Therefore, in the finite element formulation we need to refine the mesh to improve the solution for such case or – of course – correct the internal element forces as in the matrix formulation.

Example 6.7

Consider the simple frame structure shown in Figure 6.48a. The problem is to be solved both by matrix and finite element analysis. Considering the definitions shown in 6.48b which apply for both formulations:

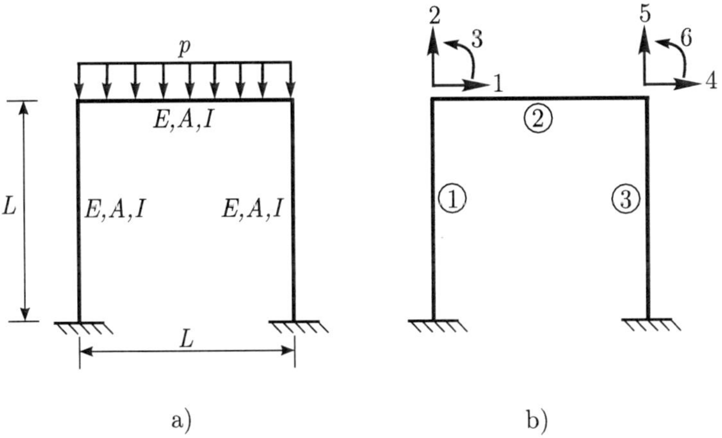

Fig. 6.48. Problem definition: $p = 400$ kN/m, $L = 2$ m, $E = 20 \times 10^6$ kN/m^2, $A = 0.12$ m^2, $I = 36 \times 10^{-4}$ m^4

(i) Calculate the load column matrix associated with the free degrees of freedom.
(ii) Find the nodal displacements.
(iii) Find the bending moment diagram for bar 2.
(iv) Consider bar 2 discretized with 10 finite elements. Find the solution for the degrees of freedom of the original discretization. Show the bending moment diagram for the domain of bar 2.

Solution

(i) For the finite element analysis

$$\mathbf{R}_a = \mathbf{R}_{B,a}$$

and for this case $\mathbf{R} = \mathbf{R}_B = \mathbf{R}_B^{(2)}$

$$\mathbf{R}_B^{(2)} = \int_0^L \mathbf{H}^{(2)^T} (-p)\ dx = -p \int_0^L \mathbf{H}^{(2)^T}\ dx.$$

Considering the free degrees of freedom

$$\mathbf{R}_a = -p \begin{bmatrix} 0 \\ \int_0^L h_2(x)\ dx \\ \int_0^L h_3(x)\ dx \\ 0 \\ \int_0^L h_5(x)\ dx \\ \int_0^L h_6(x)\ dx \end{bmatrix} = \begin{bmatrix} 0 \\ -\frac{pL}{2} \\ -\frac{pL^2}{12} \\ 0 \\ -\frac{pL}{2} \\ \frac{pL^2}{12} \end{bmatrix}. \tag{6.124}$$

For the matrix analysis, referring to Section 4.2.4, we need to evaluate $\mathbf{R}_a - \mathbf{R}_{0,a}$. Since there are no loads applied to the nodes $\mathbf{R}_a = \mathbf{0}$. Referring to Figure 4.45

$$\mathbf{f}_0^{(2)^T} = \begin{bmatrix} 0 & \frac{pL}{2} & \frac{pL^2}{12} & 0 & \frac{pL}{2} & -\frac{pL^2}{12} \end{bmatrix}$$

and for this case

$$-\mathbf{R}_{0,a}^T = -\mathbf{f}_0^{(2)^T} = \begin{bmatrix} 0 & -\frac{pL}{2} & -\frac{pL^2}{12} & 0 & -\frac{pL}{2} & \frac{pL^2}{12} \end{bmatrix}. \tag{6.125}$$

Hence, we obtained the same load column matrix for both approaches.

(ii) Since the load column matrix for the free degrees of freedom are the same for both formulations, the nodal displacements will also be the same. Solving the matrix equation we obtain the predictions for $\mathbf{U}_a$ for both approaches

$$\mathbf{U}_a^T = \begin{bmatrix} 2.717 \times 10^{-5} & -3.333 \times 10^{-4} & -6.309 \times 10^{-4} \\ -2.717 \times 10^{-5} & -3.333 \times 10^{-4} & 6.309 \times 10^{-4} \end{bmatrix}.$$

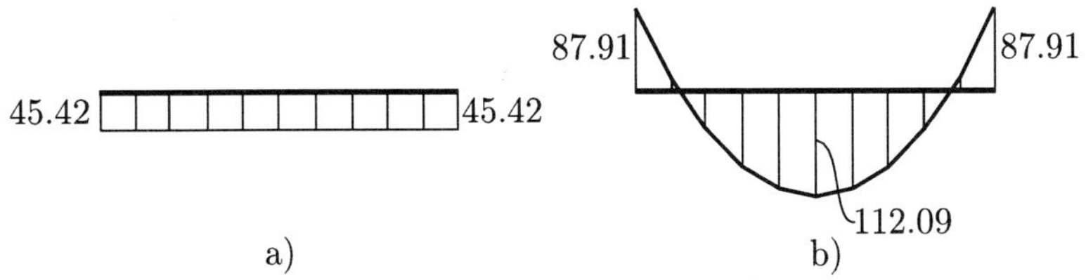

Fig. 6.49. Moment diagram for bar 2. a) Finite element analysis; b) Matrix structural analysis (moments are given in kN.m)

(iii) In Figure 6.49 we show the moment diagram for bar 2 using the finite element approach and structural matrix analysis.

We note that the matrix structural analysis gives the exact moment distribution as detailed in Section 4.2.4. Since in the finite element solution the distributed load is lumped to the nodes, we do not obtain the exact moment distribution and the differential equilibrium along the bar is violated. We also note that the moment distribution is always linear in the finite element approach since the loads are lumped to the nodes.

We recall that in matrix structural analysis to obtain the exact nodal solution when part of the external loads is applied to the bars, we used the superposition as illustrated in Example 4.14. We can proceed in the same way in the finite element solution. We first clamp each bar, calculate the fixed-end forces and moments, apply the reverse to the structure free to move, and then superimpose the two solutions.

In the present analysis, we have for bar 2, the situation summarized in Figure 6.50.

(iv) In Figure 6.51 we show the bending moment diagram for the domain corresponding to bar 2 obtained with the finite element approach without any correction when the bar is discretized with 10 elements.

Here, despite the fact the moment diagram is linear in each element, we obtain a reasonable approximation of the moment distribution directly from the finite element solution without taking recourse to the superposition described above.

□

A very important observation is that there is no matrix analysis approach that leads to the exact solution for 2-D, 3-D solids, plates and shells. One reason is that for these models we can not solve the governing differential equations defined in an element domain for the nodal displacements since these are defined only at discrete locations, not representing the complete element boundary which is a line for 2-D, plate and shell models, and a surface for 3-D models.

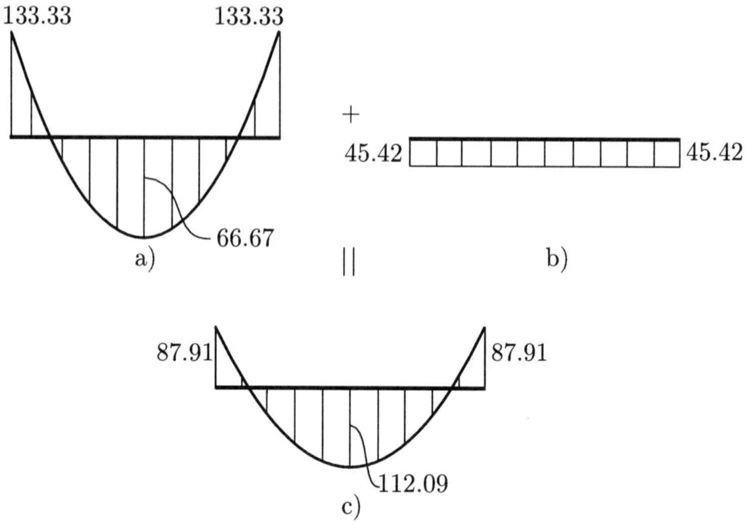

Fig. 6.50. Superposition for bar 2. a) Solution for clamped bar at both ends; b) Finite element solution (moments are given in kN.m); c) Solution obtained by superposition

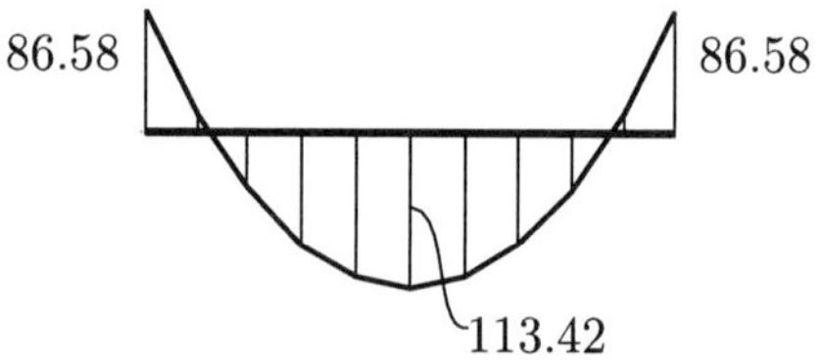

Fig. 6.51. Moment distribution for bar 2 (moments are given in kN.m)

6.4.3 Plate and shell finite elements

Referring to the brief description of the shell models presented in Section 4.4.2, a number of mathematical models could lead to shell finite element formulations. Setting as a requirement that we would like the elements to be *general*, *i.e.*, representing as many behaviors as possible, the choice is the *basic shell model*.

Due to the kinematic hypothesis, the basic shell model represents transverse shear deformations, therefore the model can be used to analyze moderately thick as well as thin shells. The fundamental idea of the shell finite element formulation is to introduce the kinematic and stress hypotheses of the basic shell model through the discretization process into a 3-D description of the shell. This discretization strategy involves the interpolation of the geometry and the displacements and leads to the so-called degenerated solid approach which was proposed by Ahmad, Irons and Zienkiewicz, 1970.

Later, Chapelle and Bathe, 2000 formulated the basic shell mathematical model whose discretization leads to the degenerated solid approach, see also Lee and Bathe, 2005.

We also note that when the midsurface of the shell is flat and the shell (plate) is subjected to transverse loading only, the degenerated solid approach corresponds to the displacement-based finite element formulation for the Reissner-Mindlin plate bending model.

When a flat shell (plate) is also subjected to in-plane loading, the solution can be obtained by superposition of the Reissner-Mindlin plate model and the plane stress model since, in this case, the in-plane and bending responses are decoupled. The degenerated solid approach implicitly models both, the in-plane and bending behaviors. In summary, the shell finite element formulation based on the basic shell model contains an effective plate bending element formulation and the plane stress element formulation as special cases.

Consider typical plate and shell structures as shown in Figure 6.52a. In Figure 6.52b we show the plate and shell midsurfaces and typical elements defined on the midsurfaces. The shell element is also shown in Figure 6.53 where we define for any generic node k a unit vector ${}^0\mathbf{V}_n^k$, called director vector, and a shell thickness denoted by a_k.

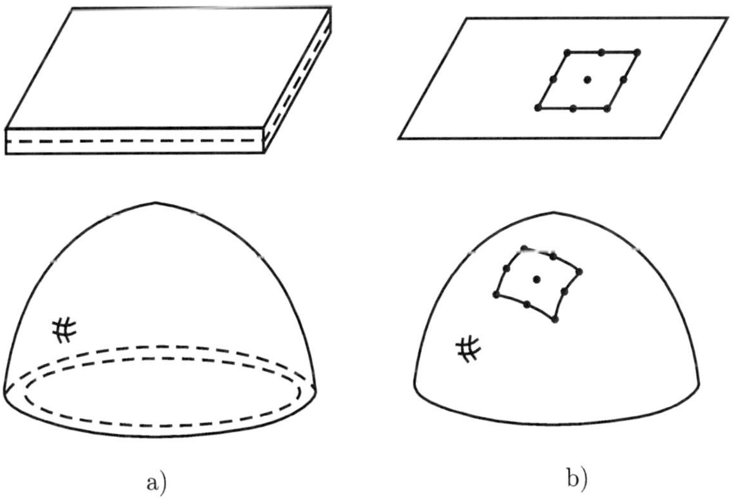

Fig. 6.52. Generic representation of a shell, its midsurface and a shell finite element

We define the interpolation of the geometry corresponding to a generic element by

$$ {}^0\mathbf{x}(r,s,t) = \sum_{k=1}^{q} h_k(r,s)\,{}^0\mathbf{x}_k + \frac{t}{2}\sum_{k=1}^{q} a_k h_k(r,s)\,{}^0\mathbf{V}_n^k. \tag{6.126} $$

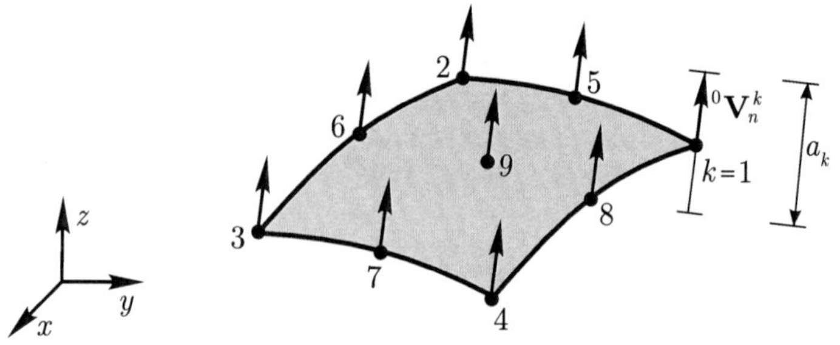

Fig. 6.53. Director vector and thickness definitions

Here we are using the isoparametric mapping and the interpolation functions $h_k(r,s)$ are the usual two-dimensional interpolation functions of a q-node element. We can identify two terms in equation (6.126). The first term gives the interpolation of the midsurface, where ${}^0\mathbf{x}_k$ represents the position vector of node k in the undeformed configuration. Of course, when the isoparametric coordinates r and s span the intervals $-1 \leq r \leq 1$ and $-1 \leq s \leq 1$, the first term gives the position vector of all points on the interpolated shell element midsurface. The second term is needed to obtain the geometry interpolation for the points out of the midsurface. In fact, considering a fixed r, s, say r^*, s^*, the second term gives the points on a straight line which crosses the midsurface at the point defined by the first term when evaluated at $r^*$and s^*. For $t = -1$ we have the point of this straight line located at the lower outer surface, point A, and when $t = +1$ at the upper outer surface, *i.e.*, point B. In Figure 6.54 we summarize the definitions given above.

We note that the left superscript 0 which was used in expression (6.126) indicates that these quantities are referred to the undeformed configuration of the shell and, therefore, (6.126) gives the interpolation of the undeformed configuration.

Now, if we denote by ${}^1\mathbf{x}_k$, ${}^1\mathbf{V}_k^n$ the analogous quantities for the deformed configuration, we can define this configuration for a generic element by

$$ {}^1\mathbf{x}(r,s,t) = \sum_{k=1}^{q} h_k(r,s)\,{}^1\mathbf{x}_k + \frac{t}{2}\sum_{k=1}^{q} a_k h_k(r,s)\,{}^1\mathbf{V}_n^k \qquad (6.127) $$

which complies with the kinematic hypothesis of the basic shell model. Indeed, equations (6.126) and (6.127) synthesize what we meant by introducing the kinematic hypothesis of the basic shell model through the discretization process. The displacements can be written as

$$ \mathbf{u} = {}^1\mathbf{x} - {}^0\mathbf{x} = \sum_{k=1}^{q} h_k(r,s)\mathbf{u}_k + \frac{t}{2}\sum_{k=1}^{q} a_k h_k(r,s)\left({}^1\mathbf{V}_n^k - {}^0\mathbf{V}_n^k\right) \qquad (6.128) $$

where $\mathbf{u}_k$ are the nodal point displacements. The director vector ${}^1\mathbf{V}_n^k$ is actually ${}^0\mathbf{V}_n^k$ after a rigid rotation. Let us define an orthonormal coordinate

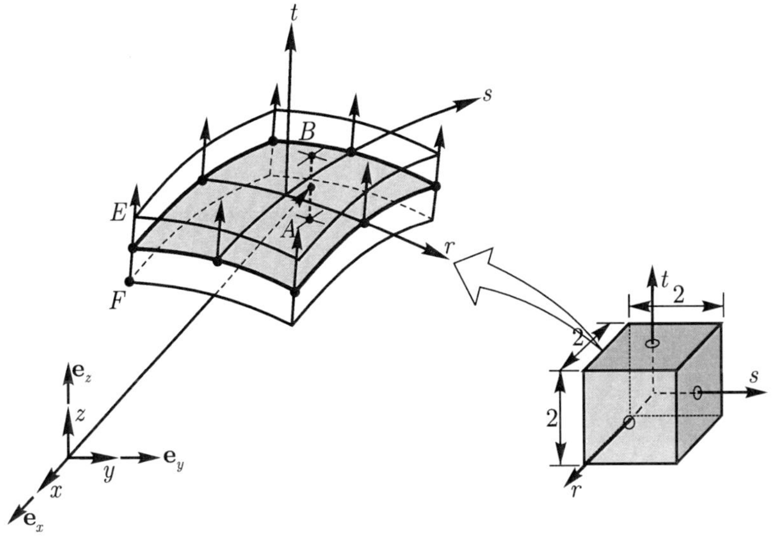

Fig. 6.54. Interpolation of shell geometry

system whose base vectors are $(\mathbf{V}_1^k, \mathbf{V}_2^k, \mathbf{V}_n^k)$ as shown in Figure 6.55a which are characterized by

$$ {}^0\mathbf{V}_1^k = \frac{\mathbf{e}_y \times {}^0\mathbf{V}_n^k}{\|\mathbf{e}_y \times {}^0\mathbf{V}_n^k\|} \tag{6.129} $$

where $\mathbf{e}_y$ is a unit vector in the direction of the y axis (when ${}^0\mathbf{V}_n^k$ is parallel to $\mathbf{e}_y$, we can use ${}^0\mathbf{V}_1^k = \mathbf{e}_z$), and

$$ {}^0\mathbf{V}_2^k = {}^0\mathbf{V}_n^k \times {}^0\mathbf{V}_1^k. \tag{6.130} $$

Of course, equations (6.129) and (6.130) are not the only possibility of defining an orthonormal basis for which $\mathbf{V}_n^k$ is the third base vector. But, it is certainly a valid way. Since the rotation of ${}^0\mathbf{V}_n^k$ is assumed to be infinitesimally small, we have

$$ {}^1\mathbf{V}_n^k - {}^0\mathbf{V}_n^k = \beta_k \mathbf{V}_1^k - \alpha_k \mathbf{V}_2^k \tag{6.131} $$

where α_k and β_k are the rotation components as summarized in Figure 6.55b. Equation (6.128) can now be re-written using (6.131) as

$$ \mathbf{u} = \sum_{k=1}^{q} h_k \mathbf{u}_k + \frac{t}{2} \sum_{k=1}^{q} a_k h_k \left(\beta_k \mathbf{V}_1^k - \alpha_k \mathbf{V}_2^k \right) $$

which implicitly defines the element displacement interpolation. The element nodal displacement column matrix is given by

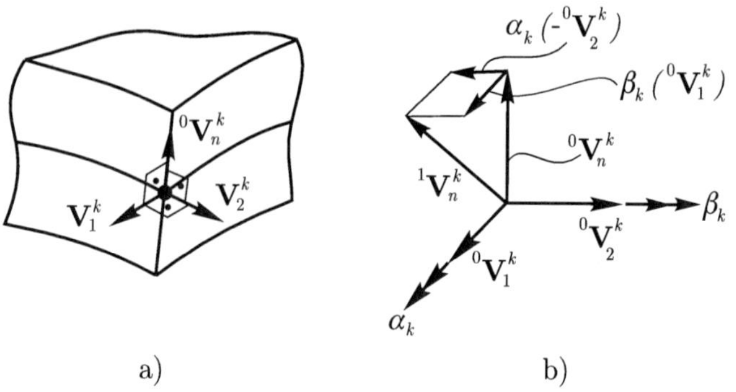

Fig. 6.55. a) A local orthonormal basis; b) Kinematics for the director vector at node k

$$\hat{\mathbf{u}}^T = \left[\begin{array}{ccccc|c|ccccc} u_1 & v_1 & w_1 & \alpha_1 & \beta_1 & \cdots\cdots\cdots & u_q & v_q & w_q & \alpha_q & \beta_q \end{array} \right]$$

and therefore the interpolation matrix $\mathbf{H}$ is

$$\mathbf{H} = \left[\begin{array}{ccccc|c} h_1 & 0 & 0 & -\frac{t}{2} a_1 h_1 V_{2x}^1 & \frac{t}{2} a_1 h_1 V_{1x}^1 & \\ 0 & h_1 & 0 & -\frac{t}{2} a_1 h_1 V_{2y}^1 & \frac{t}{2} a_1 h_1 V_{1y}^1 & \cdots\cdots \\ 0 & 0 & h_1 & -\frac{t}{2} a_1 h_1 V_{2z}^1 & \frac{t}{2} a_1 h_1 V_{1z}^1 & \end{array} \right.$$

$$\left. \begin{array}{c|ccccc} & h_q & 0 & 0 & -\frac{t}{2} a_q h_q V_{2x}^q & \frac{t}{2} a_q h_q V_{1x}^q \\ \cdots\cdots & 0 & h_q & 0 & -\frac{t}{2} a_q h_q V_{2y}^q & \frac{t}{2} a_q h_q V_{1y}^q \\ & 0 & & h_q & -\frac{t}{2} a_q h_q V_{2z}^q & \frac{t}{2} a_q h_q V_{1z}^q \end{array} \right]$$

where V_{1x}^k, V_{1y}^k and V_{1z}^k are respectively the x, y and z components of $\mathbf{V}_1^k$, and analogous definitions hold for $\mathbf{V}_2^k$. Since the displacement interpolation has been fully characterized, the stiffness matrix can be derived following the steps detailed before. We remark that the second hypothesis of the basic shell model, that the normal stress in planes parallel to the midsurface is zero, is imposed when defining the matrix $\mathbf{C}$ much like it was done for the plate and shell models in Chapter 4. Of course, in route to the stiffness matrix there are several details that need to be established. We refer the interested reader to Bathe, 1996.

We list below the main reasons that make this shell element formulation attractive:

- Since the element is based on the basic shell model it can be used to describe thin and moderately thick shells.
- When the midsurface of the shell is flat it becomes a plate element. Actually, due to the kinematic and mechanical hypotheses it corresponds to the plate element of the Reissner-Mindlin mathematical model.

- Since the midsurface is interpolated it can be used to model shells of an arbitrary geometry even those for which an analytical equation of its midsurface is not known. For example, the coordinates of the midsurface may only be given at discrete locations from the output of a geometric modeling program.
- The formulation does not contain the assumption that the ratio of thickness to radius of curvature << 1.0.
- In geometric nonlinear analysis – as introduced in Chapter 8 – the solution is obtained incrementally in load steps and for each load level a new configuration is established. Therefore, to be used in general nonlinear analysis, an element should be able to represent these incremental configurations. For example, a plate subjected to a transverse load deforms and becomes a shell as the load is increased. Considering the kinematic description of the element given in equations (6.126) and (6.127), we see that this element can be used to characterize any configuration; hence the formulation can be directly extended to general nonlinear analysis, see Bathe, 1996.
- The nodal degrees of freedom are only displacements and rotations.

Although the element formulation has all the desirable properties mentioned above, the elements are not reliable and effective, because they suffer from a difficult and detrimental numerical deficiency called locking, which will be addressed in Section 6.5. Fortunately, these elements were successfully modified to largely overcome locking while preserving the desirable characteristics outlined above.

6.4.4 Beam finite elements

If we set for beams the same requirements of generality as for shell elements, we would promptly recognize that no beam mathematical model of Chapter 4 provides this desired level of generality which could be outlined as:

- The beam axis can be straight, a planar curve or a three-dimensional curve. The axis could even be described only by a set of discrete locations.
- The beam should model transverse shear deformations.
- The beam should model cross-sectional variations along its axis.
- The beam should not contain the assumption that the ratio of thickness to radius of curvature << 1.0, as used in Section 4.2.7.

As for shells, we can identify basic hypotheses of beam behavior characterizing a basic beam model and reflect these hypotheses in the beam finite element formulation through the discretization process. The hypotheses are:

Kinematic hypothesis: Plane sections remain plane during deformations.

Mechanical hypothesis: Normal stresses acting on planes normal to the cross-sectional plane are zero. Also, the in-plane shear stresses at the cross-section are zero.

We next present the formulation of a beam element which complies with the above requirements. To characterize the beam element, we define the cross-section of the beam at the nodal locations. In Figure 6.56 we show these definitions for a 3-node beam element of rectangular cross-section; of course different cross-sections could be used. Note that at each nodal point

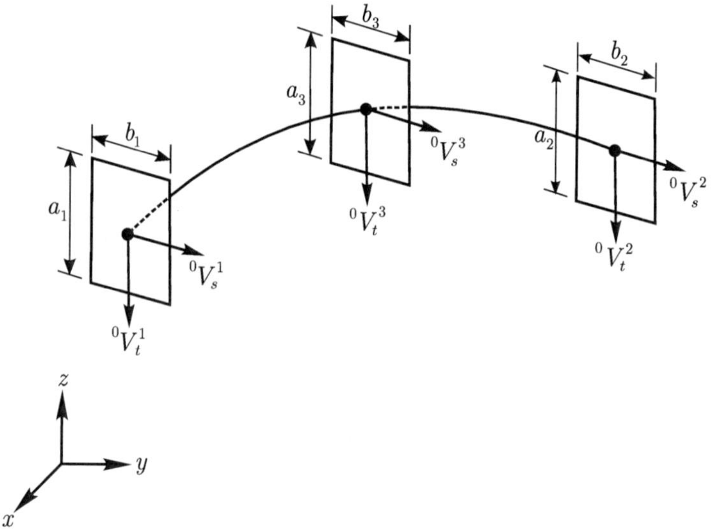

Fig. 6.56. Definition of beam geometric properties at nodal locations

the unit vectors ${}^0\mathbf{V}_s^k$, ${}^0\mathbf{V}_t^k$ position the cross-section in 3-D space. The beam axis is implicitly defined by the interpolation of the geometry given by

$$ {}^0\mathbf{x}(r,s,t) = \sum_{k=1}^{q} h_k(r)\,{}^0\mathbf{x}^k + \frac{t}{2}\sum_{k=1}^{q} a_k h_k(r)\,{}^0\mathbf{V}_t^k + \frac{s}{2}\sum_{k=1}^{q} b_k h_k(r)\,{}^0\mathbf{V}_s^k \quad (6.132) $$

which is written for the q-node element. The nodal points are given by ${}^0\mathbf{x}^k$, $h_k(r)$ are the 1-D interpolation functions for the q-node element and the reference cube with the isoparametric coordinates $(r,\ s,\ t)$ (see for example Figures 6.45 and 6.54) is mapped to the beam element geometry by equation (6.132). The deformed configuration can be written in analogous form

$$ {}^1\mathbf{x}(r,s,t) = \sum_{k=1}^{q} h_k(r)\,{}^1\mathbf{x}^k + \frac{t}{2}\sum_{k=1}^{q} a_k h_k(r)\,{}^1\mathbf{V}_t^k + \frac{s}{2}\sum_{k=1}^{q} b_k h_k(r)\,{}^1\mathbf{V}_s^k $$

where the left superscript 1 identifies $\mathbf{x}_k$, $\mathbf{V}_t^k$, $\mathbf{V}_s^k$ for the deformed configuration.

The displacements are obtained by

$$\mathbf{u} = {}^1\mathbf{x} - {}^0\mathbf{x} = \sum_{k=1}^{q} h_k(r)\,\mathbf{u}^k + \frac{t}{2}\sum_{k=1}^{q} a_k h_k(r)\,\mathbf{V}_t^k + \frac{s}{2}\sum_{k=1}^{q} b_k h_k(r)\,\mathbf{V}_s^k \quad (6.133)$$

where $\mathbf{u}^k$ are the nodal displacements and

$$\mathbf{V}_t^k = {}^1\mathbf{V}_t^k - {}^0\mathbf{V}_t^k, \quad \mathbf{V}_s^k = {}^1\mathbf{V}_s^k - {}^0\mathbf{V}_s^k.$$

Since the displacements are assumed to be infinitesimally small, we can write

$$\mathbf{V}_t^k = \theta_k \times {}^0\mathbf{V}_t^k \quad (6.134)$$

$$\mathbf{V}_s^k = \theta_k \times {}^0\mathbf{V}_s^k \quad (6.135)$$

where θ_k is the rotation vector of the section located at node k which can be written in the global reference system as

$$\theta_k = \begin{bmatrix} \theta_x^k \\ \theta_y^k \\ \theta_z^k \end{bmatrix}.$$

Therefore, there are six degrees of freedom per node and the nodal displacement column matrix can be written as

$$\hat{\mathbf{u}}^T = \begin{bmatrix} u_1 & v_1 & w_1 & \theta_x^1 & \theta_y^1 & \theta_z^1 | & \cdots & | u_q & v_q & w_q & \theta_x^q & \theta_y^q & \theta_z^q \end{bmatrix}.$$

The $\mathbf{H}$ matrix can be obtained from (6.133), (6.134) and (6.135). The constitutive equation is given by

$$\begin{bmatrix} \tau_{\eta\eta} \\ \tau_{\eta\xi} \\ \tau_{\eta\zeta} \end{bmatrix} = \begin{bmatrix} E & 0 & 0 \\ 0 & Gk & 0 \\ 0 & 0 & Gk \end{bmatrix} \begin{bmatrix} \varepsilon_{\eta\eta} \\ \gamma_{\eta\xi} \\ \gamma_{\eta\zeta} \end{bmatrix}$$

where (η, ξ, ζ) is a convected coordinate system such that at a cross-section $\tau_{\eta\eta}$ is the normal stress and $\tau_{\eta\xi}$, $\tau_{\eta\zeta}$ are the transverse shear stresses with ξ and ζ parallel to the sides of the rectangular cross-section, and k is the shear correction factor. With these definitions $\mathbf{B}$ and $\mathbf{k}^{(m)}$ can be evaluated. We refer to Bathe, 1996 for the details of these matrices and their derivations. Here warping displacements were not considered, but can also be introduced (see Bathe and Chaudhary, 1982).

These elements, as the shell elements in Section 6.4.3 are not yet effective – they are prone to locking – and need to be modified to arrive at reliable and hence useful elements (see Section 6.5).

6.5 Effective finite elements

In the previous sections we introduced a displacement-based finite element formulation for beams, plates and shells. The associated elements seem attractive due to their simple formulations and, apparently, general applicability. However, these elements[4] suffer from a detrimental numerical deficiency in that they can be much too stiff when bending situations are modeled – for which these elements are to be used extensively! This deficiency has been termed "locking" since the "numerical stiffening" can lead to very small displacements – and in the limit to zero deformations.

A similar locking phenomenon is also seen when the 2-D (plane strain and axisymmetric) and 3-D solid displacement-based elements presented above are used to model incompressible or almost incompressible behavior.

A tremendous research effort has been directed to overcome the locking deficiency in order to arrive at finite elements which are general and at the same time reliable and efficient. Nowadays, some such elements are known and we call them effective elements.

In the following sections, we first present the basic convergence properties of displacement-based finite element formulations and introduce the deleterious effects of locking. Then, in the simple setting of the 2-node Timoshenko beam element, we examine the source of locking and how to overcome it. Finally, the main characteristics of locking for the remaining mathematical models earlier discussed are summarized and a list of effective elements is given with references to their formulations.

6.5.1 Convergence of displacement-based finite element formulations and the effects of locking

We considered in Section 6.2 the convergence of 1-D finite element solutions and much of that discussion is also applicable to 2-D and 3-D solutions of solids. Let us summarize below the convergence results for the displacement-based formulations for the analysis of a solid.

Let[5] $\mathbf{V}$ be the admissible displacement space defined by

$$\mathbf{V} = \left\{\mathbf{v},\ \mathbf{v} : Vol \rightarrow \mathbb{R}^n \mid a(\mathbf{v},\mathbf{v}) < \infty,\ \ \mathbf{v}|_{S_u} = 0\right\}$$

where $a(\mathbf{v},\mathbf{v})$ as defined in (5.43) is twice the strain energy corresponding to $\mathbf{v}$, Vol represents the domain of the undeformed configuration and $n = 2$ for a 2-D formulation and $n = 3$ for a 3-D formulation. Merely for ease of presentation, we consider that the displacements are zero on S_u.

[4] Except the Bernoulli-Euler beam element for modeling straight beams. The reason is that shear deformations are not modeled and the beam axis is straight.

[5] Here we use $\mathbf{V}$ in boldface and soon also $\mathbf{V}_h$ since these are spaces of vector valued functions

Let $\mathbf{V}_h$ be a generic finite element space given by a mesh choice of isoparametric elements. Note that a generic element of $\mathbf{V}_h$ can be characterized elementwise from the nodal displacements using (6.121). The continuity property at element interfaces guarantees that $\mathbf{V}_h \subset \mathbf{V}$. We can construct a sequence of finite element spaces which obey (6.50) by subdividing the elements of the initial mesh which defines $\mathbf{V}_h^1$.

Let $\mathbf{u}$ be the exact solution and $\mathbf{u}_h^i$ the finite element solution obtained using $\mathbf{V}_h^i$. It can be shown that

$$\left\|\mathbf{u} - \mathbf{u}_h^i\right\|_1 \leq Ch^k \tag{6.136}$$

when $\mathbf{u}$ is smooth enough. Note the similarity between (6.136) and (6.63). Here C is a positive constant and h is the size of the largest element (see Bathe, 1996 for a precise definition of element sizes in 2-D and 3-D situations). The 1-norm for a function $\mathbf{f} : Vol \to \mathbb{R}^3$

$$\mathbf{f} = \begin{bmatrix} f_x(x,y,z) \\ f_y(x,y,z) \\ f_z(x,y,z) \end{bmatrix} \tag{6.137}$$

is given by

$$\begin{aligned} \|\mathbf{f}\|_1 &= \left\{\int_{Vol}\left[\left(\frac{\partial f_x}{\partial x}\right)^2 + \left(\frac{\partial f_x}{\partial y}\right)^2 + \left(\frac{\partial f_x}{\partial z}\right)^2 + \left(\frac{\partial f_y}{\partial x}\right)^2 + \left(\frac{\partial f_y}{\partial y}\right)^2\right.\right. \\ &\quad + \left(\frac{\partial f_y}{\partial z}\right)^2 + \left(\frac{\partial f_z}{\partial x}\right)^2 + \left(\frac{\partial f_z}{\partial y}\right)^2 + \left(\frac{\partial f_z}{\partial z}\right)^2 \\ &\quad \left.\left. + (f_x)^2 + (f_y)^2 + (f_z)^2\right] dVol\right\}^{1/2}. \end{aligned} \tag{6.138}$$

For $\mathbf{f} : Vol \to \mathbb{R}^2$ it suffices to suppress f_z in (6.137) and (6.138). Note that (6.138) is a generalization of (6.55) since now we have to account for all the function components and their partial derivatives.

To define k, we consider the Pascal triangle for 2-D shown in Figure 6.57, and note that k is given by the order of the complete polynomial which is exactly represented in each element. For example, $k = 2$ when P_2 is represented, *i.e.*, each monomial 1, x, y, x^2, xy, y^2 can be exactly and independently represented. Note that to guarantee convergence k should be at least 1. In Bathe, 1996 it is shown that every isoparametric element, even if distorted and/or curved, represents P_1 and hence the convergence of displacement-based isoparametric elements is assured. For the finite element formulation to satisfy compatibility and contain P_1 is equivalent to representing exactly the rigid body modes and the constant strain states for any patch of finite elements. This is sometimes referred to as the convergence condition for continuum elements.

	Degree of complete polynomial	corresponding space
1		
$x \quad y$	1	P_1
$x^2 \quad xy \quad y^2$	2	P_2
$x^3 \quad x^2y \quad xy^2 \quad y^3$	3	P_3
$x^4 \quad x^3y \quad x^2y^2 \quad xy^3 \quad y^4$	4	P_4
$x^5 \quad x^4y \quad x^3y^2 \quad x^2y^3 \quad xy^4 \quad y^5$	5	P_5

Fig. 6.57. Pascal triangle

For the displacement-based beam, plate and shell elements it is possible to show that error estimates such as (6.136) also hold (but the value of k may be linear than given by the Pascal triangle) and therefore their convergence properties are also guaranteed, see Chapelle and Bathe, 2010*a*.

Since it is possible to obtain error estimates like (6 136) for every displacement-based formulation studied so far, the reader might ask: "What can possibly be the deficiency of displacement-based finite element formulations which leads to locking?"

To address this question, we refer to Figure 6.21 in which the error estimate (6.136) is graphically given for $k = 1$. In problems prone to locking, the constant C in (6.136) depends on a problem parameter such as the thickness (beam, plates and shells) or ν (for the elasticity problems) and this constant may become very large. Therefore, as this parameter changes and C becomes larger the straight line in Figure 6.21 shifts position as schematically shown in Figure 6.58.

To understand the consequences of this shift, we show the pairs ($\log h$, $\log E_h$) represented by A and B for two meshes before the locking parameter was changed. We assume that these two meshes are fine enough such that accurate predictions were obtained (E_h is small enough) and that the full convergence rate has already been achieved for these meshes. If we further refine the mesh the convergence rate will not diminish. However, as the parameter is changed and the straight line shifts due to the increase of the constant C as shown in Figure 6.58 the same meshes yield points A' and B' shown in the diagram. That is, the errors increased (A compared to A') and the convergence rate decreased (A' to B' compared to A to B) since the condition now for the pairs ($\log h$, $\log E_h$) is to be below or on the upper solid line. The shift may be so large that accurate predictions may be obtained only for extremely fine meshes that are of no practical interest.

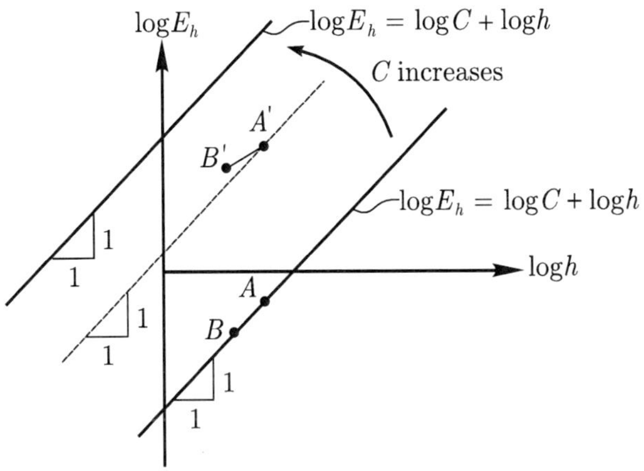

Fig. 6.58. Possible change in convergence diagram due to locking

Note that for a given problem with the applicable parameter fixed, the finite element solution converges if we sufficiently refine the mesh due to (6.136). However, due to locking and consequently a very large C, the fact that convergence is eventually achieved has no value for engineering applications since we obtain very inaccurate predictions for meshes of practical interest.

Therefore, a formulation is free of locking if the constant C in (6.136) and the order of convergence are independent of the parameter which governs locking. Then, convergence curves do not shift due to changes of the parameter and convergence rates are not affected.

Although we introduced the effects of locking, we did not address its source which is discussed in the next section by means of the Timoshenko beam element.

6.5.2 Locking for the 2-node Timoshenko beam element

The general beam formulation discussed in Section 6.4.4 degenerates to the Timoshenko beam element formulation when the axis is straight. We can use this simple beam formulation to exemplify the phenomenon of locking, see also Bathe, 1996 and Chapelle and Bathe, 2010*a*. For this purpose we rewrite the formulation as the minimization of the total potential energy

$$\begin{aligned} \Pi\left(w(x), \beta(x)\right) &= \frac{EI}{2}\int_0^L \left(\frac{d\beta}{dx}\right)^2 dx + \frac{GAk}{2}\int_0^L \left(\frac{dw}{dx} - \beta\right)^2 dx \\ &\quad - \int_0^L wp\, dx \end{aligned} \tag{6.139}$$

and refer to Section 4.2.8 for the definition of the quantities appearing in this expression. Of course, $I = \frac{bt^3}{12}$, $A = bt$ with b the width and t the thickness of the rectangular cross-section. The solution is given by the functions $w(x)$ and $\beta(x)$ which minimize the potential. The phenomenon of locking is seen as the beam becomes thin and to study it, we consider a sequence of problems. Each problem in the sequence is associated with a different thickness which is made smaller and smaller. Clearly, if (6.139) is simply used with the same load p, we must expect the displacement $w(x)$ to become infinitely large as $t \to 0$. To seek well-behaved solutions, we consider for each problem of the sequence a load given by

$$p(x) = t^3 g(x)$$

and the scaled potential

$$\begin{aligned} \tilde{\Pi} &= \frac{\Pi}{t^3} = \frac{bE}{24}\int_0^L \left(\frac{d\beta}{dx}\right)^2 dx + \frac{Gbk}{t^2}\int_0^L \left(\frac{dw}{dx} - \beta\right)^2 dx \\ &\quad - \int_0^L wg\, dx. \end{aligned} \tag{6.140}$$

Note that the functions $w(x)$ and $\beta(x)$ which minimize Π are the same functions that minimize $\tilde{\Pi}$.

Considering the minimization of $\tilde{\Pi}$ as $t \to 0$ we have that

$$\int_0^L \left(\frac{dw}{dx} - \beta\right)^2 dx \to 0 \qquad \text{as} \qquad t \to 0.$$

Otherwise, the term of $\tilde{\Pi}$ associated with the shear strain energy would go to infinity as $t \to 0$. Therefore, for a very thin beam we need to have

$$\frac{dw}{dx} - \beta \cong 0 \qquad \Rightarrow \qquad \frac{dw}{dx} \cong \beta \tag{6.141}$$

and of course (6.141) illustrates the reasonableness of the kinematic assumption of the Bernoulli-Euler beam model, where we assume $\frac{dw}{dx} = \beta$. The result given in (6.141) indicates that the Timoshenko beam solution tends to the Bernoulli-Euler beam solution as the beam thickness tends to zero.

Now, considering the finite element solution we need to have

$$\frac{dw_h(x)}{dx} - \beta_h(x) \cong 0 \qquad \text{as} \qquad t \text{ becomes small} \tag{6.142}$$

where $w_h(x)$ and $\beta_h(x)$ need to be in the finite element spaces for displacements, W_h, and rotations, B_h, corresponding to the mesh, *i.e.*, $w_h \in W_h$ and $\beta_h \in B_h$. The condition (6.142) represents a constraint for the finite element solutions $w_h(x)$ and $\beta_h(x)$. Hence, we need that W_h and B_h be such that the constraint $\frac{dw_h}{dx} = \beta_h$ can be satisfied for a sufficient number of functions in W_h and B_h. By a sufficient number of functions, we mean that these functions should be able to give a reasonable approximation of the exact solution of the mathematical model. If this is not the case, the finite element discretization will lock. We illustrate the source of the difficulty in the next example.

Example 6.8

Consider a cantilever Timoshenko beam of rectangular cross-section subjected to a tip moment as shown in Figure 6.59a and the discretization with a single two-node Timoshenko beam element as shown in Figure 6.59b. Examine the behavior of the solution as the beam thickness tends to zero.

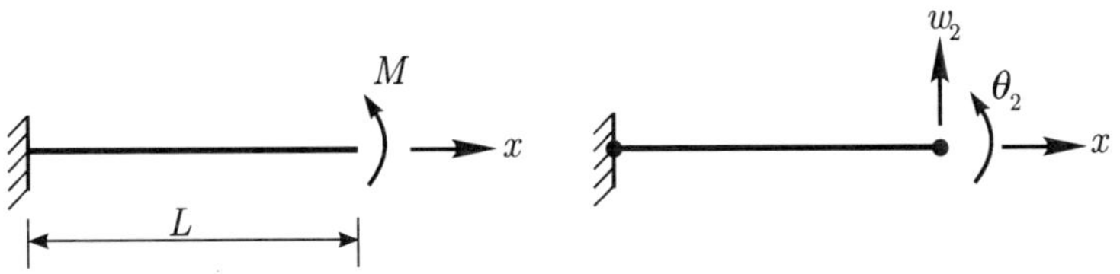

Fig. 6.59. Problem definition discretized with one Timoshenko beam element

This problem is also considered in Bathe, 1996.

Solution

Since the interpolation is linear for, both, displacement and section rotations and the displacement and section rotation are zero at the clamped end, we have

$$w_h(x) = \frac{x}{L} w_2, \qquad \beta_h(x) = \frac{x}{L}\theta_2$$

and the shear strain is given by

$$\gamma_h(x) = \frac{dw_h}{dx} - \beta_h = \frac{1}{L}\left(w_2 - x\theta_2\right). \tag{6.143}$$

The above discussion shows that as t becomes small, the constraint $\gamma_h \to 0$ is increasingly enforced and for t small enough we have

$$w_h(x) \cong 0 \qquad \text{and} \qquad \beta_h(x) \cong 0. \tag{6.144}$$

This illustrates the phenomenon of solution locking. In Table 6.4 we provide some values of w_2 and θ_2 normalized with respect to the exact solution for different ratios t/L. We can see that, indeed w_2 and θ_2 tend to zero, and even for $t/L = 1/10$ the solution is already much too stiff.

□

We have used the coarsest discretization possible in the example above, but the locking behavior is of course also seen for any fine mesh of 2-node

Table 6.4. Normalized values for tip displacements and rotations obtained with the model of Figure 6.59 using the displacement-based element

t/L	Calculated transverse displacement /exact transverse displacement	Calculated section rotation /exact section rotation
1/10	2.03×10^{-2}	2.00×10^{-2}
1/100	2.03×10^{-4}	2.00×10^{-4}
1/1000	2.03×10^{-6}	2.00×10^{-4}

Timoshenko beam elements. Since each element in the fine mesh then behaves as does the single element in the example, the locking phenomenon simply propagates through the mesh.

Mathematically, locking in general beam, plate and shell analyses means that uniform convergence of the finite element solution with respect to the thickness t of the beam (plate or shell) is not seen. That is, we do not have the desirable property

$$\frac{\|u - u_h\|}{\|u\|} \leq Ch^p \tag{6.145}$$

where $\widetilde{C}$ and p are independent of t, see Chapelle and Bathe, 2010*a*.

The most successful approach to overcome locking is the use of a mixed method based on a variational formulation analogous to the principle of virtual work. A mixed formulation for the Timoshenko beam model considered in (6.140) can be constructed by introducing a shear strain type variable in the formulation, see Bathe, 1996.
Find $w(x) \in W$, $\beta(x) \in B$ and $\zeta(x) \in Q$ such that

$$\frac{Eb}{12}\int_0^L \frac{d\bar{\beta}}{dx}\frac{d\beta}{dx}dx + Gbk\int_0^L \zeta\left(\bar{\beta} - \frac{d\bar{w}}{dx}\right)dx$$

$$= \int_0^L g\bar{w}\,dx \qquad \text{for all} \qquad \bar{\beta}(x) \in B,\ \bar{w}(x) \in W \tag{6.146}$$

$$Gb\int_0^L \bar{\zeta}\left(\beta - \frac{dw}{dx}\right)dx - t^2\int_0^L \bar{\zeta}\zeta\,dx = 0 \qquad \text{for all} \qquad \bar{\zeta} \in Q.$$

The additional function $\zeta(x)$ can be interpreted as a shear strain field. The functions $\bar{\beta}(x)$, $\bar{w}(x)$ and $\bar{\zeta}(x)$ are virtual fields and the analogy to the principle of virtual work is immediate. The functional spaces W, B and Q can be precisely defined and their choices should guarantee that the integrals above are finite.

It can be shown that the solution for $w(x)$ and $\beta(x)$ of problem (6.146) and of the minimization problem defined by (6.140) are the same. Therefore, in the continuum setting, problem (6.146) is an alternative formulation – yet more complex – to formulate the Timoshenko beam problem.

A finite element formulation of (6.146) can be obtained by choosing finite element spaces $W_h \in W$, $B_h \in B$ and $Q_h \in Q$. Consider the choice where W_h and B_h are continuous piecewise linear functions and Q_h piecewise constant functions. To see whether this choice makes sense, we consider the case $t = 0$ in (6.146) and obtain the problem.
Find $w_h(x) \in W_h$, $\beta_h(x) \in B_h$ and $\zeta_h(x) \in Q_h$ such that

$$\frac{Eb}{12}\int_0^L \frac{d\bar{\beta}_h}{dx}\frac{d\beta_h}{dx}dx + Gbk\int_0^L \zeta_h\left(\bar{\beta}_h - \frac{d\bar{w}_h}{dx}\right)dx$$

$$= \int_0^L g\bar{w}_h\, dx \text{ for all } \bar{w}_h(x) \in W_h,\ \bar{\beta}_h(x) \in B_h$$

$$\int_0^L \bar{\zeta}_h\left(\beta_h - \frac{dw_h}{dx}\right)dx = 0 \text{ for all } \bar{\zeta}_h \in Q_h.$$

The second equation above plays the role of the constraint $\int_0^L \left(\frac{dw_h}{dx} - \beta_h\right)^2 dx = 0$. However, taking into account that $\beta_h(x) - \frac{dw_h(x)}{dx}$ is linear in each element, the condition that

$$\int_0^L \bar{\zeta}_h\left(\beta_h - \frac{dw_h}{dx}\right)dx = 0$$

where $\bar{\zeta}_h$ can assume an arbitrary constant value in each element is satisfied only if

$$\left(\beta_h - \frac{dw_h}{dx}\right)\Bigg|_{\text{mid point of element } (m)} = 0 \text{ for each element } (m). \qquad (6.147)$$

The equation (6.147) is the constraint associated with the mixed formulation. This constraint is much less restrictive than that associated with the displacement-based formulation.

It is possible to show that this formulation leads to optimal error estimates for the transverse displacements and section rotations independent of the beam thickness. Such result guarantees uniform optimal convergence with respect to the beam thickness, therefore, no shear locking. For the problem of Figure 6.59a we report in Table 6.5 the solutions obtained with the 2-node mixed Timoshenko beam element.

The finite element formulation of the mixed problem given in (6.146) for any thickness t and for certain choices of W_h, B_h and Q_h can be obtained from a much simpler mathematical statement given by:
Find $w_h(x) \in W_h$, $\beta_h(x) \in B_h$ such that

Table 6.5. Normalized values for tip displacements and rotations obtained with the model of Figure 6.59 using a single mixed-formulated element

t/L	Calculated transverse displacement /exact transverse displacement	Calculated section rotation /exact section rotation
1/10	1.00	1.00
1/100	1.00	1.00
1/1000	1.00	1.00

$$\frac{Ebt^3}{12}\int_0^L \frac{d\bar{\beta}_h}{dx}\frac{d\beta_h}{dx}dx + GAk\int_0^L \bar{\gamma}_h^{AS}\gamma_h^{AS}\ dx = \tag{6.148}$$

$$\int_0^L p\bar{w}\ dx \text{ for all } \bar{\beta}_h(x) \in B_h,\ \bar{w}_h(x) \in W_h$$

where γ_h^{AS} is an assumed (ASsumed) shear strain field. We note that (6.148) is actually the principle of virtual work when we use γ_h^{AS} for the shear strain.

However, we still need to define γ_h^{AS} to fully characterize the formulation given in (6.148). It is possible to show that the choices of W_h and B_h of continuous piecewise linear functions, and Q_h of piecewise constant functions for (6.146) lead to an identical formulation as that obtained with (6.148) when we choose

$$\gamma_h^{AS} = \left(\frac{dw_h}{dx} - \beta_h\right)\bigg|_{\text{mid point of the element}} \tag{6.149}$$

which is then, of course, constant in each element.

We can use

$$\gamma_h^{DI} = \gamma_h = \frac{dw_h}{dx} - \beta_h$$

where DI (Direct Interpolation) emphasizes that the γ_h^{DI} gives the strains obtained from the interpolated kinematic variables β_h and w_h. Then (6.149) can be re-written as

$$\gamma_h^{AS} = \gamma_h^{DI}\big|_{\text{mid point of the element}}\ .$$

Since γ_h^{AS} is obtained from the kinematic variables w_h and β_h, the element stiffness matrix will correspond to nodal transverse displacement and section rotation degrees of freedom only. Therefore, the fact that the element is derived from a mixed formulation is not apparent when using the element since it has only the usual degrees of freedom.

The points where we impose $\gamma_h^{AS} = \gamma_h^{DI}$ are referred to as tying points. Higher-order mixed elements which have optimal convergence properties, *i.e.*,

Table 6.6. Mixed interpolated Timoshenko beam elements

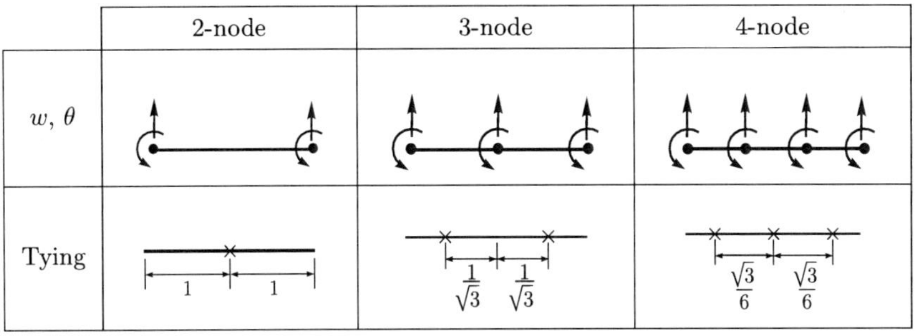

which are independent of the beam thickness, can be derived. Table 6.6 summarizes the elements, see Bathe, 1996 and Chapelle and Bathe, 2010*a*.

The tying points shown in Table 6.6 are actually Gauss points and an efficient implementation of these elements is obtained by using these Gauss points to perform the integrations. Since these points lead to integration rules which are of lower order than those usually required, their usage might be referred to as reduced integration. However, in general, we can not obtain the implementation of a mixed element formulation by means of a reduced integration rule. In fact, there are no reduced integration rules which lead to the mixed plate and shell elements we recommend.

6.5.3 Locking in a general setting

There are several mathematical models for which the displacement-based finite element formulations are prone to locking. In all such cases the potential energy of the model can, in principle, be written as

$$\Pi(\mathbf{v}) = \frac{1}{2}\left(\varepsilon^3 E_1(\mathbf{v}) + \varepsilon E_2(\mathbf{v})\right) - \varepsilon^3 P(\mathbf{v}) \tag{6.150}$$

where

$$\mathcal{U}(\mathbf{v}) = \frac{1}{2}\left(\varepsilon^3 E_1(\mathbf{v}) + \varepsilon E_2(\mathbf{v})\right)$$

is the strain energy, $P(\mathbf{v})$ the potential of the external loads, $\mathbf{v}$ the kinematic field of the model and ε is a parameter which may become very small.

Since we are interested in situations in which ε is very small, we consider a sequence of problems, each problem in the sequence associated with a value of ε. We can define for a particular problem in the sequence the normalized potential

$$\Pi_\varepsilon(\mathbf{v}) = \frac{\Pi(\mathbf{v})}{\varepsilon^3} = \frac{1}{2}\left(E_1(\mathbf{v}) + \frac{1}{\varepsilon^2}E_2(\mathbf{v})\right) - P(\mathbf{v}).$$

Table 6.7. Locking characterization for mathematical models

Mathematical model	$\frac{\varepsilon^3}{2}E_1$	$\frac{\varepsilon}{2}E_2$	ε
Timoshenko beam model	Bending strain energy	Shear strain energy	t: section height
3-D elasticity Plane strain Axisymmetric	Distortional strain energy $\times(\sqrt{1-2\nu})^3$	Volumetric strain energy $\times(\sqrt{1-2\nu})^3$	$\sqrt{1-2\nu}$ ν: Poisson´s ratio
Reissner-Mindlin Plate	Bending strain energy	Shear strain energy	t: plate thickness
Membrane-bending shell model	Bending strain energy	Membrane strain energy	t: shell thickness
Shear-membrane-bending shell model	Bending strain energy	Membrane and shear strain energy	t: shell thickness
Basic shell model	Bending strain energy	Membrane and shear strain energy	t: shell thickness

The solution for a problem in the sequence is given by that $\mathbf{u}$ which minimizes the potential Π or equivalently Π_ε. For the limit problem $\varepsilon \to 0$ the solution $\mathbf{u}$ has to satisfy

$$E_2(\mathbf{u}) = 0 \tag{6.151}$$

otherwise the term $\frac{1}{\varepsilon^2}E_2(\mathbf{u})$ would became unbounded. Therefore, when ε approaches zero (6.151) is a constraint for the mathematical model whose potential is given by (6.150). Of course, for the Timoshenko beam model this constraint is $\frac{dw}{dx} - \beta = 0$ as discussed above.

When we introduce a finite element formulation by the choice of a finite element space $\mathbf{V}_h$, the solution is given, for a problem in the sequence, by that $\mathbf{u}_h$ which minimizes the normalized potential Π_ε among all functions in $\mathbf{V}_h$. Therefore, by the same argument, when $\varepsilon \to 0$ the solution for the finite element problem should satisfy

$$E_2(\mathbf{u}_h) = 0$$

which depending on the choice of $\mathbf{V}_h$ may lead to locking.

In Table 6.7, we summarize the mathematical models which are potentially prone to locking and how they fit into the framework described above.

The overcoming of locking for these models has represented a great challenge to the finite element community which motivated a tremendous research effort. The use of mixed finite element formulations has proven to be the most successful approach to arrive at finite elements that are free of locking, reliable and, therefore, effective. There are now mixed elements which can be recommended for engineering analysis for all mathematical models listed in Table 6.7. For practical purposes, finite elements with optimal convergence properties, *i.e.*, which are independent of the value of the parameter ε, have been obtained using mixed formulations for all the mathematical models of Table 6.7 except for shell models. The reliable analysis of complex shells – considering also nonlinear analysis – represents one of the greatest challenges in all of mechanics. However, numerical experimentations, guided by deep mathematical analysis, have shown that there are also some mixed shell elements which are quite reliable for many analyses and the range of shell thickness of engineering interest, Bucalem and Bathe, 1997, Bathe and Lee, 2011.

As detailed for the Timoshenko beam model, a mixed formulation has additional unknown fields besides the displacement fields. Its finite element formulation requires the choice of finite element spaces for the kinematic fields, generically represented by $\mathbf{V}_h$, and also for the additional fields described by $\mathbf{Q}_h$. In Table 6.8 we summarize the fields of the mixed formulations which are interpolated by the choices of $\mathbf{V}_h$ and $\mathbf{Q}_h$. We also introduce the terminology for the formulations of the most effective elements. For the 2-D and 3-D incompressible or almost incompressible analyses, we refer to the displacement-pressure formulation – u/p formulation. For plates and shells we mention the MITC (Mixed Interpolation of the Tensorial Components) formulation.

For a given mixed formulation, the choices of pairs of finite element spaces $\mathbf{V}_h$, $\mathbf{Q}_h$ which lead to optimal convergence properties, independent of the parameter ε, is not trivial. Mathematically, the choice is governed by the ellipticity and inf-sup conditions (Brezzi and Bathe, 1990, Brezzi and Fortin, 1991, Bathe, 2001, Chapelle and Bathe, 2010*b*). We refer to Bathe, 1996 for a discussion of the inf-sup condition where the inf-sup condition is presented in various forms. One form of the inf-sup condition, based on our earlier discussion, see 6.2.3, is for displacement-based formulations given by

$$\|\mathbf{u} - \mathbf{u}_h\|_V \leq c\, d\,(\mathbf{u}, \mathbf{V}_h)_V$$

with c independent of the parameter ε. This result is, in essence, (6.136) with C independent of the parameter ε since

$$d\,(\mathbf{u}, \mathbf{V}_h)_V \leq \tilde{c} h^k \tag{6.152}$$

for $\mathbf{u}$ sufficiently smooth. Of course, $\tilde{c}$ does not depend on the parameter ε since (6.152) is only a functional approximation result and does not depend on the finite element solution. Therefore, the satisfaction of the inf-sup condition would render the displacement-based formulation free of locking. However, this is very difficult to achieve for the problems of interest in practical analyses, and the use of mixed formulations and elements is much more effective.

Table 6.8. Effective elements for incompressible materials and plates/shells

Mathematical model	Effective elements
Timoshenko beam model	• 2, 3 and 4-node mixed elements ref. Bathe, 1996
2-D elasticity • Plane strain • Axisymmetric	• Quadrilateral: 9-3 u/p element • Triangular: 7-3 u/p element ref. Crouzeix and Raviart, 1973, Brezzi and Bathe, 1986, Sussman and Bathe, 1987
3-D elasticity	• Hexahedral: 27-4 u/p element • Tetrahedral: 11-4 u/p element ref. Crouzeix and Raviart, 1973, Brezzi and Bathe, 1986, Sussman and Bathe, 1987
Reissner-Mindlin plate model	• Quadrilateral: MITC4, MITC9, MITC16 • Triangular: MITC7, MITC12 ref. Bathe, Bucalem and Brezzi, 1990
Basic Shell model	• Quadrilateral: MITC4, MITC9, MITC16 ref. Dvorkin and Bathe, 1984, Bathe and Dvorkin, 1986, Bucalem and Bathe, 1993, Bathe and Hiller, 2003 • Triangular: MITC3, MITC6 ref. Lee and Bathe, 2004, Kim and Bathe, 2009

In Table 6.8 we summarize the mixed elements that we recommend to be used in engineering practice[6], see also Bathe, 1996. We note that the nodal degrees of freedom of the elements listed in Table 6.8 are the same as those of their displacement-based counterparts. Therefore, these elements are in use as simple as the displacement-based elements. For beam, plate and shell elements the interpolation of the strains is achieved with the aid of tying points and the computational effort to obtain the stiffness matrix is similar to that of the displacement-based elements. For the u/p elements listed in Table 6.8, the pressure fields are discontinuous at element interfaces allowing the static condensation of the pressure degrees of freedom.

We listed in Table 6.8 only the Timoshenko beam elements free of locking. Of course, the two-node Bernoulli-Euler beam model is very effective in the

[6] Note that for plane stress analysis, the displacement-based elements can be used since the stress-strain law (in Table 4.3) enforces automatically that the thickness is adjusted to satisfy the incompressibility condition (volumetric strain is zero) when Poisson's ratio = 0.5

analysis of straight thin beams, and it is by far most frequently used in engineering analysis.

Plate bending elements for the Kirchhoff model can also be proposed. However, the Reissner-Mindlin elements are preferred due to several reasons. The Reissner-Mindlin model is hierarchically a higher-order model with respect to the Kirchhoff theory and better represents the physical behavior of plates, specially near the edges where boundary layers may develop which are well captured by the Reissner-Mindlin model (see Häggblad and Bathe, 1990). Also, the Reissner-Mindlin model predicts transverse shear deformations and can be used to model moderately thick plates. However, most importantly, the interpolations for the Kirchhoff elements should have C^1 continuity leading to relatively complex and inefficient elements that are also difficult (indeed almost impossible) to further develop for the effective analysis of shells and general nonlinear analysis.

We also recall that the elements for the basic shell model degenerate to Reissner-Mindlin plate bending elements when the shell midsurface is flat. The Reissner-Mindlin elements, which are listed in Table 6.8, have been directly derived for plate analysis and have a complete mathematical convergence analysis. However, in practice it is more efficient to use the elements of the basic shell model for, both, the analysis of plate and shell situations, to unify the analysis approach and since in nonlinear analysis the plate becomes a shell, see Lee and Bathe, 2010. These shell elements can also be hierarchically extended to 3-D shell elements to represent additional 3-D effects, see Kim and Bathe, 2008.

6.6 Finite element modeling tools

We discussed in the preceding sections how to formulate reliable finite elements for the mathematical models presented in this book. However, to take advantage of the full potential of these formulations for engineering structural analysis we still need to discuss how to efficiently construct finite element models deemed to solve practical engineering problems.

In this section we discuss several issues and techniques related to the construction of such finite element models.

6.6.1 Combining different type of elements in the same model

For an engineering problem, an efficient model is usually only obtained when different types of elements are combined. The reasons are many. For example, we may have different structural behaviors for different parts or regions of the model. Sometimes, even for the same region of the model the structural behavior may be more efficiently captured by combining elements.

As an example, we show the model of a cable-stayed bridge in a construction phase.

The bridge has two 127-meter-long spans symmetrically disposed with respect to the pylon, a 28-meter-wide deck carrying six traffic lanes, a 54-meter-tall pylon and a central stay plane with 36 cables.

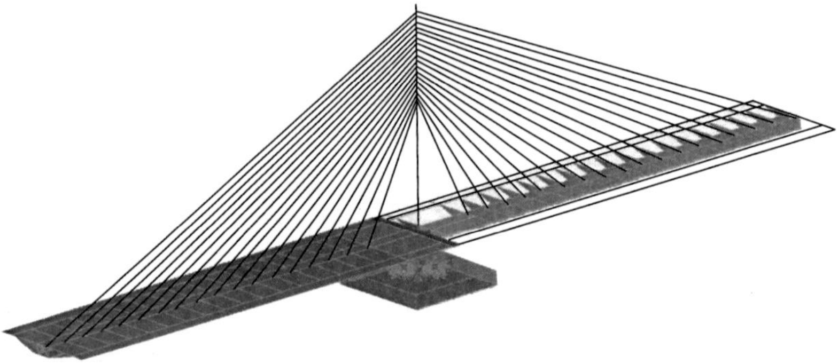

Fig. 6.60. Finite element model of cable stayed bridge

In Figure 6.60 we show the finite element model representing only the mesh outline and in Figure 6.61 a close-up region of the model. The pylon is modeled with Bernoulli-Euler beam elements whose axes are represented in Figure 6.60. The cables are modeled with pre-stressed truss elements, the foundation block with 3-D solid elements, the soil with spring elements and the bridge deck with shell elements. In Figure 6.60 we can see in the first span, the deck exterior and, in the second, its interior which gives an idea of the cellular section used.

We can appreciate that many element types are combined. Otherwise, we would not obtain an efficient model. Of course, depending on the objective of the analysis the deck could be modeled with beam elements, instead.

6.6.2 Constraint equations

In constructing a finite element model, it is sometimes effective, or even necessary, to establish a relation between the nodal degrees of freedom of the model.

A general form of this linear relation which is referred to as *constraint equation* is

$$U_i = \sum_{j=1}^{r_i} \alpha_{qj} U_{qj}$$

where U_i is a generic degree of freedom which is constrained or dependent. The degrees of freedom $U_{q1}, U_{q2}, \cdots, U_{qr_i}$ are independent degrees of freedom. As an example, consider the part of a model described in Figure 6.62a, where

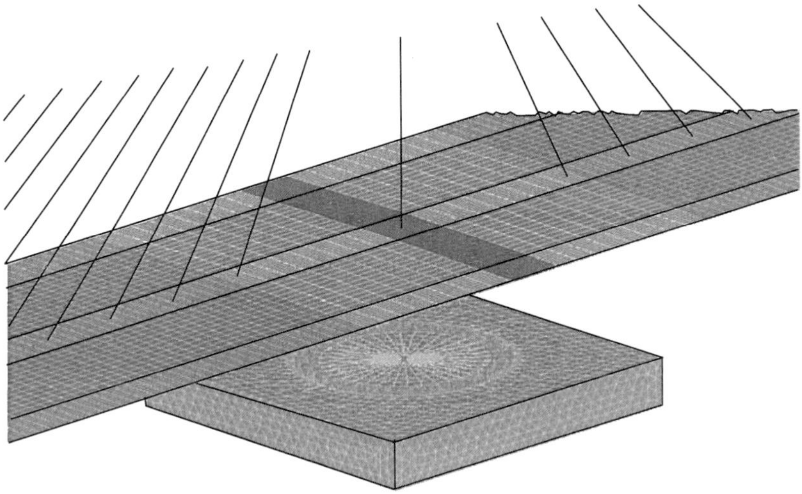

Fig. 6.61. Finite element model close-up of pylon deck intersection region

a thin structure working in bending connects to a deformable block. An appropriate modeling choice is to model the thin part with beam elements and the block with 2-D plane stress elements (here h is small). A mesh choice is shown in Figure 6.62b.

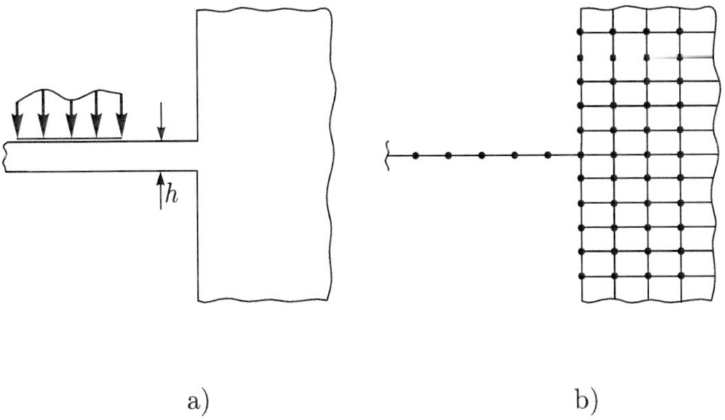

Fig. 6.62. a) Part of a structure; b) Mesh choice

In Figure 6.63a, we show a close-up view of the intersection region and in Figure 6.63b the corresponding mesh. Note that for the beam we have three degrees of freedom per node while for the 2-D elements we have only two. For the node which is common to the beam and the 2-D elements we should use three degrees of freedom, where U_r is the nodal rotation degree of freedom of

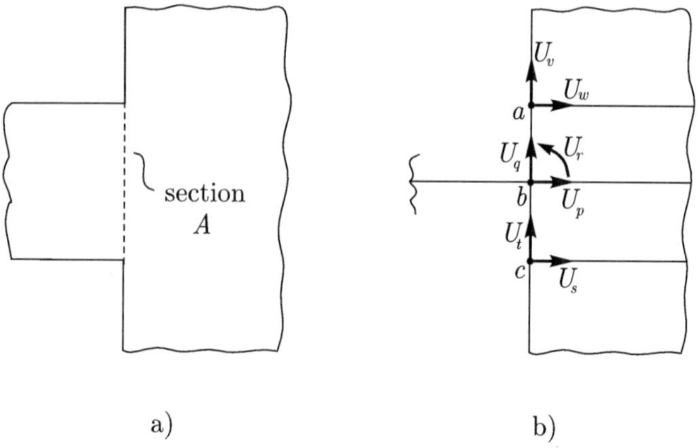

Fig. 6.63. a) Close up of the intersection region; b) Mesh choice for intersection region

the beam. However, the degree of freedom choices shown in Figure 6.63b do not yet correctly represent the behavior of the structure at the intersection region since the beam would behave as if it were pinned at node b neglecting the stiffness contribution of the block to the rotation of the Section A of the beam. The constraint equations

$$U_w = -\frac{h}{2}U_r + U_p \tag{6.153}$$

$$U_s = \frac{h}{2}U_r + U_p \tag{6.154}$$

model such stiffness contribution, and couple the rotation of the beam to the block.

6.6.3 Rigid links

A rigid link establishes constraint equations between the degrees of freedom of two nodes. One of the nodes is called the *master* node and the other the *slave* node. The degrees of freedom of the master node are independent while those of the slave node are dependent reflecting the constraint equations associated with the rigid link.

The constraint equations established by the rigid link are such that the distance between the two nodes does not change with deformation and the rotations are the same.

Rigid links are a very useful modeling tool and we exemplify their usage below. In Figure 6.64a we show part of the structure to be modeled. It is representative of a very common arrangement in structural engineering like a concrete slab and beam or a stiffened shell. An efficient modeling is obtained by using plate/shell and beam models as shown in Figure 6.64b. Of course, the beam should correspond to the geometric region below the bottom surface of the plate/shell.

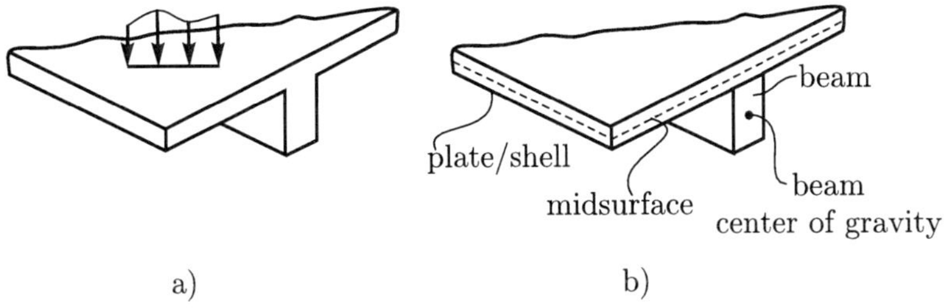

Fig. 6.64. Part of the structure to be modeled

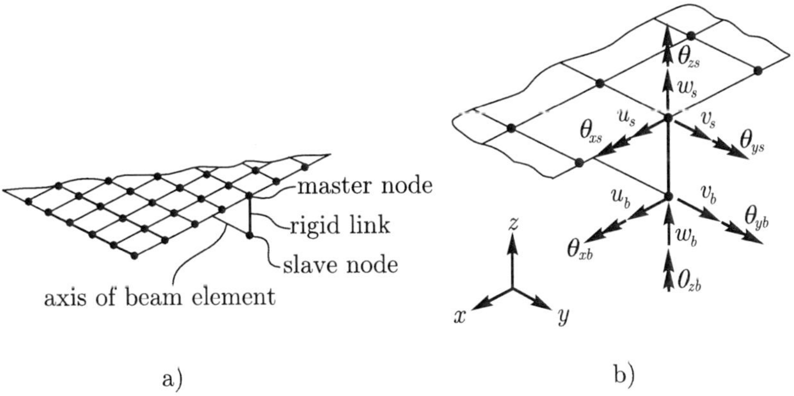

Fig. 6.65. Discretization for part of structure of Figure 6.64

In Figure 6.65a we show a typical discretization. One of the rigid links is seen connecting the master node, belonging to a shell element, and the slave node, belonging to a beam element (here, the choice of which node is to be the master node is arbitrary). In Figure 6.65b a detail is shown where we can see the degrees of freedom of the master and slave nodes.

Since the structure is assumed to be working in bending and membrane actions, the constraint is enforcing that the distance between the nodes does not change with deformation.

The constraints for the rotations deserve some detailed comments. In Figure 6.66a, we show the deformation associated with the rotation θ_x. The equality $\theta_{xb} = \theta_{xs}$ enforces that the rotation of the beam is the same as that of the director vector of the shell. These rotations induce bending deforma-

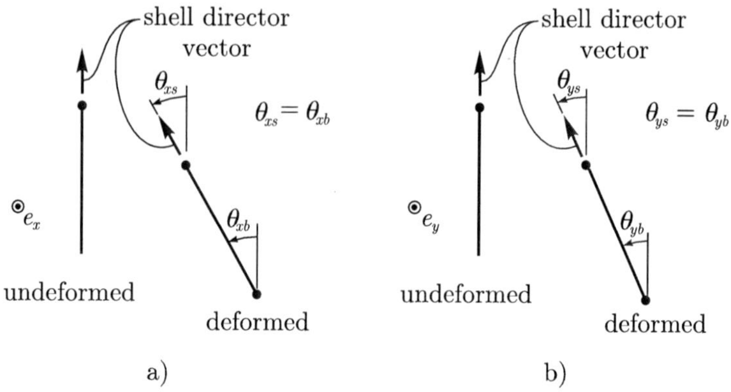

Fig. 6.66. Rotations of master and slave node around the x and y axes

tions both in the shell and in the beam. A similar behavior occurs for the rotation around the y axis which is summarized in Figure 6.66b. Note that θ_{yb} induces torsion in the beam while θ_{ys} induces bending in the shell. Finally, the rotation around z does not induce any deformation in the shell due to the kinematics of the shell element (see Section 6.4) but it induces bending in the beam. Note that the total stiffness associated with the rotation around z is only coming from the beam. Hence, this shell node should be assigned six degrees of freedom, *i.e.*, we should have θ_z as a degree of freedom for the shell node. Otherwise, we would be artificially preventing the rotation of the beam section around z.

6.6.4 Spring elements

Spring elements are very useful for modeling. Sometimes a spring element may be used to represent a whole structure when we are interested in capturing only the stiffness contribution of such structure to the current model. In another instance, it may be used to connect parts of a structure representing the stiffness of the connection. In the modeling of civil engineering structures, springs are also often used to represent the soil supporting the structure foundation.

In Figure 6.67a the most used spring arrangement is shown where the spring links a given degree of freedom U_i to the "ground" with spring constant k. Its contribution is accounted for by adding the spring constant k to the diagonal of the global stiffness matrix corresponding to degree of freedom U_i.

In Figure 6.67b, we generically represent a spring linking degrees of freedom U_i and U_j. The stiffness matrix of this spring in the local system u_1, u_2 is given by

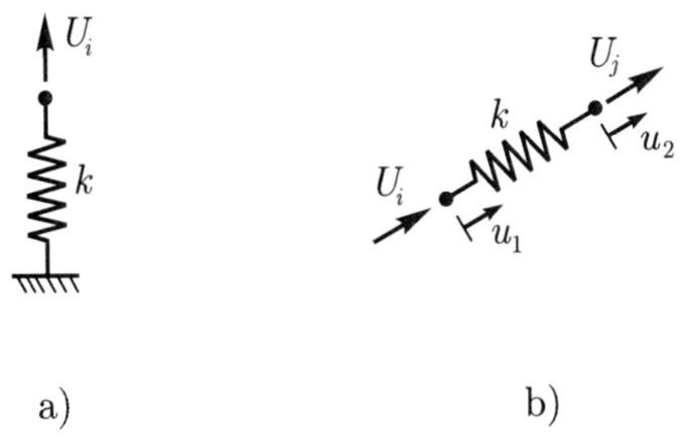

Fig. 6.67. Examples of spring arrangements

$$\mathbf{k} = \begin{bmatrix} k & -k \\ -k & k \end{bmatrix}.$$

6.6.5 Model symmetries

The use of symmetries can substantially decrease the model size, consequently, the solution and result interpretation efforts.

Plane symmetry

To exemplify the symmetry that occurs at a plane, consider the plate with a hole described in Figure 6.68a, we show the complete model and identify the planes of symmetry. The simplest way to understand symmetry is to recognize that at the plane of symmetry we may keep only one of the symmetric parts of the structure and represent all effects of the other part by means of prescribed boundary conditions. In fact, consider generic points P and P' of coordinates (y, z) and $(y, -z)$, respectively. Due to the symmetry conditions, *i.e.*, symmetric geometry and loading, we have

$$\begin{aligned} v(y,z) &= v(y,-z) \\ w(y,z) &= -w(y,-z) \end{aligned}$$

which lead, at the plane of symmetry xy, to zero displacements in the z direction and unconstrained displacements in the y direction, *i.e.*, zero tractions in the y direction. Of course, at the plane of symmetry there can not be either a positive or negative w justifying the conclusion of w being zero. Since the displacements v are the same for symmetric points located at infinitesimally small distances from the plane of symmetry there is no tendency of relative motion between the parts in the y direction leading to zero shear stress at the symmetry plane. Hence, as we extract one part, we should impose a zero traction condition in the y direction.

In Figure 6.68b, we show the boundary conditions which take into account the planes of symmetry xy and xz. Note that the boundary conditions associated to the plane of symmetry xz are justified by an analogous reasoning.

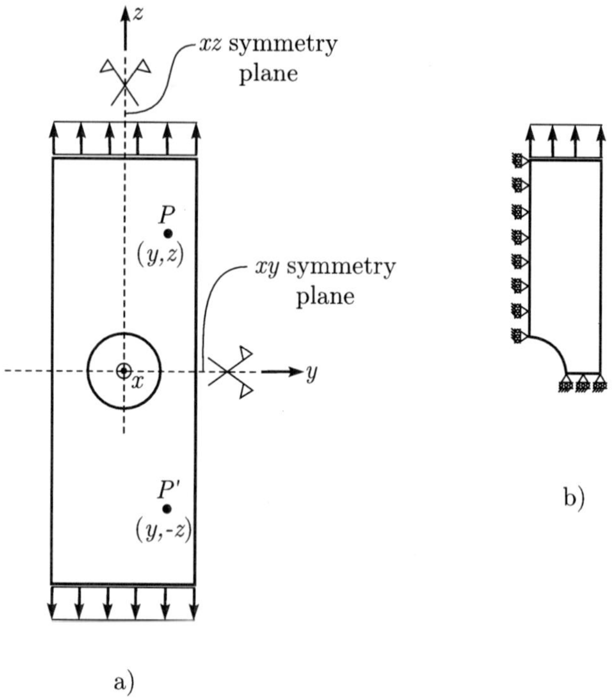

Fig. 6.68. a) Complete model; b) Model taking advantage of the symmetry

Cyclic symmetry

Consider the flywheel shown in Figure 6.69a. The flywheel can be geometrically obtained as a repetition and rotation of the part shown in Figure 6.69b. Suppose that the flywheel is rotating at constant angular velocity. Then, the deformations and stresses can be obtained by solving a static problem with the structure subjected to the inertia forces. These are body forces which act in the radial direction and also repeat themselves for a generic part such as that of 6.69b. We can establish a plane stress model of the flywheel considering the different thicknesses of its parts. Due to cyclic symmetry of the geometry and loading, we only need to model the part of Figure 6.69b with appropriate boundary conditions to take into account the cyclic symmetry as described in Figure 6.70. The justifications for these boundary conditions are as for planes of symmetry (see above).

Of course, although we presented the cyclic symmetry modeling for a specific problem, it can be used for any structure which complies with the cyclic symmetry conditions for the geometry and loading.

In case the geometry is cyclic symmetric but not the loading, we can still use solution procedures which take advantage of the cyclic symmetry of the

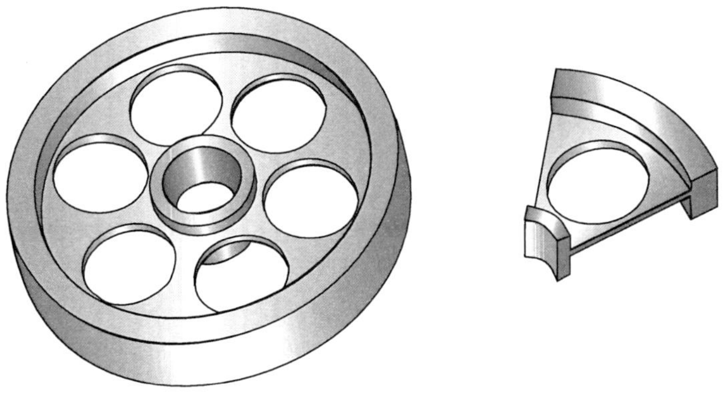

a) b)

Fig. 6.69. a) Complete flywheel; b) Cyclic symmetric part

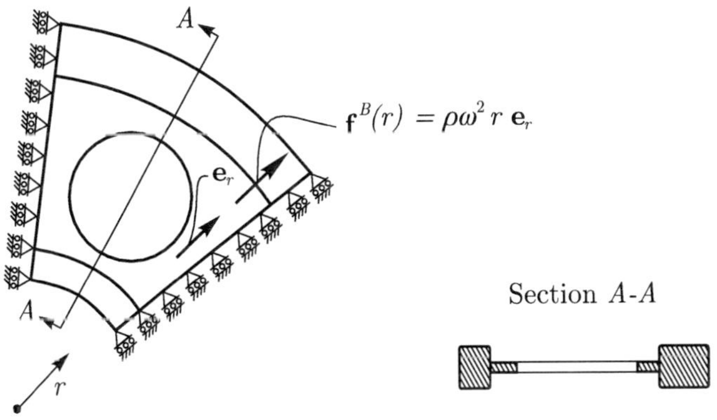

a) b)

Fig. 6.70. Plane stress model of the flywheel. Only a part needs to be modeled

geometry. These solution procedures can result into major advantages since only one cyclic part needs to be meshed and the solution times can be much smaller than those for the full problem. We refer to Zhong and Qui, 1983 for details, and ADINA, 2008.

6.6.6 Substructures

As a motivation, consider the finite element model of the cantilever with holes shown in Figure 6.71. There is a repetition pattern in the mesh used. In Figure 6.72a we show the mesh part, which when repeated leads to the complete

Fig. 6.71. Finite element model of cantilever with holes

mesh. In Figure 6.72b we indicate the nodes whose degrees of freedom are shared by more than one part of the model or for which boundary conditions are assigned.

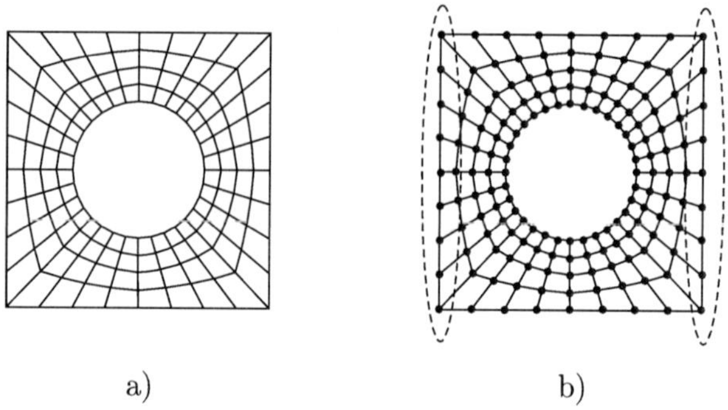

Fig. 6.72. Finite element mesh of part of the model

The substructuring technique consists of two steps. In the first step – the static condensation – the stiffness matrix of the repeating part is established for only the degrees of freedom which are shared with other parts or have assigned boundary conditions as exemplified below. This part with the reduced number of degrees of freedom may be interpreted as a macroelement.

In the second step the stiffness matrices of these macroelements are assembled to obtain the structure stiffness matrix using much fewer degrees of freedom than originally.

The computational advantages of the substructuring technique are twofold. The stiffness matrix evaluation effort is generally much reduced and the solution effort may also be significantly less since there are much less active degrees of freedom.

The static condensation part is detailed below. Let $\mathbf{U}_m$ and $\mathbf{U}_s$ be column matrices which collects the degrees of freedom that will be kept for the main structure and those that will be condensed out, respectively. Let $\mathbf{K}^p$ be the stiffness matrix of the part. Then, we can write

$$\begin{bmatrix} \mathbf{K}^p_{mm} & \mathbf{K}^p_{ms} \\ \mathbf{K}^p_{sm} & \mathbf{K}^p_{ss} \end{bmatrix} \begin{bmatrix} \mathbf{U}_m \\ \mathbf{U}_s \end{bmatrix} = \begin{bmatrix} \mathbf{R}_m \\ \mathbf{0} \end{bmatrix}$$

which leads to

$$\mathbf{K}^p_{mm}\mathbf{U}_m + \mathbf{K}^p_{ms}\mathbf{U}_s = \mathbf{R}_m \tag{6.155}$$

$$\mathbf{K}^p_{sm}\mathbf{U}_m + \mathbf{K}^p_{ss}\mathbf{U}_s = \mathbf{0}. \tag{6.156}$$

From equation (6.156)

$$\mathbf{U}_s = -\mathbf{K}^{p^{-1}}_{ss}\mathbf{K}^p_{sm}\mathbf{U}_m$$

which substituted into (6.155) yields

$$\left(\mathbf{K}^p_{mm} - \mathbf{K}^p_{ms}\mathbf{K}^{p^{-1}}_{ss}\mathbf{K}^p_{sm}\right)\mathbf{U}_m = \mathbf{R}_m$$

and therefore

$$\mathbf{K}^{pr} = \mathbf{K}^p_{mm} - \mathbf{K}^p_{ms}\mathbf{K}^{p^{-1}}_{ss}\mathbf{K}^p_{sm}$$

is the reduced stiffness matrix of the part.

Note that upon solution of this problem, $\mathbf{U}_m$ is determined for each part and from equation (6.156) $\mathbf{U}_s$ can be obtained using

$$\mathbf{U}_s = -\mathbf{K}^{p^{-1}}_{ss}\mathbf{K}^p_{sm}\mathbf{U}_m$$

and then all solution variables can be determined for the whole domain. The process can be very efficient, but is not always so, see Bathe, 1996 for more details.

6.6.7 Connecting different type of elements

Connecting beam and 2-D elements

When discussing constraint equations, we exemplified above one way of linking beam and plane elasticity elements.

In fact, the easiest way to establish constraints is to use rigid links. Referring to Figure 6.63b, we could define rigid links between nodes b and a, and b and c which would assume a rigid connection.

An alternative way is to use transition elements. In Figure 6.73 we schematically show a transition element. At one side, the transition element has the degrees of freedom of the 2-D element and at the other the degrees of freedom of the beam element, therefore, preserving displacement compatibility. Although the use of rigid links is more common in practice since it does not require a new element, the transition element concept is very attractive and provides useful elements to connect 3-D and shell elements as also

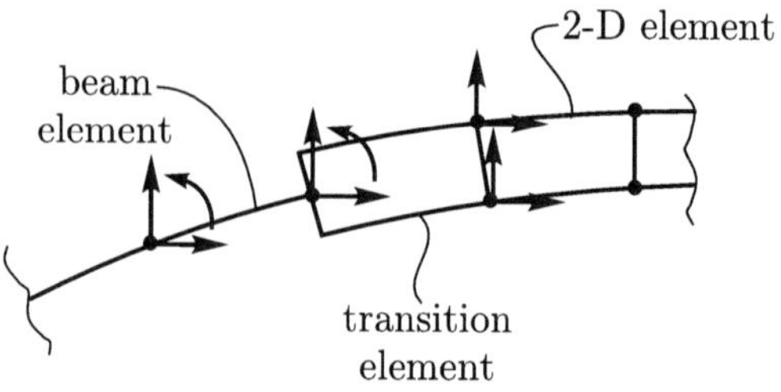

Fig. 6.73. 2-D transition element

discussed below. We refer to Bathe, 1996 for the derivation of the transition element of Figure 6.73.

Connecting beam and shell elements

We already discussed constraint equations and rigid links to connect beam and shell elements. This modeling is effective and takes into account the relative position of the beam axis with respect to the midsurface of the plate/shell.

Connecting shell and 3-D elements

As for the connection of beam and 2-D elements, we have two approaches. The first is the use of rigid links between the shell and 3-D element nodes at the interface as summarized in Figure 6.74. The second is the transition element as shown in Figure 6.75. Shell transition elements have some top and bottom nodes with only translational degrees of freedom. We refer to Bathe and Ho, 1981 for the detailing of this transition element.

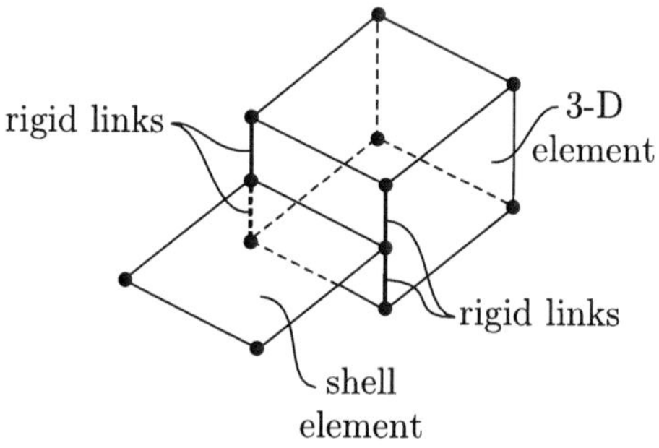

Fig. 6.74. Connecting a four-node shell element with an eight-node 3-D element through rigid links

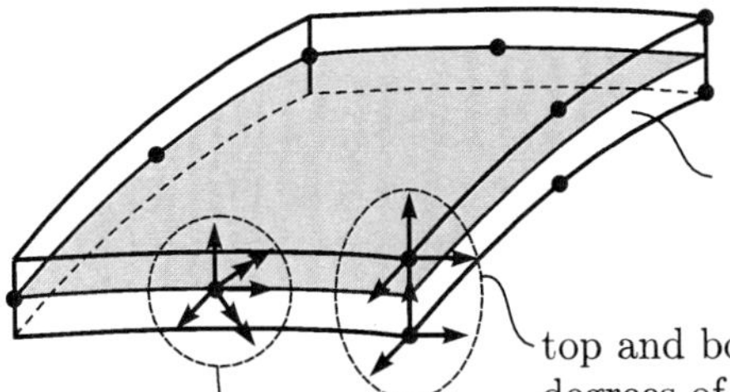

Fig. 6.75. Transition shell element

6.6.8 Skew systems

Frequently, it is necessary to have at a node a local reference system to assign displacement boundary conditions or to apply forces. For example, refer to Figure 6.70a, where we show the model of a cyclic part. If we align one of the sides for which displacement boundary conditions are specified with the global system, we need a local system for the other side. This local reference system, rotated with respect to the global system, to be assigned to nodes is referred to as a *skew system.*

Of course, the structure stiffness matrix has to be modified to account for these skew systems. These modifications are efficiently introduced at the element levels prior to the assemblage of the global stiffness matrix. In fact, let

$$\mathbf{u} = \mathbf{T}_s \tilde{\mathbf{u}}_s$$

$$\mathbf{f} = \mathbf{T}_s \tilde{\mathbf{f}}_s$$

where $\mathbf{u}$ is the element nodal displacement column matrix referred to the global system and $\tilde{\mathbf{u}}_s$ is the column matrix of element nodal displacements in which some of the degrees of freedom are referred to a skew system. Analogous definitions for the element nodal force column matrices $\mathbf{f}$ and $\tilde{\mathbf{f}}_s$ apply. Following the derivations presented in Chapter 2, namely equations (2.30) to (2.32), we obtain the stiffness matrix of the element which takes into account the skew system as

$$\mathbf{k}_s = \mathbf{T}_s^T \mathbf{k} \mathbf{T}_s. \tag{6.157}$$

Although, equation (6.157) formally gives the transformation, $\mathbf{k}_s$ is efficiently obtained operating only with the rows and columns of $\mathbf{k}$ associated with the degrees of freedom for which the skew system is specified.

6.7 Meshing issues and error assessment

Nowadays mesh generators are widely used to obtain finite element meshes in the analysis of engineering problems. In this context, the analyst has to specify some mesh parameters which control the mesh generation process. Usually, these parameters are related to the mesh density which is desired in the different regions of the model.

To fix ideas, let us consider the plate with a hole problem described in Figure 6.68. We model one quarter of the plate as shown in Figure 6.68b. It is well known that there is a stress concentration region near the hole and the highest stress gradients for the plane stress components τ_{yy}, τ_{zz}, τ_{yz} are expected to occur in this region.

As already exemplified for the 1-D problem, mesh grading is very efficient to capture high gradients of the variables to be predicted. Therefore, it is natural to specify smaller mesh densities for the hole region. In Figure 6.76a, we show the mesh densities at the lines of the geometric model and in Figure 6.76b the outcome of the mesh generator used. Due to the varying mesh density and the geometry, general quadrilateral and a few triangular elements have been generated. We generically refer to these quadrilateral elements as distorted elements since we have as reference the square element.

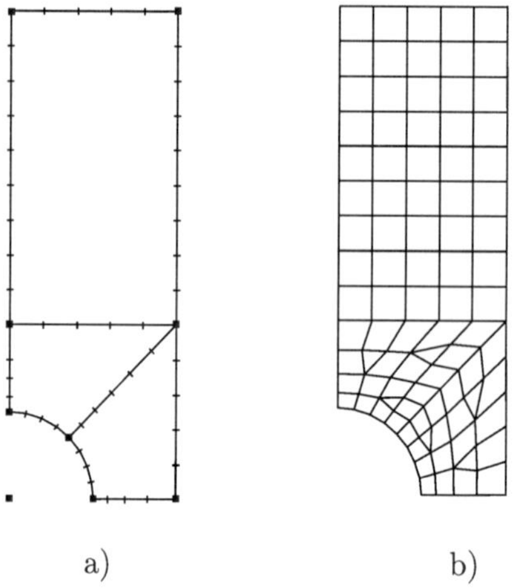

Fig. 6.76. Typical 2-D mesh generation

A natural question arises: Is the solution obtained with this mesh "good enough"?

To answer this question we refer to Figure 1.13 where the process of hierarchical modeling is described. In that figure we implicitly assumed that the solution of each mathematical model can be obtained exactly or as close to the exact solution as desired. Since we are also assuming that the finite element models used comply with the convergence requirements, we can indeed obtain a finite element solution as close to the exact solution as desired by sufficiently refining the mesh. However, from an engineering perspective overrefinement of the mesh may be very expensive and a waste of resources.

We recall that the ultimate objective of the modeling is to estimate the response of the structure. We are actually interested in some variables and we are also aware that in each of the mathematical models there are assumptions that lead to some errors in the prediction of these variables when measured with respect to the predictions of the most-comprehensive mathematical model. Therefore, as part of the hierarchical modeling process, we also need to establish the precision with which each variable should be predicted in the finite element solution of a mathematical model.

Hence the above "good enough" need not mean very close to the exact solution of the mathematical model since some level of error – determined in the hierarchical modeling process – is acceptable. Nevertheless, the essence of the question remains: is the solution obtained with this mesh "within the required precision"?

A pragmatic approach could be to keep on refining the mesh until for two consecutive solutions the variables to be predicted change by less than the required precision. This might be expensive, however.

The objective of this section is to discuss some issues related to mesh quality and solution accuracy which should aid the analyst to arrive at a finite element solution of a "good enough" accuracy.

6.7.1 Mesh grading

In general, as in the plate with a hole problem, the analyst has an expectation about the qualitative behavior of the solution and/or in which model region a more accurate prediction is required. This information should be used to guide the selection of the first mesh.

6.7.2 Element shape

We assume from the start that we are using effective elements as characterized in Section 6.5. Therefore, the elements are isoparametric with the use of mixed methods to overcome locking when required.

We would like to discuss the effect of the element shape on the accuracy of the predictions. The terminology element distortion is generically used to refer to an element shape which is not a square for 2-D quadrilateral elements and not a cube for 3-D hexahedral elements. A more precise definition of distortion types is given in Figure 6.77 (see Lee and Bathe, 1993, Bathe, 1996) for 2-D quadrilateral elements. Note that the distortions (c) and (f) are not applicable to a four-node 2-D element since these are defined based on the position of the mid-side nodes.

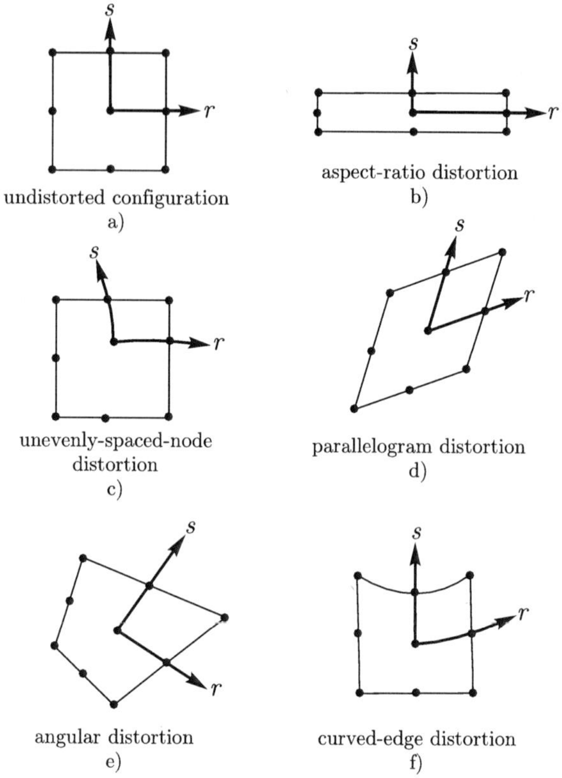

Fig. 6.77. Classification of element distortions

Considering isoparametric elements, we recall that the interpolation functions are defined for the reference element using the isoparametric coordinates r and s. The interpolations in the physical coordinates depend on the element shape. For example, consider a generic four-node element as defined in Figures 6.29 and 6.30. Equations (6.111) and (6.112) show that the interpolation functions $\tilde{h}_i(x, y)$ defined in the physical coordinates depend on the isoparametric mapping $(r(x, y), s(x, y))$. Therefore, the element shape which implicitly defines the isoparametric mapping through the element nodal coordinates affects the interpolation functions in the physical coordinates.

Of course, the approximation power of an element is linked to the monomials of the Pascal triangle that the element can actually represent in the physical coordinates.

Lee and Bathe (Lee and Bathe, 1993) have examined how element distortions affect the monomials which can be exactly represented in the physical coordinates. We summarize their findings in Table 6.9.

Table 6.9. Effect of element distortions for 2-D elements. Q_n,P_n represent all monomials of the Pascal triangle which are up to order n in each variable

Type of element	Monomials exactly represented: Undistorted	Angular distortion	Curved-edge distortion
4-node element	Q_1	P_1	–
8-node element	P_2, x^2y, xy^2	P_1	P_1
9-node Lagrangian element	Q_2	P_2	P_1
12-node element	P_3, x^3y, xy^3	P_1	P_1
16-node Lagrangian element	Q_3	P_3	P_1
25-node Lagrangian element	Q_4	P_4	P_2

We note that the 8 and 12-node elements are significantly more affected by angular distortions than the Lagrangian elements and the curved-edge distortion severely diminishes the number and order of the monomials that can be exactly represented. These observations lead to practical guidelines for mesh constructions, namely:

- When elements with angular distortion are anticipated – the usual case for engineering problems – Lagrangian elements are preferable compared to the 8 and 12-node elements.
- Unless we are modeling curved boundaries, curved edge distortions should be avoided, that is, for each edge the interior node should be placed on the straight line linking the corner nodes such that the distance between neighbouring nodes are equal.

6.7.3 Error estimation and adaptative procedures

Error estimation in finite element analysis is a broad field. In this section we would only like to briefly review some developments in this area with respect to their use in practical finite element analysis.

In Section 6.2, we discussed error estimates or bounds as given in (6.63) for 1-D and in (6.136) for 2-D and 3-D analyses. These error bounds are referred to as *a priori* error bounds since they give general bounds on the error even before any finite element solution is obtained. These bounds, of course, involve unknown constants. To obtain an error estimate for a specific finite element solution, the *a posteriori* error estimates are used.

A posteriori error estimates are useful to give the analyst an idea of how far the obtained solution is from the exact solution. Since usually the a posteriori error estimate is defined pointwise for the problem domain, it is also used to steer adaptative mesh refinement procedures (either h or p).

We note that every error estimate is based on a quantity for which the error is being estimated. In the early developments this quantity was the energy norm.

More recently error estimates are being developed for specific variables of interest. These estimates are very attractive for the hierarchical modeling process, since this process is driven by the precision with which some variables are to be predicted.

Considering the voluminous amount of material which has been published regarding error estimation and the actual usage of these developments in practical finite element analysis, we observe a substantial gap. The reasons are many, but in essence the developments have still not attained a level which permit their usage in a reliable and efficient manner for practical problems. We refer the reader to Grätsch and Bathe, 2005 for a review on this subject.

Before we close this section we would like to mention a very practical procedure used in engineering analysis to assess the accuracy of finite element solutions – the isobands of stresses – as proposed in Sussman and Bathe, 1986.

We recall that although the finite element solution always satisfies some equilibrium properties as discussed in Section 6.1, differential equilibrium is not satisfied pointwise, *i.e.*, the equations given by (3.114) are not satisfied when the stress field is obtained from the displacement finite element solution through the compatibility and constitutive relations. Also, at element boundaries, which are shared by two elements, equilibrium is, in general, violated. In fact, for a point on this boundary we can evaluate the stress (a stress component or an invariant to have a scalar value) from the displacement solution considering either element to arrive, in general, at different values. Of course, if the loads, the geometry and the material properties of the problem are smooth, there is only one stress value corresponding to the exact solution at that point and the difference in the finite element prediction is referred to as a stress jump which is a local measure of the error for the stress prediction.

By means of the stress isobands we can graphically visualize the stress jumps throughout the finite element domain. The stress isobands are constructed by plotting bands of selected stress measures (pressure, effective stress, stress components, among others) from the displacement finite element solution.

A typical situation for two neighboring elements is shown in Figure 6.78a. Each band, black or white, represents a specified range of the stress measure.

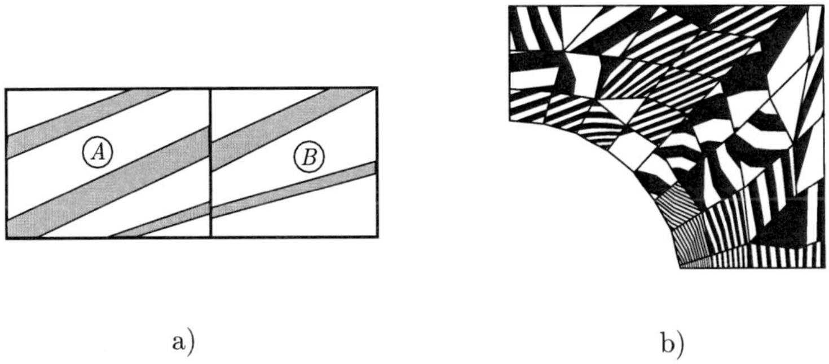

Fig. 6.78. a) Schematic isoband plot two neighbouring elements; b) Isobands of a mesh region

When we examine the common boundary of elements A and B, we identify some discontinuities. Selecting the magnitude of the band interval with respect to some relevant reference value for the problem, we can visualize how accurate the finite element solution is with respect to that stress measure. In Figure 6.78b we show a stress isoband plot for a region of a finite element model which we examine next. For the selected band interval we can identify all sorts of discontinuities revealing poor predictions.

Let us consider the plate with a hole problem described in Figure 6.68. We are interested in assessing the accuracy of the prediction for the stress τ_{zz} when using selected finite element meshes. In Figure 6.79a we show the isoband plots of τ_{zz} obtained with 4-node quadrilateral plane stress elements. In Figure 6.79b we plot for the same mesh the analogous results obtained with 9-node quadrilateral elements. We first note that the applied uniform traction has a magnitude of 25 N/mm^2 and the stress intensity factor for the component τ_{zz} predicted by the exact solution is of the order of 4 leading to values for τ_{zz} around 100 N/mm^2 near the hole. We selected 2.5 N/mm^2 as the stress band interval magnitude which is, therefore, of the order of 2.5% of the maximum τ_{zz}. Note that for the 4-node element model we can not clearly identify a band pattern and there are very large band discontinuities revealing a poor quality prediction. For the 9-node element model we have a clear band pattern and the discontinuities are small indicating good predictions for τ_{zz} when a 2.5 N/mm^2 error is acceptable.

In order to obtain better predictions with the 4-node element a much more refined mesh is required. In Figure 6.80a, we show the isobands for a

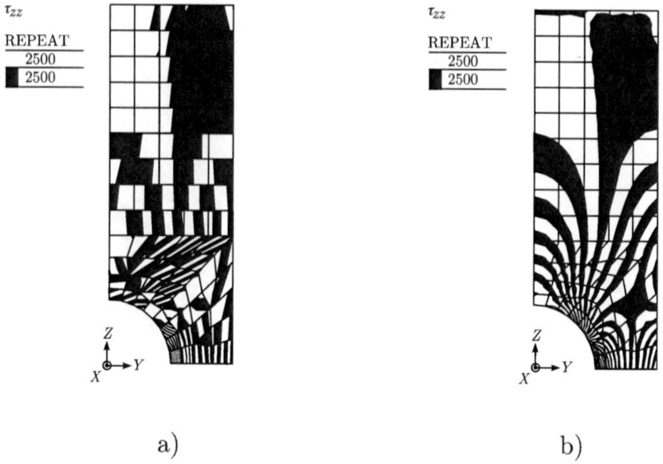

Fig. 6.79. Isoband plots for τ_{zz}. Problem data: plate width 20 mm, height 56 mm, thickness 1 mm, hole diameter 10 mm, applied traction 25 N/mm^2, $E = 7.0 \times 10^4$ N/mm^2 and $\nu = 0.25$

refined mesh of 4-node elements. Of course, since the interpolations are of low-order for the 4-node element some discontinuities are unavoidable for distorted elements. If the results are smoothed by taking average values at element boundaries we obtain the isobands of Figure 6.80b. Care must be exercised when using isoband plots of smoothed results since they are always continuous at element interfaces no matter how bad the predictions might be. Therefore, smoothed results may be used to improve the visualization only when the unsmoothed results have been examined and have revealed reasonable accuracy.

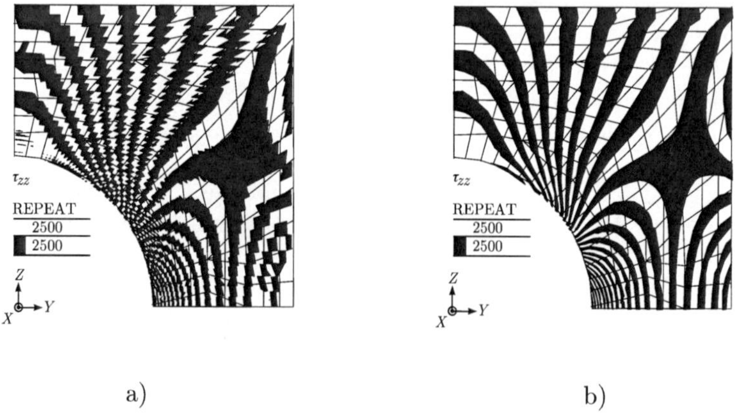

Fig. 6.80. Isoband plots for refined 4-node element mesh

Finally, we note that the stress band plots are particularly effective in identifying incompatibilities in constructed element meshes. That is, if by a data input error, displacement incompatibilities have been created by not connecting elements properly, the stress band plots will quite likely show artificial stress gradients and stress jumps, thus indicating the regions where the mesh should be checked and repaired.

6.8 Finite element model construction

Before we close this chapter we would like to mention some developments that have significantly facilitated the use of the finite element method and have therefore helped to spread the use of this technology. This task is often referred to as mesh generation. In the early stages of applications of the finite element method in industry, the mesh generation was performed by defining the geometric properties of each element and its position with respect to a coordinate system. For three-dimensional problems the generation of the mesh was a major task and consumed most of the resources involved in the finite element model construction. Therefore continuous efforts have been directed to ease this burden. Nowadays, the construction of complex 3-D finite element models takes advantage of the geometric modeling of solids and makes use of automatic mesh generation algorithms which act directly on the solid models.

In order to illustrate these developments, let us briefly describe how the construction of model 3 for the carabiner (see Chapter 1 and Section 7.3) could be aided by use of geometric modeling and automatic mesh generation algorithms.

In Figure 6.81, a 3-D representation of the carabiner, *i.e.*, a geometric model is shown. These geometric models are, in general, constructed in Com puter Aided Design (CAD) codes for other purposes than performing a finite element analysis. Their primary objective is usually linked with the production process of the part/equipment/structure. We note that this is the reason why some details are present in the geometric model shown in Figure 6.81 such as the letter imprints and the small hole for the pin of the superior closing part. For the kind of structural predictions generally sought, such details are not important and their modeling would lead to a much more refined finite element model than required. Therefore, it is necessary to modify the geometric model before generating the finite element model and implicitly the mesh. This process is called defeaturing and it corresponds to the suppression of features that have no significant impact on the prediction of the variables sought in the analysis.

In Figure 6.82, we give an illustration of the defeaturing process. We see that it is still performed at the level of the geometric model using, in general, the tools of the geometric modeling program. We show in Figure 6.83 the results of some steps of the finite element model construction that can still

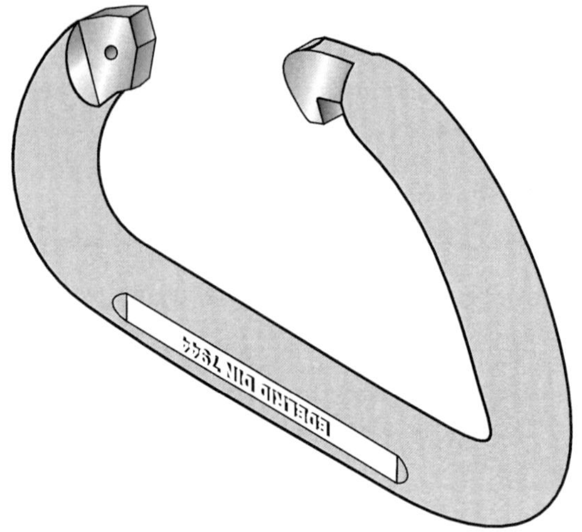

Fig. 6.81. Geometric model of the carabiner

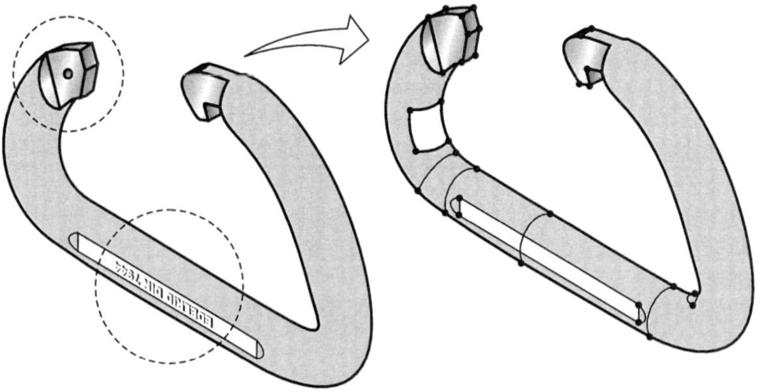

Fig. 6.82. Ilustration of the defeaturing process

be made at the geometric modeling level. The surfaces to apply the pressure and to specify the displacement boundary conditions can be defined on the geometric model. Also, the representation of only half of the model due to symmetry can be introduced at the geometric modeling level. The geometric model of Figure 6.83 with the mechanical conditions already introduced is ready for the mesh generation. In Figure 6.84 an automatically constructed mesh is presented.

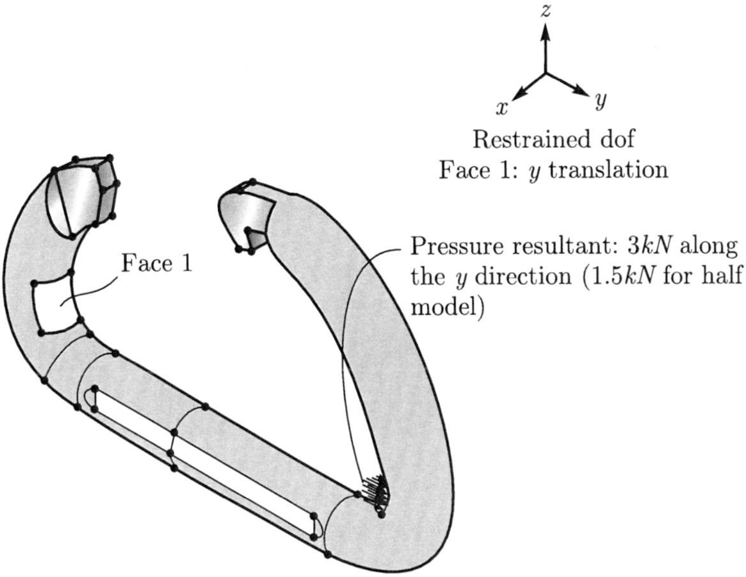

Fig. 6.83. Finite element mechanical conditions that are applied to the geometric model

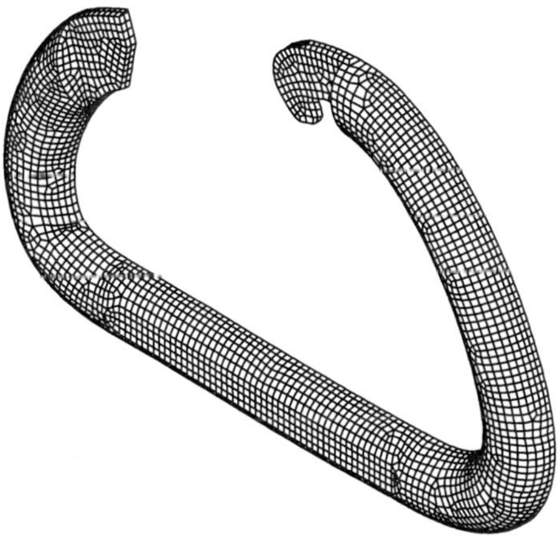

Fig. 6.84. Finite element mesh for the carabiner

6.9 A finite element modeling example

We choose to end this chapter by discussing the finite element modeling of a simple problem. Actually, this example could have been included in the next chapter as one of the modeling examples. However, placed at this point it should be interpreted as a synthesis of many concepts introduced in this chapter.

Spherical dome subjected to its own weight

This problem was studied in Section 4.4.2 and solved with the shell mathematical models presented in that section. We consider first the situation shown in Figure 4.116 for which the dome is supported tangentially. The data of the problem is that given in Example 4.21.

The sequence of mathematical models for this problem is:

1. Membrane-bending shell model
2. Basic shell model
3. 3-D elasticity model

Of course, in case the support prevents the motion along the tangential direction only, model 1 is analytically solved using the membrane theory of shells of revolution loaded axisymmetrically as detailed in Examples 4.19 and 4.20.

We resort to finite elements to solve models 2 and 3. The most efficient way to solve model 2 is to use the degenerated solid approach particularized to axisymmetric conditions. This procedure actually leads to axisymmetric shell elements based on the degenerated approach, see Bathe, 1996. In this study, however, we use shell elements to obtain the solution of model 2 and to exemplify their usage.

To solve model 3, we use axisymmetric 2-D elements. Of course, in this case, the solution of the axisymmetric 2-D elasticity model corresponds to the exact solution of the 3-D model as discussed in Section 4.1.3.

Model 2 – Shell model

We need to choose an initial mesh. We selected 20 elements along the meridian direction to discretize the curved geometry and no grading is used since due to the tangential support condition a membrane type solution is expected.

In Figure 6.85a we show a mesh which discretizes the whole shell domain with eighty elements along the circumferential direction. The elements in the shell apex region are degenerated triangular elements which have such a small width that they can not be seen with the scale of Figure 6.85a. In Figure 6.85b, we show a generic element and the degrees of freedom for two nodes. These degrees of freedom are selected such that two of their directions (2 and 3) lie on the meridian plane and direction 1 is orthogonal to it. Due to the symmetry conditions the rotations about directions 2 and 3 should be zero and the displacement along the direction 1 should also be zero. Therefore, considering these symmetry conditions we can select to discretize only a

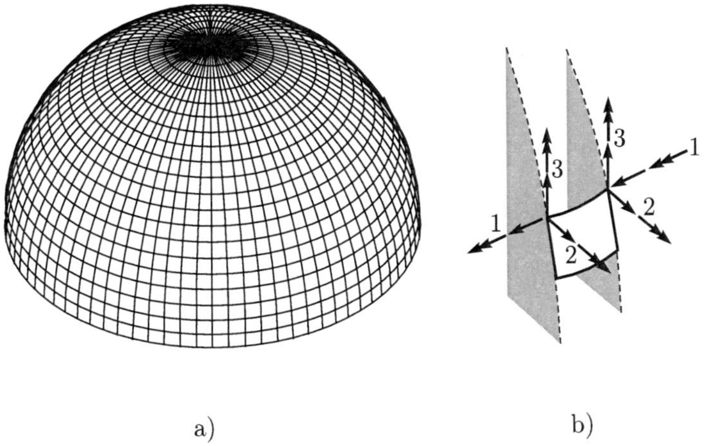

Fig. 6.85. a) Shell model; b) Degrees of freedom for a node at the meridian

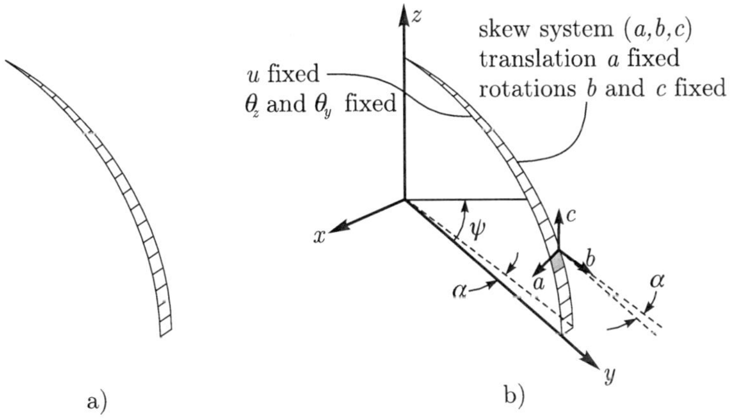

Fig. 6.86. a) Model of a shell wedge; b) Symmetry boundary conditions

wedge of the shell as shown in Figure 6.86a instead of discretizing the whole domain.

As long as we impose the symmetry boundary conditions as detailed above for every node on the meridian planes and the elements are geometrically the same as those of the mesh of Figure 6.85a for one element along the circumferential direction, we would obtain exactly the same predictions with the models of Figures 6.85a and 6.86a.

In order to impose the boundary conditions discussed above for the model of Figure 6.86a we can position the model with respect to the global system and define a skew system as shown in Figure 6.86b. Then the appropriate boundary conditions can be imposed as detailed in Figure 6.86b.

The model of Figure 6.86 was solved with MITC9 shell elements, considering first that the shell is tangentially supported. Since our model is hemispherical, we prevent the vertical displacement at the shell edge.

In Figure 6.87 we compare the membrane force predictions obtained with the model of Figure 6.86 to those of the membrane theory derived in Example 4.19.

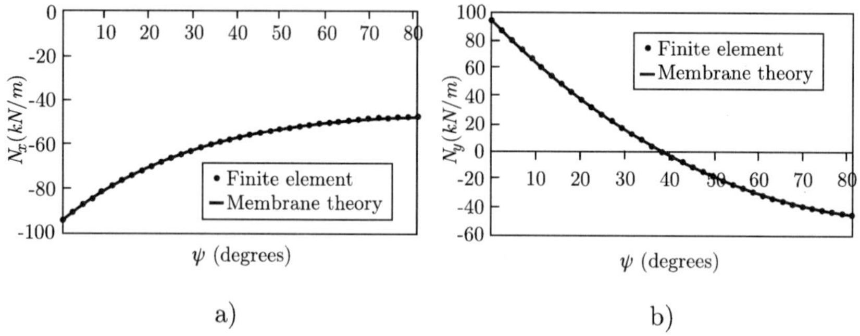

Fig. 6.87. Membrane force predictions. The angle ψ is measured from the edge as shown in Figure 4.122

We note the very close agreement between the membrane theory solution and the finite element solution. The finite element model predictions for the bending moments lead to very small values of less than 1 kN.m/m, therefore confirming the membrane behavior of the shell for the tangential support.

In Figure 6.88 the tangential u and radial w displacement predictions are shown. We again note the very close agreement between the membrane theory solutions and the finite element model predictions.

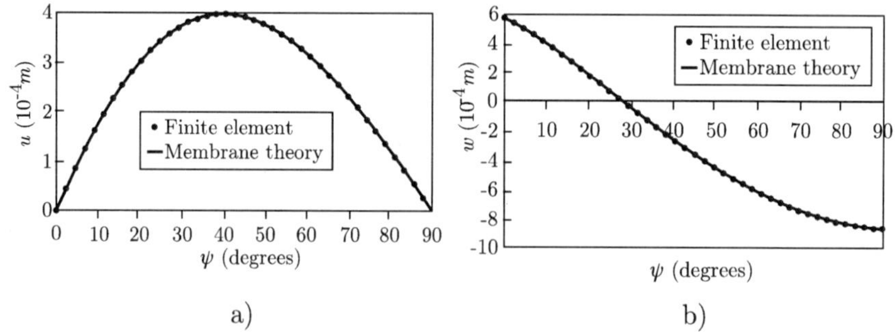

Fig. 6.88. Displacements of the midsurface of the shell. Finite element shell and membrane theory solutions; u and w measured as in Figure 4.115

Model 3 – 2-D axisymmetric model

In Figure 6.89, we show a close-up at the support regions of the finite element mesh for the tangentially supported shell where the model boundary conditions are shown. Again, a uniform mesh was used since we expect a membrane type solution. We used a total of eighty 9-node 2-D axisymmetric elements with two elements through the shell thickness.

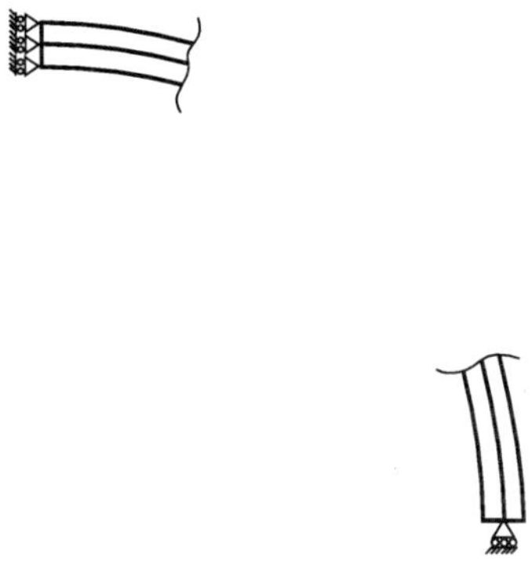

Fig. 6.89. Axisymmetric 2-D model

We compare the nodal displacements at the line which corresponds to the midsurface of the shell to the membrane theory solution. The results are shown in Figure 6.90 and the same convention used for Figure 6.88 is adopted. We note that with the scale of the graph we can barely distinguish between the model solutions.

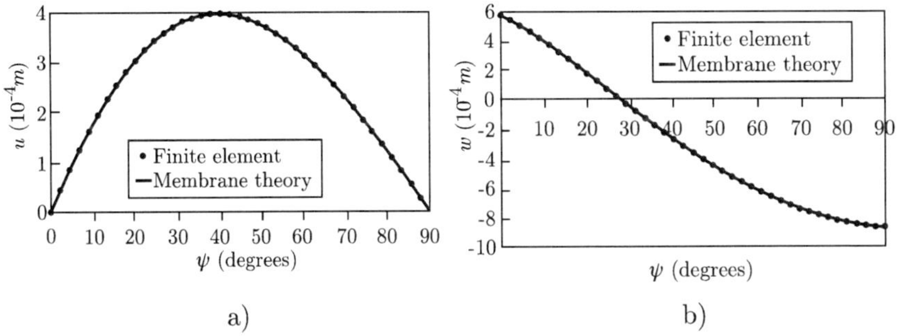

Fig. 6.90. Displacement predictions of the midsurface of the shell. Finite element 2-D and membrane theory solutions

The stress resultants such as axial forces and bending moments can be obtained through the integration of the stress fields. These resultants were evaluated leading to predicted values which are very close to those of the membrane theory.

Modeling of the clamped case

As discussed in Section 4.4.2 when a spherical dome is clamped, the shell can no longer resist the external loading through membrane internal actions alone and some bending takes place.

In Example 4.21 this problem with clamped conditions was solved. We obtained the bending moment distribution M_x reported in Figure 4.124.

Now we want to solve this problem with the shell finite element model and since we anticipate high gradients for the variables of interest near the edge, it is natural to use a graded mesh. We refined our mesh by grading it along the meridian assigning a smaller mesh density near the edge. We also refined the mesh for the circumferential direction by selecting a smaller central angle of 1° instead of 4.5° for the previous case. This is equivalent to considering a full shell model with 360 divisions in the circumferential direction.

In Figure 6.91a we show a much magnified deformed mesh and in Figure 6.91b a close-up view of the deformations near the edge.

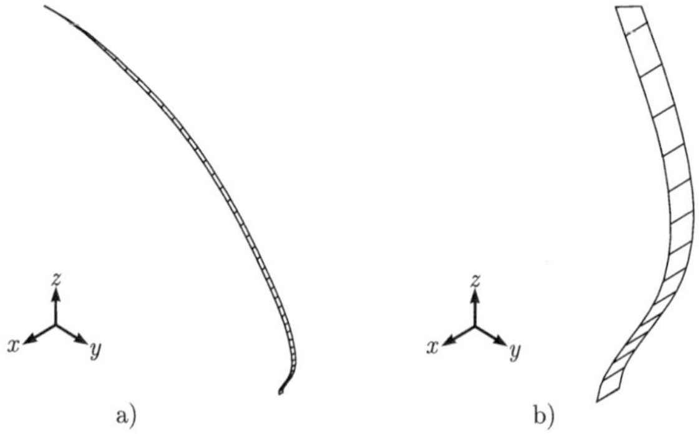

Fig. 6.91. Deformed mesh

In Figure 6.92 we compare the solution obtained with the shell model of Figure 6.91 with the shell theory solution derived in Example 4.21. We can see the close agreement obtained for the variable M_x.

This clamped case was also solved with the 2-D axisymmetric model. A mesh grading was also used and the stress resultants such as M_x were evaluated by integration; the calculated values were very close to those predicted by the shell theory and shell finite element model.

Summary

In this section we detailed the modeling of a simple shell structure. Nevertheless, several issues related to model choices and finite element mesh

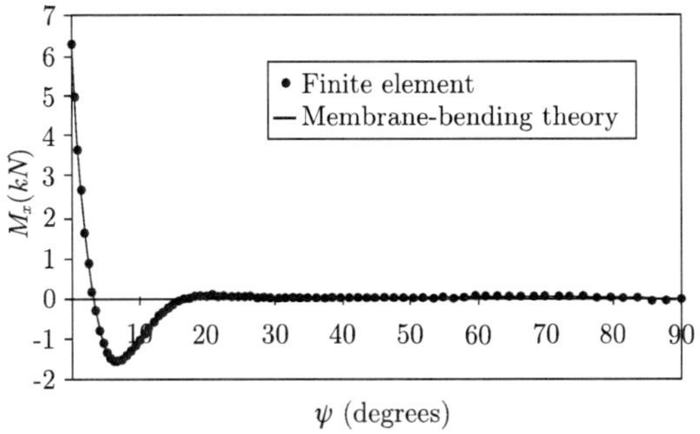

Fig. 6.92. Moment M_x predictions

constructions could be explored. We emphasize that the knowledge of the mathematical models used was an essential ingredient. Also, it was very important to be able to anticipate the nature of the solution, since this will guide the choice of mesh density. The use of symmetry and skew systems simplified a great deal the shell models, not only diminishing the computational effort but also facilitating the understanding through an easier interpretation of the results.

The three models considered led to very close predictions for displacements and internal actions showing that they are reliable for this problem.

7. Hierarchical modeling examples

The objective of this chapter is to apply the hierarchical modeling approach discussed in Chapter 1 to three selected problems using the mathematical models studied in Chapters 3 and 4, and the finite element procedures introduced in Chapter 6. Therefore, it is a synthesis chapter in which the concepts and tools discussed so far are used to perform the finite element modeling of engineering structures.

The chosen examples are relatively simple, but rich enough to highlight several modeling issues. Indeed, the discussion will show the importance of hierarchical modeling and, depending on our experience regarding analyses, also contain some surprises. While we only consider three examples, it is important to recognize that the methodology used in the modeling and solutions is quite general.

7.1 Built-in cantilever subjected to a tip load

The physical problem we would like to study is summarized in Figure 7.1.

The aluminum plate/beam is connected to the steel plate by a process that guarantees that no relative displacements between the aluminum and steel plates occur at any point at the connecting section. The block is suspended by a thin cable which is pinned at the center of gravity of the rectangular end section. The total weight of the block plus the cable is given by P. We neglect the self-weight of the plates.

We are interested in predicting the displacements of the aluminum plate/beam and the stresses at sections $x = \ell/2$ (mid-span) and $x = 0$ (built-in end) for the relations $\ell/h = 100$, $\ell/h = 10$ and $\ell/h = 5$. In order to set some values, let $h = 0.1$ m and $b = 0.01$ m which is also the thickness of the steel plate. The load P is scaled such that the maximum normal stress at the built-in end predicted by the Bernoulli-Euler beam model is $0.9\tau_y$, where $\tau_y = 223$ MPa is the yield stress of the aluminum used.

The sequence of models used in the hierarchical process is:

- Model 1: Bernoulli-Euler beam model
- Model 2: Timoshenko beam model
- Model 3 to 6: Plane stress models

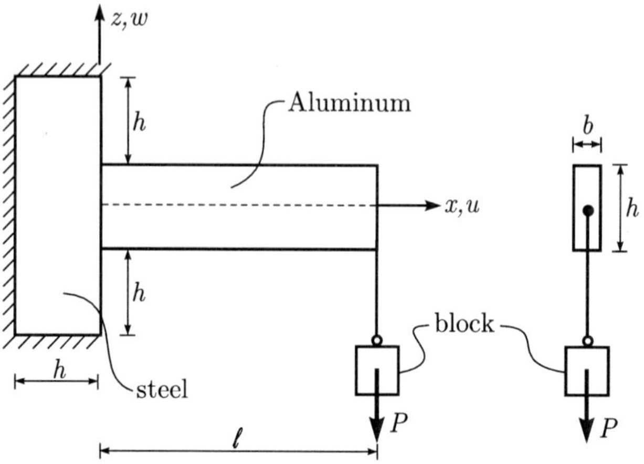

Fig. 7.1. Description of built-in cantilever. Material properties of the aluminum are $E = 70$ GPa, $\nu = 0.33$ and of the steel are $E = 210$ GPa, $\nu = 0.3$

We consider several plane stress models. However, their precise characterization is given as the modeling process unfolds.

7.1.1 Bernoulli-Euler beam model

The conditions that represent, in the context of the Bernoulli-Euler beam theory, the displacement restrictions and loading of the physical problem are summarized in Figure 7.2. Of course, in this model we are implicitly assuming that the steel plate is rigid.

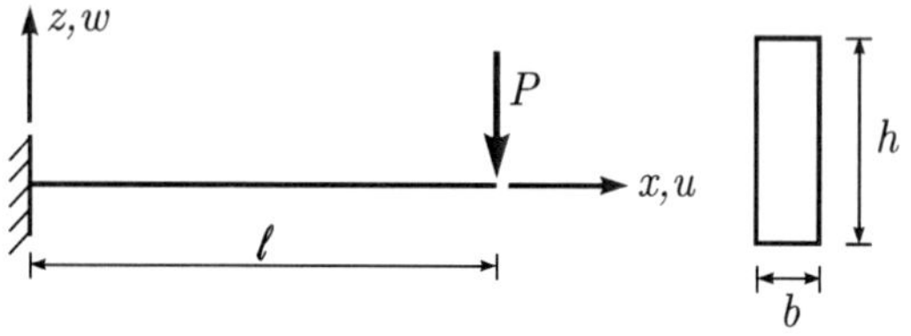

Fig. 7.2. Problem description for Bernoulli-Euler beam theory

These boundary conditions are

$$w(0) = 0, \ \frac{dw}{dx}(0) = 0 \tag{7.1}$$

$$M(\ell) = 0, \ V(\ell) = P. \tag{7.2}$$

The governing differential equation is

$$\frac{d^4w}{dx^4} = 0 \tag{7.3}$$

since there is no distributed transverse load along the beam axis. The force boundary conditions described in (7.2) can be expressed in terms of the displacements by

$$M(\ell) = EI\frac{d^2w}{dx^2}(\ell) = 0 \;\Rightarrow\; \frac{d^2w}{dx^2}(\ell) = 0 \tag{7.4}$$

$$V(0) = EI\frac{d^3w}{dx^3}(\ell) = P. \tag{7.5}$$

The integration of equation (7.3), considering the boundary conditions given in (7.1), (7.4) and (7.5) leads to the solution

$$w(x) = \frac{P}{6EI}(x^3 - 3\ell x^2). \tag{7.6}$$

Of course, from the above solution the moments and the shear forces can be obtained. The longitudinal displacement $u(x, z)$ can be directly derived from (7.6) considering the Bernoulli-Euler hypothesis that plane sections remain plane and orthogonal to the deformed beam axis. Therefore

$$u(x, z) = -z\frac{dw}{dx} = -\frac{Pz}{2EI}(x^2 - 2\ell x).$$

7.1.2 Timoshenko beam model

The characterization of the physical problem for the Timoshenko beam theory can also be described by Figure 7.2. Referring to the differential equations (4.238) and (4.239) which were derived in Section 4.2.8 and introducing $p = 0$ we obtain

$$-GA\left[\frac{d}{dx}\left(\frac{dw}{dx} - \beta\right)\right] = 0 \tag{7.7}$$

$$EI\frac{d^2\beta}{dx^2} = -GA\left(\frac{dw}{dx} - \beta\right). \tag{7.8}$$

The boundary conditions are given by

$$\beta(0) = 0 \tag{7.9}$$

$$w(0) = 0 \tag{7.10}$$

$$M(\ell) = 0 \tag{7.11}$$

$$V(\ell) = P. \tag{7.12}$$

Taking the derivatives with respect to x in equation (7.8) and using (7.7), we arrive at

$$EI\frac{d^3\beta}{dx^3} = 0$$

which can be integrated leading to

$$\beta(x) = C_2\frac{x^2}{2} + C_1 x + C_0. \tag{7.13}$$

Considering equation (7.11) and that for the Timoshenko beam model $M = EI\frac{d\beta}{dx}$, we obtain

$$\frac{d\beta}{dx}(\ell) = 0. \tag{7.14}$$

Introducing (7.9) and (7.14) into (7.13), we obtain

$$\beta(x) = C_2\frac{x^2}{2} - C_2\ell x. \tag{7.15}$$

Recalling that for the Timoshenko model $V = -GA\left(\frac{dw}{dx} - \beta\right)$, the boundary condition (7.12), together with equation (7.8), gives

$$-GA\left(\frac{dw}{dx} - \beta\right)\bigg|_{\ell} = P = EI\frac{d^2\beta}{dx^2}\bigg|_{\ell}.$$

Therefore from (7.15)

$$C_2 = \frac{P}{EI}$$

and

$$\beta(x) = \frac{P}{2EI}\left(x^2 - 2\ell x\right). \tag{7.16}$$

Introducing the above equation into (7.8)

$$\frac{dw}{dx} = \frac{P}{2EI}\left(x^2 - 2\ell x\right) - \frac{P}{GA}$$

which can be integrated to give

$$w(x) = \frac{P}{EI}\left(\frac{x^3}{6} - \frac{\ell x^2}{2}\right) - \frac{P}{GA}x + C.$$

Imposing the boundary condition (7.10), we obtain

$$w(x) = \frac{P}{6EI}\left(x^3 - 3\ell x^2\right) - \frac{P}{GA}x.$$

Introducing the shear correction factor for a rectangular cross-section (see discussion in Chapter 4) and using $A = bh$, we arrive at

$$w(x) = \frac{P}{6EI}\left(x^3 - 3\ell x^2\right) - \frac{12 + 11\nu}{10(1+\nu)}\frac{P}{Gbh}x.$$

The longitudinal displacement $u(x,z)$ can be obtained considering that for the Timoshenko beam the sections remain plane and rotate by the angle $\beta(x)$. Therefore from (7.16)

$$u(x,z) = -z\beta = -\frac{Pz}{2EI}\left(x^2 - 2\ell x\right).$$

7.1.3 Plane stress solution

Consider the plane stress model of Figure 7.3 in which we also assume that the steel plate is rigid.

Fig. 7.3. Problem description in the context of the 2-D plane stress model

The displacement boundary conditions are

$$u(0,z) = 0 \qquad (7.17)$$

$$w(0,z) = 0. \qquad (7.18)$$

The force boundary conditions are given pointwise by

$$\mathbf{Tn} = \mathbf{f}^S$$

which for the free boundaries leads to

$$\tau_{zz}(x, \frac{h}{2}) = \tau_{zz}(x, -\frac{h}{2}) = 0$$

$$\tau_{xz}(x, \frac{h}{2}) = \tau_{xz}(x, -\frac{h}{2}) = 0.$$

For the edge defined by $x = \ell$, we have

$$\tau_{xx}(\ell, z) = f_x^S(\ell, z) = 0$$

$$\tau_{xz}(\ell, z) = f_z^S(\ell, z)$$

and $f_z^S(\ell, z)$ should represent the action of the load P. In other words, we need to distribute the load along a portion of the edge ($x = \ell$) denoted by L_P such that the resultant of the shear tractions $\left(f_z^S\right)$ acting on L_P is equal to P, *i.e.*,

$$\int_{L_P} -f_z^S(\ell, z) b \, dL = P.$$

The choice of L_P and the functional distribution of f_z^S over L_P is actually a modeling consideration and depends on how the loading P is actually applied in the physical situation.

Since to impose the displacement boundary conditions given in equations (7.17) and (7.18) and to find an analytical solution for this problem is not an easy task, we seek instead the solution of the problem described in Figure 7.4.

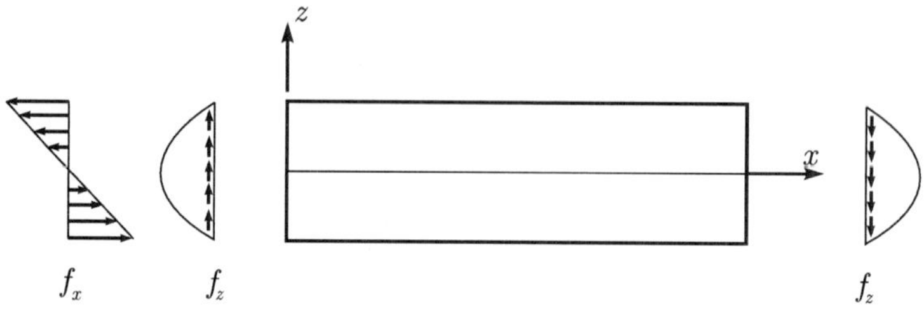

Fig. 7.4. Simplified 2-D plane stress problem, $L_P = h$

To relate the problems described in Figures 7.3 and 7.4 we need to impose at the edge $x = \ell$

$$\int_{-\frac{h}{2}}^{\frac{h}{2}} f_z^S(\ell, z) b \, dz = -P \tag{7.19}$$

and at the edge $x = 0$

$$\int_{-\frac{h}{2}}^{\frac{h}{2}} f_z^S(0, z) b \, dz = P \tag{7.20}$$

$$\int_{-\frac{h}{2}}^{\frac{h}{2}} f_x(0, z) z b \, dz = P\ell. \tag{7.21}$$

Of course, equation (7.20) guarantees the equilibrium in the vertical direction and equation (7.21) enforces global moment equilibrium. We choose the functional form of the traction distribution based on the shear stress distribution in beam models as previously discussed.

The derivations given below were reported in many textbooks and are repeated here merely for the convenience of the reader. We follow the classical textbook of Timoshenko and Goodier, 1970.

To derive the solution, we assume that the stress distributions are the same as that for a beam in bending. Hence

$$\tau_{xx}(x,z) = \frac{P(\ell - x)}{I} z = \frac{P(\ell - x)}{\frac{bh^3}{12}} z = \frac{12P(\ell - x)}{bh^3} z \tag{7.22}$$

$$\tau_{xz}(x,z) = -\frac{3P}{2bh}\left(1 - \frac{4z^2}{h^2}\right) = -\frac{P}{2I}\left(\frac{h^2}{4} - z^2\right) \tag{7.23}$$

$$\tau_{zz}(x,z) = 0.$$

This stress field satisfies all traction boundary conditions. Since there are no body forces, the differential equilibrium equations are (see equations (4.45) and (4.46))

$$\begin{aligned} \frac{\partial \tau_{xx}}{\partial x} + \frac{\partial \tau_{xz}}{\partial z} &= 0 \\ \frac{\partial \tau_{xz}}{\partial x} + \frac{\partial \tau_{zz}}{\partial z} &= 0 \end{aligned}$$

which are satisfied by the stress field given in (7.22) and (7.23).

Using the appropriate constitutive equation (4.44) we obtain the strains

$$\begin{aligned} \varepsilon_{xx} &= \frac{\sigma_{xx}}{E} = \frac{P(\ell - x)}{EI} z \\ \varepsilon_{zz} &= -\frac{\nu}{E}\sigma_{xx} = \frac{-\nu P(\ell - x)}{EI} z \\ \gamma_{xz} &= \frac{\tau_{xz}}{G} = -\frac{P}{2GI}\left(\frac{h^2}{4} - z^2\right). \end{aligned}$$

The strain-displacement relations are imposed next and used to obtain, by integration, a single valued displacement field[1] that satisfies the displacement boundary conditions. Referring to equations (4.47) to (4.49), we write

$$\varepsilon_{xx} = \frac{\partial u}{\partial x} = \frac{P}{EI}(\ell - x)z$$

hence

[1] We note that not all stress fields that satisfy equilibrium generate, through the use of the constitutive equations, compatible strain fields, *i.e.*, strain fields which can be integrated to lead to a single valued displacement field, see Section 3.2.2. Therefore, methods of solution that use as primary variables the stresses, such as the Airy stress function method, must enforce, besides the equilibrium conditions, the strain compatibility relations written in terms of the stresses

$$u(x,z) = \frac{P}{EI}(\ell x - \frac{x^2}{2})z + f(z) \tag{7.24}$$

where $f(z)$ is a function of z only. Also

$$\varepsilon_{zz} = \frac{\partial w}{\partial z} = \frac{-\nu P}{EI}(\ell - x)z$$

$$w(x,z) = \frac{-\nu P}{EI}(\ell - x)\frac{z^2}{2} + g(x) \tag{7.25}$$

where $g(x)$ is a function of x only. We also have

$$\gamma_{xz} = \frac{\partial u}{\partial z} + \frac{\partial w}{\partial x} = -\frac{P}{2GI}\left(\frac{h^2}{4} - z^2\right). \tag{7.26}$$

Substituting equations (7.24) and (7.25) into equation (7.26), we obtain

$$\frac{P}{EI}(\ell x - \frac{x^2}{2}) + \frac{df(z)}{dz} + \frac{\nu P}{2EI}z^2 + \frac{dg(x)}{dx} = -\frac{P}{2GI}\left(\frac{h^2}{4} - z^2\right). \tag{7.27}$$

Defining

$$G(x) = \frac{P}{EI}(\ell x - \frac{x^2}{2}) + \frac{dg(x)}{dx} \tag{7.28}$$

$$F(z) = \frac{\nu P}{2EI}z^2 - \frac{P}{2GI}z^2 + \frac{df(z)}{dz} \tag{7.29}$$

$$K = -\frac{Ph^2}{8GI}$$

we can re-write (7.27) as

$$G(x) + F(z) = K. \tag{7.30}$$

Since K is a constant and $G(x)$ and $F(z)$ are functions of x and z only, we conclude that $G(x)$ and $F(z)$ must be constant functions, *i.e.*,

$$G(x) = e$$

and

$$F(z) = d$$

where e and d are real constant values. Therefore (7.28) and (7.29) become

$$\begin{aligned}
\frac{dg(x)}{dx} &= -\frac{P}{EI}(\ell x - \frac{x^2}{2}) + e \\
\frac{df(z)}{dz} &= -\frac{\nu P}{2EI}z^2 + \frac{P}{2GI}z^2 + d
\end{aligned}$$

leading to

$$g(x) = -\frac{P}{EI}(\ell\frac{x^2}{2} - \frac{x^3}{6}) + ex + j \tag{7.31}$$

$$f(z) = -\frac{\nu P}{6EI}z^3 + \frac{P}{6GI}z^3 + dz + i \tag{7.32}$$

where j and i are also real constant values. Substituting (7.31) and (7.32) into (7.24) and (7.25) leads to

$$u(x,z) = -\frac{P}{2EI}(x^2 - 2\ell x)z - \frac{\nu}{6EI}Pz^3 + \frac{Pz^3}{6GI} + dz + i \tag{7.33}$$

$$w(x,z) = -\frac{\nu P}{2EI}(\ell - x)z^2 + \frac{P}{6EI}(x^3 - 3\ell x^2) + ex + j. \tag{7.34}$$

Since in the model of Figure 7.4 we are applying force boundary conditions along all boundaries, we need to suppress the rigid body motions while representing as close as possible the displacement boundary conditions for the problem of Figure 7.3. This condition allows us to determine the constants that appear in equations (7.33) and (7.34). We impose

$$u(0,0) = 0$$

and

$$w(0,0) = 0$$

that is, we fix the mid-point at the built-in end. These conditions lead to $i = j = 0$. However, the rotation about this point has still to be prevented and how to enforce this condition is not obvious. Referring to the problem of Figure 7.4, we choose to enforce

$$\frac{\partial u}{\partial z}(0,0) = 0 \tag{7.35}$$

that is we impose that a vertical infinitesimal fiber with origin at the point (0,0) remains vertical after the deformation has taken place. We note that other infinitesimal fibers along the edge $x = 0$ do not remain vertical, as detailed later. Another choice would be to impose $\frac{\partial w}{\partial x}(0,0) = 0$, that is, to enforce that a horizontal infinitesimal fiber, at the built-in end, along the x axis, remains horizontal. Since the distortion $\gamma_{xz}(0,0)$ is different from zero we can not satisfy both conditions simultaneously and, considering the physical problem, to prevent the rotation of the vertical fiber corresponds to an appropriate modeling assumption. We note that to fix the horizontal fiber corresponds to neglecting the effect on the transverse displacements of the distortion at point $(0,0)$. Considering the boundary condition given in equation (7.35) and using equation (7.33), we obtain

$$\frac{\partial u}{\partial z}(0,0) = d = 0$$

and from (7.30) we obtain $e = -\frac{Ph^2}{8GI}$. Hence, the displacements are

$$u(x,z) = -\frac{P}{2EI}(x^2 - 2\ell x)z - \frac{\nu}{6EI}Pz^3 + \frac{Pz^3}{6GI} \quad (7.36)$$

$$w(x,z) = \frac{P}{6EI}(x^3 - 3\ell x^2) - \frac{\nu P}{2EI}(\ell - x)z^2 - \frac{Ph^2}{8GI}x. \quad (7.37)$$

Let us examine the longitudinal displacements at the built-in end, *i.e.*, $u(0,z)$. From equation (7.36)

$$u(0,z) = \frac{(2+\nu)}{6EI}Pz^3 \quad (7.38)$$

and the shear strain at point $(0,0)$ can also be evaluated from equations (7.36) and (7.37)

$$\gamma_{xz}(0,0) = -\frac{Ph^2}{8GI}. \quad (7.39)$$

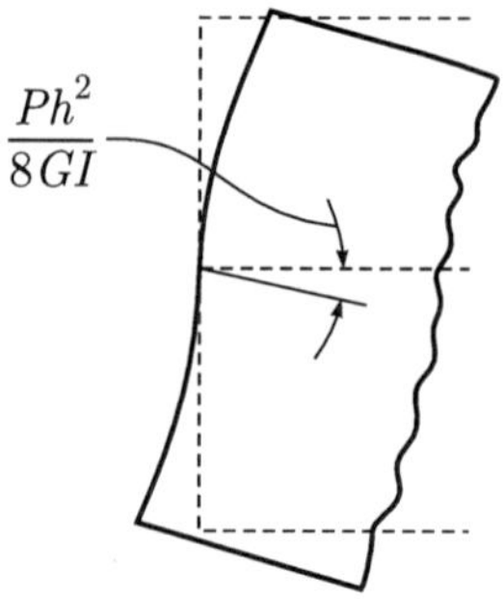

Fig. 7.5. Detail of the magnified displacements at built-in end

In Figure 7.5 we show the deformed and undeformed configurations at the built-in end for the data $E = 70$ GPa, $\nu = 0.33$, $P = 3345$ N, $h = 0.1$ m, $b = 0.01$ m magnified by a factor of 30. We note that the built-in end condition of the physical problem is not exactly satisfied as given by equations (7.17) and (7.18). We also show in Figure 7.5 the shear strain at the point $(0,0)$.

The natural choice for the next mathematical model in our sequence would be the 2-D plane stress model with the boundary conditions at the built-in end enforced exactly, *i.e.*, $u(0,z) = 0$ and $w(0,z) = 0$. However, there is no closed form analytical solution available under these conditions. Hence, we

Table 7.1. Displacement solutions for the first three models in the sequence

Model	u
Bernoulli-Euler	$-\frac{zP}{2EI}(x^2 - 2\ell x)$
Timoshenko	$-\frac{zP}{2EI}(x^2 - 2\ell x)$
Plane stress	$-\frac{zP}{2EI}(x^2 - 2\ell x) + \frac{Pz^3}{6EI}(2+\nu)$

Model	w
Bernoulli-Euler	$\frac{P}{6EI}(x^3 - 3\ell x^2)$
Timoshenko	$\frac{P}{6EI}(x^3 - 3\ell x^2) - \frac{12+11\nu}{10(1+\nu)}\frac{P}{Gbh}x$
Plane stress	$\frac{P}{6EI}(x^3 - 3\ell x^2) - \frac{3}{2}\frac{P}{Gbh}x - \frac{\nu P}{2EI}(\ell - x)z^2$

take recourse to finite element modeling. Before doing so, let us obtain more insight into the problem by exploring the analytical solutions of the three first models of our sequence. We are interested in examining the difference in predictions both qualitatively and quantitatively of these models. In particular, we would like to see the effect on the predicted variables for different ratios between the length and height of the aluminum plate/beam.

In Table 7.1 we summarize the longitudinal and transverse displacement solution for the first three models in our sequence. We note that for the longitudinal displacement $u(x, z)$ both the Bernoulli-Euler and Timoshenko models give exactly the same predictions which correspond to a linear variation along the section height. This is, of course, linked to the kinematic hypothesis used in both models that plane sections remain plane. This linear variation is also predicted by the 2-D model which displays, in addition, a cubic term in z. At the built-in end, this is the only term which gives non-vanishing displacements and it is responsible for the violation of the boundary condition at this end as discussed earlier. In order to have an estimate of the magnitude of the displacement associated with this term when compared with the displacement predicted by the linear term, let us examine the displacement at the tip, *i.e.*, at $x = \ell$. We can write

$$u(\ell, z) = \frac{zP\ell^2}{2EI}\left(1 - \left(\frac{z}{\ell}\right)^2\left(\frac{2+\nu}{3}\right)\right)$$

which shows that the relative importance of the displacement associated with the cubic term with respect to the linear term is decreasing for increasing length to height ratios. In Table 7.2 we give some values for different length to height ratios. These reported values indicate that for usual beams ($10 \leq \ell/h \leq 100$) the contribution of the cubic term is very small.

Considering the transverse displacements, there is a common term to all models which corresponds to the effect of bending. For the Timoshenko and plane stress models, the second term represents the effect of shearing. We note that these shear terms are comparable since they are both linear with

Table 7.2. Displacement ratios for linear and cubic terms ($\nu = 0.33$)

ℓ/h	u_{cubic}/u_{linear} $(x = \ell,\ z = h/2)$
5	7.767×10^{-3}
10	1.942×10^{-3}
100	1.942×10^{-5}

respect to x and taking, for example, $\nu = 0.33$ the coefficients are $1.18\frac{P}{Gbh}$ for the Timoshenko model and $1.5\frac{P}{Gbh}$ for the plane stress model.

A quadratic variation of the transverse displacement through the beam height is predicted by the plane stress model which is linked to Poisson's ratio and corresponds to a zero contribution at points on the beam axis.

It is of interest to compare the relative importance of the bending and shear deformations for the resulting transverse displacements. Hence, we write the expressions for the transverse displacement of the axis as $\bar{w}(x)$ in a single expression for the three models

$$\bar{w}(x) = \frac{2P\ell^3}{Ebh^3}\left[\left(\frac{x}{\ell}\right)^3 - 3\left(\frac{x}{\ell}\right)^2 - k\left(\frac{h}{\ell}\right)^2\left(\frac{x}{\ell}\right)\right] \tag{7.40}$$

where $k = 0$ for the Bernoulli-Euler model, $k = \frac{12+11\nu}{10(1+\nu)}$ for the Timoshenko model and $k = \frac{3(1+\nu)}{2}$ for the plane stress model. It is easy to see that the relative contribution of shear to the transverse displacement is proportional to the ratio $\left(\frac{h}{\ell}\right)^2$. Let us define

$$r_w = \left|\frac{\bar{w} - w_{BE}}{w_{BE}^{\max}}\right| \times 100 \tag{7.41}$$

which gives the percent of the transverse displacement due to shear measured and normalized with respect to the maximum displacement of the Bernoulli-Euler model. In expression (7.41) $\bar{w}$ represents the total displacement of either the Timoshenko or the plane stress model and w_{BE} the transverse displacement of the Bernoulli-Euler model. Simplifying expression (7.41) yields

$$r_w = \frac{k}{2}\left(\frac{h}{\ell}\right)^2 \frac{x}{\ell} \times 100$$

which shows that the percent difference for a given point on the beam axis is proportional to the factor $\frac{k}{2}\left(\frac{h}{\ell}\right)^2$. Since k is independent of the ratio between the height and the length of the beam, the relative importance of shear deformations is given by the factor $\left(\frac{h}{\ell}\right)^2$. In Table 7.3, r_w is shown for $x = \ell$ (beam tip). We note, as expected, that the shear deformation contribution to the total transverse displacement is very small and that even for the relatively large ratio $h/\ell = 0.2$ the increase of transverse displacement is less than 4%.

Table 7.3. r_w for some ℓ/h ratios using $\nu = 0.33$

	$r_w\,(x=\ell)$		
Model	$\ell/h = 5$	$\ell/h = 10$	$\ell/h = 100$
Timoshenko	3.12600	0.78150	0.007815
Plane stress	3.99000	0.99750	0.009975

We recall that the stress distribution of the plane stress model is exactly that of the beam models since in the derivation of the plane stress solution we assumed from the onset that the stress distribution would be that given by the beam models (refer to equations (7.22) and (7.23). Of course, the displacement solutions reported in Table 7.1 lead to these stresses through the compatibility and constitutive equations.

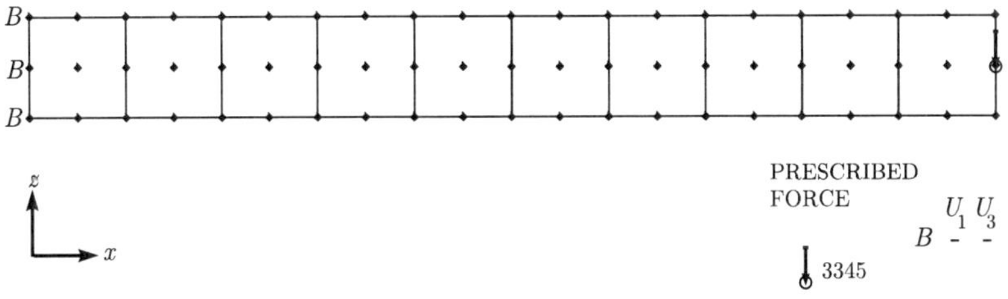

Fig. 7.6. Coarse finite element model (1 × 10), $\ell/h = 10$; $\nu = 0.33$

We return now to study the model described in Figure 7.3 and consider $\ell/h = 10$ and Poisson's ratio $\nu = 0.33$. As mentioned before, we seek the solution of this model using finite elements. Since our objective is to discuss the solution of this mathematical model, we use a mesh refinement strategy to guarantee that our (final) finite element solution is "very" close to the solution of the mathematical model. We start with the model shown in Figure 7.6 in which we use ten 9-node displacement-based plane stress elements and refer to it as the 1 × 10 element model. We consider successive mesh refinements by subdividing each of the elements into four elements to obtain the 2 × 20, 4 × 40 and 8 × 80 element models, and also give the results obtained using an extremely fine mesh of 64 × 640 elements. The last mesh results are really not needed but included to underline the convergence behavior. The 8 × 80 element model is shown in Figure 7.7.

In Figure 7.8, we report the strain energy values of the finite element solutions. We can clearly identify a convergent behavior since the absolute difference in predicted strain energy between two consecutive solutions is decreasing and eventually the difference is small (considering the two finest models).

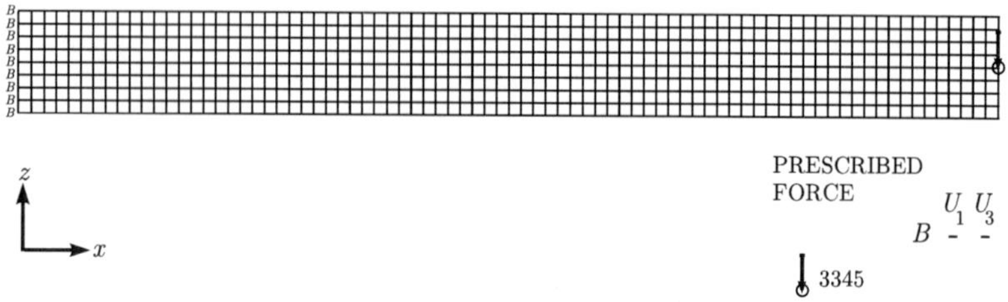

Fig. 7.7. (8 × 80) finite element model

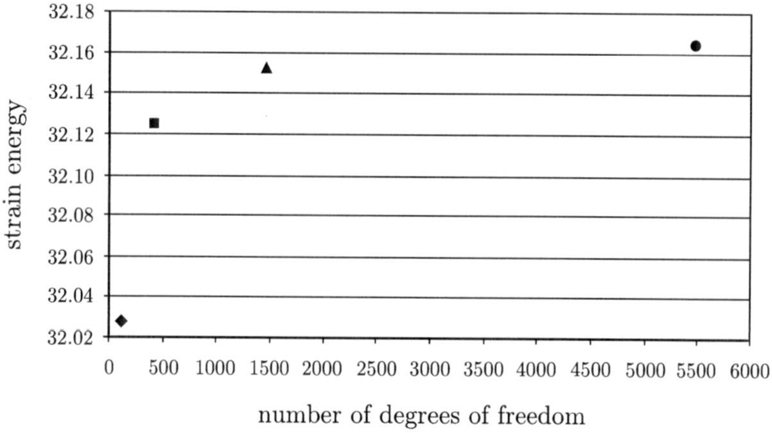

Fig. 7.8. Convergence in strain energy ($N.m$)

In Table 7.4, we report the predictions for the tip transverse displacement for all models. As expected, the convergent behavior of the finite element predictions is also observed. Note that the finite element solutions converge to a slightly smaller tip displacement value than given by the plane stress model studied earlier. The difference is quite small, but in the plane stress model studied analytically, the displacement boundary conditions do not represent exactly the built-in condition leading to a more flexible model.

In order to study the stress predictions, we select two sections: the built-in section and a section located at mid-span, *i.e.*, $x = \ell/2$. Let us first examine the predictions obtained with the 2×20 element model which already gives relatively accurate displacement predictions. The stress predictions[2] are shown in Figures 7.9 to 7.12. We note that the predictions for the mid-span section are reasonably close to those of the solution of models 1 to 3 (Model 1: Bernoulli-Euler beam model, see Section 7.1.1; Model 2: Timoshenko beam model, see Section 7.1.2; Model 3: Plane stress model, see Section 7.1.3).

[2] Here we also use the parabolic shear stress distribution for the Bernoulli and Timoshenko beam elements as the analytical solutions, see Sections 4.2.2 and 4.2.8.

Table 7.4. Transverse displacement at tip for various models (in meters)

Model	w at tip	normalized w.r.t. the plane stress model
Bernoulli-Euler	-0.01911	0.99012
Timoshenko	-0.01926	0.99786
Plane stress	-0.01930	1.00000
FE plane stress (1x10)	-0.01915	0.99195
FE plane stress (2x20)	-0.01921	0.99497
FE plane stress (4x40)	-0.01922	0.99583
FE plane stress (8x80)	-0.01923	0.99620
FE plane stress (64x640)	-0.01924	0.99690

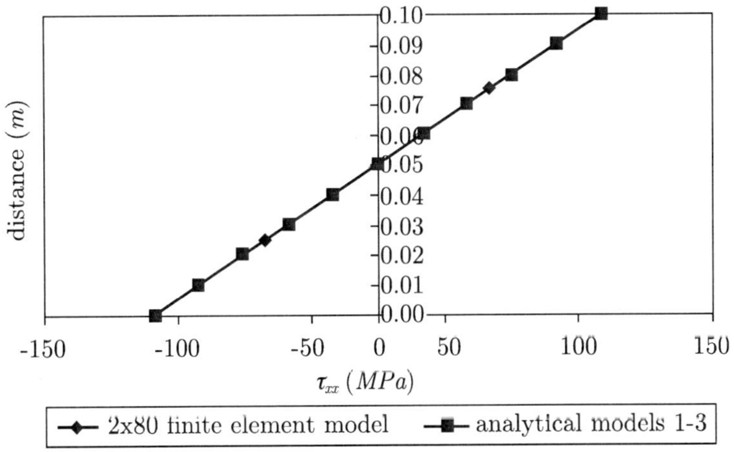

Fig. 7.9. Normal stress predictions at mid-span

However, the normal stress predictions at the built-in end shows some deviation from the solutions of models 1 to 3 and the shear stress predictions is completely different.

We report in Figures 7.13 to 7.16 the analogous results obtained with the 8×80 element model. The same trend as observed for the 2×20 element model can be seen. Now, however, the mid-span stress predictions agree very well with those of models 1 to 3 and the difference in predictions for the stresses at the built-in end are even more pronounced. If we denote by A the top point at the built-in section, *i.e.*, $(x = 0,\ z = h/2)$, we note that the normal stress prediction for this point is much higher than that obtained with models 1 to 3. Regarding the shear stress at the built-in section, we note that although the shape of the shear stress curve is similar to that obtained with

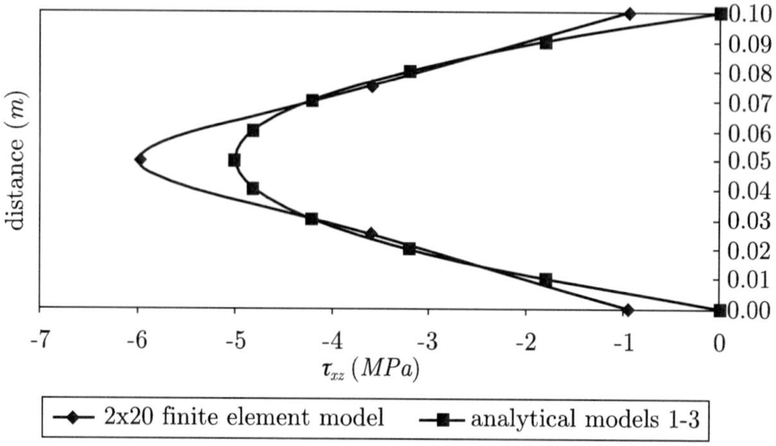

Fig. 7.10. Shear stress predictions at mid-span

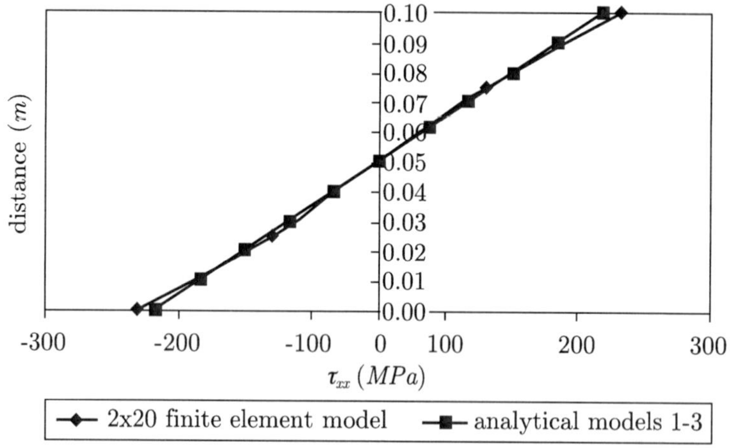

Fig. 7.11. Normal stress predictions at built-in section

the 2×20 element model, the values are substantially different, especially at point A.

The results reported indicate that as we refine the mesh the stresses increase in the vicinity of point A. In fact, for the boundary conditions described in Figure 7.3, it can be shown that the stresses which correspond to the exact solution of the mathematical model are singular at point A, *i.e.*, they tend to infinity (see Timoshenko and Goodier, 1970). Considering the hierarchical process, it is crucial to realize that this model is only one model in our hierarchical sequence and our primary objective is to arrive at good predictions for the actual physical problem at hand. Therefore it is essential at this point to use the concept of the very-comprehensive mathematical model introduced

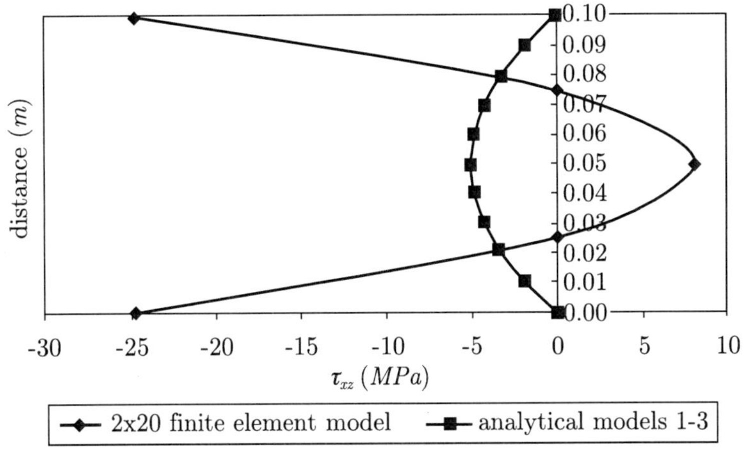

Fig. 7.12. Shear stress predictions at built-in section

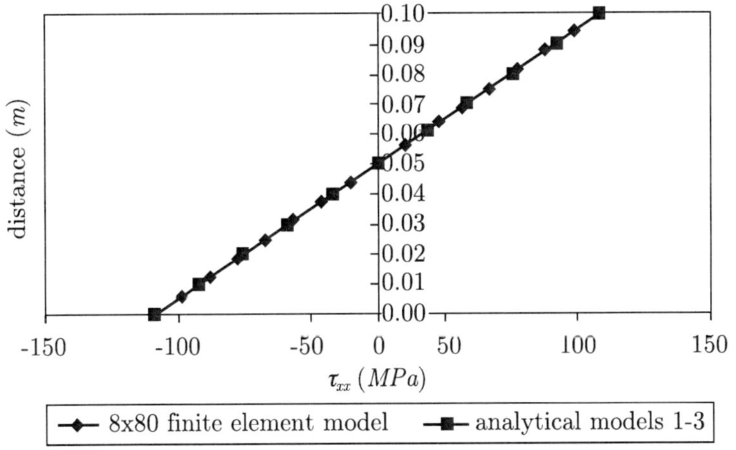

Fig. 7.13. Normal stress predictions at mid-span

in Chapter 1. Of course, in such a model the stresses would not tend to infinity as predicted by the elasticity model for the problem in Figure 7.3, since in the very-comprehensive mathematical model inelastic deformations would occur as the stresses exceed some limit values. Moreover, the sharp corner at point A is "not really sharp" in the physical problem and in the very-comprehensive mathematical model this fact would be represented by some geometric rounding of the corner. Another consideration that should be reflected in the very-comprehensive mathematical model is the fact that the support is "not really rigid" and the boundary conditions shown in Figure 7.3 are not really representative of what happens in the physical problem. In order to improve the model in this respect, a portion of the adjoining domain

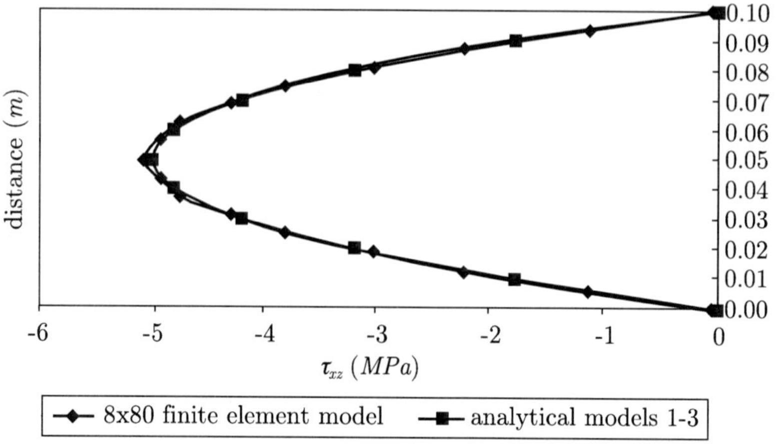

Fig. 7.14. Shear stress predictions at mid-span

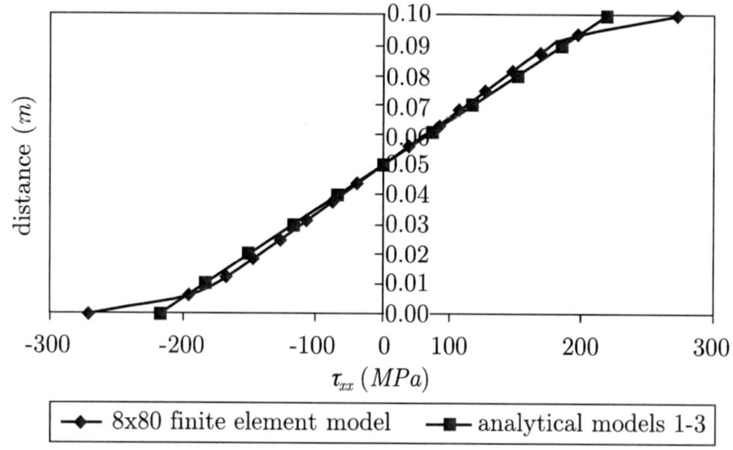

Fig. 7.15. Normal stress predictions at built-in section

at the built-in end should be included in the model representing the flexibility of the material in that region.

In light of the above discussion, we introduce two additional models in our hierarchical sequence. First we consider some rounding of the corners at the built-in end and next we model the adjoining region to the built-in end, which is made of steel. We could also introduce elasto-plastic effects. However, we choose not to do so here since nonlinear formulations are discussed only in the next chapter.

In Figure 7.17, the geometric model with the rounding is described. We note that the rounding is a very small feature and it was introduced adding material in such way that point A, shown in the detail of Figure 7.17, is

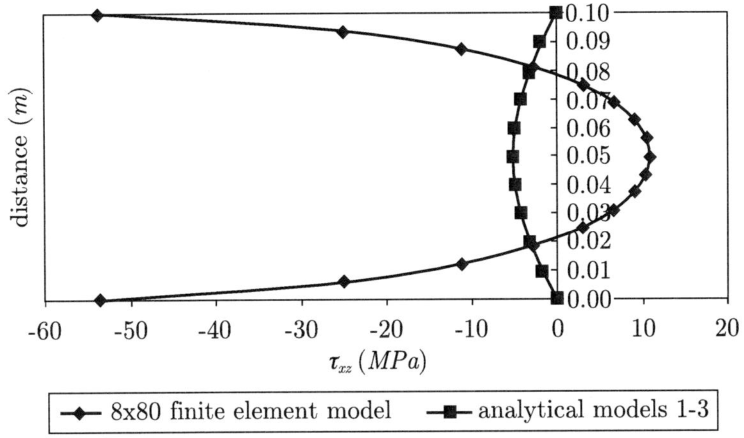

Fig. 7.16. Shear stress predictions at built-in section

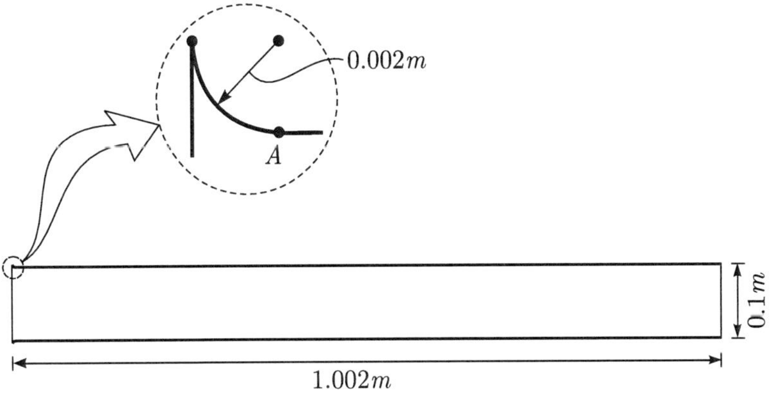

Fig. 7.17. Geometric model with the rounding

at the same geometric location as before, *i.e.*, $x = 0$, $z = h/2$. We study the stresses at the section defined by point A and since the stress resultants at this section are the same as those of the built-in section of the model described in Figure 7.3 comparisons are allowed. Since we expect to have removed the singularity at point A due to the rounding, the solution to the mathematical model should display a vanishing shear stress at point A due to the horizontal free edge. However, to predict this vanishing shear stress along free edge an extremely fine mesh needs to be used because of the very small radius of curvature of the feature.

In Figures 7.18 and 7.19 we report the normal and shear stress predictions respectively. We can see that the shear stress drops back to zero as point A is approached. In Figure 7.20, a detail of the variation of the shear stress in the vicinity of point A is shown. We can see the high stress gradients.

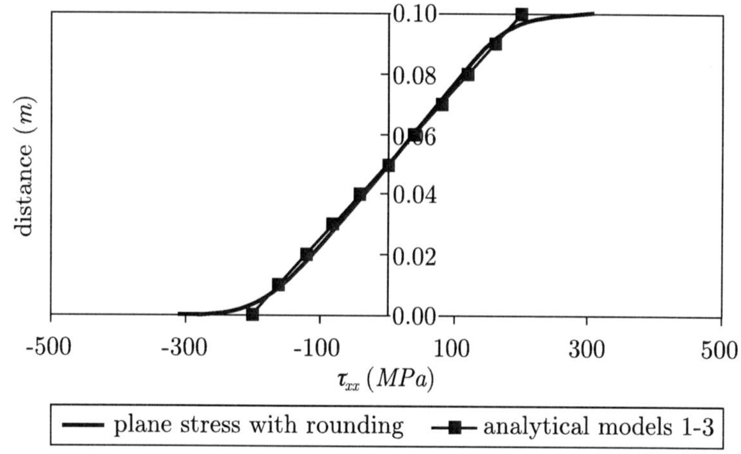

Fig. 7.18. Normal stress predictions at section that passes through point A

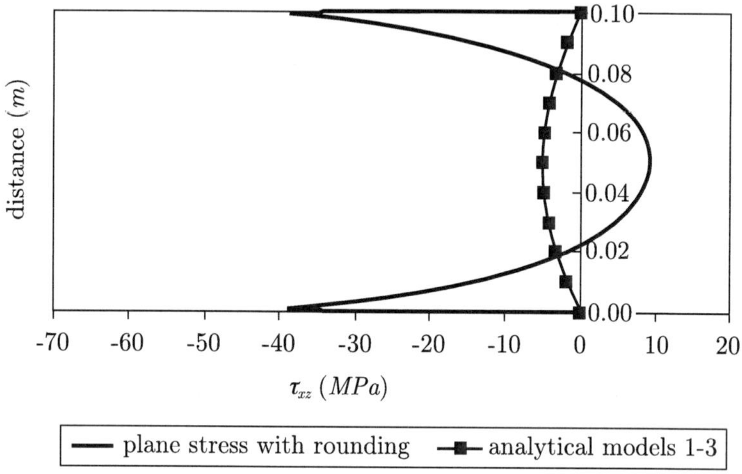

Fig. 7.19. Shear stress predictions at section that passes through point A

As a final model improvement, still in the context of linear analysis, we introduce in the model the portion made of steel. In Figure 7.21, the geometric model is summarized. We note that we have also introduced the rounding in this model. In Figures 7.22, 7.23, 7.24 we show the stress curves predicted at the section which contains point A analogous to those reported for the previous model. We see that the peak values have increased. This increase in stress may be a surprise, and is due to the fact that the aluminum material now punches into the steel support. However, the overall shapes of these curves are the same as before, retaining the qualitative behavior of the stress predictions at this section.

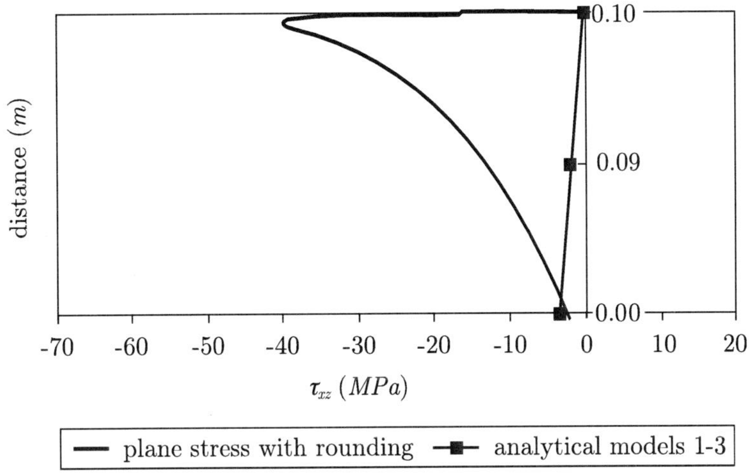

Fig. 7.20. Detail of shear stress predictions at section that passes through point A

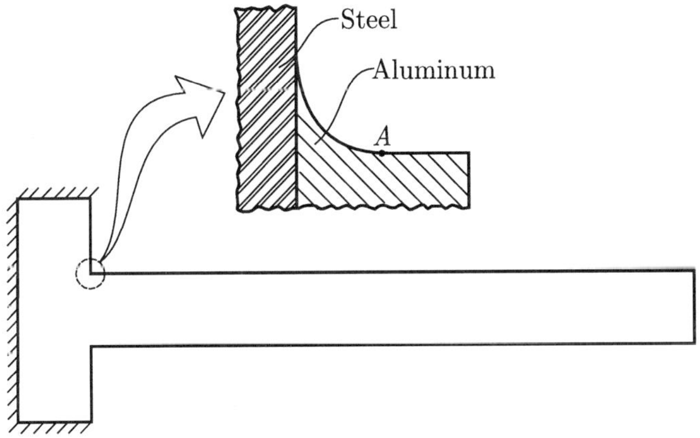

Fig. 7.21. Geometric model which includes the steel portion. Material properties of steel are $E = 210$ GPa and $\nu = 0.3$

We might say that this last model captures all the basic phenomena which can be represented in linear analysis. Of course, some improvement could still be obtained by introducing an elasto-plastic constitutive law for the materials involved. Since the yield stress of the aluminum considered is $\tau_y = 223\ MPa$ we expect a decrease in the stress prediction near the vicinity of point A. Nevertheless, such modifications should not qualitatively change the distribution of the stresses at the section which contains point A. Therefore, the stress distributions shown in Figures 7.22 to 7.24 should be relatively close to those of the very-comprehensive mathematical model. Note that the shear stress distribution at the section which contains point A is qualitatively

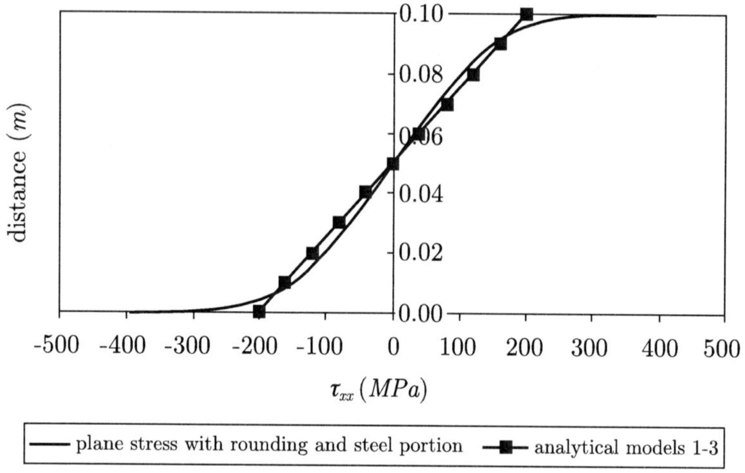

Fig. 7.22. Normal stress predictions at section that passes through point A

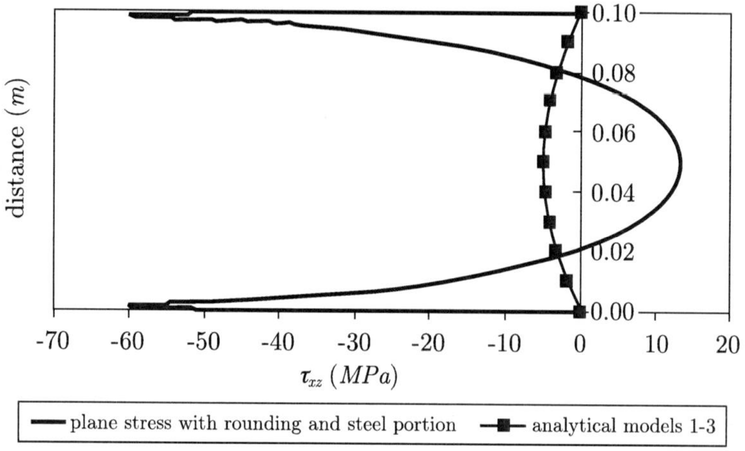

Fig. 7.23. Shear stress predictions at section that passes through point A

(and hence also quantitatively) completely different from the predictions of models 1 to 3 which are routinely used for such kind of analyses. This result illustrates the importance of hierarchical modeling for this case if accurate stress predictions are required for this section.

7.2 Machine Tool Jig

In Figure 7.25 the machine tool jig to be analyzed is schematically shown. We assume that the displacement boundary conditions and the loading of the

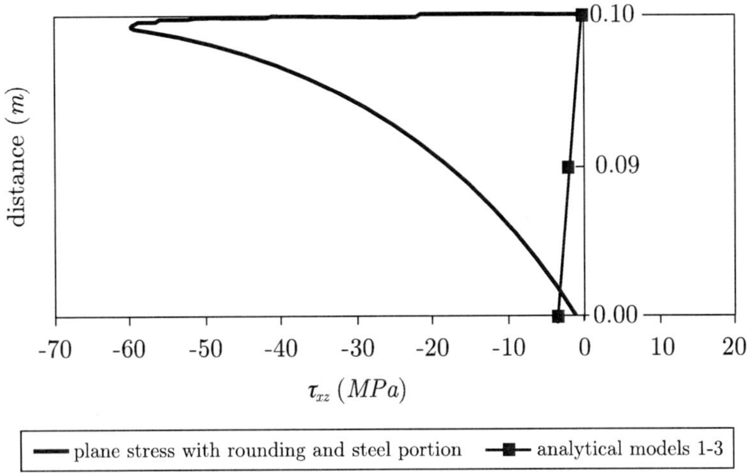

Fig. 7.24. Detail of shear stress predictions at section that passes through point *A*

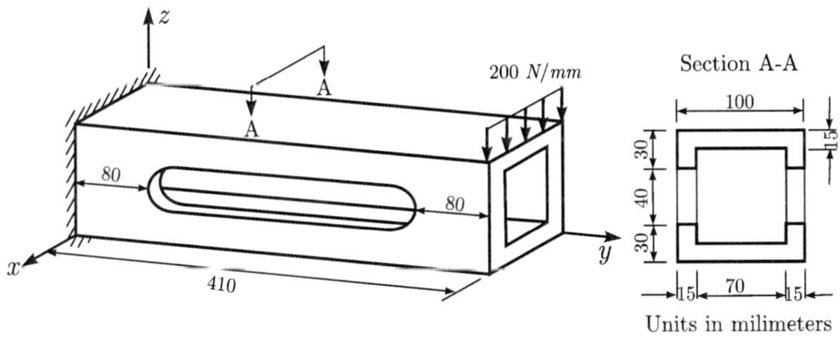

Fig. 7.25. Machine Tool Jig problem description

physical problem are accurately modeled by the idealization represented in Figure 7.25. Therefore, we concentrate on the structural modeling of this idealization.

The variables to be predicted are the transverse displacement and the stresses. These variables are chosen, since there are limit values for the transverse displacement, due to operational requirements, and for the stresses, to guarantee structural integrity.

Our objectives are to obtain a qualitative understanding of the structural behavior of the part and to investigate which models would predict the transverse displacement with a precision of around 15 % and the stresses within a margin of 20 % error.

Our sequence of hierarchical models is:

1) Model 1: Beam model

2) Model 2: Shell model
3) Model 3: Three-dimensional elasticity model.

We detail below the models used.

7.2.1 Beam Model

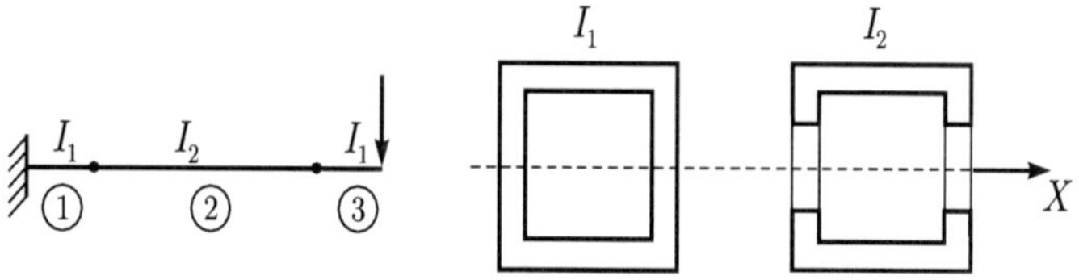

Fig. 7.26. Uniaxial beam model, in each case the moment of inertia is, as in Section 4.2.2, taken about the X-axis

The simplest beam model is shown in Figure 7.26 which is obtained by considering Bernoulli-Euler beam elements with different cross-sections. This model is included here with the only purpose of showing, in a very simple context, an example of bad modeling. The mid-beam element, represented in Figure 7.26 by number 2, does not represent the basic mechanical behavior of the part in this region. The reason is that due to the presence of the side openings, there is no physical connection between the top and bottom parts of any section in this region. Therefore, these top and bottom parts do not behave as geometric subdomains of the same beam section, and the fundamental hypothesis of the Bernoulli-Euler beam theory, that beam sections remain plane (with the shear force resisted on a connected section) is violated. Some results obtained with this model are shown later to give an idea of how far off these solutions may be when compared to solutions obtained with properly selected models.

The beam model described in Figure 7.27 should be able to capture the qualitative behavior of the part. In this model, the top and bottom parts of the tool in the opening region are modeled with separate beams, represented by elements 3 and 4, with sections shown also in Figure 7.27. The rigid links as discussed in Section 6.6.3 can be understood as constraints which enforce that the rotations of nodes 1, 3 and 5 should be the same as well as those of nodes 2, 4 and 6. This modeling assumption is justified since the parts of the model which have a box type section, represented by elements 1 and 2, do behave approximately as beam sections having moment of inertia I_1. Note that these beam elements with box type section are quite rigid when compared to the beam elements of the open region. Therefore, it is a consistent modeling hypothesis to assume that, at the geometric section where elements 3 and 4 meet with element 1 at the left side, and with element 2 at the right side,

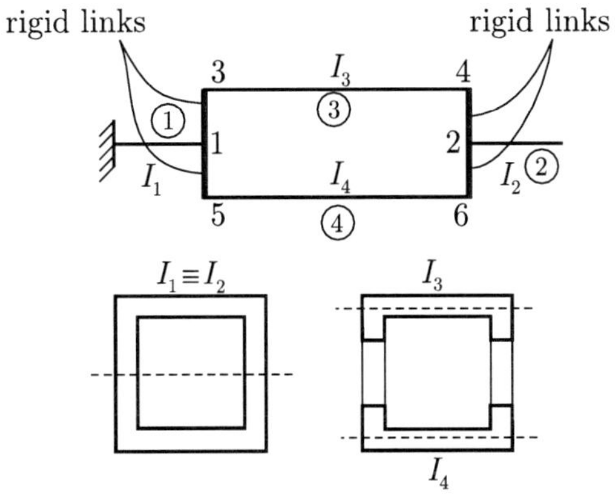

Fig. 7.27. Beam model with rigid links

the rotations are the same since the box type sections remain plane during deformations.

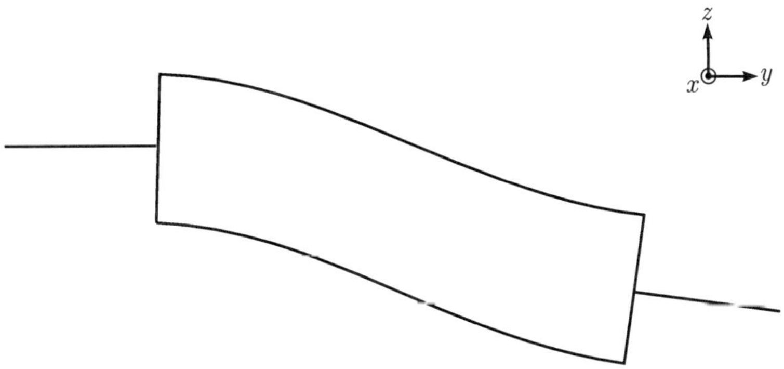

Fig. 7.28. Magnified deformed configuration of beam model with rigid links

To obtain more insight into the behavior of this model we show in Figure 7.28 a magnified deformed configuration representing only the bar axes. We can easily recognize that most of the deformation occurs in the central part of the model which behaves as a frame that is being sheared and subjected to an external moment. In Figure 7.29 the bending moment and axial force diagrams are shown confirming this frame type behavior. It is interesting to contrast the behavior of this beam model with that described in Figure 7.26 which is deemed to represent a "bad" modeling. For this purpose, we select a section of the model as shown in Figure 7.30 and compare qualitatively the internal forces and stresses. The distribution of normal stresses shows totally different predictions. While in the model presented in Figure 7.30b, the top and bottom parts bend individually, the model of Figure 7.30a predicts that the section bends as a whole. Consequently, the stress distributions, schematically shown in Figure 7.30 for section $B - B$, reflect these distinct behaviors leading to totally different predictions.

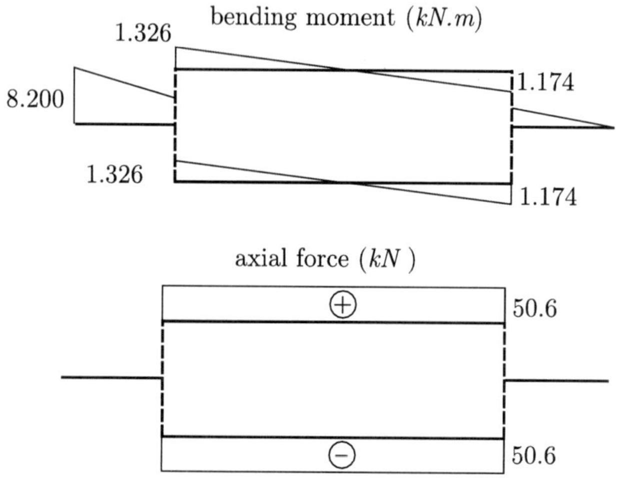

Fig. 7.29. Bending moment and normal force diagrams for beam model with rigid links

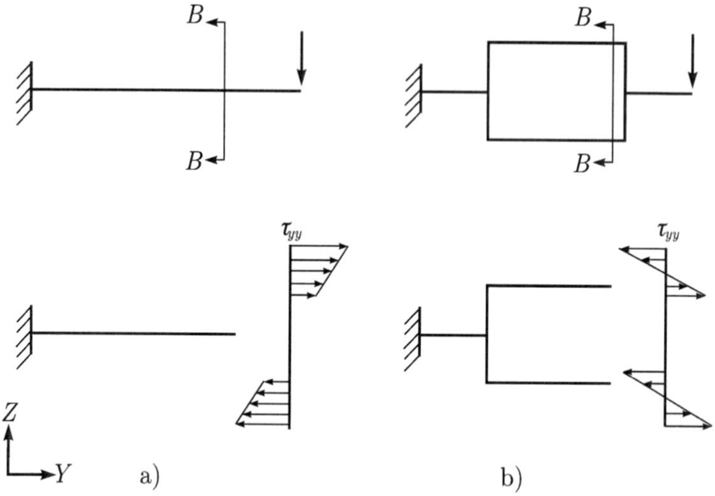

Fig. 7.30. Schematic longitudinal normal stress distributions for beam models. a) Model of Figure 7.26; b) Model of Figure 7.27

7.2.2 The Shell Model

This model represents the part as a shell structure. We note that the thickness/length ratio of this shell model is given by 15/100 = 15% if we take the width to be the representative length size. Therefore, if the behavior is to be represented by a shell structure, we need to consider at least a moderately thick shell model. Even using such a model, if we look at the corners of the open regions, as indicated in Figure 7.31, and consider the actual dimensions of the problem, we might expect the behavior to be mostly three-dimensional in these regions. This fact partially justifies the use of model 3 of the sequence. Nevertheless, the shell model should provide a better description of the behavior of the open regions (the rounded part) than the beam model and predict spatial variations of the variables along the width direction which are, of course, not predicted by the beam model.

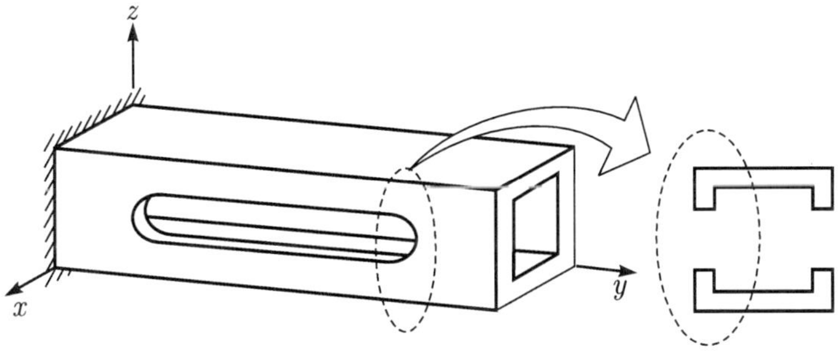

Fig. 7.31. Detail near end of open region

The solution of the shell model is obtained in a finite element analysis. The elements used are solving the (mathematical) basic shell model (see Section 4.4.2) and, therefore, the predictions include the effects of shear deformations. A sufficiently fine mesh, shown in Figure 7.32, of eight-node elements is used to ensure that the solution obtained is close to the solution of the mathematical model. We present the results of this analysis in Sections 7.2.4 and 7.2.5.

7.2.3 Three-Dimensional Elasticity Model

A motivation for the use of this model was already given when presenting the shell model, *i.e.*, the structural behavior of the part is, *e.g* in the corner regions, of a three-dimensional nature due to the geometric and loading conditions. Therefore, an enhancement in predicted behavior is expected when considering a three-dimensional model. Again, a sufficiently fine mesh, now of twenty-node hexahedral elements, is considered as shown in Figure 7.33.

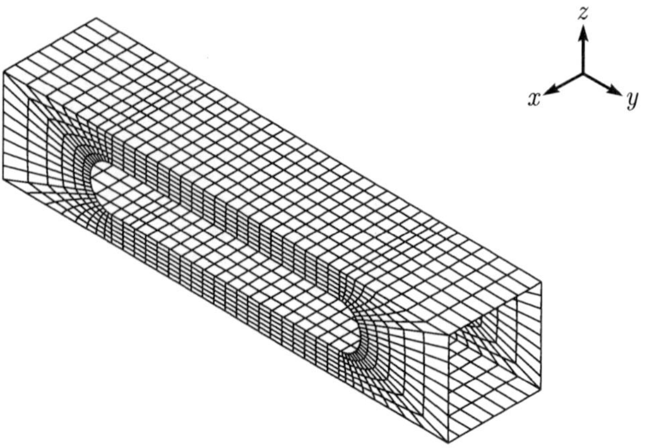

Fig. 7.32. Finite element mesh for the shell model

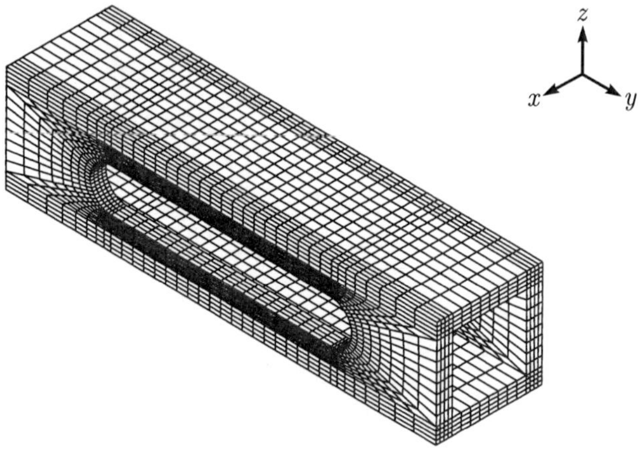

Fig. 7.33. Finite element mesh for 3-D model

A magnified deformed configuration of the 3-D model is shown in Figure 7.34. We observe that the gross behavior is consistent with that of the beam model with rigid links.

7.2.4 Qualitative Analysis

The qualitative behavior of the displacements was, to some extent, already examined. Both end parts are very rigid and behave as box sections. The central part – the open region – has a frame type behavior in shear and the top and bottom parts behave as beams.

In Figure 7.35, we select some lines of the part to examine the predicted results. In Figure 7.36, we show the transverse displacement, *i.e.*, the dis-

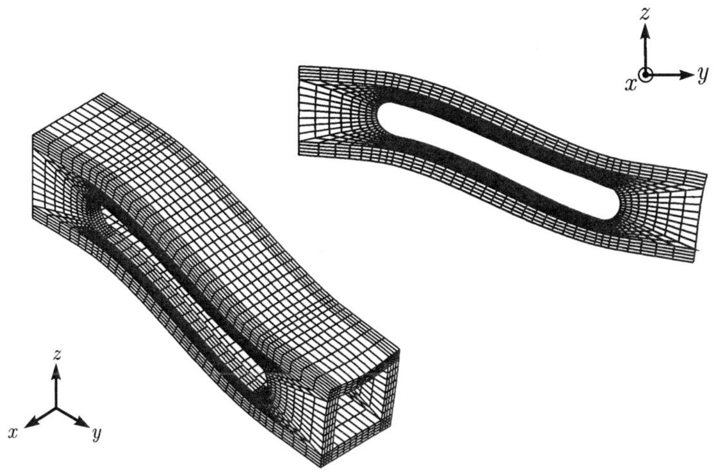

Fig. 7.34. Magnified deformed configuration for the 3-D model

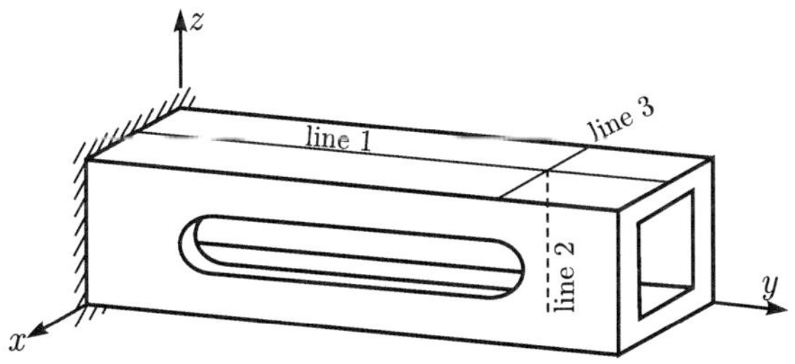

Fig. 7.35. Selected model lines

placement in the z-direction, along line 1. We note that the gross behavior is captured by the three valid models. Naturally, the beam model with rigid links is the most stiff while the 3-D model is the least stiff. As might be expected, or may be a surprise if Figure 7.30 has not been studied, the beam model of Figure 7.26 does not capture the behavior at all, being much too stiff. In Figure 7.37 we show the stress τ_{yy}, which is the normal stress associated with bending, along line 1. We note that in the central open region $120 < y < 270$ the three valid models lead to quite close predictions. However, the behavior is complex and the peak stresses are not well predicted by the beam model with rigid links. We also include the results obtained with the beam model of Figure 7.26. As now expected considering Figure 7.36, this model completely misses the behavior of the tool jig with respect to stresses.

The stresses along line 2 are shown in Figure 7.38. This figure reveals that the qualitative behavior of the top and bottom parts is predicted by

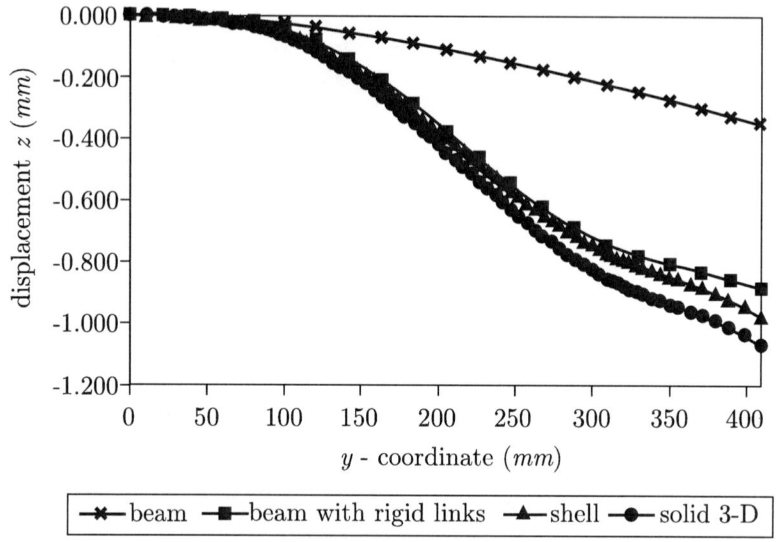

Fig. 7.36. Transverse displacement along line 1

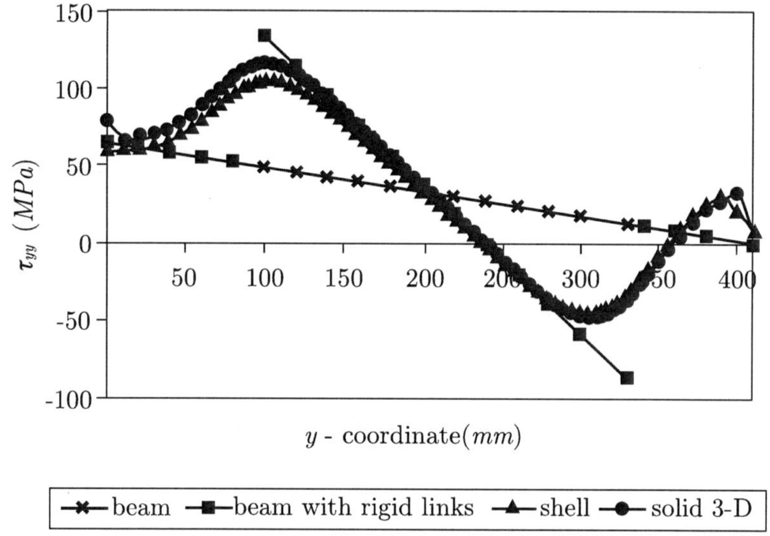

Fig. 7.37. Longitudinal normal stress along line 1

all models except by the beam model of Figure 7.26 which leads to wrong predictions.

The results along line 3 are reported in Figure 7.39. Of course the beam models predict constant stress values along this line. The 3-D and shell models can predict the variation along the width. We point out that the results

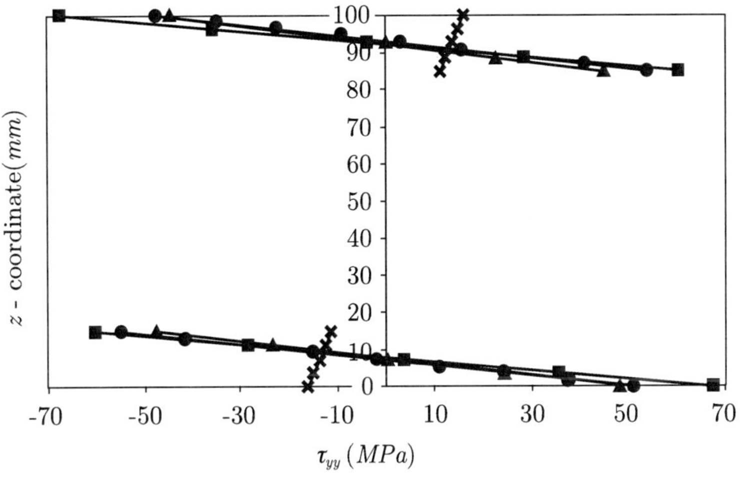

Fig. 7.38. Longitudinal normal stress along line 2

Table 7.5. Transverse displacement at selected points (in mm)

Model	P1	P2	P3
Beam	-0.355	-0.227	-0.110
Beam with rigid links	-0.884	-0.747	-0.375
Shell	-0.976	-0.772	-0.400
Solid 3-D	-1.067	-0.853	-0.442

reported for the shell model do not extend to the lateral faces of the part, since, at the corner, the shell node is located at the midsurface implying that the geometric domain in the corner region is not fully represented in the shell model.

7.2.5 Quantitative Analysis

We do not perform a comprehensive quantitative analysis. Our purpose is to only show the numerical results at some selected points, defined in Figure 7.40, to obtain an estimate of the accuracy of the results.

In Table 7.5 we report the transverse displacement results and in Table 7.6 the same results are shown normalized with respect to the results of the 3-D model, our highest order model. We note that the hierarchical order of the selected models is confirmed by the displacement predictions, *i.e.*, the beam model with rigid links displays an "error" of around 15% and the shell

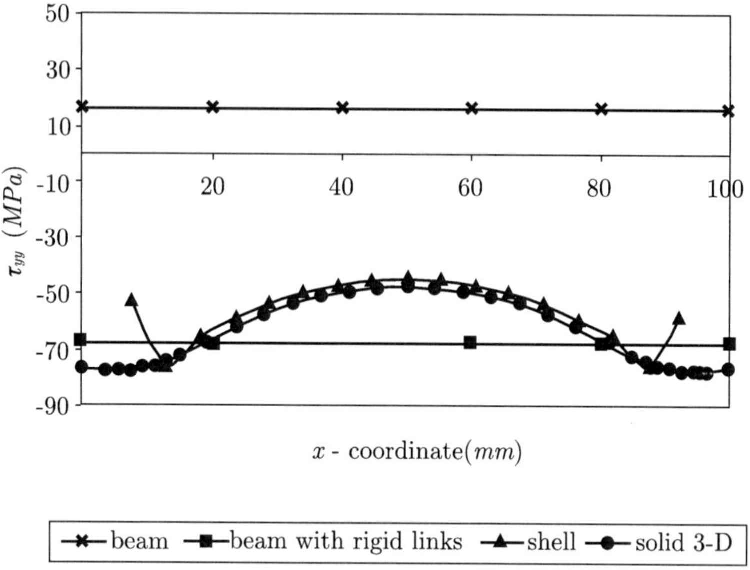

Fig. 7.39. Longitudinal normal stress along line 3

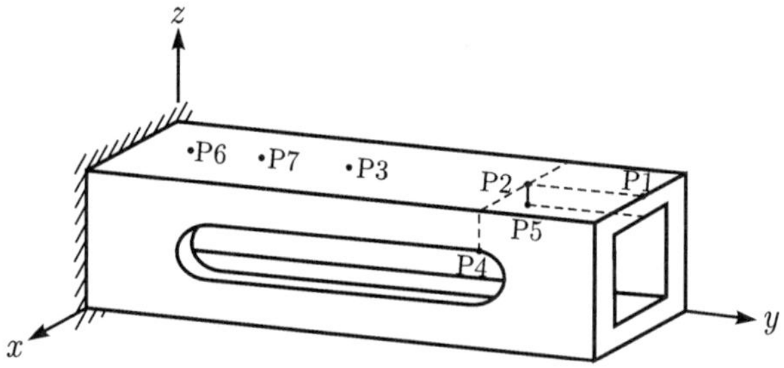

Fig. 7.40. Selected model points

model an error of about 10% when compared with the 3-D model results.

Table 7.6. Normalized transverse displacement at selected points

Model	P1	P2	P3
Beam	0.333	0.266	0.249
Beam with rigid links	0.828	0.876	0.848
Shell	0.915	0.905	0.905
Solid 3-D	1.000	1.000	1.000

Table 7.7. Longitudinal normal stress at selected points (in MPa)

Model	P2	P4	P5	P6	P7
Beam	16.20	6.48	11.34	61.40	46.82
Beam with rigid links	-67.35	187.97	60.31	61.40	113.66
Shell	-44.48	190.76	45.02	61.37	99.52
Solid 3-D	-47.54	206.12	53.94	69.19	108.98

Table 7.8. Normalized longitudinal normal stress at selected points

Model	P2	P4	P5	P6	P7
Beam	-0.341	0.031	0.210	0.887	0.430
Beam with rigid links	1.417	0.912	1.118	0.887	1.043
Shell	0.936	0.925	0.835	0.887	0.913
Solid 3-D	1.000	1.000	1.000	1.000	1.000

The uniaxial beam model of Figure 7.26, due to the wrong representation of the kinematics, is much too stiff.

In Table 7.7 we show the stress τ_{yy} at selected points and in Table 7.8 the same predictions are reported normalized with respect to the 3-D model.

The analysis of the results for the stresses is more intricate, since the quality of the predictions for the same model may significantly vary from region to region. For example, point P7 falls in the central region for which the beam model with rigid links captures the behavior of the structure quite well, leading to a quite accurate prediction. As discussed earlier, points P2, P4 and P5 are located in a region of complex behavior for which we should not expect accurate predictions from the beam model with rigid links. In fact, the point P2 prediction by this model is off by around 40% with respect to the 3-D model. We note that this result is also seen in Figure 7.39, and is due to the variation of the stress along the width direction.

Of course, the structure being analyzed is statically determinate. Therefore, the resultant forces that are transmitted by any section are the same for all models considered. This is the reason why the stress prediction for point P6 is relatively accurate even for the uniaxial beam model: since at the section of point P6 the structure behaves as a beam with a box section, and since the stress resultants are exactly predicted we obtain reasonable predictions for the stresses. The situation would be quite different if the structure were supported at a point (e.g. $y = 410$) preventing the transverse displacement at this point. In such a case, the structure would be statically indeterminate and the accuracy of the resultant forces would depend on the accuracy of the displacement predictions as well.

We can summarize the conclusions regarding the analysis of the structural idealization of Figure 7.25 as follows:

- The beam model with rigid links, the shell and 3-D models capture the qualitative behavior of the structure for displacements and stresses. The uniaxial beam model fails to represent the behavior in the open region and should not be used.
- The beam model with rigid links, being the simplest model that properly captures the behavior, could be used to understand the behavior of the structure when performing parametric studies involving variations of the geometric characteristics.
- Only the shell and 3-D models gave predictions that have the required accuracy (15% for displacements and 20% for stresses). Here it is implicitly assumed, as a modeling consideration, that the predictions of the 3-D elasticity model are very close to the predictions of the very-comprehensive model for the regions of interest. Of course, this is the case only if the loading is limited to a value which leads to stresses that are well below some limit values, namely those which would cause yielding of the material and/or lead to geometric nonlinear behavior.

7.3 Modeling of a carabiner

The modeling of the carabiner was discussed in Chapter 1 to introduce the main concepts of the hierarchical modeling process. The description of the physical problem is given in Section 1.2.1 and is not repeated here. As in Section 7.2, our objective is to calculate the displacements and stresses to an accuracy of about 15% and 20%, respectively. Referring to Figure 1.10 we consider in this section the following hierarchical models:

1) Model 1: Straight bar model
2) Model 2: Curved bar model
3) Model 3: Three-dimensional elasticity model.

7.3.1 Straight bar model

The model is constructed using straight Bernoulli-Euler beam elements. The end-points of each beam element are placed on the axis of the geometric model of the carabiner. An elliptical cross-section is chosen to describe the part. In Figure 7.41 we show the model where the diamond symbols indicate the end-points of the elements. We see that in the curved regions many elements were chosen to properly describe the geometry.

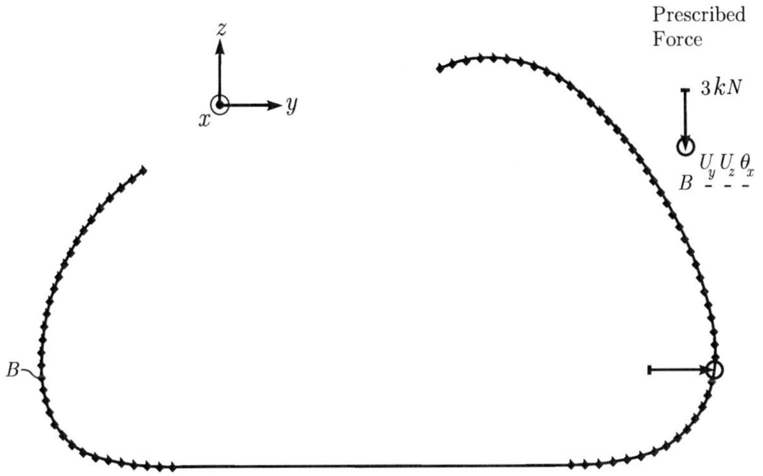

Fig. 7.41. Finite element mesh for beam models. The material of the carabiner is aluminum with $E = 70$ GPa, $\nu = 0.33$

7.3.2 Curved bar model

For this model we use general curved 3-node beam elements (See Section 6.4.4). We recall that in this beam element formulation the geometry of the element is interpolated from the cross-sections at the nodes, and therefore the mesh represents, in an approximate manner, the curved geometry. The discretization used is defined by the nodal positions shown in Figure 7.41, *i.e.*, both beam models have the same nodes but, of course, not the same elements since in this model we have 3-node elements. Note that these curved beam elements can model thick beams and include the effect of shear deformations.

7.3.3 Three-dimensional elasticity model

The three-dimensional finite element model is constructed from the solid model shown in Figure 7.42a. We note that this solid model has a number of features such as the hole for the pin and letter imprints, which, if considered in the finite element model, would render the model very large, adding little relevant information for the structural behavior of the part. Therefore, a defeaturing of the geometric model was undertaken leading to the geometry shown in Figure 7.42b (see Section 6.8). We note that the geometric model shown in Figure 7.42b was constructed for the finite element discretization procedure, *i.e.*, only half of the geometry is considered due to symmetry conditions and surfaces are conveniently created for the application of the boundary conditions which are directly imposed on the geometric model, as shown in Figure 7.43. An automatically generated and sufficiently fine mesh is displayed in Figure 7.44. We use this mesh, instead of the mesh in Figure 6.84, because such tetrahedral element meshes are typically used in practice.

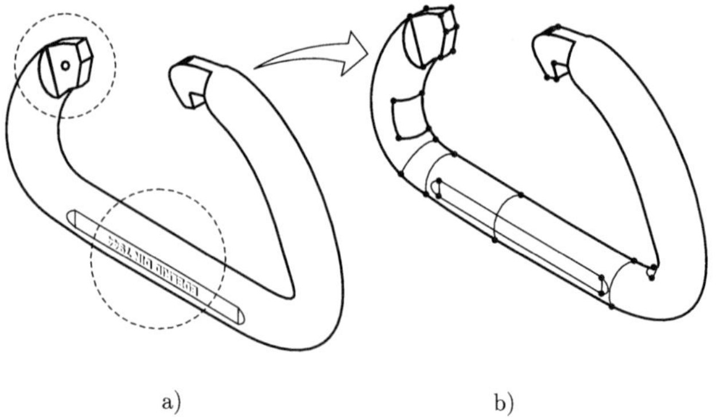

Fig. 7.42. Defeaturing process

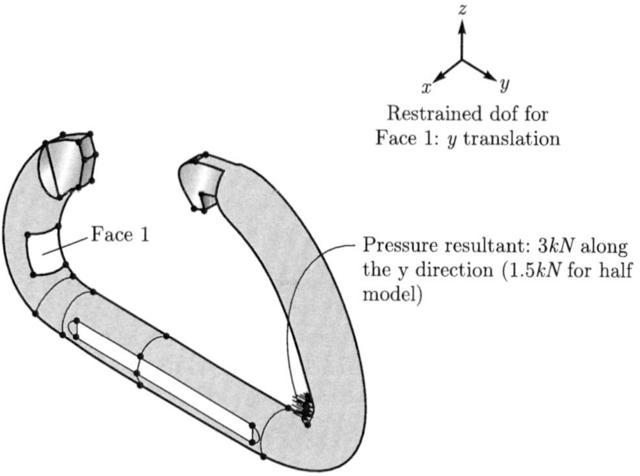

Fig. 7.43. Geometric model prepared for finite element mesh construction

7.3.4 Qualitative analysis

The qualitative behavior of this structural part is very simple and it should be captured by all models. Both, displacements and stresses are mainly due to bending actions. In order to obtain insight into this behavior, let us refer to Figure 7.45 where we schematically show the moment diagram that we should expect for a bar representation of the part. Due to the statically determinate nature of the problem, the moment distribution is that reported in Figure 7.45 for all models considered. Hence, the induced displacement and stress patterns should be qualitatively similar to each other for the models studied since they are primarily due to bending. In other words, for a given section, the bending moment, which is equal for all models, is transferred by normal stresses. However, considering the bending moment stresses

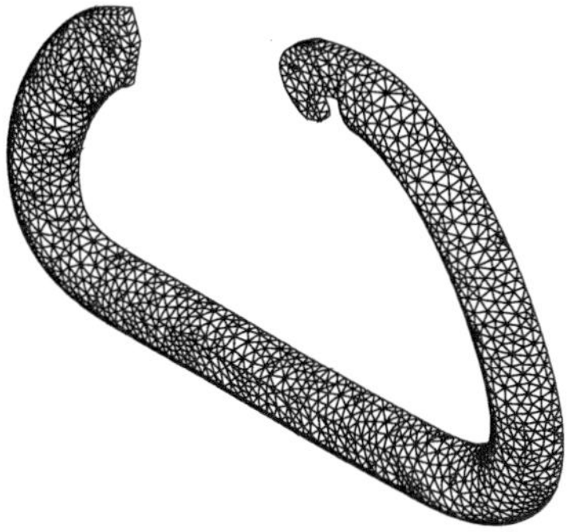

Fig. 7.44. Finite element mesh for 3-D model

more closely, in the regions where the geometric axis of the part is straight we expect a linear stress distribution through the thickness for all models. But in the regions where the axis is curved we *only* have a linear normal stress distribution for the Bernoulli-Euler beam model, since each element is straight, and that distribution irrespective of how many straight elements are used to represent a curved region. Hence a surprise may be that convergence to the stresses of a strongly curved thick beam can never be reached with this model. Considering the other two models, the normal stress distribution will deviate significantly from linearity. For curved bar models in which the height of the beam section is large when compared with the radius of curvature of the axis, the normal stress distribution is hyperbolic (see Section 4.2.7). Therefore, significant differences in predictions are primarily expected in the curved regions of the structure, and for the peak normal stress.

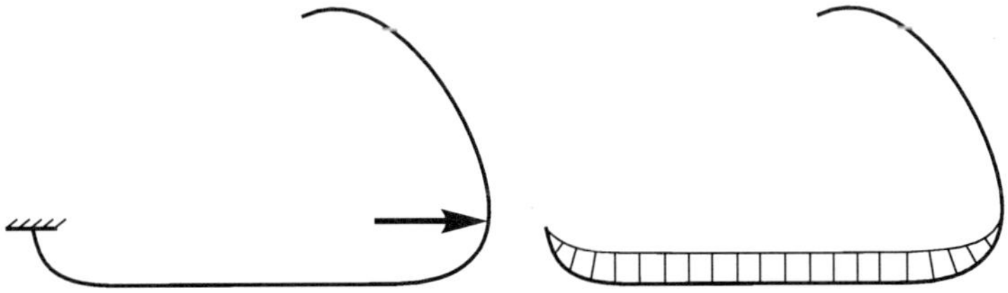

Fig. 7.45. Schematic bending moment diagram

7.3.5 Quantitative analysis

In Figure 7.46 we show some selected points for which numerical predictions are reported below. In Table 7.9 the displacements in the Y direction for points P_1 and P_2 are shown. The values are normalized with respect to

Table 7.9. Displacement in the y direction at selected points

	Displacement y (mm)		Normalized displacement y	
Model	P1	P2	P1	P2
Straight bar	2.976	0.967	0.921	0.949
Curved bar	2.952	0.971	0.914	0.953
3-D solid	3.230	1.019	1.000	1.000

Table 7.10. Displacement in the z direction at selected points

	Displacement z (mm)			Normalized displacement z		
Model	P1	P2	P3	P1	P2	P3
Straight bar	-0.881	-2.530	-1.712	0.936	0.912	0.926
Curved bar	-0.900	-2.527	-1.704	0.956	0.911	0.922
3-D solid	-0.941	-2.775	-1.848	1.000	1.000	1.000

the 3-D model. The same kind of results are reported in Table 7.10 for the displacements in the z direction for points P_1, P_2 and P_3.

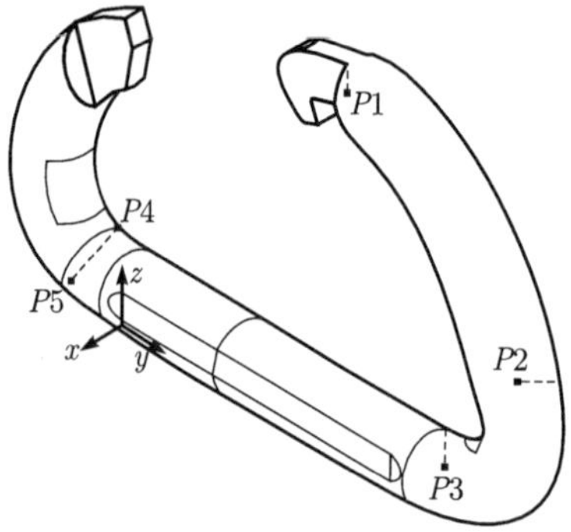

Fig. 7.46. Selected model points

Considering that the displacements at P_1, P_2 and P_3 are representative of the overall displacement pattern of the part, we might say that the three models give relatively close displacement predictions. We can see that the beam models give very close predictions which differ, at most, by about 2%. The 3-D model gives, as expected, a more flexible behavior. The difference in prediction when compared to the beam models is relatively small falling in the range of 10%. Hence both beam models, reported upon in the tables, give results within the desired margin of error.

Let us examine the normal stress predictions for points P_4 and P_5 which are extreme points of a section in a curved region. These results are reported in Table 7.11. We see that the predictions obtained with the straight beam

Table 7.11. Normal stresses at selected points

	Normal stress (MPa)		Normalized normal stress	
Model	**P4**	**P5**	**P4**	**P5**
Straight bar	230.75	-181.30	0.569	1.417
Curved bar	402.78	-105.29	0.992	0.823
3-D solid	405.86	-127.91	1.000	1.000

model are quite far from those obtained with the 3-D model. In order to obtain insight into the normal stress behavior, we show in Figure 7.47 the normal stress distribution due to the bending moment only obtained for a straight and a curved bar model for the section in which points P_4 and P_5 are located. Of course, for the curved bar model, the radius of curvature is an important variable which for this part is given by $R = 9.85$ mm with the ratio of the section height to the radius $h/R = 1.254$. Note that the differences in predictions are significant and that the same trend as observed in the numerical predictions for points P_4 and P_5 is displayed in the results reported in Figure 7.47.

It is interesting to verify, as anticipated, that as the radius of curvature is increased the differences in predictions between the straight and curved bar models should become smaller. Indeed, the results shown in Figure 7.48 confirm this anticipated behavior.

Overall we see that the curved bar model gives a much better stress prediction, measured on the 3-D solid model results, than the straight bar model, and the stresses are almost within the desired margin of error. But for this very large ratio of h/R a 3-D solid model is more appropriate if the stresses need to be predicted very accurately.

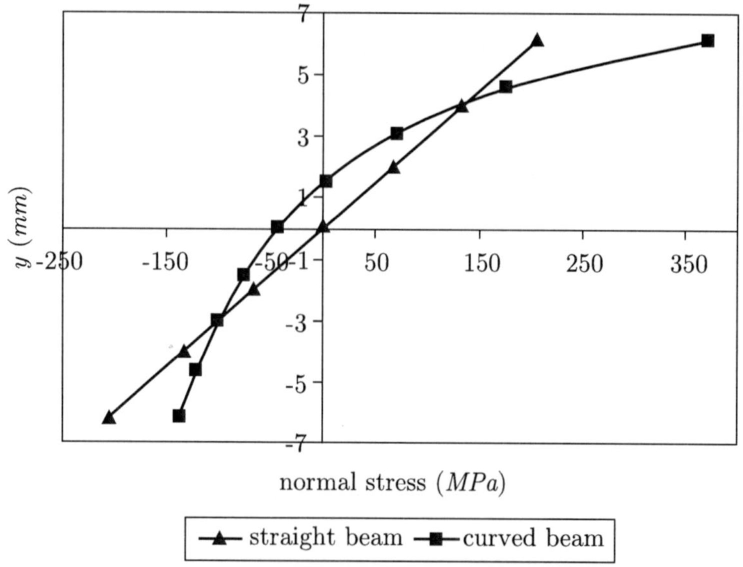

Fig. 7.47. Comparison of normal stress predictions for straight and curved beam models. For the curved model the radius of curvature is $R = 9.85\ mm$ and the section height is $h = 12.35\ mm$. The moment is 37 $kNmm$

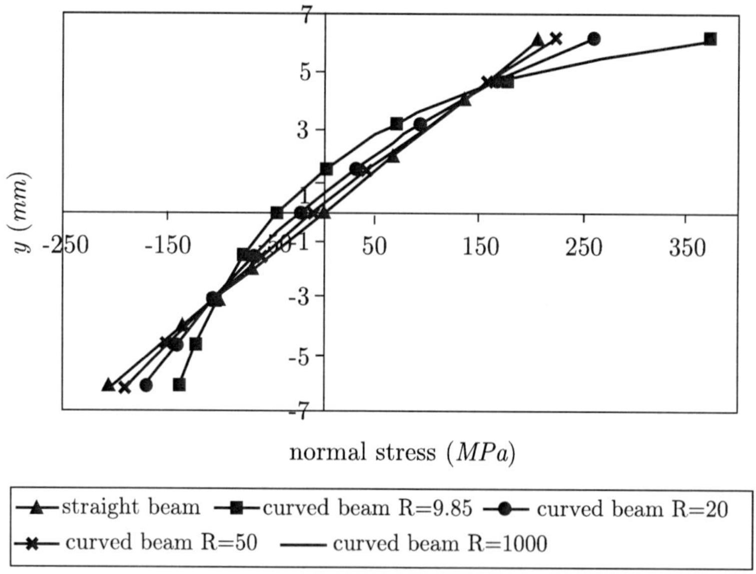

Fig. 7.48. Normal stress distribution for various radii of curvature; the $R = 1000$ results are practically equal to the straight beam results

8. Modeling for nonlinear analysis. An aperçu

Our objective in this chapter is to introduce some main concepts of nonlinear structural analysis. We will focus on the fundamental issues.

We choose the analysis of truss structures as a means to discuss in a simple setting the basic concepts of nonlinear analysis. Let us recall that in Chapter 2 truss structures were used to exemplify the application of the fundamental steps in structural mechanics leading to the matrix formulation. Later, several more complex mathematical models and finite element formulations were presented. Despite the complexity of these models, we could always find a clear analogy between their formulations and those for truss structures.

Of course, it is out of the scope of this book to discuss nonlinear formulations for the 2-D, 3-D, beam and shell mathematical models detailed in Chapters 3 and 4 (for such formulations see Bathe, 1996). However, we do endeavour to make the presentation for truss structures sufficiently comprehensive for the reader to gain insight into the important general aspects of nonlinear analysis of complex problems. In our presentation we use the notation of Bathe, 1996.

8.1 Sourccs of nonlinearity

We can classify many nonlinearities encountered in structural analysis as geometric, material and contact nonlinearities (see Bathe, 1996). Of course, in a general nonlinear analysis, these nonlinearities may arise together. We give below an introductory discussion of these nonlinearities.

Geometric nonlinearities

A most important assumption of linear analysis – introduced in Chapter 2 – is that the displacements are assumed to be infinitesimally small. This assumption is adequate for most applications of structural engineering. However, there are situations in which it does not apply. Consider, for example, the trusses described in Figure 8.1.

The truss of Figure 8.1a corresponds to the truss defined in Figure 2.16 with $R_1 = R_2 = -\frac{R\sqrt{2}}{2}$ and the cross-sectional areas $A = A_1 = A_2 =$

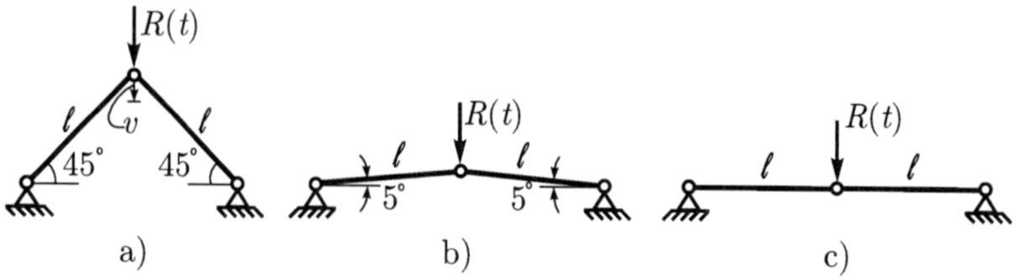

Fig. 8.1. Selected truss structures ($\ell = 2$ m)

1.5×10^{-3} m. The displacement v under the load can be readily evaluated considering linear analysis and is given by

$$v = \frac{R\ell}{2EA\sin(45°)^2} = 0.000508 \text{ m}$$

which is very small compared to the dimensions of the truss. Therefore, the linear assumption – imposing equilibrium in the undeformed configuration – is adequate. For the truss of Figure 8.1b the linear analysis leads to

$$v = \frac{R\ell}{2EA\sin(5°)^2} = 0.033434 \text{ m}$$

which is a somewhat large value when compared to the height $H = \ell \sin 5° = 0.17431$ m and, hence, imposing the equilibrium in the undeformed configuration might not be an appropriate assumption.

In order to solve the problem allowing for any magnitude of displacement we need to guarantee that equilibrium is satisfied in the deformed configuration which is not known a priori. We solve this problem in Example 8.1 below and valuable insight in the nonlinear formulation will be gained from this solution.

Finally, we note that the truss of Figure 8.1c can display infinitesimally small rigid motions and therefore can not be solved in linear analysis.

In nonlinear analysis, it is usual to consider that the load is increased from zero to a given value. We can describe such load variation by a function

$${}^tR = R(t)$$

where t is the time variable. Since we pursue a static analysis the loads are assumed to be applied slowly, so that inertia effects are negligible. Therefore, the time variable is merely used to define how the load varies during the analysis. The nonlinear solution will then give the deformed configuration for every time t in the range of interest. Of course, equilibrium will be enforced in these deformed configurations.

Example 8.1[1]

Consider the arch described in Figure 8.2a which can be modeled as a one bar truss structure. Obtain the deformed configurations associated with ${}^tR$

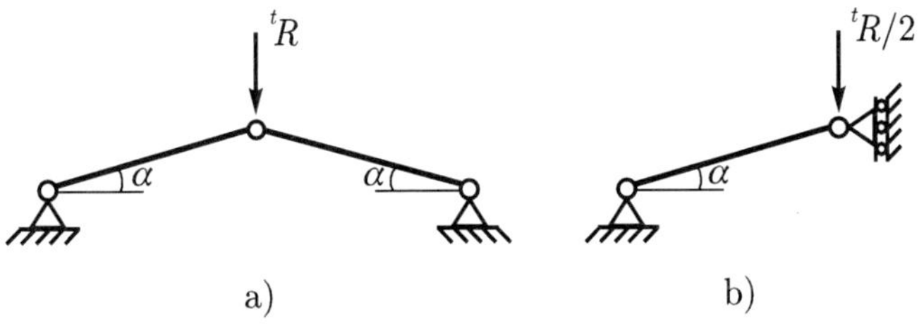

Fig. 8.2. a) Arch structure; b) Model using symmetry

which is increased from zero. Assume that α is small such that the resulting strain is also small and, therefore, Hooke's law applies.

Solution

The kinematics of the structure is summarized in Figure 8.3a from which we can write

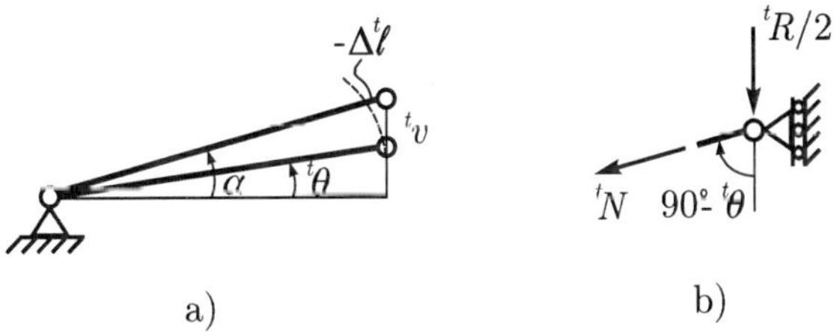

Fig. 8.3. a) Kinematics of the truss; b) Normal force for a deformed configuration

$$ {}^0\ell = \ell, \qquad \Delta^t\ell = {}^t\ell - {}^0\ell $$

$$ {}^t\ell \cos {}^t\theta = {}^0\ell \cos \alpha $$

$$ {}^t\ell \sin {}^t\theta = {}^0\ell \sin \alpha - {}^tv $$

which yields

[1] This example is also discussed in (Bathe, 1996, Example 6.3)

$$\Delta {}^{t}\ell = \sqrt{({}^{0}\ell)^2 - 2\,{}^{0}\ell\,{}^{t}v \sin\alpha + ({}^{t}v)^2} - {}^{0}\ell \tag{8.1}$$

and

$$\sin {}^{t}\theta = \frac{{}^{0}\ell \sin\alpha - {}^{t}v}{{}^{0}\ell + \Delta {}^{t}\ell}. \tag{8.2}$$

Referring to Figure 8.3b, equilibrium gives

$$\frac{{}^{t}R}{2} = -\,{}^{t}N \sin {}^{t}\theta \tag{8.3}$$

where ${}^{t}N$ is the axial force of the bar at time t. Since the strains are small we can use Hooke's law to obtain

$${}^{t}N = EA\frac{\Delta {}^{t}\ell}{{}^{0}\ell}. \tag{8.4}$$

From equations (8.1), (8.2), (8.3) and (8.4), we obtain

$$\frac{{}^{t}R}{2EA} = \left(\frac{1}{\sqrt{1 - 2\frac{{}^{t}v}{{}^{0}\ell}\sin\alpha + \left(\frac{{}^{t}v}{{}^{0}\ell}\right)^2}} - 1\right)\left(\sin\alpha - \frac{{}^{t}v}{{}^{0}\ell}\right). \tag{8.5}$$

We note that equation (8.5) allows to obtain the deformed configurations in which ${}^{t}R$ is in equilibrium with the bar forces in the deformed configurations.

In Figure 8.4a the load-displacement relation is given when $\alpha = 5°$. We note that for small ${}^{t}R$ the response is almost linear (up to point C). For the load level between points C and A the response is truly nonlinear. At the configuration corresponding to point A, with a "tiny" load increase the structure "snaps" and reaches an equilibrated[2] configuration corresponding to point B. The configurations corresponding to points A and B are schematically presented in Figure 8.4b and, although the load increase was "tiny", the structure could not reach an equilibrated configuration in the vicinity of the configuration corresponding to point A. The configurations between those of points A and B can only be reached by imposing displacements. The load level R_{cr} corresponding to the *limit point* A is called the *critical load*.

□

Material nonlinearities

The discussion presented so far assumed that the material has a linear elastic behavior. However, there are situations for which the use of a nonlinear material behavior is essential for correct modeling. Nonlinear elastic and elasto-plastic behaviors are common examples.

[2] Since the structural motion corresponding to the "snap" occurs in a short time period, in physical reality, the structure would oscillate around the configuration given by point B, before this configuration is finally reached due to energy dissipation

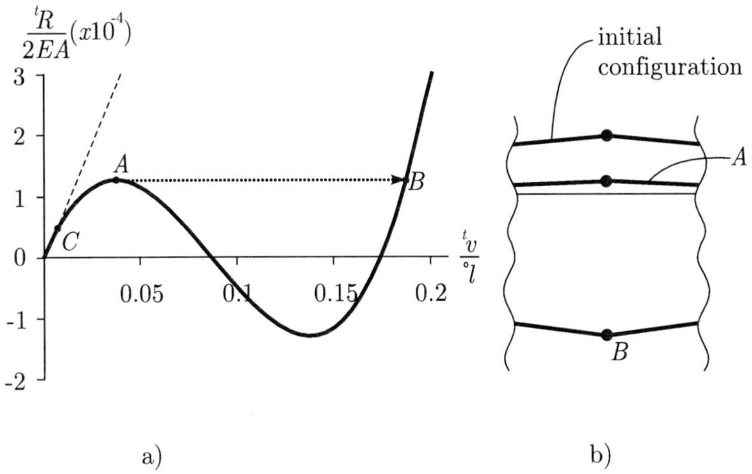

Fig. 8.4. a) Load-displacement curve. Linear response shown by the dashed line; b) Selected configurations around middle node

In the discussion below we address only one-dimensional states of stress and strain which are those relevant to the truss formulation and which allow us to introduce the main concepts.

We recall the linear elastic behavior in Figure 8.5a and introduce the non-linear elastic behavior in Figure 8.5b. The essential ingredient of an elastic behavior is that the deformations are immediate and reversible. The deformation being immediate means that there is no time lag between the development of a strain state given a stress state. The reversibility means that the stress state depends only on the current strain state, *i.e.*, not on any history of stress and strain. For example, for point A shown on the diagrams of Figure 8.5, loading and unloading may have occurred many times before the stress and strain states associated with A have been reached.

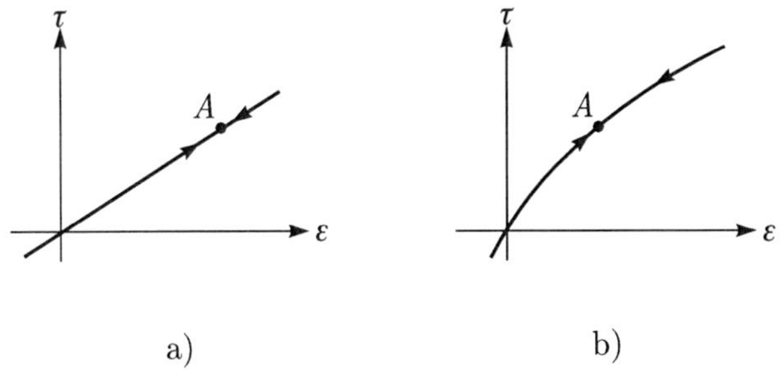

Fig. 8.5. a) Linear elastic; b) Nonlinear elastic

Let us consider an elasto-plastic behavior as an example in which inelastic deformations take place. In Figure 8.6a typical test data from a uniaxial tension laboratory experiment of a metallic material is presented. The arrows indicate the loading and unloading paths. We note the strongly nonlinear behavior and that upon unloading a residual strain ε^p– the plastic strain – is present. In Figure 8.6b a model representation of the data of Figure 8.6a is shown. In this bilinear model the behavior is purely elastic until the yield stress τ_y is reached at point 1. Above this stress level, the increments in stress and strain are related by the modulus E_T (hardening modulus) and unloading, like from point 2, occurs according to the Young's modulus E of the material. Total unloading will lead to a residual plastic deformation ε^p. Reloading from this point (point 3), the behavior is purely elastic up to point 4, indicating that the yield stress has increased (strain hardening), and then loading continues like to point 5.

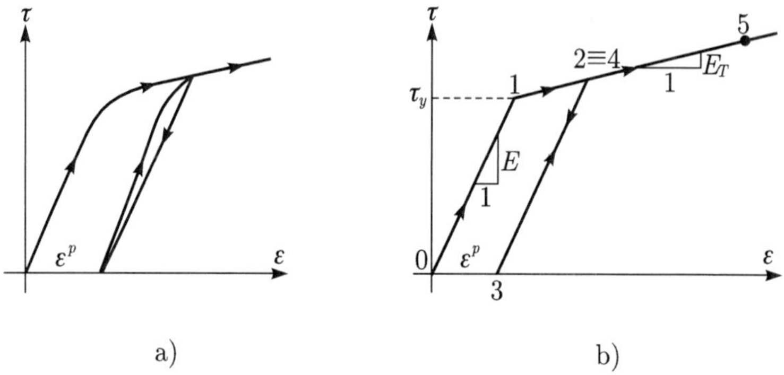

Fig. 8.6. a) Stress-strain response from a laboratory experiment; b) Model stress-strain curve

If we assume $E_T = 0$ there is no strain hardening and the material is referred to as elastic perfectly plastic.

Of course, the modeling of plasticity is a very large field in solid mechanics (see *e.g.* Kojic and Bathe, 2005) but the behavior mentioned above already allows us to present a valuable analysis example.

Example 8.2

Evaluate the load displacement response of the truss of Figure 8.7a; increase the load P from zero up to start of collapse of the structure and then totally unload the structure. The material is elastic perfectly plastic as described in Figure 8.7b.

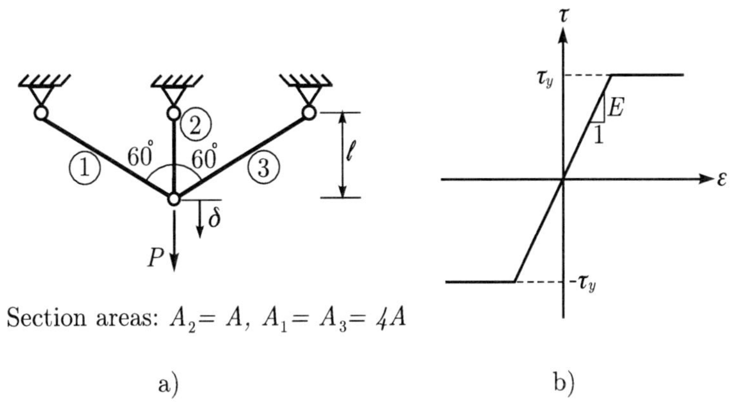

Fig. 8.7. a) Truss structure; b) Stress-strain curve adopted

Solution

We suppose that the displacements are small enough such that equilibrium is appropriately enforced in the undeformed configuration, and the nonlinear behavior is due only to the material behavior. We refer to such analyses as materially–nonlinear–only (M.N.O.) analyses.

The initial response of the structure is elastic. Due to symmetry conditions $N_1 = N_3$ and equilibrium yields

$$2N_1 \cos 60^\circ + N_2 = P \qquad \Rightarrow \qquad N_1 + N_2 = P.$$

Referring to Figure 8.8, compatibility yields

$$\Delta\ell_2 = 2\Delta\ell_1$$

and since

$$\Delta\ell_1 = \frac{N_1\ell_1}{E_1A_1} = \frac{N_1\ell}{2EA} \qquad \text{and} \qquad \Delta\ell_2 = \frac{N_2\ell_2}{E_2A_2} = \frac{N_2\ell}{EA}$$

we obtain

$$N_1 = N_2 = \frac{P}{2}.$$

During the elastic phase, the stresses in the bars are given by

$$\tau_1 = \frac{P}{2A_1} = \frac{P}{8A} \qquad \text{and} \qquad \tau_2 = \frac{P}{2A_2} = \frac{P}{2A}.$$

Therefore, bar 2 attains its plastic limit first at the load P_I which is given by

$$P_I = 2\tau_y A$$

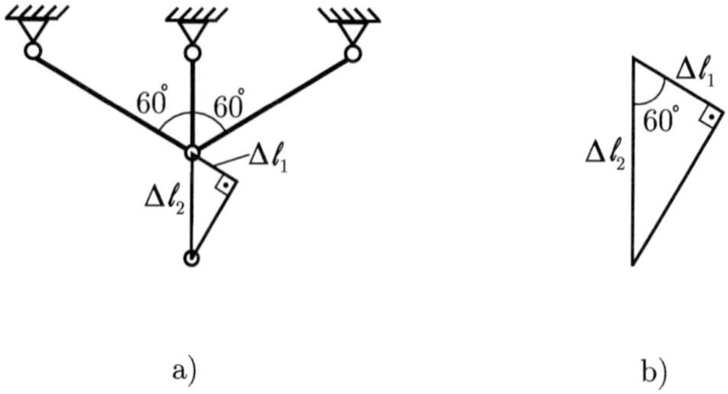

Fig. 8.8. a) Compatibility of displacements; b) Geometric relations

and the stress at bar 1 for this load is

$$\tau_1 = \frac{P_I}{2A_1} = \frac{\tau_y}{4}.$$

For increments of load beyond P_I, only bars 1 and 3 have stiffness and bar 2 will deform, without increase in stress, to maintain compatibility. From equilibrium

$$2\Delta N_1 \cos 60^\circ = \Delta P \qquad \Rightarrow \qquad \Delta N_1 = \Delta P.$$

The collapse of the structure is attained when the stress in bars 1 and 3 is equal to τ_y. Then the structure is no longer able to sustain load increments. Therefore, the load P_{II} at the collapse of the structure is given by

$$P_{II} = P_I + \Delta P^{II}$$

where

$$\frac{\Delta P^{II}}{4A} + \frac{\tau_y}{4} = \tau_y$$

leading to

$$P_{II} = 5\tau_y A.$$

The unloading is initially elastic for all three bars. However, bar 2 might attain its compression yield stress before total unloading. In fact, the total unloading requires

$$\Delta P = -5\tau_y A$$

leading to

$$\Delta\tau_2 = \frac{\Delta N_2}{A} = \frac{\Delta P}{2A} < -2\tau_y$$

Therefore, the elastic part of the unloading phase associated with ΔP^e can be evaluated by imposing

$$\Delta\tau_2 = \frac{\Delta N_2}{A} = \frac{\Delta P^e}{2A} = -2\tau_y$$

leading to

$$\Delta P^e = -4\tau_y A.$$

For the final unloading phase only bars 1 and 3 change stress.

Since bars 1 and 3 remain elastic throughout the loading and unloading phases they characterize the truss deformation. We note that

$$\delta = \Delta\ell_2 = 2\Delta\ell_1 = \frac{N_1\ell}{2EA}.$$

Let point A be associated with the yielding of bar 2 in tension; point B with

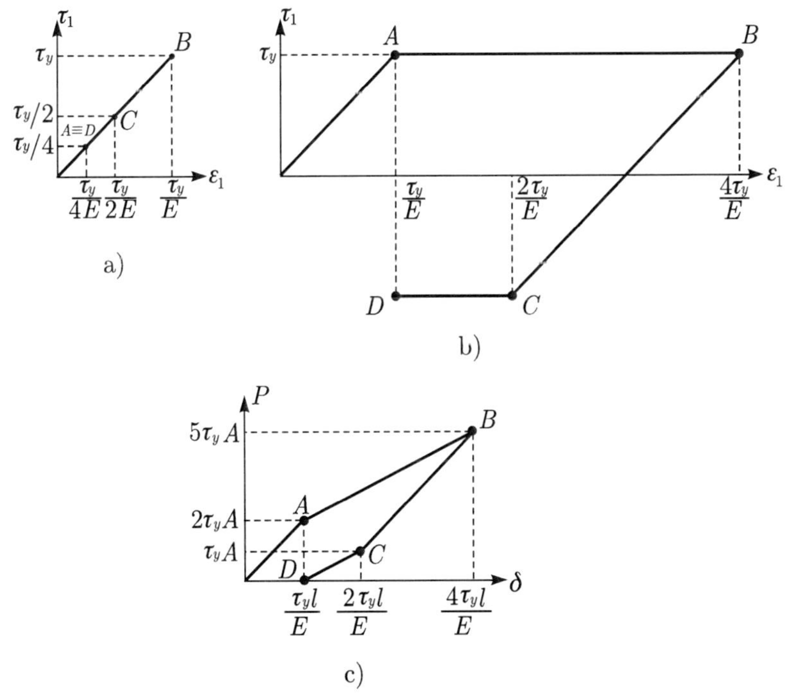

Fig. 8.9. a) Response of bars 1 and 3; b) Response of bar 2; c) Response of the structure

the imminence of collapse; point C with the yielding of bar 2 in compression and point D with total unloading. Then, Figure 8.9 summarizes the response of the structure.

We should note the path dependency of the response of the structure and that at the final configuration, without any external load, bars 1 and 3 are in tension and bar 2 is in compression, all with residual stresses.

□

Contact nonlinearities

Contact nonlinearities arise when during the deformation process a body establishes contact with other bodies or with itself. In essence, contact conditions correspond to a change of boundary conditions on the body coming into contact. A typical situation for a simple truss is presented in Figure 8.10.

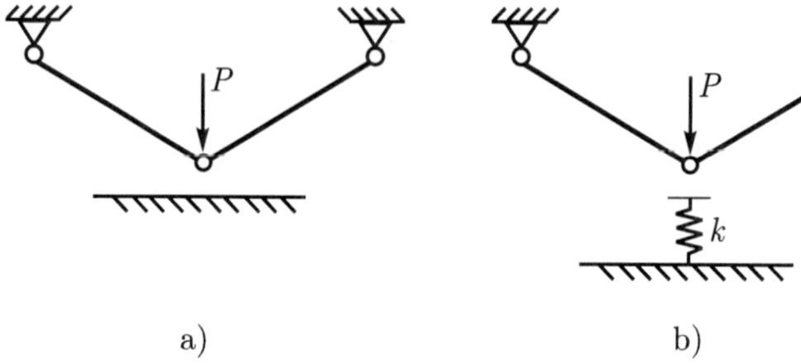

Fig. 8.10. a) Contact implying a change of boundary condition; b) Spring modeling the contact

Other sources of nonlinearities

In addition to the common sources of nonlinearities mentioned above, there are other instances. For example, during construction or manufacturing processes, structural parts may be added or suppressed which affect the subsequent structural response. Also, problems with fluid-structure interactions and thermo-mechanical interactions lead in general to a nonlinear response, see *e.g.* Bathe, 2005.

8.2 Incremental formulation for nonlinear analysis

A basic result of the matrix analysis of truss structures derived in Chapter 2 was that for a given $\mathbf{R}$, the column matrix of external nodal loads, the equilibrium conditions are given by

$$\mathbf{F} = \mathbf{R}$$

where $\mathbf{F}$ is column matrix of the nodal forces corresponding to the element stresses. We also call these nodal point forces corresponding to the element stresses, internal nodal point forces. In linear analysis, introducing the compatibility conditions and constitutive behaviors, we obtain

$$\mathbf{F} = \mathbf{KU} \tag{8.6}$$

which leads to the final matrix equation

$$\mathbf{KU} = \mathbf{R}.$$

In nonlinear analysis equation (8.6) is no longer valid since the nodal forces $\mathbf{F}$ depend nonlinearly on the nodal displacements.

Let us assume that the external nodal loads do not depend on the nodal displacements and are given by ${}^t\mathbf{R}$ where t is the time variable representing the load and the configuration at time t.

The equilibrium conditions at time t can be expressed as

$${}^t\mathbf{F} = {}^t\mathbf{R} \tag{8.7}$$

where ${}^t\mathbf{F}$ are the nodal forces corresponding to the element stresses for the configuration at time t, *i.e.*,

$${}^tF_i = {}^tF_i\left({}^tU_1, {}^tU_2, \cdots, {}^tU_N\right) \qquad \text{for } i = 1, \cdots, N \tag{8.8}$$

which in matrix notation is given by

$${}^t\mathbf{F} = {}^t\mathbf{F}\left({}^t\mathbf{U}\right) \tag{8.9}$$

where

$${}^t\mathbf{U}^T = \begin{bmatrix} {}^tU_1 & {}^tU_2 & \cdots & {}^tU_N \end{bmatrix}$$

gives the nodal displacements at time t. In a nonlinear problem, the ${}^tF_i$ of equation (8.8) depend nonlinearly on ${}^tU_j$, $j = 1, \cdots, N$. Therefore, to solve directly (8.7) is, in general not possible. This fact prompts us to introduce the incremental procedure as an effective approach to solve (8.7).

The fundamental idea of the incremental procedure is to suppose that the solution is known at time t and develop a methodology to obtain the solution at time $t + \Delta t$, where Δt is a finite time increment referred to as "the time step" corresponding to positive or negative "increments" in forces and displacements. With this methodology available, we can obtain, starting from time $t = 0$, the solution for any time t by applying repeatedly this time-step solution scheme (see also Bathe, 1996).

We may write

$$ {}^{t+\Delta t}\mathbf{F} = {}^{t}\mathbf{F} + \mathbf{F} \tag{8.10} $$

where $\mathbf{F}$ is the unknown internal nodal force increment such that

$$ {}^{t+\Delta t}\mathbf{F} = {}^{t+\Delta t}\mathbf{R}. \tag{8.11} $$

Substituting (8.10) into (8.11) yields

$$ \mathbf{F} = {}^{t+\Delta t}\mathbf{R} - {}^{t}\mathbf{F}. \tag{8.12} $$

Let dF_i be an infinitesimally small internal nodal force increment for degree of freedom i which is given by

$$ dF_i = \sum_{j=1}^{N} \frac{\partial\, {}^{t}F_i}{\partial\, {}^{t}U_j} dU_j \tag{8.13} $$

where the dU_j are infinitesimally small nodal displacement increments. Defining

$$ {}^{t}\mathbf{K} = \begin{bmatrix} \frac{\partial\, {}^{t}F_1}{\partial\, {}^{t}U_1} & \frac{\partial\, {}^{t}F_1}{\partial\, {}^{t}U_2} & \cdots & \frac{\partial\, {}^{t}F_1}{\partial\, {}^{t}U_N} \\ \vdots & \ddots & & \vdots \\ \vdots & & \ddots & \vdots \\ \frac{\partial\, {}^{t}F_N}{\partial\, {}^{t}U_1} & \cdots & \cdots & \frac{\partial\, {}^{t}F_N}{\partial\, {}^{t}U_N} \end{bmatrix} \tag{8.14} $$

we can re-write (8.13) as

$$ \mathbf{dF} = {}^{t}\mathbf{K}\ \mathbf{dU} \tag{8.15} $$

where the $\mathbf{dF}$ and $\mathbf{dU}$ are column matrices which collect the increments dF_i, dU_j defined above. The matrix ${}^{t}\mathbf{K}$ defined by equation (8.14) is called the tangent stiffness matrix at time t with the obvious interpretation given by (8.15) that it relates the infinitesimally small increments of nodal displacements to the infinitesimally small increments of internal nodal point forces.

Let us define, considering equation (8.12), a first estimate $\Delta\mathbf{U}^{(1)}$ for the increment of nodal displacements from time t to time $t + \Delta t$ by

$$ {}^{t}\mathbf{K}\ \Delta\mathbf{U}^{(1)} = {}^{t+\Delta t}\mathbf{R} - {}^{t}\mathbf{F} \tag{8.16} $$

where we naturally use (8.15), but for larger and complex load changes. We note that the right hand side of equation (8.16) is the out-of-balance load when we change the external load from ${}^{t}\mathbf{R}$ to ${}^{t+\Delta t}\mathbf{R}$. The column matrix $\Delta\mathbf{U}^{(1)}$ is, in general, only an approximation of the exact nodal displacement increment. A first estimate of the nodal displacements at time $t + \Delta t$ is therefore given by

$$^{t+\Delta t}\mathbf{U}^{(1)} = {}^{t}\mathbf{U} + \Delta\mathbf{U}^{(1)}$$

and

$$^{t+\Delta t}\mathbf{F}^{(1)} = {}^{t+\Delta t}\mathbf{F}\left({}^{t+\Delta t}\mathbf{U}^{(1)}\right)$$

defines the internal nodal point forces corresponding to the nodal displacements $^{t+\Delta t}\mathbf{U}^{(1)}$. We can evaluate the out-of-balance load $^{t+\Delta t}\mathbf{R}- {}^{t+\Delta t}\mathbf{F}^{(1)}$ associated with the intermediate configuration (between t and $t+\Delta t$) given by $^{t+\Delta t}\mathbf{U}^{(1)}$, and then obtain a second nodal displacement increment $\Delta\mathbf{U}^{(2)}$ from

$$^{t+\Delta t}\mathbf{K}^{(1)}\ \Delta\mathbf{U}^{(2)} = {}^{t+\Delta t}\mathbf{R}- {}^{t+\Delta t}\mathbf{F}^{(1)}.$$

Hence we obtain

$$^{t+\Delta t}\mathbf{U}^{(2)} = {}^{t+\Delta t}\mathbf{U}^{(1)} + \Delta\mathbf{U}^{(2)}$$

and

$$^{t+\Delta t}\mathbf{F}^{(2)} = {}^{t+\Delta t}\mathbf{F}\left({}^{t+\Delta t}\mathbf{U}^{(2)}\right).$$

Note that $^{t+\Delta t}\mathbf{K}^{(1)}$ is the tangent stiffness matrix for the intermediate configuration given by $^{t+\Delta t}\mathbf{U}^{(1)}$, *i.e.*,

$$^{t+\Delta t}K_{ij}^{(1)} = \left.\frac{\partial\, {}^{t+\Delta t}F_i}{\partial\, {}^{t+\Delta t}U_j}\right|_{{}^{t+\Delta t}\mathbf{U}^{(1)}}.$$

This iterative procedure is repeated until the out-of-balance load $^{t+\Delta t}\mathbf{R}- {}^{t+\Delta t}\mathbf{F}^{(k)}$ is sufficiently small. We refer the reader to Bathe, 1996 for more details and a discussion of the convergence of the iteration. When we have found $^{t+\Delta t}\mathbf{U}^{(k)}$ for which convergence has been achieved, we set

$$^{t+\Delta t}\mathbf{U} = {}^{t+\Delta t}\mathbf{U}^{(k)}$$

and the next time step is considered.

The iteration procedure for each time step can be summarized by

$$\begin{aligned} ^{t+\Delta t}\mathbf{K}^{(i-1)}\ \Delta\mathbf{U}^{(i)} &= {}^{t+\Delta t}\mathbf{R}- {}^{t+\Delta t}\mathbf{F}^{(i-1)} \\ ^{t+\Delta t}\mathbf{U}^{(i)} &= {}^{t+\Delta t}\mathbf{U}^{(i-1)} + \Delta\mathbf{U}^{(i)} \end{aligned}$$

with the initial conditions

$$^{t+\Delta t}\mathbf{U}^{(0)} = {}^{t}\mathbf{U}, \qquad {}^{t+\Delta t}\mathbf{K}^{(0)} = {}^{t}\mathbf{K}, \qquad {}^{t+\Delta t}\mathbf{F}^{(0)} = {}^{t}\mathbf{F}.$$

The tangent stiffness matrix is a fundamental ingredient for the incremental procedure and hence we need to address how to obtain $^{t}\mathbf{K}$ for a truss structure. As in linear analysis, we obtain $^{t}\mathbf{K}$ by summing up the stiffness contributions of the truss elements of the assemblage. Therefore, if we have the tangent stiffness matrix ${}^t\mathbf{k}$ for a generic truss element, we can obtain ${}^t\mathbf{K}$ as in linear analysis (see Section 2.3).

Nonlinear stiffness matrix of a truss element

We follow in this derivation Example 6.16 of Bathe, 1996 and consider the configurations at time 0 and t of a generic truss element as summarized in Figure 8.11a. For simplicity of the derivations and interpretations, the truss

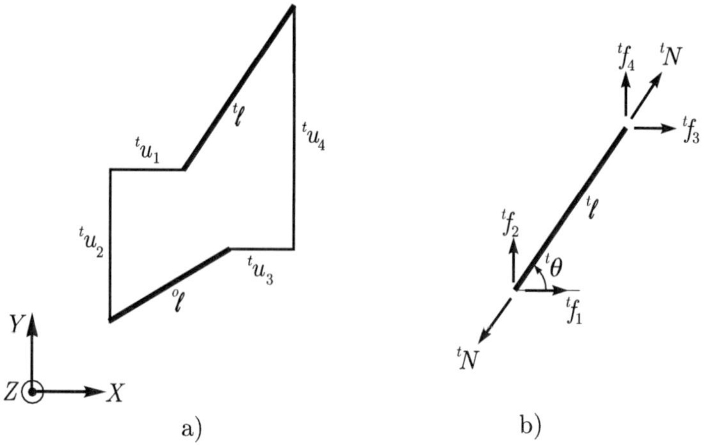

Fig. 8.11. a) Nodal displacement for a truss element; b) Nodal forces for a truss element

element is supposed to be in the XY plane throughout the motion. In Figure 8.11a we show the nodal displacements for time t and in Figure 8.11b the nodal forces where we also show the axial force ${}^tN$ and the angle ${}^t\theta$ that the bar axis makes with respect to the global X axis, both for time t.

The generic term of the tangent stiffness matrix of the truss element at time t is given by

$$
{}^tk_{ij} = \frac{\partial\, {}^tf_i}{\partial\, {}^tu_j}.
$$

The column matrix of nodal displacements and nodal forces at time t are given by

$$
{}^t\mathbf{u} = \begin{bmatrix} {}^tu_1 \\ {}^tu_2 \\ {}^tu_3 \\ {}^tu_4 \end{bmatrix}, \qquad {}^t\mathbf{f} = \begin{bmatrix} {}^tf_1 \\ {}^tf_2 \\ {}^tf_3 \\ {}^tf_4 \end{bmatrix}.
$$

Referring to Figure 8.11b, we can write

$$
{}^t\mathbf{f} = \begin{bmatrix} -\,{}^tN\cos{}^t\theta \\ -\,{}^tN\sin{}^t\theta \\ {}^tN\cos{}^t\theta \\ {}^tN\sin{}^t\theta \end{bmatrix}. \tag{8.17}
$$

Therefore, to evaluate ${}^t\mathbf{k}$ we need to calculate the derivatives of ${}^tN$ and ${}^t\theta$ with respect to the nodal displacements ${}^tu_i$. Referring to Figure 8.12,we

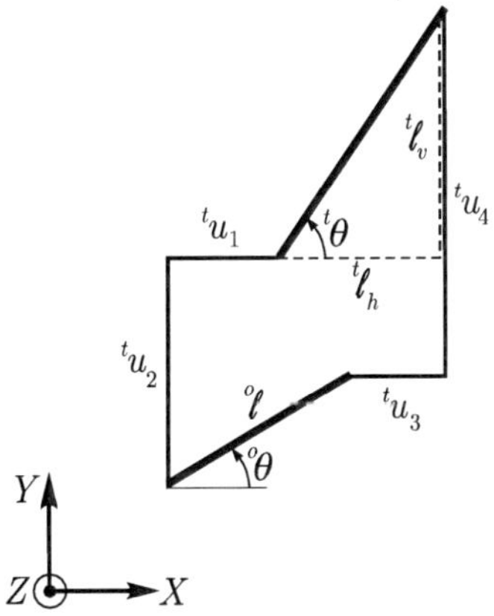

Fig. 8.12. Additional definitions related to the truss element kinematics

define the horizontal and vertical projections of the bar length at time t by

$$
{}^t\ell_h = {}^t\ell\cos{}^t\theta = {}^0\ell\cos{}^0\theta + {}^tu_3 - {}^tu_1 \tag{8.18}
$$

$$
{}^t\ell_v = {}^t\ell\sin{}^t\theta = {}^0\ell\sin{}^0\theta + {}^tu_4 - {}^tu_2 \tag{8.19}
$$

and defining

$$
{}^t\ell = {}^0\ell + \Delta^t\ell
$$

we can re-write (8.18) and (8.19) as

$$
{}^t\ell_h = \left({}^0\ell + \Delta^t\ell\right)\cos{}^t\theta \tag{8.20}
$$

$$
{}^t\ell_v = \left({}^0\ell + \Delta^t\ell\right)\sin{}^t\theta. \tag{8.21}
$$

A generic element of ${}^t\mathbf{k}$ can be written as

$$
{}^tk_{ij} = \frac{\partial\,{}^tf_i}{\partial\,{}^tu_j} = \frac{\partial\,{}^tf_i}{\partial\,{}^t\ell_h}\frac{\partial\,{}^t\ell_h}{\partial\,{}^tu_j} + \frac{\partial\,{}^tf_i}{\partial\,{}^t\ell_v}\frac{\partial\,{}^t\ell_v}{\partial\,{}^tu_j}. \tag{8.22}
$$

Note that we can easily obtain $\frac{\partial\, ^t\ell_h}{\partial\, ^tu_j}$ and $\frac{\partial\, ^t\ell_v}{\partial\, ^tu_j}$ from equations (8.18) and (8.19). Therefore, let us focus on the terms $\frac{\partial\, ^tf_i}{\partial\, ^t\ell_h}$ and $\frac{\partial\, ^tf_i}{\partial\, ^t\ell_v}$.

From equations (8.20) and (8.21), we obtain

$$\begin{bmatrix} \frac{\partial}{\partial(\Delta^t\ell)} \\ \frac{\partial}{\partial\, ^t\theta} \end{bmatrix} = \begin{bmatrix} \cos\, ^t\theta & \sin\, ^t\theta \\ -\left(^0\ell + \Delta^t\ell\right)\sin\, ^t\theta & \left(^0\ell + \Delta^t\ell\right)\cos\, ^t\theta \end{bmatrix} \begin{bmatrix} \frac{\partial}{\partial\, ^t\ell_h} \\ \frac{\partial}{\partial\, ^t\ell_v} \end{bmatrix}$$

which upon inversion leads to

$$\begin{bmatrix} \frac{\partial}{\partial\, ^t\ell_h} \\ \frac{\partial}{\partial\, ^t\ell_v} \end{bmatrix} = \begin{bmatrix} \cos\, ^t\theta & -\frac{\sin\, ^t\theta}{^0\ell + \Delta^t\ell} \\ \sin\, ^t\theta & \frac{\cos\, ^t\theta}{^0\ell + \Delta^t\ell} \end{bmatrix} \begin{bmatrix} \frac{\partial}{\partial(\Delta^t\ell)} \\ \frac{\partial}{\partial\, ^t\theta} \end{bmatrix}.$$

Therefore,

$$\begin{bmatrix} \frac{\partial\, ^tf_i}{\partial\, ^t\ell_h} \\ \frac{\partial\, ^tf_i}{\partial\, ^t\ell_v} \end{bmatrix} = \begin{bmatrix} \cos\, ^t\theta & -\frac{\sin\, ^t\theta}{^0\ell + \Delta^t\ell} \\ \sin\, ^t\theta & \frac{\cos\, ^t\theta}{^0\ell + \Delta^t\ell} \end{bmatrix} \begin{bmatrix} \frac{\partial\, ^tf_i}{\partial(\Delta^t\ell)} \\ \frac{\partial\, ^tf_i}{\partial\, ^t\theta} \end{bmatrix} \tag{8.23}$$

together with equation (8.22) give us the element of $^t\mathbf{k}$.

Let us detail the evaluation of one term. For example, to obtain $^tk_{11}$ we need

$$\frac{\partial\, ^tf_1}{\partial\, \Delta^t\ell} = -\frac{\partial\, ^tN}{\partial\, \Delta^t\ell}\cos\, ^t\theta, \qquad \frac{\partial\, ^tf_1}{\partial\, ^t\theta} = {}^tN\sin\, ^t\theta$$

$$\frac{\partial\, ^t\ell_h}{\partial\, ^tu_1} = -1, \qquad \frac{\partial\, ^t\ell_v}{\partial\, ^tu_1} = 0.$$

Using the above relations together with (8.23) and (8.22), we obtain

$$k_{11} = \frac{\partial\, ^tN}{\partial\, \Delta^t\ell}\cos^2\, ^t\theta + \frac{^tN}{^0\ell + \Delta^t\ell}\sin^2\, ^t\theta. \tag{8.24}$$

This equation can be written in an alternative form using the derivative rule of the quotient of functions

$$\frac{\partial\, ^tN}{\partial\, \Delta^t\ell} = \frac{\partial}{\partial\, \Delta^t\ell}\left(\frac{^tN}{^0\ell + \Delta^t\ell}\right)\left(^0\ell + \Delta^t\ell\right) + \frac{^tN}{^0\ell + \Delta^t\ell}$$

which substituted in (8.24) yields

$$k_{11} = \frac{\partial}{\partial\, \Delta^t\ell}\left(\frac{^tN}{^0\ell + \Delta^t\ell}\right)\left(^0\ell + \Delta^t\ell\right)\cos^2\, ^t\theta + \frac{^tN}{^0\ell + \Delta^t\ell}.$$

The other terms of the stiffness matrix can be obtained analogously leading to

$$
{}^t\mathbf{k} = \frac{\partial}{\partial\, \Delta^t\ell}\left(\frac{{}^tN}{{}^0\ell + \Delta^t\ell}\right)({}^0\ell + \Delta^t\ell)
$$

$$
\begin{bmatrix}
\cos^2 {}^t\theta & \sin {}^t\theta \cos {}^t\theta & -\cos^2 {}^t\theta & -\sin {}^t\theta \cos {}^t\theta \\
\sin {}^t\theta \cos {}^t\theta & \sin^2 {}^t\theta & -\sin {}^t\theta \cos {}^t\theta & -\sin^2 {}^t\theta \\
-\cos^2 {}^t\theta & -\sin {}^t\theta \cos {}^t\theta & \cos^2 {}^t\theta & \sin {}^t\theta \cos {}^t\theta \\
-\sin {}^t\theta \cos {}^t\theta & -\sin^2 {}^t\theta & \sin {}^t\theta \cos {}^t\theta & \sin^2 {}^t\theta
\end{bmatrix}
\tag{8.25}
$$

$$
+ \frac{{}^tN}{{}^0\ell + \Delta^t\ell}
\begin{bmatrix}
1 & 0 & -1 & 0 \\
0 & 1 & 0 & -1 \\
-1 & 0 & 1 & 0 \\
0 & -1 & 0 & 1
\end{bmatrix}
= {}^t\mathbf{k}_L + {}^t\mathbf{k}_{NL}
$$

where we partitioned ${}^t\mathbf{k}$ into ${}^t\mathbf{k}_L$, the linear part, and ${}^t\mathbf{k}_{NL}$, the nonlinear part of the tangent stiffness matrix (see also (8.26)).

Let us now assume that the displacements are large but the strains are small, and that Hooke's law applies. Then

$$
{}^tN = \frac{EA}{{}^0\ell}\,\Delta^t\ell.
$$

Note that, for such case,

$$
\frac{\partial}{\partial\, \Delta^t\ell}\left(\frac{{}^tN}{{}^0\ell + \Delta^t\ell}\right)({}^0\ell + \Delta^t\ell) = \frac{EA}{{}^0\ell + \Delta^t\ell}
$$

which substituted in equation (8.25) gives

$$
{}^t\mathbf{k} = \frac{EA}{{}^0\ell + \Delta^t\ell}
\begin{bmatrix}
\cos^2 {}^t\theta & \sin {}^t\theta \cos {}^t\theta & -\cos^2 {}^t\theta & -\sin {}^t\theta \cos {}^t\theta \\
\sin {}^t\theta \cos {}^t\theta & \sin^2 {}^t\theta & -\sin {}^t\theta \cos {}^t\theta & -\sin^2 {}^t\theta \\
-\cos^2 {}^t\theta & -\sin {}^t\theta \cos {}^t\theta & \cos^2 {}^t\theta & \sin {}^t\theta \cos {}^t\theta \\
-\sin {}^t\theta \cos {}^t\theta & -\sin^2 {}^t\theta & \sin {}^t\theta \cos {}^t\theta & \sin^2 {}^t\theta
\end{bmatrix}
$$

$$
+ \frac{{}^tN}{{}^0\ell + \Delta^t\ell}
\begin{bmatrix}
1 & 0 & -1 & 0 \\
0 & 1 & 0 & -1 \\
-1 & 0 & 1 & 0 \\
0 & -1 & 0 & 1
\end{bmatrix}
$$

with

$$
{}^t\mathbf{k}_L = \frac{EA}{{}^0\ell + \Delta^t\ell}\begin{bmatrix} \cos^2 {}^t\theta & \sin {}^t\theta \cos {}^t\theta & -\cos^2 {}^t\theta & -\sin {}^t\theta \cos {}^t\theta \\ \sin {}^t\theta \cos {}^t\theta & \sin^2 {}^t\theta & -\sin {}^t\theta \cos {}^t\theta & -\sin^2 {}^t\theta \\ -\cos^2 {}^t\theta & -\sin {}^t\theta \cos {}^t\theta & \cos^2 {}^t\theta & \sin {}^t\theta \cos {}^t\theta \\ -\sin {}^t\theta \cos {}^t\theta & -\sin^2 {}^t\theta & \sin {}^t\theta \cos {}^t\theta & \sin^2 {}^t\theta \end{bmatrix}
$$

$$
{}^t\mathbf{k}_{NL} = \frac{{}^tN}{{}^0\ell + \Delta^t\ell}\begin{bmatrix} 1 & 0 & -1 & 0 \\ 0 & 1 & 0 & -1 \\ -1 & 0 & 1 & 0 \\ 0 & -1 & 0 & 1 \end{bmatrix}. \tag{8.26}
$$

Comparing ${}^t\mathbf{k}_L$ with the stiffness matrix of the linear truss element given in equation (2.33), we can see that ${}^t\mathbf{k}_L$ corresponds to the stiffness matrix of a linear truss element which is geometrically identical to the nonlinear truss element at time t. Therefore, the interpretation of ${}^t\mathbf{k}_L$ – the linear part of the tangent stiffness matrix – is that it corresponds to a linear stiffness matrix which incorporates the change of geometry of the element due to the deformations.

The matrix ${}^t\mathbf{k}_{NL}$ – the nonlinear part of the tangent stiffness matrix – gives the stiffness contribution due to the force ${}^tN$ in the element.

Since we assumed that the strains are small we can generally set ${}^0\ell + \Delta^t\ell = {}^0\ell$ in (8.26).

Example 8.3

Interpret the coefficients of the truss nonlinear stiffness matrix (8.26) imposing unit end displacements.

Solution

For convenience, suppose that the truss element is horizontal at time t, as shown in Figure 8.13a. In Figure 8.13b, we show the unit end displacement

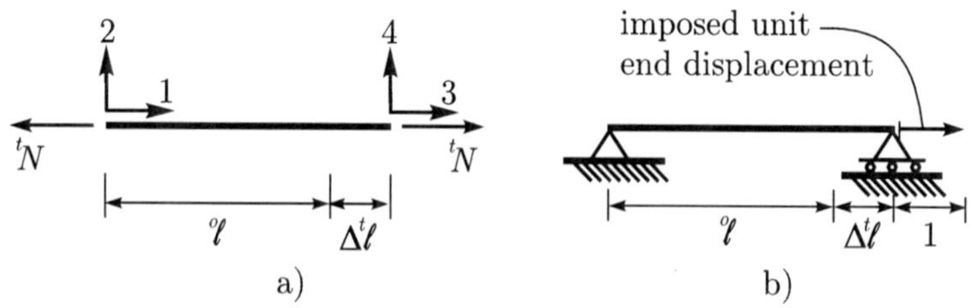

Fig. 8.13. a) Horizontal truss element at time t; b) Unit end displacement imposed

of magnitude 1 which is imposed at the local degree of freedom 3 while the remaining degrees of freedom are fixed. The axial force acting on the bar after the imposition of the unit end-displacement is given by

$$\frac{EA}{{}^0\ell}\left(\,\Delta^t\ell+1\right)=\frac{EA}{{}^0\ell}\,\Delta^t\ell+\frac{EA}{{}^0\ell}={}^tN+\frac{EA}{{}^0\ell}.$$

Since ${}^tN$ was already acting on the bar at time t, the force necessary to introduce the unit end displacement is

$${}^tk_{33}=\frac{EA}{{}^0\ell}.$$

Let us verify that this is the ${}^tk_{33}$ which is obtained from (8.26). Considering that ${}^t\theta=0$, (8.26) gives

$${}^tk_{33}={}^tk_{L_{33}}+{}^tk_{NL_{33}}=\frac{EA}{{}^0\ell+\,\Delta^t\ell}+\frac{{}^tN}{{}^0\ell+\,\Delta^t\ell}. \tag{8.27}$$

Introducing the relation ${}^tN=\frac{EA}{{}^0\ell}\,\Delta^t\ell$ into (8.27), we obtain

$${}^tk_{33}=\frac{EA}{{}^0\ell+\,\Delta^t\ell}+\frac{EA\,\Delta^t\ell}{{}^0\ell\,({}^0\ell+\,\Delta^t\ell)}=\frac{EA}{{}^0\ell+\,\Delta^t\ell}\left(1+\frac{\Delta^t\ell}{{}^0\ell}\right)=\frac{EA}{{}^0\ell}. \tag{8.28}$$

Note that from equilibrium

$${}^tk_{31}=-\frac{EA}{{}^0\ell},\qquad {}^tk_{32}=0\qquad\text{and}\qquad {}^tk_{34}=0$$

which can be easily obtained from (8.26).

Now consider the unit end displacement imposed for the local degree of freedom 4 as shown in Figures 8.14a and 8.14b. The forces required to guarantee equilibrium after the introduction of the unit end displacement lead to

$${}^tk_{14}=0,\qquad {}^tk_{24}=-\frac{{}^tN}{{}^0\ell+\,\Delta^t\ell},\qquad {}^tk_{34}=0\qquad\text{and}\qquad {}^tk_{44}=\frac{{}^tN}{{}^0\ell+\,\Delta^t\ell}.$$

Note that ${}^tk_{14}={}^tk_{34}=0$ and that the stiffness coefficients above are those of ${}^t\mathbf{k}_{NL}$, see (8.26).

□

Example 8.4

Consider that the truss structure of Figure 8.15a is subjected to the imposed horizontal displacement that leads to the configuration at time t as shown in Figure 8.15b. Evaluate the tangent stiffness associated with the vertical displacement δ.

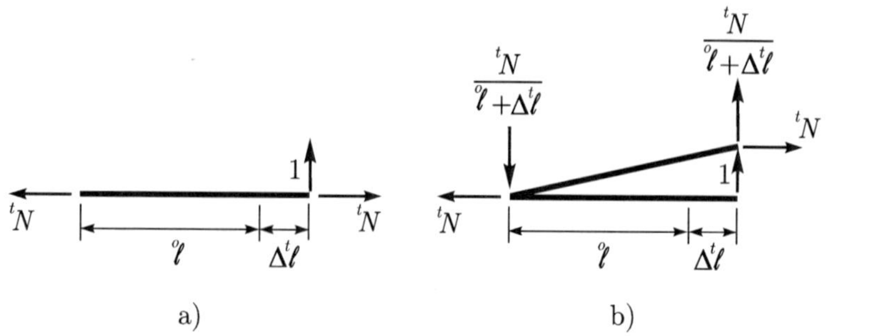

Fig. 8.14. a) Horizontal truss element at time t; b) Equilibrium considering the unit end transverse displacement

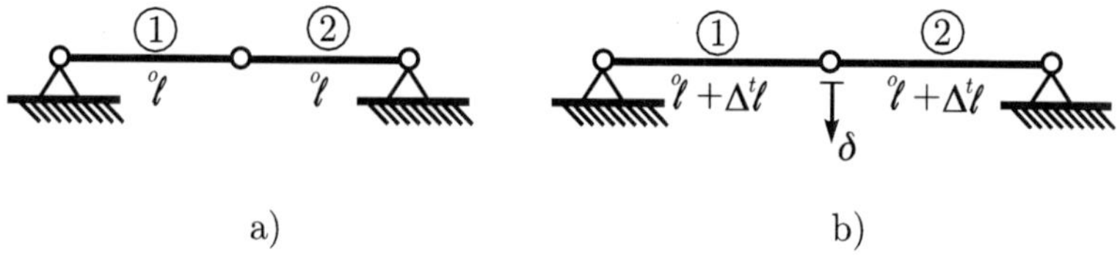

Fig. 8.15. a) Configuration at time 0; b) Configuration at time t

Solution

We can directly obtain the stiffness from the truss bar stiffness matrix derived above,

$$\begin{aligned} {}^t k &= {}^t k_{44}^{(1)} + {}^t k_{22}^{(2)} = {}^t k_{NL_{44}} + {}^t k_{NL_{22}} \\ &= \frac{{}^t N}{{}^0\ell + \Delta^t\ell} + \frac{{}^t N}{{}^0\ell + \Delta^t\ell} = \frac{2\,{}^t N}{{}^0\ell + \Delta^t\ell}. \end{aligned}$$

We note that ${}^t k$ is only due to ${}^t k_{NL}$ since the bars in the configuration at time t are still horizontal, and that the stiffness is proportional to the axial force in the bars.

□

8.3 Determination of ultimate loads leading to structural collapse

Considering a spatial load distribution acting on a structure, it is of great interest to calculate the value of the load multiplier that causes the structure to lose its capability of sustaining a further increase in load, *i.e.*, which causes the structural collapse.

Structural collapse is a very involved subject and our objective in this section is to only discuss briefly some basic facts.

In Section 8.1 we have already encountered situations in which the structure lost its ability to sustain a further increase in load, namely Examples 8.1 and 8.2.

We examine, through the following example, another situation of interest.

□

Example 8.5

Consider the structure shown in Figure 8.16a, where k is a rotational spring. The bar is assumed to be rigid. Find the equilibrium configurations when ${}^tR$ is increased.

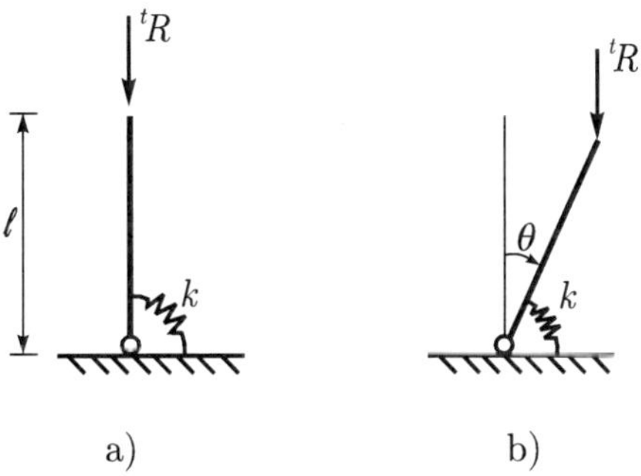

Fig. 8.16. a) Initial configuration; b) Potential deformed configuration

Solution

Let us examine possible equilibrium configurations. The bar remaining vertical gives a configuration which is always possible and, of course, in such a case the axial force in the bar is

$$ {}^tN = -\,{}^tR. $$

Consider the equilibrium configuration shown in Figure 8.16b. Moment equilibrium requires

$$ -\,{}^tR\ell \sin\theta + M_k = 0 $$

where M_k is the moment associated with the rotational spring which is given by

$$ M_k = k\theta $$

leading to

$$-\,{}^tR\ell\sin\theta + k\theta = 0$$

which is satisfied by the undeformed configuration ($\theta = 0$) and by the configuration which corresponds to

$${}^tR = \frac{k\theta}{\ell\sin\theta}.$$

Since

$$\frac{\theta}{\sin\theta} > 1 \text{ for } \theta \neq 0$$

an equilibrium configuration with $\theta \neq 0$ is only possible for

$${}^tR > \frac{k}{\ell}.$$

The equilibrium configurations which are possible are summarized in Figure 8.17a: Up to the load ${}^tR = \frac{k}{\ell}$ there is only one equilibrium configuration possible and it corresponds to the vertical position of the bar. At the load ${}^tR = \frac{k}{\ell}$ there is a branching of the equilibrium paths, which is known as a bifurcation. For load levels greater than the bifurcation load, there are three possible equilibrium configurations as schematically shown in Figure 8.17a by B, B' and B''.

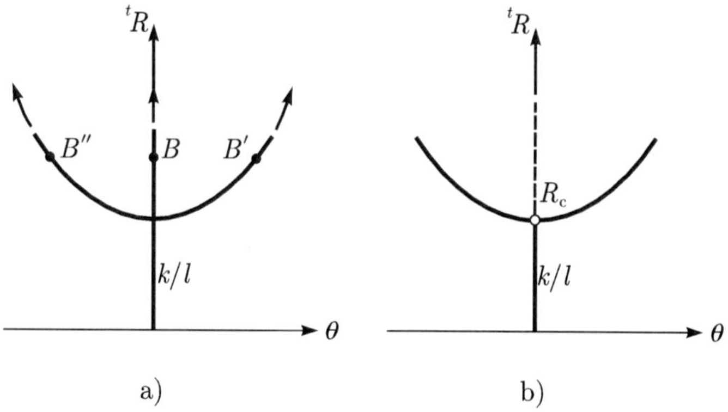

Fig. 8.17. a) Possible equilibrium configurations; b) Possible equilibrium configurations and their stability: the solid line represents stable equilibrium configurations while the dashed line unstable equilibrium configurations

Now consider that we increase the load "slowly" from zero to a load greater than the bifurcation or critical load $R_c = \frac{k}{\ell}$. We may then ask which equilibrium path will be followed and this question prompts the discussion below. □

Stability of an equilibrium configuration

The basic idea associated with the concept of an equilibrium configuration being stable is that if we introduce a small perturbation, for example through any small imposed force acting during a short period of time, the resulting motion will remain close to the equilibrium configuration, just like in linear analysis. Of course, there are many rigorous definitions of stability and formulations to verify whether an equilibrium configuration of a system/structure is stable or not (see *e.g.* Thompson and Hunt, 1973). Since our objective here is only to discuss the subject in an introductory and conceptual manner, we adopt the above concept and verify – in our small example above – the stability of the equilibrium state by identifying if the total potential energy (see Section 5.3) is at a local minimum for the configuration. That is, we use the result that for a conservative system an equilibrium configuration is stable if the total potential energy is at a local minimum[3].

Let us examine the stability of the equilibrium configurations for the problem described in Figure 8.16. Since the bar is rigid, the strain energy is given by

$$\mathcal{U}(\theta) = \frac{1}{2}k\theta^2$$

and, therefore, the total potential energy is given by

$$\Pi(\theta, {}^tR) = \frac{1}{2}k\theta^2 - {}^tR\ell\,(1 - \cos\theta)\,.$$

Verifying the stability as defined above, we have that for $0 < {}^tR < R_{cr}$ the equilibrium configurations are stable. For ${}^tR > R_{cr}$ and $\theta = 0$ the equilibrium configurations are unstable and for $\theta \neq 0$ they are stable. The configuration that defines the bifurcation is itself unstable.

In Figure 8.17b we redraw Figure 8.17a using the solid lines to represent stable configurations and the dashed line to represent those which are unstable. Based on these conclusions, if we increase ${}^tR$ from zero, very slowly, the "deformed" configuration corresponds to $\theta = 0$ for ${}^tR < R_{cr}$. For a given ${}^tR > R_{cr}$ we could theoretically have a deformed configuration corresponding to $\theta > 0$, $\theta = 0$ or $\theta < 0$. However, any perturbation would lead to a configuration corresponding to either $\theta > 0$ or $\theta < 0$.

The bifurcation phenomenon is also referred to as buckling due to the fact that, conceptually, it is the same kind of instability that causes a straight column subjected to a compressive load to adopt an alternative curved, *i.e.* buckled, equilibrium configuration, when the magnitude of the load reaches a certain value, *i.e.*, the critical or buckling load.

We summarize in Table 8.1 the basic stability behaviors of 1-D structural systems. Problem 1 is equivalent to that of Example 8.1 since the spring is

[3] We refer the reader to the Lagrange-Dirichlet theorem which establishes rigorously a sufficient condition for the stability of an equilibrium configuration

equivalent to the truss element with $k = \frac{EA}{\ell}$. In the response of the systems, we have identified in Table 8.1 the stability status of the equilibrium configurations using the same convention as adopted for Figure 8.17b. The point for which the system changes its stability is called a *singular* or *critical point* and the associated load is the *critical load.* These points are identified for all problems described in Table 8.1. The critical point of Problem 1 is defined as a *limit point.*

The limit point is a critical point for which there is no bifurcation of the equilibrium paths and for which there is no equilibrium configuration in its neighborhood for a load greater than the load associated with the limit point.

For Problems 2 to 4 the critical points are bifurcation points since we can clearly identify the branching of the equilibrium paths. They receive specific names, as shown in Table 8.1, that are connected to the characteristics of the post-critical (post-buckling) equilibrium paths. It can be shown that, for example, for Problem 3, if the load is increased very slowly from zero to the critical load and slightly beyond with a tiny horizontal load as a perturbation, a dynamic response initiates. Of course, the objective of presenting the problems in Table 8.1 is to give the reader insight into classical stability behaviors which serve as reference behaviors for more complex models.

The response of the systems summarized in Table 8.1 are for the *perfect system*, *i.e.*, with no imperfection in geometry or load application. Since no real structure or system to be modeled is totally free from imperfections, the response of the *imperfect system* is of interest. This, in particular, because the behavior may drastically change when we go from the perfect to the imperfect system, as discussed below.

In Table 8.2 we show the response of the problems of Table 8.1 when geometrical imperfections are considered. For Problem 1, although there is a change in the magnitude of the critical load, the qualitative behavior remains the same. For Problem 2, there is a change in the qualitative behavior of the system, that is, the system no longer displays a bifurcation point. The same happens for Problem 3. However, for this case, the response of the imperfect system has a limit point, instead of the bifurcation, with a critical load magnitude lower than that of the perfect system. Note that depending on the magnitude of the imperfection, the critical load may be significantly lower than that of the perfect system. For Problem 4, depending on the sense of the imperfection the system behaves either as 2 or 3. We observe that when the imperfection leads to a sharp decrease in the critical load, the structure is said to be sensitive to imperfections.

An important conclusion reached by studying the behavior of the imperfect systems is that the response may qualitatively change. Most importantly, however, is that the critical load may be significantly lower leading to collapse much earlier than predicted for the perfect system.

Table 8.1. Stability behavior of selected models

Problem	Perfect system
1) 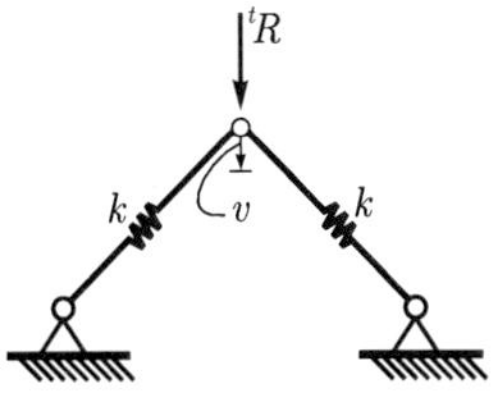	Limit point 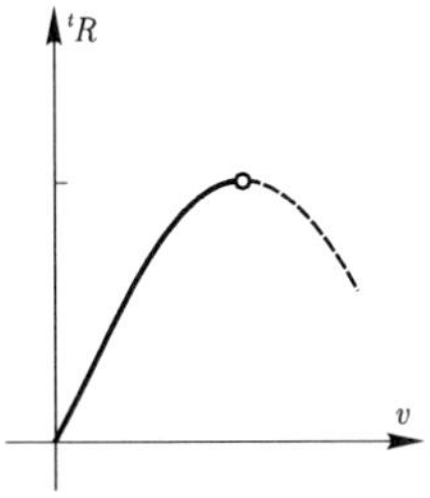
2) 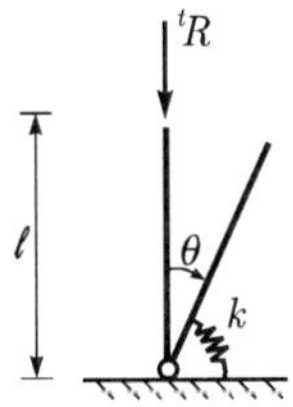	Stable symmetric point of bifurcation 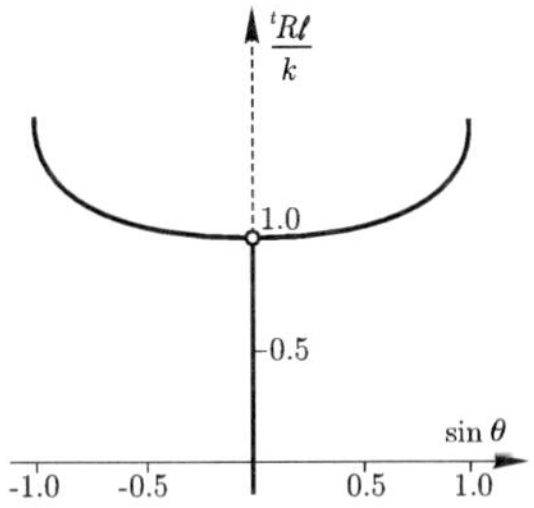
3) 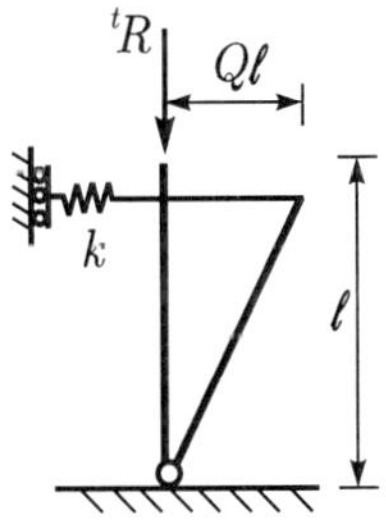	Unstable symmetric point of bifurcation 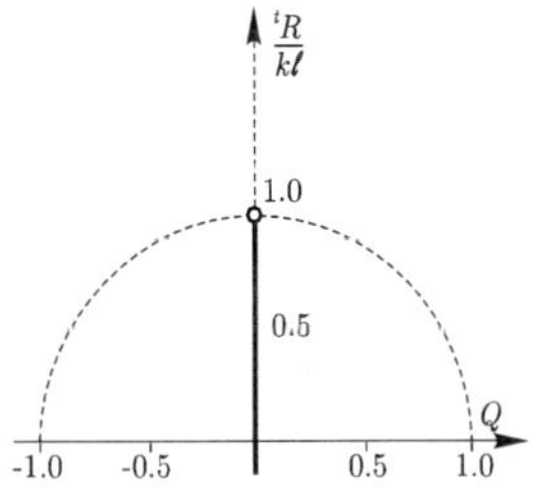
4) 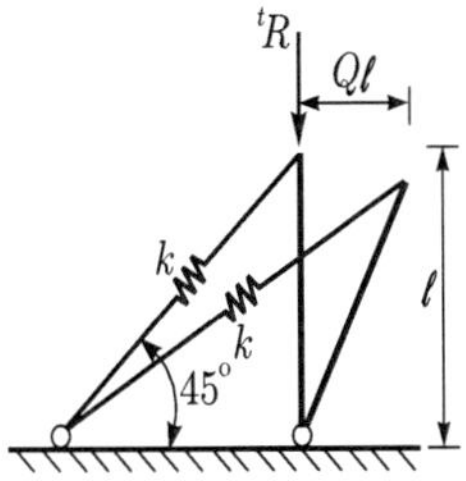	Asymmetric point of bifurcation 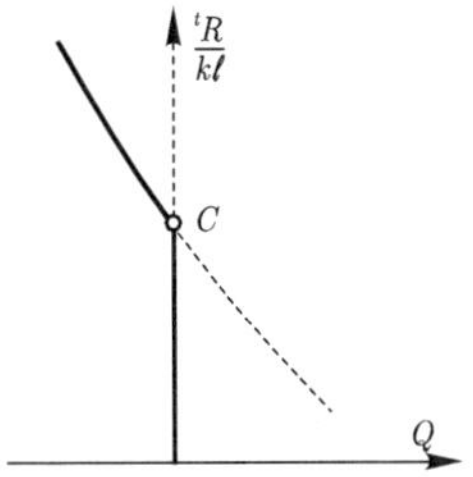

Table 8.2. Behavior of selected models subjected to geometrical imperfections. The horizontal axes of displacements include the imperfections

Problem	System with imperfection

1)

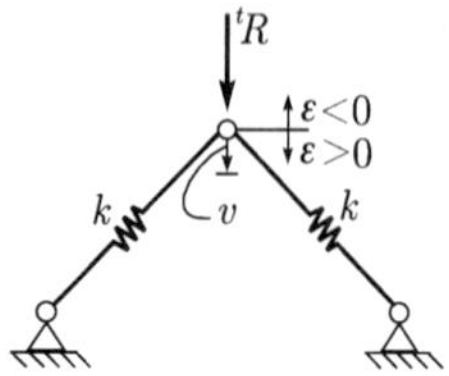

Limit point

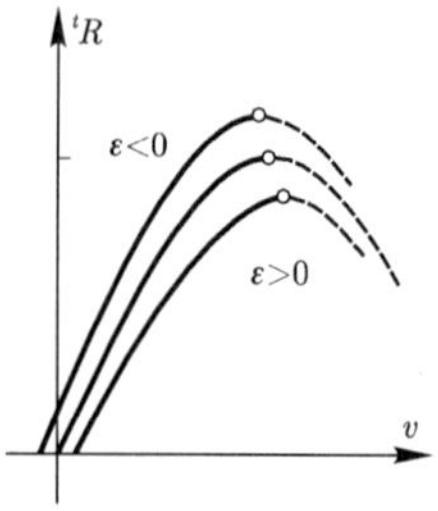

2)

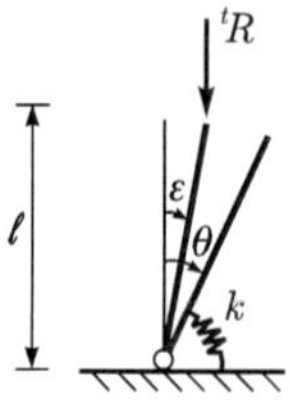

Stable symmetric point of bifurcation

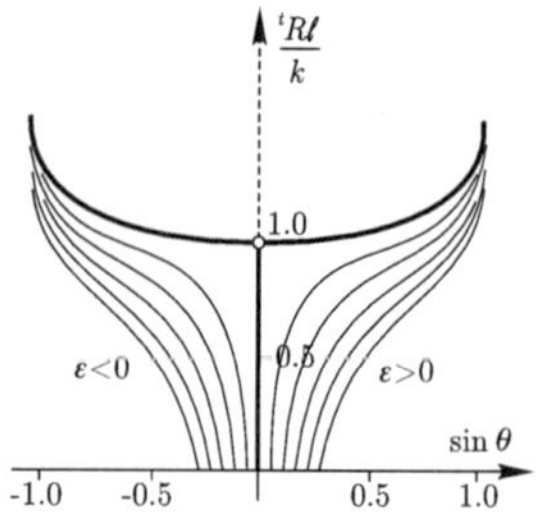

3)

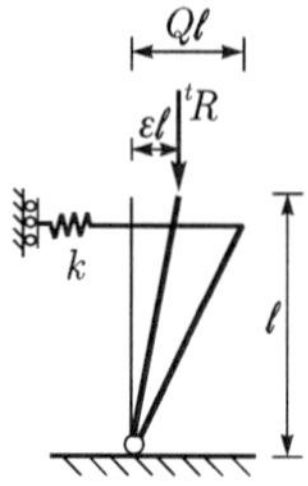

Unstable symmetric point of bifurcation

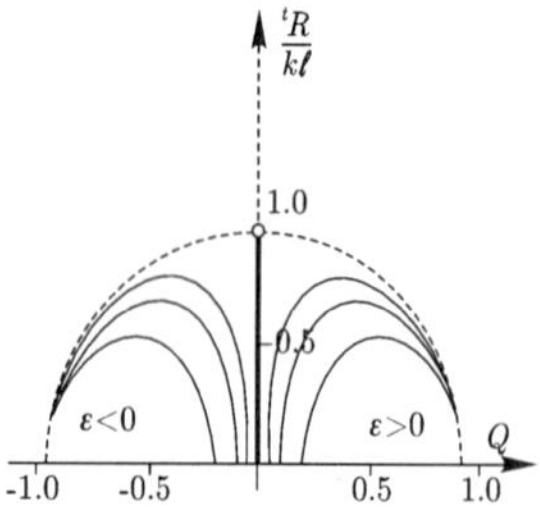

4)

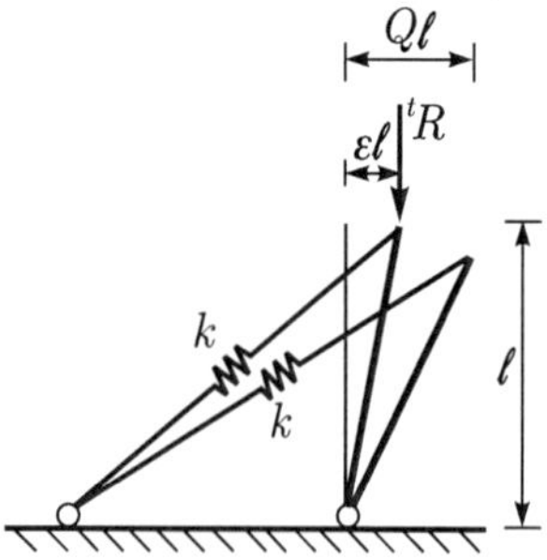

Asymmetric point of bifurcation

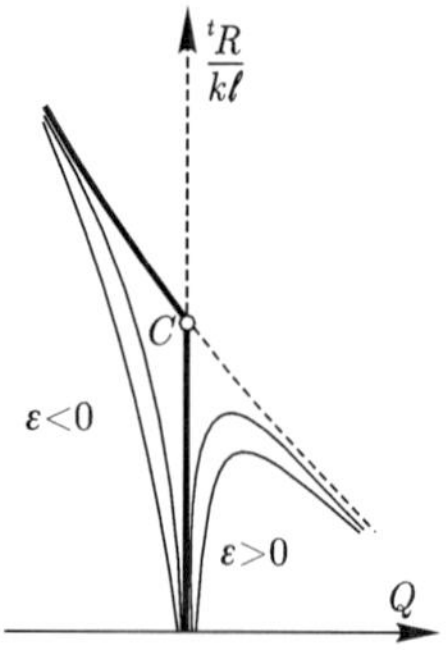

Now let us consider the Problem 3 of Table 8.1, but for a flexible bar, *i.e.*, a truss bar. The problem is summarized in Figure 8.18 where we have placed the bar horizontally.

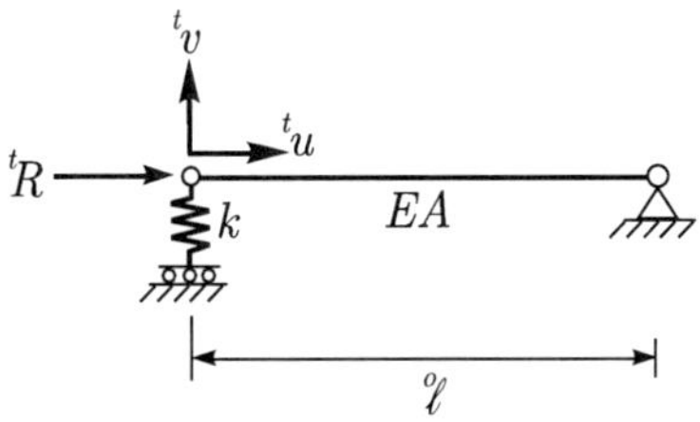

Fig. 8.18. Problem 3 of Table 8.1 with flexible bar

Considering the equilibrium configuration for a horizontal configuration of the bar, we can write

$$\begin{aligned} {}^tR &= -{}^tN \\ \Delta{}^t\ell &= \frac{{}^tN\,{}^0\ell}{EA} \\ {}^tu &= -\Delta{}^t\ell = -\frac{{}^tN\,{}^0\ell}{EA} = \frac{{}^tR\,{}^0\ell}{EA} \\ {}^tv &= 0. \end{aligned}$$

The tangent stiffness matrix is given by

$${}^t\mathbf{K} = \frac{EA}{{}^0\ell + \Delta{}^t\ell}\begin{bmatrix} 1 & 0 \\ 0 & 0 \end{bmatrix} + \frac{{}^tN}{{}^0\ell + \Delta{}^t\ell}\begin{bmatrix} 1 & 0 \\ 0 & 1 \end{bmatrix} + \begin{bmatrix} 0 & 0 \\ 0 & k \end{bmatrix}$$

where we have presented above the ${}^t\mathbf{k}_L$ and the ${}^t\mathbf{k}_{NL}$ matrices of the truss bar and the contribution of the spring. We can re-write the tangent stiffness matrix as

$${}^t\mathbf{K} = \begin{bmatrix} \frac{EA}{{}^0\ell} & 0 \\ 0 & k - \frac{{}^tR}{{}^0\ell + \Delta{}^t\ell} \end{bmatrix}.$$

The incremental displacements can be evaluated by solving

$${}^t\mathbf{K}\,\Delta\mathbf{u} = \Delta\mathbf{R} \tag{8.29}$$

where $\Delta\mathbf{R}^T = \begin{bmatrix} \Delta R & 0 \end{bmatrix}$. Therefore $\Delta\mathbf{u}^T = \begin{bmatrix} \Delta u & 0 \end{bmatrix}$ is a solution of (8.29) and if $\det{}^t\mathbf{K} \neq 0$ it is the unique solution. Note that

$$\det {}^t\mathbf{K} = \frac{EA}{{}^0\ell}\left(k - \frac{{}^tR}{{}^0\ell + \Delta^t\ell}\right)$$

where

$${}^tR = k\left({}^0\ell + \Delta^t\ell\right) = k\,{}^t\ell$$

leads to $\det {}^t\mathbf{K} = 0$ and characterizes a singular or critical point.

The simple idea presented above can be generalized for structures with many degrees of freedom. Let us assume that the load vector is defined by

$${}^t\mathbf{R} = {}^t\lambda\mathbf{R}_0$$

where $\mathbf{R}_0$ is a reference load vector and ${}^t\lambda$ is the load multiplier. In addition, suppose that we are using the incremental procedure to obtain the structural response. For a given time t, we seek the incremental displacements using

$${}^t\mathbf{K}\,\Delta\mathbf{U} = \Delta\lambda\,\mathbf{R}_0 \tag{8.30}$$

and as long as $\det {}^t\mathbf{K} \neq 0$ we can solve for $\Delta\mathbf{U}$. When ${}^t\mathbf{K}$ becomes singular, *i.e.*, $\det {}^t\mathbf{K} = 0$, we have reached a singular or critical point. Conceptually, we have an analogous situation to the 1-D problem. We may have either a bifurcation of the equilibrium path or a limit point. In the limit point case there is no branching of the equilibrium path and an equilibrium configuration in the vicinity of the critical point is associated with a decrease in ${}^t\lambda$. In this case it is still possible to obtain the post-critical response of the structure by means of an incremental solution. However, adequate procedures should be used since the (immediate) post-critical path is associated with a decrease in ${}^t\lambda$. There are incremental strategies which have as unknowns not only the incremental displacements $\Delta\mathbf{u}$ but also the increment $\Delta\lambda$ of the load multiplier parameter which may be negative, see Bathe, 1996.

In many cases of practical interest the critical point may be taken as the ultimate configuration at which the structure can no longer support a further increase of the external load. Therefore, in such cases the determination of the critical point and associated loading characterizes the ultimate load carrying capacity of the structure.

An incremental procedure always determines only one equilibrium path. Therefore, when a critical point is reached, to identify the nature of the critical point, the basic behaviors summarized in Tables 8.1 and 8.2 need be considered. For perfect systems, the nature of the critical point may indicate the kind of imperfection to be considered in the model in order to avoid an overestimation of the actual physical collapse load of the structure.

In the above approach, the determination of a critical point requires that the incremental solution be undertaken up to the critical point. Sometimes it is convenient to have an estimate of the critical load without performing the full incremental solution. We describe such approach below.

Linearized buckling analysis

The basic assumption of a linearized buckling analysis is that the tangent stiffness matrix varies linearly with respect to the load parameter λ as

$$^{\tau}\mathbf{K} = {}^{t-\Delta t}\mathbf{K} + \lambda \left({}^{t}\mathbf{K} - {}^{t-\Delta t}\mathbf{K} \right) \tag{8.31}$$

where τ is the time associated with

$$^{\tau}\mathbf{R} = {}^{t-\Delta t}\mathbf{R} + \lambda \left({}^{t}\mathbf{R} - {}^{t-\Delta t}\mathbf{R} \right). \tag{8.32}$$

We are interested in determining the value of λ which makes $^{\tau}\mathbf{K}$ singular, *i.e.*,

$$\det {}^{\tau}\mathbf{K} = 0 \tag{8.33}$$

or equivalently

$$^{\tau}\mathbf{K}\phi = \mathbf{0} \tag{8.34}$$

with $\phi \neq \mathbf{0}$. The values of λ_i that satisfy (8.33) or (8.34) can be found by solving an eigenvalue problem as detailed in Bathe, 1996 which also gives ϕ_i as the eigenvectors.

The eigenvalues λ_i define through equation (8.32) the buckling loads and the associated eigenvectors define the buckling modes, since by equation (8.34) we can see that a buckling mode corresponds to a valid displacement solution associated with no increment in the load.

The quality of the predictions obtained with the linearized buckling analysis depends primarily on whether the approximation adopted for the tangent stiffness matrix given by (8.31) is sufficiently accurate.

In general, we take $^{t-\Delta t}\mathbf{K} = {}^{0}\mathbf{K}$, *i.e.*, the stiffness matrix prior to the application of external loads. When the magnitude of the displacements which take place prior to the critical point is small, the assumption implicitly contained in (8.31) is adequate. Then, the linearized buckling analysis is quite useful.

Finally, we note that the buckling modes may be used to generate imperfections. When properly scaled, buckling modes can be added to the initial geometry of the structure generating an imperfect initial geometry, which may correspond to a significantly lower critical load than that of the perfect structure. An example of this procedure can be found in Bathe, 1996.

8.4 Modeling nonlinear problems

As mentioned at the beginning of this chapter, the finite element nonlinear formulations of the mathematical models of Chapters 3 and 4 are out of the scope of this book and we refer the reader to Bathe, 1996 where such formulations are given. We note, however, that the nonlinear formulation presented for the truss element is a valuable aid for conceptually understanding the nonlinear finite element matrices of more complex formulations. In fact, we note that there are analogies between the nonlinear formulation of the truss element and that of every finite element presented in Chapter 6. Namely:

- The incremental formulation is the same as that presented in Section 8.2.
- For each element the tangent stiffness matrix can be decomposed into a ${}^t\mathbf{k}_L$ and a ${}^t\mathbf{k}_{NL}$ with analogous interpretations as for the truss element, *i.e.*, ${}^t\mathbf{k}_L$ accounts for the additional straining of the element and ${}^t\mathbf{k}_{NL}$ for the effect that there are already internal forces/stresses in the element.

However, there are many complex additional considerations when considering nonlinear 2-D, 3-D, beam and shell analyses, and these are particularly complex because the details of the formulations matter a great deal in order to obtain reliable, accurate and effective solutions, see Bathe, 1996.

References

ADINA 2008, ADINA—A finite element program for automatic dynamic incremental nonlinear analysis, Technical report, ADINA R&D, Watertown, MA. www.adina.com.

Ahmad, S., B. M. Irons and O. C. Zienkiewicz 1970, 'Analysis of thick and thin shell structures by curved finite elements', *International Journal for Numerical Methods in Engineering* **2**, 419–451.

Arnold, D. N. and R. S. Falk 1990, 'The boundary layer for the Reissner-Mindlin plate model', *SIAM Journal on Mathematical Analysis* **21**(37), 281–312.

Bathe, K. J. 1996, *Finite element procedures*, Prentice-Hall, Inc., Englewood Cliffs.

Bathe, K. J. 2001, 'The inf–sup condition and its evaluation for mixed finite element methods', *Computers & Structures* **79**, 243–252.

Bathe, K. J. 2009, *Wiley Encyclopedia of Computer Science and Engineering*, Wiley-Interscience, chapter The finite element method, pp. 1253–1264. Wah, B. W. ed.

Bathe, K. J. and A. Chaudhary 1982, 'On the displacement formulation of torsion of shafts with rectangular cross-sections', *International Journal for Numerical Methods in Engineering* **18**, 1565–1568.

Bathe, K. J., D. Chapelle and P. S. Lee 2003, 'A shell problem "highly-sensitive" to thickness changes', *International Journal for Numerical Methods in Engineering* **57**, 1039–1052.

Bathe, K. J. and E. Dvorkin 1986, 'A formulation of general shell elements — the use of mixed interpolation of tensorial components', *International Journal for Numerical Methods in Engineering* **22**, 697–722.

Bathe, K. J., ed. 2005, *Computational fluid and solid mechanics 2005*, Elsevier Science Publishers.

Bathe, K. J. and L. W. Ho 1981, Some results in the analysis of thin shell structures, *in* Wunderlich et al., eds, 'Nonlinear finite element analysis in structural mechanics', Springer-Verlag, Berlin, pp. 72–100.

Bathe, K. J., M. L. Bucalem and F. Brezzi 1990, 'Displacement and stress convergence of our MITC plate bending elements', *Engineering Computations* **7**(4), 291–302.

Bathe, K. J., N. S. Lee and M. L. Bucalem 1990, 'On the use of hierarchical models in engineering analysis', *Computer Methods in Applied Mechanics and Engineering* **82**, 5–26.

Bathe, K. J. and P. S. Lee 2011, 'Measuring the convergence behavior of shell discretization schemes', *Computers & Structures*, in press.

Bathe, K. J., P. S. Lee and J. F. Hiller 2003, 'Towards improving the MITC9 shell element', *Computers & Structures* **81**, 477–489.

Brezzi, F. and K. J. Bathe 1986, Studies of finite element procedures — the inf-sup condition, equivalent forms and applications, *in* K. J. Bathe and D. R. J.

Owen, eds, 'Reliability of Methods for Engineering Analysis', Pineridge Press, Swansea, pp. 197–219.

Brezzi, F. and K. J. Bathe 1990, 'A discourse on the stability conditions for mixed finite element formulations', *J. Computer Methods in Applied Mechanics and Engineering* **82**, 27–57.

Brezzi, F. and M. Fortin 1991, *Mixed and hybrid finite element methods*, Springer-Verlag.

Bucalem, M. L. and K. J. Bathe 1993, 'On higher-order MITC general shell elements', *International Journal for Numerical Methods in Engineering* **36**, 3729–3754.

Bucalem, M. L. and K. J. Bathe 1997, 'Finite element analysis of shell structures', *Archives of Computational Methods in Engineering* **4**, 3–61.

Chapelle, D. and K. J. Bathe 2000, 'The mathematical shell model underlying general shell elements', *International Journal for Numerical Methods in Engineering* **48**, 289–313.

Chapelle, D. and K. J. Bathe 2010*a*, *The finite element analysis of shells – fundamentals*, second edn, Springer-Verlag.

Chapelle, D. and K. J. Bathe 2010*b*, 'On the ellipticity condition for model-parameter dependent mixed formulations', *Computers & Structures* **88**, 581–587.

Ciarlet, P. G. 1978, *The finite element method for elliptic problems*, North-Holland, Amsterdam.

Cowper, G. R. 1966, 'The shear coefficient in Timoshenko's beam theory', *Journal of Applied Mechanics* **33**, 335–340.

Crandall, S. H., N. C. Dahl and T. J. Lardner 1978, *An introduction to the mechanics of solids*, second edn, MacGraw-Hill.

Crouzeix, M. and P. A. Raviart 1973, 'Conforming and nonconforming finite element methods for solving the stationary Stokes equations', *RAIRO, Anal. Numer.* **R3**, 33–76.

Dvorkin, E. and K. J. Bathe 1984, 'A continuum mechanics based four-node shell element for general nonlinear analysis', *Engineering Computations* **1**, 77–88.

Grätsch, T. and K. J. Bathe 2005, 'A posteriori error estimation techniques in practical finite element analysis', *Computers & Structures* **83**, 235–265.

Häggblad, B. and K. J. Bathe 1990, 'Specifications of boundary conditions for Reissner-Mindlin plate bending finite elements', *International Journal for Numerical Methods in Engineering* **30**, 981–1011.

Hiller, J. F. and K. J. Bathe 2003, 'Measuring convergence of mixed finite element discretizations: an application to shell structures', *Computers & Structures* **81**, 639–654.

Jones, R. M. 1975, *Mechanics of composite materials*, MacGraw-Hill.

Kim, D. N. and K. J. Bathe 2008, 'A 4-node 3D-shell element to model shell surface tractions and incompressible behavior', *Computers & Structures* **86**, 2027–2041.

Kim, D. N. and K. J. Bathe 2009, 'A triangular six-node shell element', *Computers & Structures* **87**, 1451–1460.

Kojic, M. and K. J. Bathe 2005, *Inelastic analysis of solids and structures*, Springer-Verlag.

Lee, N. S. and K. J. Bathe 1993, 'Effects of element distortions on the performance of isoparametric elements', *International Journal for Numerical Methods in Engineering* **36**, 3553–3576.

Lee, P. S. and K. J. Bathe 2002, 'On the asymptotic behavior of shell structures and the evaluation in finite element solutions', *Computers & Structures* **80**, 235–255.

Lee, P. S. and K. J. Bathe 2004, 'Development of MITC isotropic triangular shell finite elements', *Computers & Structures* **88**, 945–962.

Lee, P. S. and K. J. Bathe 2005, 'Insight into finite element shell discretizations by use of the basic shell mathematical model', *Computers & Structures* **83**, 69–90.

Lee, P. S. and K. J. Bathe 2010, 'The quadratic MITC plate and MITC shell elements in plate bending', *Advances in Engineering Software* **41**, 712–728.

Love, A. E. H. 1934, *The mathematical theory of elasticity*, Cambridge University Press.

Murray, N. W. 1985, *Introduction to the theory of thin walled structures*, Claredon Press.

Sussman, T. and K. J. Bathe 1986, 'Studies of finite element procedures. Stress band plots and the evaluation of finite element meshes', *Engineering Computations* **3**, 178–191.

Sussman, T. and K. J. Bathe 1987, 'A finite element formulation for nonlinear incompressible elastic and inelastic analysis', *Computers & Structures* **26**, 357–409.

Thompson, J. M. T. and G. W. Hunt 1973, *A General theory of elastic stability*, Wiley, London.

Timoshenko, S. and J. M. Gere 1961, *Theory of elastic stability*, MacGraw-Hill.

Timoshenko, S. and J. N. Goodier 1970, *Theory of elasticity*, third edn, MacGraw-Hill.

Timoshenko, S. and S. Woinowsky-Krieger 1959, *Theory of plates and shells*, third edn, MacGraw-Hill.

Truesdell, C. A. 1977, *A first course in rational continuum mechanics*, Academic Press.

Zhong, W. and C. Qui 1983, 'Analyis of symmetric or partially symmetric structures', *Computer Methods in Applied Mechanics and Engineering* **38**, 1–18.

Index